10th Topical Conference on Gas Utilization 2010

Topical Conference at the 2010 AIChE Spring Meeting and 6th Global Congress on Process Safety

San Antonio, Texas, USA
21 – 25 March 2010

ISBN: 978-1-61738-441-7

Printed from e-media with permission by:

Curran Associates, Inc.
57 Morehouse Lane
Red Hook, NY 12571

CURRAN ASSOCIATES INC.
proceedings.com

Some format issues inherent in the e-media version may also appear in this print version.

Printed by Curran Associates, Inc. (2010)

AIChE
3 Park Avenue
New York, NY 10016-5991

Phone: (203) 702-7660
Fax: (203) 775-5177

www.aiche.org

Additional copies of this publication are available from:

Curran Associates, Inc.
57 Morehouse Lane
Red Hook, NY 12571 USA
Phone: 845-758-0400
Fax: 845-758-2634
Email: curran@proceedings.com
Web: www.proceedings.com

TABLE OF CONTENTS

(7a) Goals and Achievements of the US National Program On Hydrogen Storage Applied R&D

Ned Stetson, U.S. Department of Energy, Washington, DC

Grace Ordaz, Office of Hydrogen, Fuel Cells and Infrastructure Technologies, U.S. Department of Energy, Washington, DC

Carole Read, Office of Hydrogen, Fuel Cells and Infrastructure Technologies, U.S. Department of Energy, Washington, DC

Monterey Gardiner, Office of Hydrogen, Fuel Cells & Infrastructure Technologies, U.S. Department of Energy, Washington, DC

Sara Dillich, Office of Hydrogen, Fuel Cells & Infrastructure Technologies, US Department of Energy, Office of Energy Efficiency and Renewable Energy, Washington DC, DC

Robert Bowman, Oak Ridge National Lab, Franklin, OH

The US Department of Energy Fuel Cell Technologies Program's mission is to reduce oil use and carbon emissions by enabling clean, reliable fuel cell technologies for stationary, auxiliary, and portable power as well as transportation applications.

The requirements for hydrogen storage continue to be some of the most technically challenging barriers to the widespread commercialization of hydrogen and fuel cell products. The DOE-EERE hydrogen storage activities primarily focus on applied research and development of materials and engineering to provide low-pressure storage options that meet packaging, cost, safety and performance requirements to allow fuel cell systems to be competitive with current technologies.

This presentation summarizes the program goals, current status, recent accomplishments and current performance gaps of hydrogen storage technologies for fuel cell applications. Material R&D projects are focused in three main areas: metal hydrides; chemical hydrogen storage materials and hydrogen sorbents. New efforts under the Hydrogen Storage Engineering Center of Excellence will provide a coordinated approach to engineering research and development of hydrogen storage systems.

See more of this Session: Advancements in Hydrogen Technology I
See more of this Group/Topical: Topical 5: Clean Fuels and Energy Efficient Processes

Au/CeO_2 CATALYSTS FOR WATER GAS SHIFT REACTION

Camilla Galletti, Brenno A. Lenite, Stefania Specchia, Vito Specchia
Politecnico di Torino, Torino, Italy

Abstract

Thanks to their high energy efficiency and low emissions, PEM-FCs are the preferred candidates for electric power generation in mobile applications. Considering the actual lack of H_2 infrastructures for its distribution on the road, the most probable fuel for PEM-FCs is H_2-rich gas coming from reformed hydrocarbons; at reformer outlet it contains about 10% CO. The performance of Pt, used as catalytic element for H_2 oxidation in PEM-FCs, is seriously depressed from CO; therefore the latter must be removed through water gas shift (WGS) reaction (to about 0.5–1% with simultaneous increase of H_2) and then below 10 ppmv by CO preferential oxidation (CO-PROX) or selective methanation (CO-SMET). Gold supported catalysts revealed to be extremely active at low temperatures in different reactions; strict requirement was Au highly dispersion on suitable carriers. Dispersion, in fact, strongly influences activity and selectivity. An efficient deposition method able to guarantee an Au intimate dispersion in nanometric clusters on the carrier is thus a decisive issue to provide good catalytic performance. In this work the influence of synthesis parameters on Au dispersion on ceria, used as a support, and the activity of the obtained catalysts toward the WGS reaction was evaluated. Catalytic activity tests were performed on 3% Au-CeO_2 catalysts, prepared at different pH and Au precursor solution molarity M. The best catalytic activity in terms of CO abatement was obtained with the catalyst prepared at pH = 8.5 and $M = 0.2 \cdot 10^{-3}$.

Introduction

Due to the increasing attention towards the emissions from the internal combustion engines, fuel fell (FC) technology is now standing out as a ''clean'' alternative for diminishing automotive pollution. The polymer electrolyte membrane FC (PEM-FC) is currently considered as the most suitable technology for vehicles application by the majority of the automotive companies due to its high power density, mechanical robustness, low weight, compactness, long stack life, fast start-up, dynamic response characteristics. Nevertheless, one of the major problems for the application of low temperature PEM-FC as power source for electrically operated vehicles is the delivery of "nearly-CO-free" feed gas, which becomes problematic whenever H_2 is generated from fossil fuels. As the FC anodes can be poisoned even by trace impurities of CO, this contaminant has to be removed to a level below 50 ppm for state-of-art Pt-Ru anode electrocatalysts [1-3] and below 10 ppmv for Pt anode electrocatalysts [4-6]. On the other hand, in the guideline document "Multi-Year Research, Development and Demonstration Plan: Planned Program Activities for 2005-2015 – Technical Plan Fuel Cells", published on DOE web site on 2007 [7] and updated on April 2009, the CO tolerance (target 2005) is 10 ppm in steady-state and 100 ppm in transient, for a durability of 20,000 h, that will be reduced (target 2015) to 1 ppm in steady-state and 10 ppm in transient, for a durability of 40,000 h. Then, maintaining the CO concentration at low values, the PEM-FC durability is surely favored.

Considering a Fuel Processor Units (FPUs) reforming fossil fuels to produce a H_2-rich stream to be fed to PEM-FCs, in order to increase the H_2 rate in the reformate gas the CO clean-up process is based on the water gas shift (WGS) step which releases a CO concentration in the 0.5-1% range. As the final clean-up step, a reasonable selection may be made between the preferential CO oxidation (CO-PROX) and selective CO methanation (CO-SMET). The first preferred process for CO removal is based on the WGS reaction:

$$CO + H_2O \leftrightarrow CO_2 + H_2 \qquad \Delta H = -41\ kJ\cdot mol^{-1}$$

which not only reduces the CO concentration from about 10% to about 0.5-1% [8] but simultaneously increases the H_2 concentration. Since at high temperatures CO conversion is equilibrium limited and at low temperatures is kinetically controlled, normally two different WGS catalysts are commercially used. One for high temperature shift (HT-WGS), based on Fe and Cr oxides, reduces at 400-500 °C the CO concentration to about 2-5%. The other for low temperature shift (LT-WGS), with Cu and ZnO on Al_2O_3 catalyst is used at 200-400 °C for cutting the CO concentration to about 0.5-1

Commercially a combination of the two stages (HT- and LT-WGS) is used with a cooling step in between. However, commercial catalysts pose some disadvantages (low activity and some thermodynamic limitations at high temperatures for the HT catalyst; pyrophoricity for the LT one) that render themselves unsuitable for residential or automotive applications. Here, in fact, fast start-ups, carried out by heating-up the reactors chain with suitably produced flue gases, dictate the use of low reactor volumes, equipped with very active and non-pyrophoric catalytic materials, the latter feature mandatory for the presence of residual O_2 in the start-up heating medium [9].

Gold has long been regarded as a poorly active catalyst. However, when it is deposited as nanoparticles on metal oxides, it exhibits surprisingly high catalytic activity [10–15]. In particular, Au has been considered as a potentially useful element for CO abatement in WGS reaction at low temperature [10–18]. The state and structure of the carrier strongly influence the catalytic activity and selectivity of the Au-based catalysts [19-21]. The selection of an efficient carrier is thus a crucial factor to provide desirable contact between Au and the carrier itself and, consequently, a good catalytic performance [22]. In the last years, Au nanoparticles supported on CeO_2 and mixed oxides have been studied, evaluating support and preparation method effects [23-28]. Therefore, in this paper Au catalysts supported on own prepared CeO_2 were characterized and tested, to evaluate the influence of the preparation method characteristic parameters on the catalytic activity toward WGS reaction, particularly when a reformate composition typical of LT-WGS clean-up step is used.

Experimental

The CeO_2 to be used as Au carrier was on purpose prepared through the solution combustion synthesis (SCS) method [29]. Au was added by deposition-precipitation method, trying to obtain the smallest particle size, necessary for a satisfactory catalytic activity. The aqueous solution of Au precursor ($HAuCl_4$) was prepared at 80 °C and adjusting the pH to the desired value; the powder CeO_2 was then added, stirring on a heated plate. After 1 h the suspension was filtered, washed and dried under vacuum. Different catalysts were prepared, varying both pH (7-10) and molarity M of solution ($0.2\cdot10^{-3}$ - $1\cdot10^{-3}$). All the prepared catalytic materials were calcinated in calm air at 400 °C for 3 h.

X-ray diffraction analysis (XRD, Philips PW1710 apparatus equipped with a monochromator for the CuKα radiation) was performed on the prepared catalysts before the catalytic activity tests in order to verify their effective composition and derive a qualitative indication of the presence of comparatively large Au metal crystallite, if its peaks visible. The obtained Au-based catalysts were also examined by high-resolution transmission electron microscopy (HRTEM, Jeol JEM 3010 UHR apparatus) in order to investigate of the active metal both the clusters size and the dispersion on the carrier surface. The BET specific surface area of the catalysts was determined by means of N_2 adsorption analysis (Micromeritics ASAP 2020 C apparatus).

A fixed bed micro-reactor (a quartz tube of 4 mm I.D.), heated up by a PID regulated oven and containing 0.3 g of catalyst in powder, diluted with 0.5 g of SiO_2, held in place by flocks of quartz wool, was used for the WGS reaction. A K-type thermocouple was inserted into the reactor to measure the temperature of the catalytic bed. The inlet gas was fed at a flow rate of 100 Nml min^{-1} (weight space velocity WSV = 0.33 Nl min^{-1} g_{cat}^{-1}) and with different number of gaseous components, starting with a mixture of only CO and H_2O (H_2O/CO = 4), and then adding CO_2 and H_2. The 4-component mixture was representative of a typical reformate composition at the outlet of a HT-WGS reactor, where the CO concentration was partially reduced. The outlet gas stream from the WGS micro-reactor was analyzed through a micro gas-chromatograph (Varian CP-4900) equipped with a thermal conductivity detector. The CO detection limit was 2 ppmv.

The CO conversion, calculated on the basis of inlet and outlet concentrations, was evaluated in the temperature range 150-550 °C; WGS equilibrium curve at the various temperatures and gas feeding compositions considered was estimated by using the following equilibrium constant [30]:

$$K_p = \exp\left(\frac{4{,}577.8}{T} - 4.33\right)$$

Results and discussion

A preliminary characterization of the as-prepared Au-based catalysts was carried out with XRD analysis; the spectrum related to the 3% Au-CeO_2 catalyst synthesized at slightly basic pH and lowest M at is shown in Fig. 1. Only the peaks related to CeO_2 were enlightened, a sign that Au was probably homogeneously dispersed. Also the spectra of catalysts prepared with different M and pH values (not reported here) showed similar crystalline structure without Au peaks. The characteristic Au (111) metallic gold peak was detected only by deeply scanning around its characteristic peak at 2θ = 38.184 with larger time for each scan step (see insert in Fig. 1).

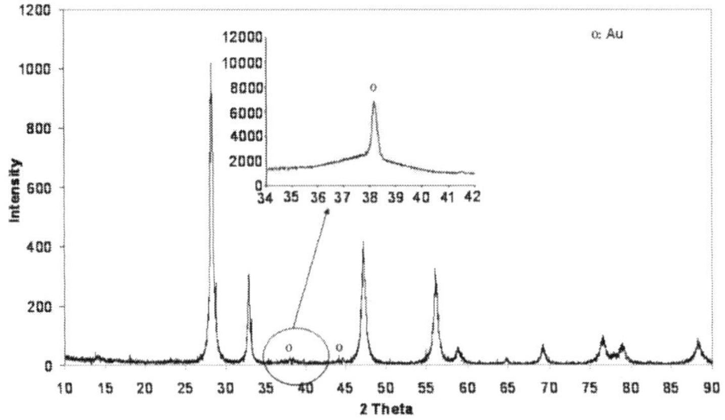

Figure 1: XRD spectrum for 3% Au-CeO_2 catalyst prepared at pH = 8.5 and M = $0.2 \cdot 10^{-3}$

The CeO_2 carrier showed a BET specific surface area of 66 m^2 g^{-1}; for all the prepared catalysts this value slightly decreased to about 62 m^2 g^{-1}, after Au deposition.

HRTEM investigation (Fig. 2) completed the catalysts physical characterization. The 3% Au-CeO_2 catalyst synthesized at pH = 8.5 and M = $0.2 \cdot 10^{-3}$ (Fig. 2.A) showed very small Au particles (1.5-2 nm) and a homogeneous distribution on the carrier. By increasing M to $1 \cdot 10^{-3}$ (Fig. 2.B) during preparation, particles appeared slightly larger, with an average size of 3 nm, and more distributed on the carrier border grain (like a decorated catalyst). Catalyst prepared at pH = 10 and M = $1 \cdot 10^{-3}$ (Fig. 2.C) showed small Au particles (~ 3 nm) coupled with larger ones (10-15 nm), and a not satisfactorily homogeneous distribution on the carrier. The reduction to pH = 7 (Fig. 2.D) caused the formation of Au particles with size about 20 nm and the agglomeration of the smallest ones. Then, preparation at slightly basic pH and low M permitted to deposit Au in small particles and with a satisfactorily homogeneous distribution on the ceria carrier, fundamental prerequisites for a good catalytic activity.

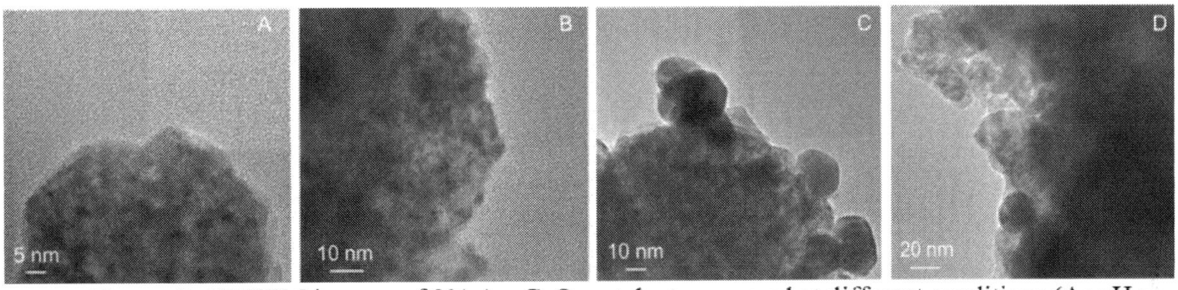

Figure 2: HRTEM images of 3% Au-CeO_2 catalysts prepared at different conditions (A: pH = 8.5, M = $0.2 \cdot 10^{-3}$; B: pH = 8.5, M = $1 \cdot 10^{-3}$; C: pH = 10, M = $1 \cdot 10^{-3}$; D: pH = 7, M = $1 \cdot 10^{-3}$)

Anyway, a preliminary screening of the catalytic performance of Au-based catalysts, in terms of CO conversion (ξ_{CO}), was carried out on the catalysts prepared at different pH and M = $1 \cdot 10^{-3}$.

Firstly, tests were performed by feeding only CO (5%) and H_2O (H_2O/CO = 4) in N_2; the obtained results are shown in Fig. 3. The catalyst prepared at pH = 10 reached ξ_{CO} = 0.9 at 440 °C, with residual CO concentration of 0.35%. Reducing the pH to 8.5, maximum ξ_{CO} decreased to 0.86 at higher temperature (500 °C), but in the range 150-440 °C showed higher ξ_{CO} values, compared to catalyst at pH = 10. A further reduction of pH to 7 depressed globally ξ_{CO}, with a maximum of 0.83 at 550 °C.

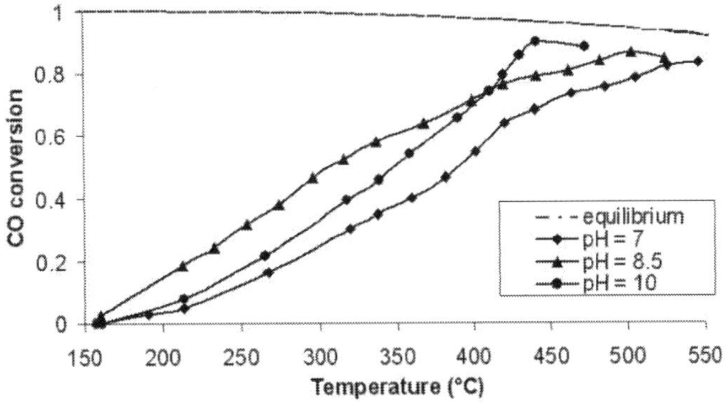

Figure 3: ξ_{CO} vs. T for 3% Au-CeO_2 catalysts prepared at M = $1 \cdot 10^{-3}$ and different pH (feed gas composition: 5% CO, 20% H_2O in N_2; WSV = 0.333 $Nl \cdot min^{-1} \cdot g^{-1}$).

The catalyst synthesized at lower pH gave the worst performance, enlightening the agreement with HRTEM analysis: large Au particles (Fig. 2.D) reduced the catalytic performance.

The real reformate gas to which apply WGS process contains also H_2 and CO_2. Then, to be closer to reality, a deeper study on the catalytic activity of the three catalysts was carried out by adding to the initial investigated mixture only CO_2 (11%, Fig 4.A), afterwards only H_2 (40%, Fig. 4.B) and finally both CO_2 and H_2 (standard feed, as representative of a real reformate mixture, Fig. 4.C), in order to evaluate the effect on ξ_{CO} of each single component addition.

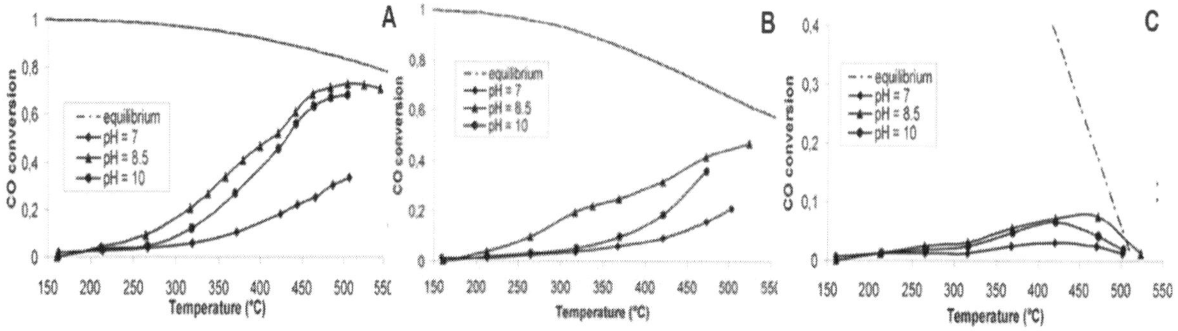

Figure 4: ξ_{CO} vs. T for 3% Au-CeO_2 catalysts prepared at $M = 1\cdot10^{-3}$ and different pH at various feed gas composition (A: 5% CO, 20% H_2O, 11% CO_2; B: 5% CO, 20% H_2O, 40% H_2; C: "standard composition" 5% CO, 20% H_2O, 11% CO_2, 40% H_2; WSV = 0.333 $Nl\cdot min^{-1}\cdot g^{-1}$).

For all the catalysts, the WGS reaction with only CO and H_2O in the feed produced the highest ξ_{CO} values, reaching nearly the corresponding equilibrium curve (see Fig. 3). This result was not surprising because with CO-H_2O mixture no parasite reactions (such as reverse water gas shift or CO methanation) would occur allowing only CO conversion to take place. The addition of CO_2, H_2 and H_2+CO_2 to the feed gas composition, in fact, reduced ξ_{CO} that dramatically worsened in comparison with the one obtained with the CO-H_2O mixture (Fig. 3).

In particular, adding only CO_2 (Fig. 4.A), the CO conversion decreased dramatically for the catalyst prepared at pH = 7 ($\xi_{CO,max}$ = 0.34), whereas the best performance was reached by the one synthesized at pH = 8.5 ($\xi_{CO,max}$ = 0.73), that showed the highest ξ_{CO} values in the whole temperature range, approaching equilibrium curve at about 525 °C. For all the three catalysts ξ_{CO} worsened much more by adding H_2 to the initial CO + H_2O mixture (Fig. 4.B): for T > 520 °C ξ_{CO} was still far from the equilibrium curve (also the latter lowered due to H_2 in the feed mixture), by reaching a $\xi_{CO,max}$ = 0.47 at pH = 8.5 and $\xi_{CO,max}$ = 0.36 at pH = 10, respectively. Finally, by feeding the standard composition (CO-H_2O-CO_2-H_2, Fig. 4.C), the performance of all the catalysts was really very poor, reaching very low $\xi_{CO,max}$ values (lower than 0.1) between 420 and 470 °C, and the equilibrium curve was never approached. For all the catalysts, at T higher than that of $\xi_{CO,max}$, the outlet CO concentration increased again, with a simultaneous decrease of both H_2 and CO_2 concentrations: since CH_4 in the outlet stream was not detected, most probably the reverse water gas shift reaction took place in a significant extent at higher temperature.

As the 3% Au-CeO_2 catalyst synthesized at pH = 8.5 showed in any case slightly higher performance compared to the other two, a deeper study was carried out by synthesizing new 3% Au-CeO_2 catalysts maintaining the same pH value of 8.5 and reducing the molarity M of the precursor solution to $0.5\cdot10^{-3}$ and $0.2\cdot10^{-3}$. The first tests were performed by feeding only CO-H_2O and the obtained results are shown in Fig. 5. By decreasing M, the shape of ξ_{CO} curves did not changed so much; the lower molarity permitted to reach a slightly higher conversion, up to $\xi_{CO,max} = 0.89$ at 500 °C, practically approaching the equilibrium curve.

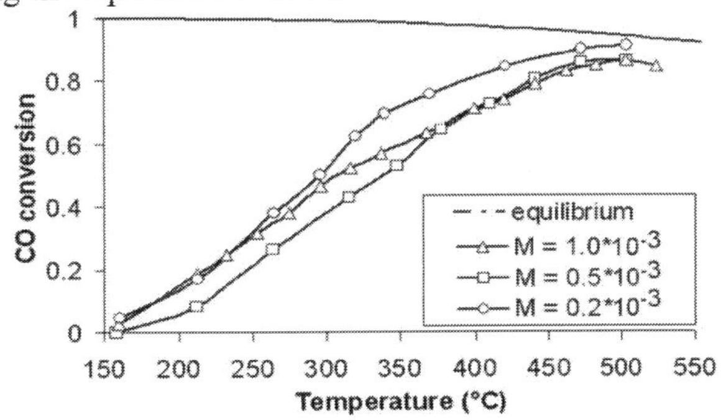

Figure 5: ξ_{CO} vs. T for 3% Au-CeO_2 catalysts prepared at pH = 8.5 and different M (feed gas composition: 5% CO, 20% H_2O in N_2; WSV = 0.333 Nl·min^{-1}·g^{-1}).

Summarizing, the best preparation conditions for the 3% Au-CeO_2 were pH = 8.5 and M = $0.2 \cdot 10^{-3}$: by feeding the simple CO-H_2O mixture, it was possible to obtain the highest CO conversion, well approaching equilibrium curve. Moreover, also with the standard feed this catalyst showed a better performance compared to the other ones prepared at different conditions: $\xi_{CO,max}$ was equal to 0.12 at 420 °C (results not shown), compared to $\xi_{CO,max} = 0.07$ at 470 °C of 3% Au-CeO_2 prepared with the same pH and M = $1 \cdot 10^{-3}$ (Fig. 4.C). These results were in agreement with HRTEM analysis, as the catalyst prepared at slightly basic pH and lowest molarity showed the smallest gold particles (1.5-2 nm) and a more homogeneous clusters distribution (Fig. 2.A), that better promoted catalytic activity. Anyway, all the catalysts showed their highest activity around 500 °C, revealing to be mostly medium temperature (MT) rather than LT WGS catalytic materials, thus possibly better exploitable with higher CO concentration, like those from the reforming reactor.

Furthermore, this catalyst was operated with the simple mixture CO-H_2O for 100 h, with temperature varying from 200 to 500 °C: its performance after this ageing time was compared (see Fig. 6) with that in the fresh status. The equilibrium curve was approached at about the same temperature, with a slightly lower $\xi_{CO,max} = 0.85$ (compared to $\xi_{CO,max} = 0.89$ in the fresh status); besides, on average, the ξ_{CO} curve remained in a lower position for the whole temperature range, thus showing some deactivation effects, probably due to some catalyst sintering.

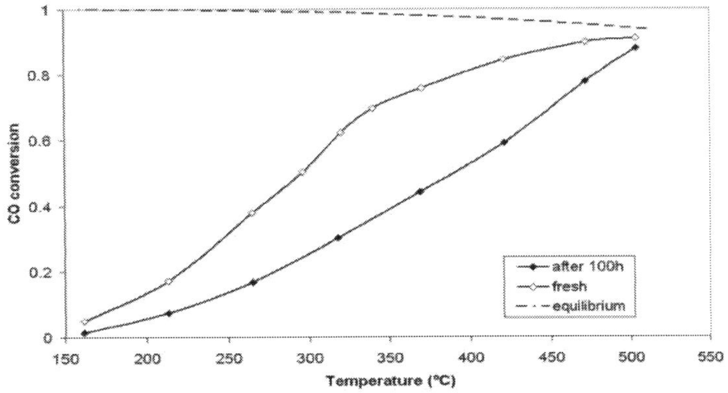

Figure 6: ξ_{CO} vs. T for 3% Au-CeO_2 catalyst prepared at pH = 8.5 and M = $0.2 \cdot 10^{-3}$, fresh and after 100 h of operation (feed gas composition: 5% CO, 20% H_2O in N_2; WSV = 0.333 Nl·min^{-1}·g^{-1}).

Conclusions

Au-based catalysts supported on CeO_2 for CO abatement via WGS reaction were synthesized starting from precursor solutions at different pH and molarity M. A screening of their performance using various synthetic gas mixture compositions was carried out. 3% Au-CeO_2 catalyst prepared with slightly basic pH (8.5) and quite low M ($0.2 \cdot 10^{-3}$) showed the best performance and the smallest Au nano-particles size. By feeding only CO and H_2O, with H_2O/CO ratio equal to 4, catalysts prepared at different pH and M = $1 \cdot 10^{-3}$ approached satisfactorily the equilibrium curve, in particular when pH = 8.5 was used. The performance worsened by increasing or decreasing pH (10 and 7, respectively), in accordance with HRTEM results: catalyst with smaller Au clusters size gave better performance. Catalytic activity tests performed with a realistic reformate feed (containing also H_2 and CO_2), showed fairly low CO conversions, as the undesired reverse water gas shift reaction became more significant. The most promising catalyst was also aged for 100 h of operation and a slight deactivation resulted.

References

1. Fujiwara N., Yasuda K., Ioroi T., Siroma Z., (2002), "Preparation of Pt-Ru onto solid PEM and the application to a DMFC anode", *Electrochimica Acta*, vol. 47, pp. 4079-4084.
2. Farrell C.G., Gardner C.L., Ternan M., (2007), "Experimental and modelling studies of CO poisoning in PEM fuel cells", *Journal of Power Sources*, vol. 171, pp. 282-293.
3. Cheng X., Shi Z., Glass N., Zhang L., Zhang J., Song D., Liu Z.-S., Wang H., Shen J., (2007), "A review of PEM H_2 FC contamination: Impacts, mechanisms, and mitigation", *Journal of Power Sources*, vol. 165, pp. 739-756.
4. Divisek J., Oetjen H.-F., Peinecke V., Schmidt V. M., Stimming U., (1998), "Components for PEM FC systems using H_2 and CO containing fuels", *Electrochimica Acta*, vol. 43, pp. 3811-3815.
5. Passalacqua E., Lufrano F., Squadrito G., Patti A., Giorni L., (2000), "CO tolerance of Pt-W electrocatalysts for PEM FCs", *Journal of New Materials for Electrochemical Systems*, vol. 3, pp. 141-146.
6. Bowers B.J., Zhao J.L., Ruffo M., Khan R., Dattatraya D., Dushman N., Beziat J.-C., Boudjemaa F., (2007), "On-board fuel processor for PEM FC vehicles", *International Journal of Hydrogen Energy*, vol. 32, pp. 1437-1442.
7. DOE guidelines (2009). Multi-Year Research, Development and Demonstration Plan: Planned Program Activities for 2005-2015. Technical Plan Fuel Cells. 3.4, pp. 1-42, http://www1.eere.energy.gov/hydrogenandfuelcells/mypp/
8. Luengnaruemitchai, S. Osuwan, E. Gulari, (2003), "Comparative studies of low-T WGS reaction over Pt/CeO_2, Au/CeO_2, and Au/Fe_2O_3 catalysts", *Catalysis Communication,* vol. 4, pp. 215-221.
9. Ghenciu A. F., (2002), "Review of Fuel Processing Catalysts for Hydrogen Production in PEM Fuel Cell Systems", *Curr. Op. Solid State Mater. Sci.*, vol. 6, pp. 389-399.
10. Haruta M., Yamada N., Kobayashi T. and Iijima S. (1989), "Gold Catalysts Prepared by Coprecipitation for Low-Temperature Oxidation of H_2 and of CO", *Journal of Catalysis*, vol. 115, pp. 301-309.
11. Haruta M, Tsubota S, Kobayashi T, Kageyama H, Genet MJ, Delmon B (1993) "Low-Temperature Oxidation of CO over Gold Supported on TiO_2, α-Fe_2O_3, and Co_3O_4", *Journal of Catalysis*, vol. 144, pp. 175-192.

12. Andreeva D., Idakiev V., Tabakova T. and Andreev A. (1996), "Low-temperature water-gas shift reaction on Au/a-Fe_2O_3 catalyst", *Applied Catalysis A: General*, vol. 134, is. 2, pp. 275-283.
13. Sakurai H., Ueda A., Kobayashi T. and Haruta M. (1997), "Low-temperature water–gas shift reaction over gold deposited on TiO_2", *Chemical Communications*, vol. 3, pp. 271-272.
14. Tabakova T., Idakiev V., Andreeva D. and Mitov I. (2000), "Influence of the microscopic properties of the support on the catalytic activity of Au/ZnO, Au/ZrO_2, Au/Fe_2O_3, Au/Fe_2O_3–ZnO, Au/Fe_2O_3–ZrO_2 catalysts for the WGS reaction", *Applied Catalysis A: General*, vol. 202, is. 1, pp. 91-97.
15. Haruta M. and Daté M. (2001), "Advances in the catalysis of Au nanoparticles", *Applied Catalysis A: General*, vol. 222, pp. 427-437.
16. Haruta M. (1997), "Novel catalysis of gold deposited on metal oxides", *Catalysis Surveys from Japan*, vol. 1, pp. 61-73.
17. Andreeva D., Idakiev V., Tabakova T. and Andreev A. (1996), "Low-Temperature Water–Gas Shift Reaction over Au/a-Fe_2O_3", *Journal of Catalysis*, vol. 158, is. 1, pp. 354-355.
18. Andreeva D., Idakiev V., Tabakova T. and Giovanoli R. (1998), "Low-temperature WGS Reaction on Au/TiO_2, Au/Fe_2O_3 and Au/Co_3O_4", *Bulgarian Chemical Communications*, vol. 30, pp. 59-68.
19. Baiker A., Kilo M., Maciejewski M., Menzi S. and Wokaun A. (1993), In: *Proceedings of the 10th ICC*, 19–24 July 1992, Budapest. Elsevier, Amsterdam, p. 1257.
20. Kozlova A., Sugiyama S., Kozlov A., Asakura K. and Iwasawa Y. (1998), "Iron-Oxide Supported Gold Catalysts Derived from Gold-Phosphine Complex $Au(PPh_3)(NO_3)$: State and Structure of the Support", *Journal of Catalysis*, vol. 176, is. 2, pp. 426-438.
21. Yuan Y., Asakura K., Kozlova A., Wan H., Tsai K. and Iwasawa Y. (1998), "Supported gold catalysis derived from the interaction of a Au–phosphine complex with as-precipitated titanium hydroxide and titanium oxide", *Catalysis Today*, vol. 44, is. 1-4, pp. 333-342.
22. Haruta M. (2002), "Catalysis of Au nanoparticles deposited on metal oxides", *Cattech*, vol. 6, is. 3 pp. 102-115.
23. Casaletto M.P., Longo A., Venezia A.M., Martorana A. and Prestianni A. (2006), "Metal-support and preparation influence on the structural and electronic properties of gold catalysts", *Applied Catalysis A: General*, vol. 302, is. 2, pp. 309-316.
24. Sandoval A., Gomez-Cortes A., Zanella R., Dıaz G. and Saniger J.M. (2007), "Au nanoparticles: Support effects for the WGS reaction", *Journal of Molecular Catalysis A: Chemical*, vol. 278, is. 1-2, pp. 200-208.
25. Andreeva D., Ivanov I., Ilieva L., Sobczak J.W., Avdeev G. and Tabakova T. (2007), "Nanosized gold catalysts supported on ceria and ceria-alumina for WGS reaction: Influence of the preparation method", *Applied Catalysis A: General*, vol. 333, is. 2, pp. 153-160.
26. Idakiev V., Tabakova T., Tenchev K., Yuan Z.Y., Ren T.Z. and Su B.L. (2007), "Gold nanoparticles supported on ceria-modified mesoporous titania as highly active catalysts for low-temperature water-gas shift reaction", *Catalysis Today*, vol. 128, is. 3-4, pp. 223-229.
27. Chang L.H., Sasirekha N., Rajesh B. and Chen Y.-W. (2007), "CO oxidation on CeO_2- and MnO_2-supported Au catalysts", *Separation and Purification Technology*, vol. 58, is. 1, pp. 211-218.
28. Galletti C., Specchia S., Saracco G., SpecchiaV. (2009), "Gold-supported catalysts for Medium Temperature-Water Gas Shift reaction", *Topics in Catalysis*, vol. 52, is. 6-7, pp.688-692.
29. Patil, K.C., Aruna, S. T.; Mimani, T., (2002), "Combustion synthesis: an update", *Current Opinion in Solid State Materials Science*, vol. 6, pp. 507-512.
30. D.S. Newsome (1980), "The Water-Gas Shift Reaction", *Catalysis Reviews Science and Engineering*, vol. 21, is. 2, pp. 275-318.

(7c) Sulfur Poisoning Resistant Palladium/Gold Alloy Membranes for High Temperature Hydrogen Separation

Gokhan O. Alptekin, TDA Research, Inc., Wheat Ridge, CO

Sarah DeVoss, TDA Research, Inc., Wheat Ridge, CO

Stephen N. Paglieri, TDA Research, Inc., Wheat Ridge, CO

J. Douglas Way, Chemical Engineering, Colorado School of Mines, Golden, CO

TDA Research, Inc., in collaboration with Colorado School of Mines (CSM) is developing a water-gas-shift membrane reactor that combines a sulfur tolerant, PdAu alloy composite membrane and a sour shift catalyst. The membrane reactor converts sulfur-containing CO-rich synthesis gas into a pure hydrogen permeate and a high pressure CO2-rich retentate stream.

This paper discusses the potential of using the PdAu membranes in the novel membrane water-gas-shift reactor. PdAu alloys have been shown to resist poisoning by H2S. Our fabrication process allows preparation of a composite membrane with a micron thick palladium alloy layer, providing high flux and acceptable selectivity at the minimum cost. The effect of the water-gas-shift components on the performance of the PdAu membranes was examined at the conditions of interest. In bench-scale tests, poisoning and/or inhibition effects of CO, CO2, H2O and H2S on separation properties were examined at 430 kPa and at 350-450°C temperature range.

See more of this Session: Advancements in Hydrogen Technology I
See more of this Group/Topical: Topical 5: Clean Fuels and Energy Efficient Processes

(7d) Separation Efficiencies of Hydrogen From Steam Gasification Products – Thermodynamic Evaluation

Bhanu Kaini, Department of Mechanical Engineering and Energy Processes, Southern Illinois University, Carbondale, IL

Kanchan Mondal, Department of Mechanical Engineering and Energy Processes, Southern Illinois University, Carbondale, IL

The paper presents a procedure for the comparison of various technologies for the production of hydrogen from carbon based materials based on First and Second Law Analysis. All known technologies in publicly available literature are analyzed based on calculations of the energetic and exergetic analysis of the whole process as well as the sub processes. Special emphasis has been given to chemical separations. The thermodynamic limits to the separation efficiencies for any given hydrogen concentration has been identified. Methane was chosen as a model compound for evaluating single stage separation. The incremental changes in efficiencies as a function of process parameters such as temperature, steam amount, chemical type and amount were estimated. Energy and exergy losses associated with generation of syngas, separation of hydrogen from COx as well as exergetic loss associated with emissions are presented. The optimal conditions for each process by minimizing these losses are presented. Based on the results, the most energy efficient process for a given hydrogen purity is suggested along with the optimal conditions for operation. The primary chemical processes that were considered were simple steam gasification, steam gasification with CO2 capture using chemical looping of a reactive carbon dioxide removal agent, steam gasification with oxygen carrier, steam gasification with Boudouard reaction catalyst and steam gasification with Boudouard reaction catalyst and reactive carbon dioxide removal agent in a chemical looping system. The information presented will clearly demonstrate the differences in efficiencies between the various chemical looping processes for hydrogen generation. The data obtained from the theoretical studies will be compared with data from experimental research available in literature and in our laboratories to establish the degree of inefficiency during practical applications. Finally, the protocol will be used for other hydrogen production systems such as membrane separations, pressure swqing absorption etc.

See more of this Session: Advancements in Hydrogen Technology I
See more of this Group/Topical: Topical 5: Clean Fuels and Energy Efficient Processes

Combustion Testing of Hydrogen-Enriched Natural Gas (HENG)

P. Gogolek, P. Hughes, A. Caverly, O. Ramadan
CanmetENERGY, Natural Resources Canada, Ottawa, ON, Canada
T. Whidden
Altantic Hydrogen Inc., Fredricton, NB, Canada

Introduction

Hydrogen-enriched natural gas (HENG) fuel provides pre-combustion removal of carbon from the natural gas and other potential benefits such as reduced emissions of NO_X and improved furnace efficiency by changing the thermal radiation and flame stability characteristics. The objective of this work was to provide a scoping investigation of the change in combustion performance as the amount of hydrogen in the fuel was increased. Controlled combustion trials were performed in the CanmetENERGY research tunnel furnace (RTF). Details of the RTF are in Gogolek et al. (2009) and Hughes et al. (2009).

Equipment

Burner

The chosen burner was a commercial self-regenerative burner (SRB). Each regenerator nozzle pair has a honeycomb type ceramic insert and the nozzles alternate between blowing air and sucking furnace gas. Inlet air temperature is within 100°C of the furnace temperature and hot air and fuel do not mix until a large amount of furnace gas has been entrained. Combustion reactions are broadly distributed in the furnace, approximating a well-stirred reactor. This avoids high flame temperatures, reducing levels of thermal NO_X and making the prompt NO_X formation mechanism proportionally more important.

Fuel Supply System

The fuel was a mixture of utility natural gas and industrial grade hydrogen. The hydrogen was introduced into the natural gas stream along the centre-line at the back of the burner. The fuel line to the burner tip was over 20 pipe diameters long, providing good mixing of the two gas streams.

Furnace

The Research Tunnel Furnace (RTF) is shown in Figure 1. The entrance of the furnace is lined with refractory wool to provide an adiabatic region for the burner. This section has an opening on one side to allow probes to be inserted. There is an exposed heat exchange coil located 1.9 m from the burner face providing a thermal load for the furnace. The inlet and outlet temperatures and flow rate are measured and logged.

Probes and other measurement equipment

A suction pyrometer measured the furnace gas temperature at the centre line of the furnace, midway between the burner and the dividing wall, using. In this pyrometer the furnace gas is drawn over a thermocouple within the probe that is shielded from the thermal radiation in the furnace, minimizing radiation errors in measurements of the gas temperature. Probe ports are at 15 cm, 30 cm, 45 cm and 60 cm from the burner face. The air temperature for the regenerative burner was measured by a thermocouple in the regenerator just before the entrance to the furnace. The nominal furnace temperature was measured with a thermocouple close to the wall near the burner face. The incident radiant heat flux at the inside wall of the furnace was measured using a 2π radiometer calibrated with a black body furnace (Figure 2). The probe has an entrance acceptance angle of ~2π steradians. The composition of gases exiting the furnace was measured with continuous analyzers for O_2, CO_2, NO_x, and CO. The electronic data (flow rates, temperatures, analyzer measurements) were logged every 2 seconds.

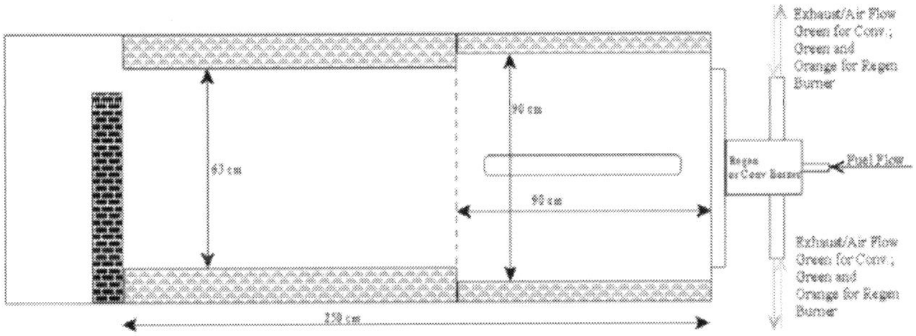

Schematic of the Interior of the Research Tunnel Furnace

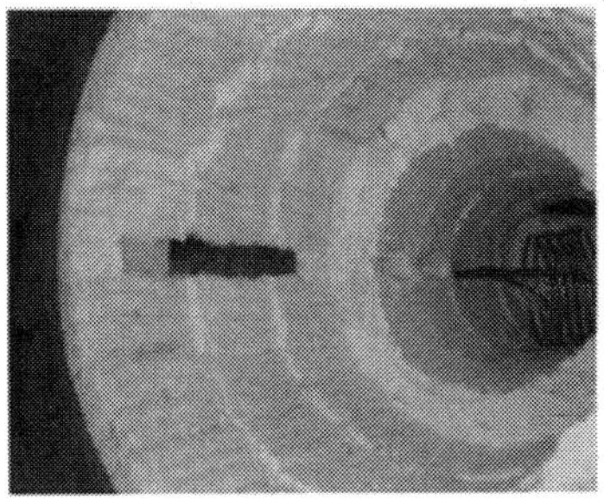

Photograph of the Interior of the Research Tunnel Furnace Showing the Insulation and Heat Exchanger

Figure 1: Interior of the Research Tunnel Furnace showing the cylindrical cavity formed in the furnace.

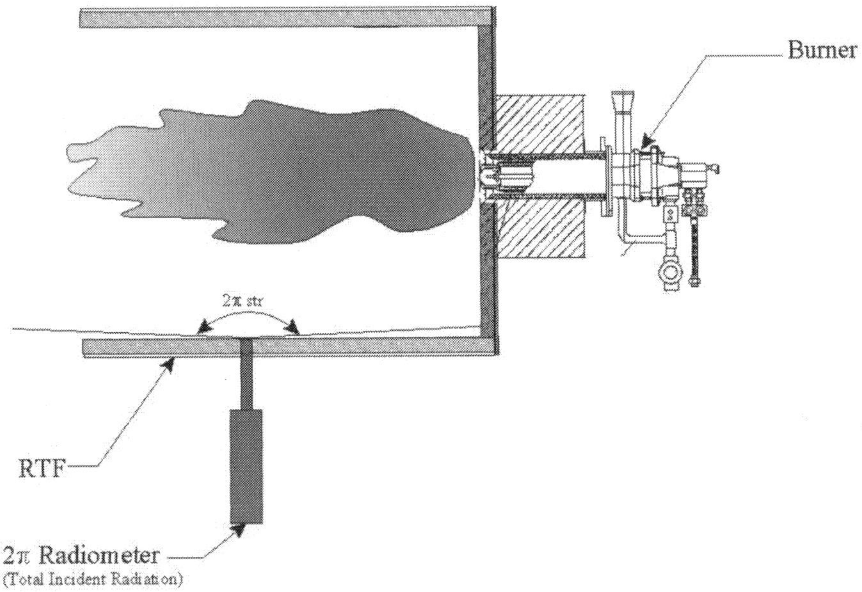

Figure 2: Orientation of the 2π radiometers in the test furnace.

Results and Discussion

The tests evaluated the effect of HENG on furnace efficiency and emissions of NO_X. The tests employed a firing rate of 200 kW with hydrogen contents from 2%-v to 20%-v. Excess air, which affects the rate of formation of NO_X, was also evaluated as an operating variable. More air produces lower furnace temperature, reducing both the furnace efficiency and the emissions of NO_X.

Tables 1 and 2 show the run conditions for these tests for 10% excess air and 20% excess air respectively. The average flows and the standard deviations (σ) of the time series are listed. The standard deviations are reported as the percentage of the average. Note that the standard deviations are all less than 1%, indicating that the runs were well-controlled.

The stack gas concentrations are given in Tables 3 and 4. The standard deviations of the time series (as a percentage of the average) for O_2, CO_2, and NO_X are small. The standard deviation of the time series for CO is large. That is to be expected with the alternating "suck and blow" of the SRB. The perturbation caused by this change in flow direction lasts less than 1 second during the 20 second cycle time. The theoretical oxygen, the expected stack oxygen concentration for the given flow rates of air and fuel, is 2% for the 10% excess air trials and 3.7% for the 20% excess air trials. The measured concentrations of oxygen were 2.1% and 1.8% higher on average respectively, caused by uncontrolled air infiltration into the furnace.

NO_x emissions

Figures 3 and 4 present the NO_X emissions from the long duration tests with 10% excess air and 20% excess air respectively. The error bars are at 3% and 3.4% of the measurement points for the two figures, these two percentages are the averages of the standard deviations for the NO_X (see Tables 3 and 4) for the 10% excess air and 20% excess air trials respectively. The baseline condition of natural gas without hydrogen addition is included. These baselines were at the beginning of the test day, with the lowest furnace temperature. The results with 20% excess air show higher NO_X that at 10% excess air. In both cases, the tests with hydrogen addition produced lower NO_X emissions. The emissions of NO_X can be presented as an emission factor with units of g/MJ fired. This avoids the difficulty of comparing concentration levels at different levels of dilution by combustion air and infiltrated air. The emission factors for NO_X are plotted against hydrogen content in Figure 5. Doubling the excess air level from 10% to 20% almost doubles the emission factor for NO_X. It is also clear that the addition of even 2% hydrogen reduces the emission factor for NO_X.

Flame properties and furnace efficiency

The SRB does not produce an intense visible flame as would be found with a conventional burner. The temperature and gas composition is approximately uniform over the furnace volume. There are three measurements that provide an indication of the impact of the addition of hydrogen on furnace efficiency: the gas temperature; the incident radiation at the wall; and the heat extracted via the heat exchange coil in the furnace. The furnace gas temperature with the SRB is much lower than the adiabatic flame temperature, much closer to the temperature of a well-stirred reactor. The gas temperature is approximately 200°C higher than the nominal furnace temperature and in the range where thermal NO_X will be formed, although at much lower levels than with a conventional burner. Figure 6 shows the gas temperature plotted against the hydrogen content in the fuel gas. There is no discernable correlation.

Table 1 - Actual run conditions for trials with 10% excess air, with the standard deviation expressed as a percentage of the average.

Run #	Natural Gas	σ	Hydrogen	σ	Air	σ	H_2 content
	kg/h	%	kg/h	%	kg/h	%	%
1	13.6	0.04	0.08	0.74	245.8	0.94	5.0
2	12.8	0.06	0.37	0.47	242.8	0.97	19.8
3	13.1	0.04	0.25	0.36	242.6	0.92	14.0
4	13.6	0.05	0.15	0.40	242.6	0.89	8.8
5	13.7	0.07	0.03	0.87	243.3	0.89	2.0

Table 2- Actual run conditions for trials with 20% excess air, with the standard deviation expressed as a percentage of the average.

Run #	Natural Gas	σ	Hydrogen	σ	Air	σ	H_2 content
	kg/h	%	kg/h	%	kg/h	%	%
1	12.90	0.05	0.084	0.74	261.3	0.87	5.2
2	13.22	0.05	0.373	0.47	261.3	0.83	19.4
3	13.69	0.04	0.252	0.36	264.5	0.85	13.5
4	13.61	0.05	0.155	0.40	264.2	0.87	8.8
5	13.38	0.05	0.033	0.87	264.0	0.84	2.0

Table 3 - The stack gas composition with averages and standard deviations as a percentage of the average for trials with 10% excess air.

Run #	Average concentration				σ			
	O_2	CO_2	CO	NO	O_2	CO_2	CO	NO
	%	%	ppm	ppm	%	%	%	%
1	4.3	9.0	73.7	35.4	1.6	1.3	22.5	3.1
2	4.3	8.6	119.7	37.1	3.9	1.5	39.0	4.4
3	4.1	8.9	48.6	36.6	2.2	1.4	36.0	3.1
4	3.8	9.2	166.5	34.8	1.8	1.3	31.6	2.7
5	3.7	9.4	274.4	35.1	2.2	1.4	14.3	2.9

Table 4- The stack gas composition with averages and standard deviations as a percentage of the average for trials with 20% excess air.

Run #	Average concentration				σ			
	O_2	CO_2	CO	NO	O_2	CO_2	CO	NO
	%	%	ppm	ppm	%	%	%	%
1	5.4	7.9	4.3	57.3	7.4	3.4	25.0	7.8
2	5.1	8.2	5.1	57.9	1.3	1.2	15.1	2.9
3	5.4	8.4	5.8	60.0	1.2	1.1	24.6	2.7
4	5.1	8.5	7.6	57.2	1.1	1.0	37.7	2.7
5	5.2	8.3	6.3	59.4	1.3	1.1	37.9	2.5

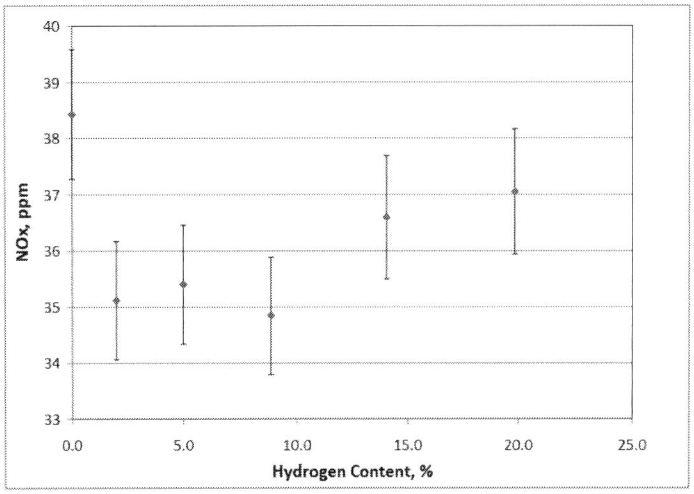

Figure 3 - NO_x emissions versus hydrogen content for the tests with 10% excess air. The error bars are 3% of the measurement.

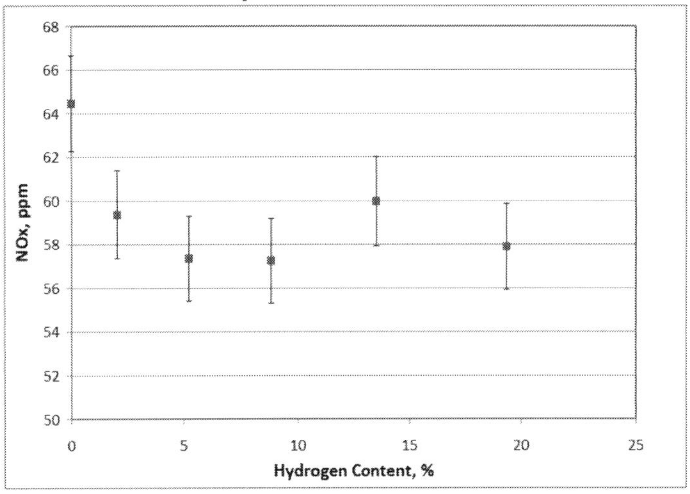

Figure 4 - NO_x emissions versus hydrogen content for the tests with 20% excess air. The error bars are 3.4% of the measured value.

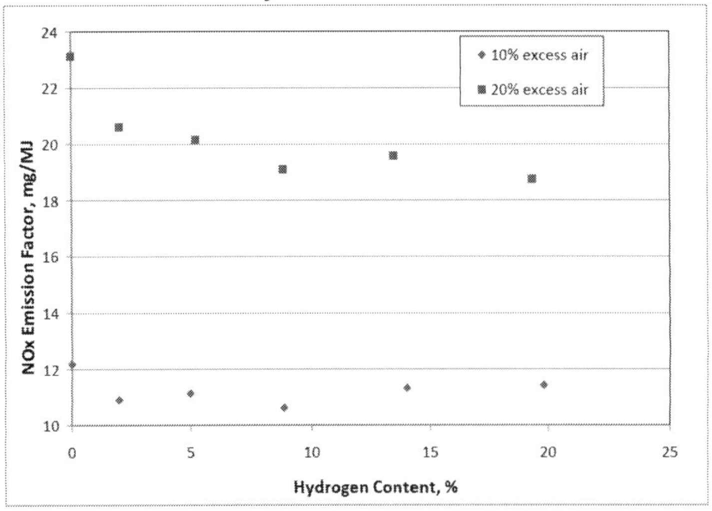

Figure 5 - NO_X emission factor on energy basis including baseline results with no hydrogen addition.

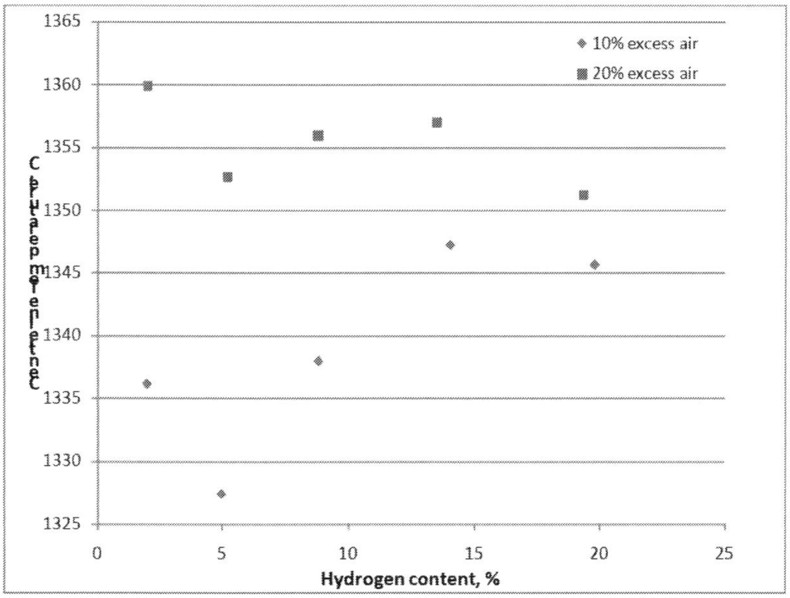

Figure 6 – Centre-line gas temperature plotted against the hydrogen content in the fuel gas.

The incident radiation at the wall is expected to be related primarily to the furnace temperature and not the gas temperature. Figure 7 plots the incident radiation at the wall against the nominal furnace temperature and there is a strong correlation, because the SRB operates in flameless combustion mode and the dominant emitter of thermal radiation is the furnace walls. There appears to be an anomalously low measurement at hydrogen content of 5%-v in Figures 6 and 8. This is the unfortunate result of using the same sequence of tests in the two different excess air levels. The hydrogen level of 5%-v was the first test of each day and therefore with the lowest furnace temperature.

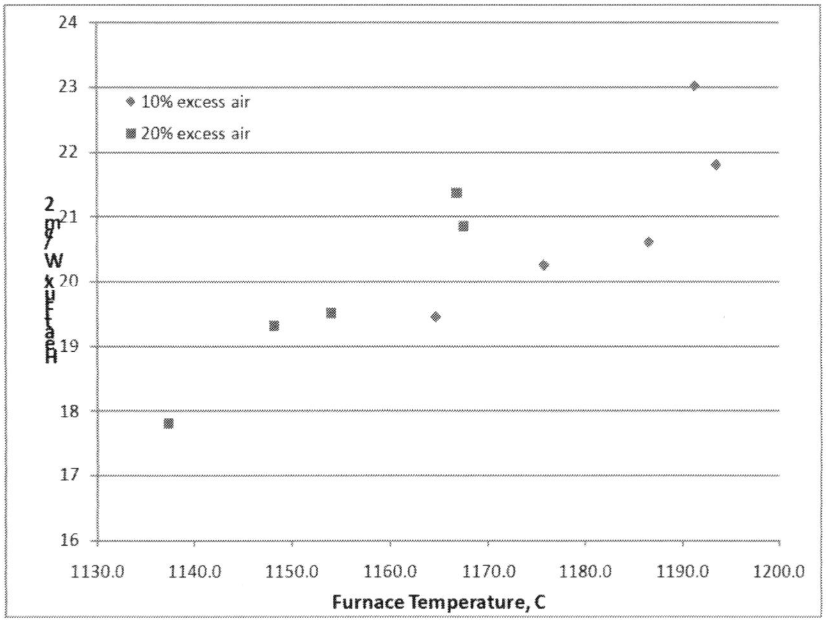

Figure 7 - Radiation at the wall as measured by the radiometer correlates with the furnace temperature.

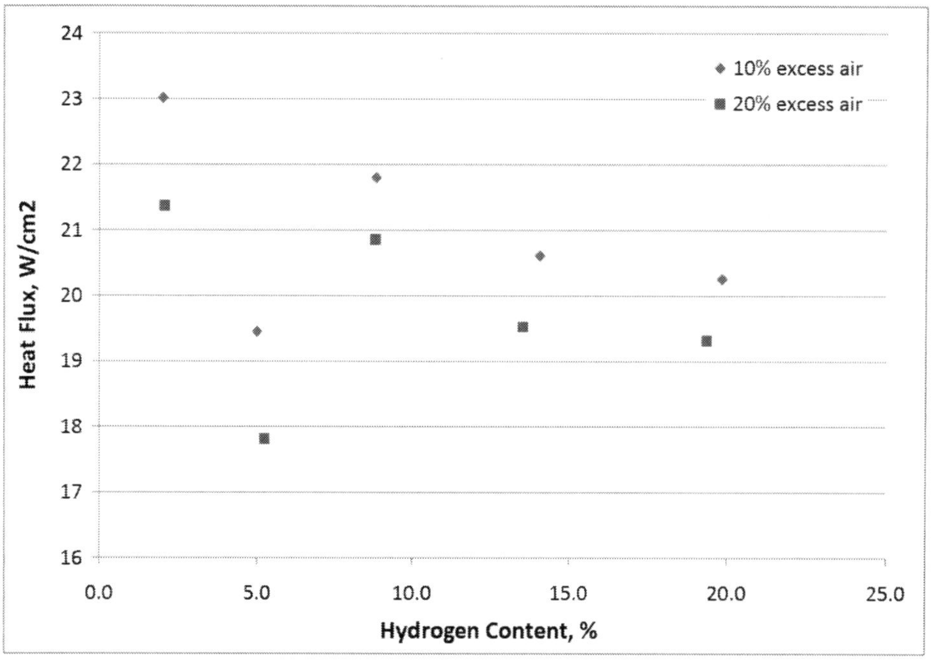

Figure 8 - The radiation incident at the wall decreases with increasing hydrogen content in the fuel.

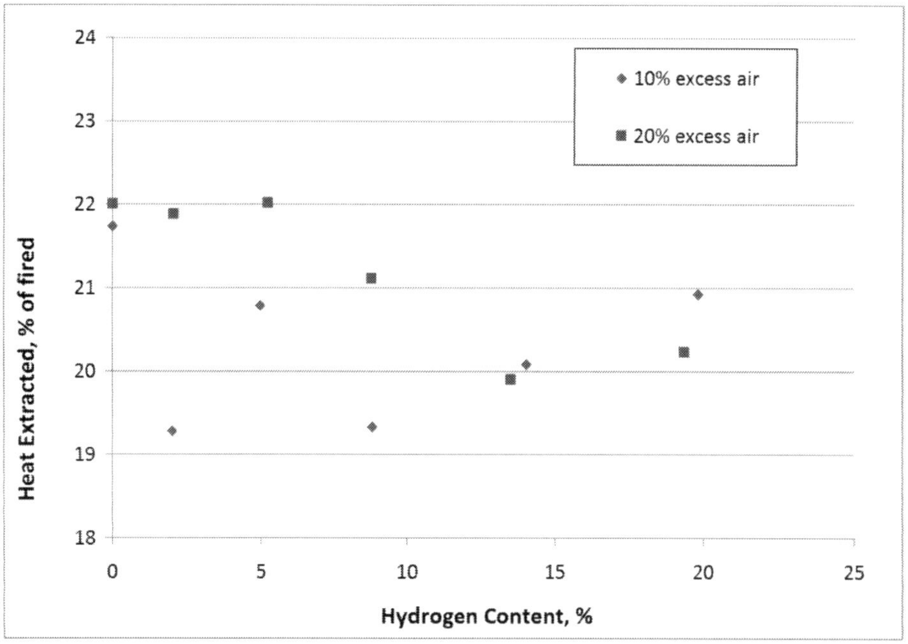

Figure 9 - Heat extracted via the heat exchange coil as a percentage of total firing rate as the hydrogen content changed.

Conclusion

The heat extracted via the heat exchange coil is our third measure of the furnace efficiency. The heat exchange coil is the model for the thermal load and receives both convective and radiative heat transfer. The heat extracted, as a fraction of the total heat input, is a good measure of the effectiveness of the furnace and the burner in getting heat into the load, and is plotted in Figure 9. There is a very small decrease in the heat extraction efficiency with increased hydrogen content. This can be attributed to the decrease in incident radiation as shown by the radiometer measurements above. The conclusion of these measurements is that the increased hydrogen content has a small negative impact on the thermal efficiency of the furnace.

Testing of the combustion properties of hydrogen-enhanced natural gas was performed in the Research Tunnel Furnace of CanmetENERGY. The firing rate was fixed at 200 kW. The hydrogen contents tested were 1%, 4%, 9%, 14% and 20%. Excess air level was 10% and 20%. A commercial self-regenerative burner (SRB) was used for the testing. Gas temperature was measured using a suction pyrometer on the centre-line of the furnace. A radiometer was used to measure the radiant heat flux at the wall. A heat exchange coil provided a model thermal load for the furnace.

It was found that NO_X emissions decreased with hydrogen addition at both excess air levels. The heat extraction efficiency, as determined by the percentage of the heat input extracted in the heat exchange coil, displayed a small decrease with increasing hydrogen content in the fuel. Taken together, these measurements show that a small addition of H_2 produces a significant reduction of NO_X emissions compared to natural gas and had no further effect as the amount of H_2 was increased. This small addition of H_2 had very little effect on the efficiency of the furnace.

References

Gogolek, P., A. Caverly, O. Ramadan, P. Hughes, "Combustion Testing of Hydrogen Enriched Natural Gas", Report to AHI, September, 2009.

Hughes, P. M. Lacelle, R. J., Idris, A., Legere, M., Percy, D., Wong, J., Parameswaran, T., "CARS and Heat Flux Measurements in Regenerative and Conventional Industrial-Scale Burners", Industrial Combustion, International Flame Research Foundation on-line Journal, May, 2009.

Ramadan, O., P. Gogolek, A. Caverly, P. Hughes "Combustion of Natural Gas Enhanced with Hydrogen – Literature Review", Report to AHI, July, 2009.

Acknowledgements

All the testing was performed by A. Idris, with the expert assistance of O. Ramadan, R. Lycett, and M. Legere. Drs. O. Ramadan and C. Lam made many useful comments to improve the report.

(7f) Ultrasmall-Angle X-Ray Scattering (USAXS) Studies of the Effect of Transition Metal (Ti2+, Ti3 +, V3+, Fe3+, and Zr4+) Dopants in NaAlH4 Powders

Tabbetha A. Dobbins, Institute for Micromanufacturing/Physics, Louisiana Tech University, Ruston, LA

Transition metal dopants added to complex metal hydrides, specifically to sodium aluminum hydride ($NaAlH_4$), by high energy ball milling enhances dehydrogenation kinetics and induces dehydrogenation reaction reversibility. This study uses ultrasmall angle x-ray scattering (USAXS) to elucidate differences in $NaAlH_4$ particle morphology as dopant type and mill time is varied. Five dopant types were used. Two dopant types were used to represent the best kinetic enhancements having high desorption rates (e. g. $TiCl_2$, $TiCl_3$) and were compared with two dopant types which do not perform as well (e.g. $ZrCl_4$, $FeCl_3$ and VCl_3). USAXS data for the doped hydrides were compared with undoped and milled $NaAlH_4$ powders. Mill times used were 0 minutes (blended), 1 minute, 5 minutes, and 25 minutes. As indicated by USAXS power-law scattering data, the undoped $NaAlH_4$ powders are comprised of primary particles having a high surface area (surface fractal geometries) and of smooth colloidal particles. The powders, when catalyzed with a single transition metal cation, and milled for 1 minute, 5 minutes, and 25 minutes show uniformly decreasing hydride particle surface area with mill time. surface area using measured particle sizes and volume fractions. The variation in desorption rates in those co-doped systems were correlated with changes in powder surface area.

See more of this Session: Advancements in Hydrogen Technology I

See more of this Group/Topical: Topical 5: Clean Fuels and Energy Efficient Processes

(10a) Future Role and Characteristics of the Fischer-Tropsch Technology

Rafael Espinoza, Emerging Fuels Technology, Tulsa, OK

Kenneth Agee, Emerging Fuels Technology, Tulsa, OK

In the past, the development of the FT technology took many years and hundreds of millions of dollars were spent by companies such as Sasol, Shell, Exxon, Conoco, Syntroleum, Rentech, etc.

The traditional approach was the development of the FT technology oriented towards large plants (i.e., in excess of 30 000 bpd, which limited the number of players and opportunities in this field.

Today, the price of oil is in the $80's per barrel and it is generally agreed that the days where the oil price is consistently below the $50's are over.

Following the higher oil price, the FT development over the past few years has been oriented towards smaller commercial plants and on non-slurry bed reactors. Currently, the FT technology is profitable at scales much smaller than before. Plants of 500 to 1000 bpd are now economically feasible, depending upon its location and application.

In the author's opinion, the technological monopoly of the traditional FT players is over. This "new" FT technology will allow smaller companies to enter the FT field, including those using non-traditional sources of carbon like biomass and municipal waste.

It is envisaged that during the next decade the expansion of the FT industry will include a large number of small to medium size FT plants, flexibly designed for specific applications.

The lowering of the FT plants size greatly increases the number of potential opportunities in the US. The strategies for the plants operation, catalyst maintenance, FT primary products transformation and handling, IBL designs, will have to be modified accordingly.

See more of this Session: Coal, Biomass and Natural Gas to Liquids I
See more of this Group/Topical: Topical 6: 10th Topical Conference on Gas Utilization

(10b) A Study of Radial Heat Transfer in a Tubular Fischer-Tropsch Synthesis Reactor

Xiaowei Zhu, Center of Material and Process Synthesis, University of the Witwatersrand, Johannesburg, South Africa

Diane Hildebrandt, Center of Material and Process Synthesis, University of the Witwatersrand, Johannesburg, South Africa

David Glasser, Center of Material and Process Synthesis, University of the Witwatersrand, Johannesburg, South Africa

Summary

A series of heat transfer experiments based on TiO_2 supported Fischer-Tropsch (FT) Co catalyst with and without reaction were performed in a relatively large ID tubular reactor. The first group of experiments was carried out with a heater in the center of the catalytic bed to simulate different FT reaction conditions and the second group of experiments was run at typical low temperature FTS conditions. The temperature profiles in the reactor with a heater in the center of the catalytic bed and with FT reaction in the reactor were measured under a range of conditions. The aim of this work is to derive a radial heat transfer model for tubular FTS reactor which can be applied for FTS reactor design.

Keywords

Fischer-Tropsch Synthesis, radial heat transfer, tubular reactor, heat transfer in packed beds.

Introduction

Due to the global increase in energy costs, the Fischer-Tropsch Synthesis (FTS) reaction, which is a well known catalytic reaction that yields synthetic fuels from a mixture of CO and H_2, has received renewed industrial and academic interest for approximately the last two decades[1, 2]. At the same time, the fixed bed reactor is widely used in the chemical industry because of certain advantages when is compared to other kinds of reactors. When it is used for FTS, the applicable tube diameter is limited due to the large amount of heat generated and the poor heat transfer in the catalyst bed.

The FTS reaction is highly exothermic reaction and has an adiabatic temperature rise of up to 1750°C. Because of this, efficient heat removal is of crucial importance because too high temperatures may lead to a loss in selectivity, possible thermal runaways and rapid deactivation of the catalyst.

Experimental

In this study, a tubular reactor with an inner diameter of 23 mm is used to investigate the radial heat transfer in a fixed bed FTS reactor. Thermo-wells were built into the reactor at different radial positions. Thermocouples could be moved up and down to measure the temperature at different axial positions. A heating coil was wrapped around the reactor and the heat input to this controlled to give a fixed value of wall temperature which was controlled using a temperature controller. Two subsidiary heaters were placed at the two ends of the reactor and these were adjusted to give a flat axial temperature profile (<1C) under no flow conditions.

Two groups of experiments were carried out in the same catalytic bed. For the first group of experiments, the reactor was fed with N_2 instead of reactant gases. The middle thermocouple was used as a heater to generate various amounts of heat to obtain different temperature profiles in the bed. For the second group of experiments, syngas was fed into reactor and FTS reaction was performed at typical low temperature FTS conditions. Different control temperatures and flow rates were chosen. Fig.1 and Fig.2 show typical measured temperature profiles with a heater in the center of the catalytic bed and the FT reaction in the reactor respectively.

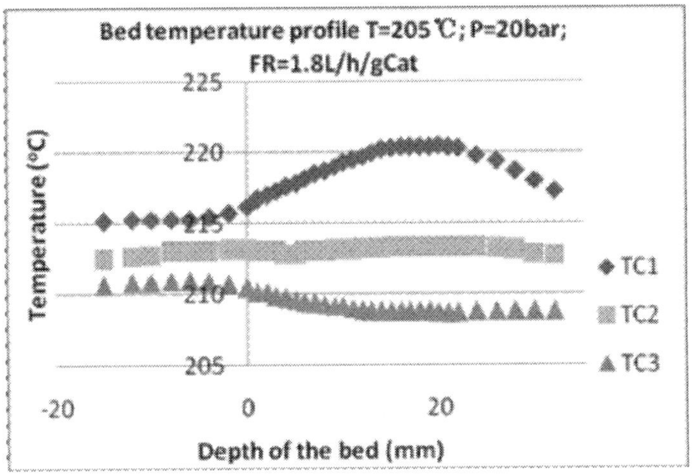

Fig.1 the temperature profile in the bed with the heater in the center, the bed depth from 0 to 22 is the catalytic bed and the other part is the bed support which is loaded with ceramic balls; the radius for TC1, TC2 and TC3 are at radius of 3.5, 6.6 and 10.9mm respectively measured from the inside wall.

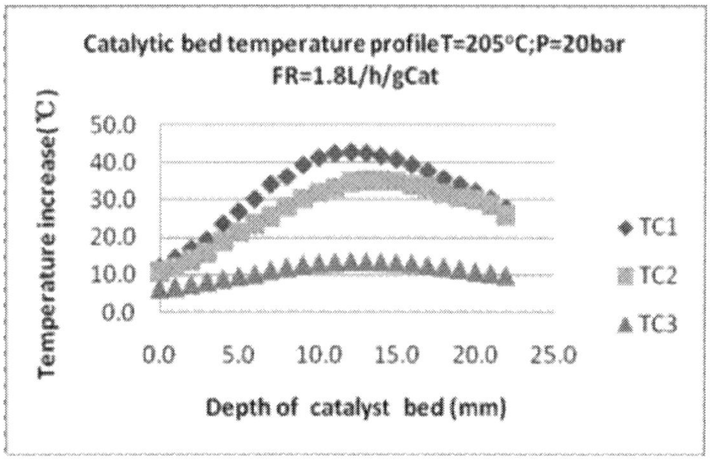

Fig.2 the catalytic bed temperature profile when the FT reaction occurs in the reactor (the temperature increase across the bed can be as high as 42°C)

Discussion

Experiments were performed for a range of flow-rates with only nitrogen flowing and different heat inputs. The experiments were repeated using similar conditions but with a feed of syngas which then caused reaction to happen in the bed.

An effective radial heat transfer coefficients for the bed with nitrogen flow was estimated using the experimental results. Using these values and measured rates of reaction in the reacting system, the heat transfer model was checked for accuracy. The results from this analysis will be discussed with reasons for discrepancies highlighted.

References

ADDIN EN.REFLIST (1) Dry, M. E. Present and future applications of the Fischer¨CTropsch process. *Applied Catalysis* **2004,** *276*, 1.

ADDIN EN.REFLIST (2) Schulz, H. Short history and present trends of Fischer¨CTropsch synthesis. *Applied Catalysis* **1999,** *186*, 3.

See more of this Session: Coal, Biomass and Natural Gas to Liquids I
See more of this Group/Topical: Topical 6: 10th Topical Conference on Gas Utilization

(10c) A Thermodynamic Approach to the Olefin Products Distribution in Fischer-Tropsch Synthesis

Xiaojun Lu, Center of Material and Process Synthesis, University of the Witwatersrand, Johannesburg, South Africa

Diane Hildebrandt, Center of Material and Process Synthesis, University of the Witwatersrand, Johannesburg, South Africa

David Glasser, Center of Material and Process Synthesis, University of the Witwatersrand, Johannesburg, South Africa

The Fischer-Tropsch Synthesis (FTS) is a network of parallel and consecutive reactions that take place within catalyst pores filled with waxy liquid hydrocarbon products. The product distribution follows the so called ASF (Anderson-Schulz-Flory) model. Although product distribution models have been developed using a kinetic approach, the kinetic models can only describe the product distribution but cannot explain why they follow this particular distribution. On the other hand, researchers have also tried to explain the formation of each single group of products from a catalytic point of view, but this approach is also not fully able to describe the experimentally found product distributions. This work tries to explain a small but very important part of the FTS reaction namely the olefin distribution that is found in both a CSTR(Autoclave Engineers, 100ml in volume) and a PFR(Autoclave Engineers, 8mm in ID). Conventional FT experiments were carried out both in a CSTR and a PFR at typical low temperature FTS conditions. The olefin to paraffin ratios for short chain hydrocarbons (C2-C5) were found to change time even when reaction conditions are kept constant in the CSTR, while the ratio of adjacent olefins (for example C4H8/C3H6) remained unchanged (see Fig). The ratios for $C_nH_{2n}/C_{n-1}H_{2(n-1)}$ from two different temperatures(465K and 485K) for the same carbon number n are found to be more of less the same. For the PFR, the temperature was varied from 485K to 523K and flow rate (FR) was varied from 1.8 to 5.4 NL/h/gcat and the ratio of adjacent olefins was also found to be constant. Furthermore, the ratio for the same carbon number n in the CSTR and the PFR are the same. A simple reaction network for FTS is assumed and studied based on the reaction characteristic: $C_{n-1}H_{2(n-1)}$ C_nH_{2n} $C_{n-1}H_{2(n-1)+2}$ The thermodynamic model for olefin distribution in this work is developed based on: $C_{n-1}H_{2(n-1)}+C_{n+1}H_{2(n+1)}=2C_nH_{2n}$ The thermodynamic equilibrium constant derived from thermodynamic calculations and experiments are compared. The results from experiments suggest that the olefin distribution in FTS might be described by thermodynamic equilibrium.

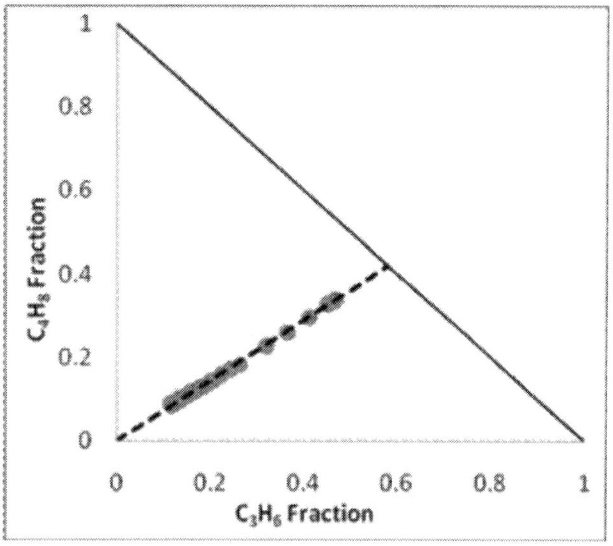

See more of this Session: Coal, Biomass and Natural Gas to Liquids I
See more of this Group/Topical: Topical 6: 10th Topical Conference on Gas Utilization

(10d) Application of Compact Heat Exchange Reactor in Fischer-Tropsch Reaction

Zhijun Jia, Chart Energy & Chemicals, La Crosse, WI

Larry Stryker, Chart Energy & Chemicals, La Crosse, WI

Dennis E. Sparks, Center for Applied Energy Research, University of Kentucky, Lexington, KY

Burtron H. Davis, Center for Applied Energy Research, University of Kentucky, Lexington, KY

Chart's compact heat exchange reactor has been used in various applications especially for highly exothermic or endothermic reactions. FT reaction is one of the applications. The reactor consists of multiple reaction and cooling layers alternatively arranged. Each reaction layer contains multiple mini-channels; each of them is in mm scale. Catalysts, provided by Burt Davis's group at University of Kentucky's Center for Applied Energy Research, are packed in those mini-channels. The tested reactor is made of aluminum which provides an excellent heat transfer character. The operating conditions are 220°C at 15 bars. The reactors were continuously running the reactor for three weeks. The results indicated that the compact heat exchange reactor provided unique features to improve the process efficiency and productivity.

See more of this Session: Coal, Biomass and Natural Gas to Liquids I

See more of this Group/Topical: Topical 6: 10th Topical Conference on Gas Utilization

(24a) Energy Security and Affordability in the 21st Century

John Hofmeister, Citizens for Affordable Energy, Houston, TX

Hofmeister, former President of Shell Oil Company, and founder of the non-profit association, Citizens for Affordable Energy, tackles two of the most critical concerns facing industry today – energy availability and affordability. While green energy is helpful, it's inadequate for the demands of the future. Technology and innovation are important elements, but do they offer realistic solutions right now? Add in an aging infrastructure as a potential disabler to economic competitiveness and the picture is bleak. But it's not all doom and gloom, and it seems the answers are well within our reach. According to Hofmeister, investment in the traditional energy base is the path to a competitive affordable energy supply. And as an added bonus, necessary environmental protections can be value adding to the economy if managed wisely.

See more of this Session: Energy Leadership Keynote Address - John Hofmeister, Founder, Citizens for Affordable Energy - Sponsored by the AIChE Fuels and Petrochemicals Division
See more of this Group/Topical: Liaison Functions

(30a) Effects of Catalytic Supports On the Structure and Dehydrogenation of Ammonia Borane

Fernando Soto, Engineering, Louisiana Tech University, Ruston, LA

Aditya Khandavelli, Chemical Engineering, Louisiana Tech University, Ruston, LA

Scott A. Gold, Institute for Micromanufacturing, Louisiana Tech University, Ruston, LA

Daniela S. Mainardi, Department of Chemical Engineering/ Institute for Micromanufacturing, Louisiana Tech University, Ruston, LA

Hydrogen gas is encouraged as a fuel as it is more energy efficient and environmental friendly when compared to other conventional fuels [1]. However, the main concern has been the storage of hydrogen for its effective usage as a fuel. Recently, hydrogen storage using metal hydrides and chemical hydrides has been receiving great attention due to their higher gravimetric storage. Among the chemical hydrides, B-N compounds are much attractive as both B and N are lighter elements and have an ability to bear multiple hydrogen atoms [1]. Ammonia Borane (AB), which has a gravimetric and volumetric hydrogen storage densities up to 19.6 wt. % and 0.16 kg-H_2/L upon complete dissociation of AB [2, 3], is considered to be one of the best materials for hydrogen storage.

The present work includes computational studies of the structure and the dehydrogenation of AB. Based on the experimental studies performed, hydrogen release is observed when the AB solution came in contact with the Zirconia catalytic support. Computational studies are performed to interpret the experimental results of the above mentioned reaction to gain a better understanding of the reaction thermodynamics and mechanisms, which would help in design of new catalysts. Quantum Mechanics (QM) and Molecular Dynamics (MD) simulations are performed using CASTEP programming package to analyze structure, stability and dehydrogenation of AB.

References:

1. C. W. Hamilton, R. T. Baker, A. Staubitzc, I. Manners, "B–N compounds for chemical hydrogen storage," *Chemical Society Reviews*, 38, 279-293, 2009.

2. C. R. Miranda and G. Ceder, "Ab initio investigation of ammonia-borane complexes for hydrogen storage," *Journal of Chemical Physics*, 126, 1847031-11, 2007.

3. M. E. Bowden, G. J. Gainsford, and W. T. Robinson, "Room temperature structure of ammonia borane," *Australian Journal of Chemistry*, 60, 149-153, 2007.

See more of this Session: Advancements in Hydrogen Technology II
See more of this Group/Topical: Topical 6: 10th Topical Conference on Gas Utilization

(30b) Hydrogen Storage Via Sodium Borohydride/ Metaborate Recycling

H. Bruce Li, 21-Century Silicon, Inc., Richardson, TX

A patent-pending process will allow us to perform low-cost recycling of sodium metaborate into sodium borohydride by adding four moles of hydrogen gas per each mole of sodium metaborate. This process permits us to avoid the high cost associated with the Brown-Schlessinger process. In our process, the primary cost is the cost of producing hydrogen via electrolysis of water using photovoltaic electricity, which is the primary business of 21-Century Silicon, Inc.

Thirteen metallic catalysts are able to produce sodium borohydride from sodium metaborate and hydrogen. Only 321 KJ/mol of endothermic change of Gibb's free energy occurs. The reaction is carried out using high pressure plasma. Water formed during the reaction is captured by metallic oxides to suppress the reverse reaction. Vacuum distillation of sodium borohydride is carried out afterwards.

See more of this Session: Advancements in Hydrogen Technology II
See more of this Group/Topical: Topical 6: 10th Topical Conference on Gas Utilization

Study of the dimensioning of ATEX formed in the case of leakage of H_2-N_2 mixtures from pressurized ducts

Jacques CHAINEAUX[a], Sébastien EVANNO[a], Bernard GAUTIER[b]

[a] INERIS Parc Technologique Alata BP 2, Verneuil-en-Halatte, F-60550. prénom.nom@ineris.fr

[b] EDF-SEPTEN – DPI/DIN SEPTEN 12-14 Avenue Dutrievoz, Villeurbanne, F-69628. bernard.gautier@edf.fr

Abstract

The methods implemented by EDF-SEPTEN (Electricité de France – Department of Thermic and Nuclear Studies and Project) of Basic Nuclear Installations (BNI) in Nuclear Electricity Power Stations (NEPS) requires the implementation of pressurized hydrogen.

Within the scope of a study on safety, hypothetical leaks are investigated, leading to the formation of explosive atmosphere (ATEX); the intention here is to evaluate their dangerousness.

The present publication deals with the establishment of a simplified model in order to determine the x_{LEL} [distance on the axis of the leak jet, where a concentration is reached equal to the Lower Explosive Limit (LEL)] and V_{ATEX} (volume of the locus of the spatial points where the concentration is greater than or equal to LEL) parameters for typical H_2-N_2 mixture leaks.

Key-words: ATEX, EXPLOJET software, Explosion, H_2, LEL, PHAST software, V_{ATEX}, x_{LEL}.

1. Introduction

EDF-SEPTEN is confronted with the problem of assessing the risks of ATEX formation during exploitation of various networks of ducts, which convey a pressurized gas (pure hydrogen or hydrogen-nitrogen mixture with known composition) and which are set up in ventilated locals with various volumes of a few hundred cubic meters. Within the scope of studies on safety considering major aggressions (e.g.: earthquake), major leaks of the guillotine rupture type are taken as a hypothesis on these networks, with the goal of controlling their consequences [1].

Under the assumption of loss of the seal of a member of the network, the gas contained in the network would be discharged into the locals through a leak orifice and would mix with air in order to form an ATEX [2], at least in the proximity of this orifice.

EDF-SEPTEN requested INERIS to evaluate the volume of ATEX which would be formed inside these locals [3], [4].

Our study is an integral part of a request from the French Nuclear Safety Authority (NSA) which dealt with various cases of hydrogen leaks, supposed to occur in certain locals and originating from various pressurized ducts, bringing into play pure hydrogen or H_2-N_2 mixtures with a defined H_2 content, C_{H2-N2} (vol. %), in order to evaluate the ATEX volume (V_{ATEX}) and the distance to the LEL (x_{LEL}).

In this study, a simplified model may be established in order to determine for defined H_2-N_2 mixture leaks, the following quantities which characterize the dimensions of the ATEX resulting from the leak :

- the distance to the LEL, x_{LEL} (m),
- the volume of formed ATEX, V_{ATEX} (m^3),

from the following parameters defining a leak :

- the data relating to the leaking duct:
  - its nominal diameter, D (m),
  - its outer diameter, D_{ext} (m),
  - its thickness, e (mm),
- the data relating to the leak orifice supposed to be circular :
  - the diameter of the leak orifice d (m); under the assumption of a guillotine rupture of this duct, d is related to D_{ext} and to e by the relationship $d = (D_{ext} - 2e)$.
  - the diameter of the leak orifice, d (m); under the assumption of a half-guillotine rupture of this duct, d is related to D_{ext} and to e by the relationship $d = (D_{ext} - 2e)/\sqrt{2}$.
- the H_2 concentration of the H_2-N_2 mixture, C_{H2-N2} (%),
- the absolute pressure in the duct, P (absolute bars).

2. Methodology used

2.1 Phenomenology of gas leaks

During a leak on a pressurized network, the compressed gas will mix with ambient air through a most often turbulent jet. The conditions under which the discharged gas and air mix depend on many parameters.

However, if the leak occurs in free air and while it is delivering, it is possible to state that :

- it is always pure gas which is present in the plane of the leak orifice,
- on the contrary, there always exists an area of the space sufficiently away from the leak orifice where the air concentration of discharged gas remains very low or even zero.

The leak therefore generates a concentration field. For this reason and under the assumption that the discharged gas is flammable, there always exists an area of the space where the gas concentration in the air belongs to the explosivity domain and where the air-gas mixture therefore forms an ATEX. Beyond this area, there no longer exists any possibility of ATEX.

If, on the contrary, the leak occurs in a confined medium (for example in locals), a concentration field also appears within the locals by accumulation. Its development over time, as well as the possible increase in the formed ATEX volume depend :

- on the volume of the locals as compared with the gas volume (a priori limited) which is able to be discharged through the leak orifice,
- on the air renewal rate of the locals, regardless of whether it is equipped with mechanical or natural ventilation.

Thus, under the assumption of a hydrogen-nitrogen mixture leak in locals, the problem is therefore not of knowing whether an ATEX will be formed, but rather of knowing whether the volume of this ATEX will be such that its explosion would cause significant damages.

An explosive atmosphere of more than ten liters present in a constant amount in closed locals is always considered as dangerous independently of the dimensions of the locals [5]. It is possible to proceed with an approximate evaluation by applying the empirical rule according to which explosive atmospheres present in closed locals are dangerous when the amounts are greater than one ten thousandth (1/10 000) of its volume. However this does not mean that the whole space should be described as a dangerous location; only the portion of the locals where a dangerous atmosphere is likely to form should be considered as a dangerous location.

Under the assumption of sufficiently ventilated locals where accumulation remains low, the evaluation of the risks of explosion related to the formed ATEX in the case of a leak is therefore directly related to that of the maximum volume of formed ATEX on the jet.

2.2 Approach for evaluating the formed ATEX volume: identification of the influent parameters

In order to evaluate the maximum volume of formed ATEX, all the parameters which have an influence on this volume should be identified first.

2.2.1 Parameters related to the leak conditions

Among the parameters which have a determining influence on the maximum volume of formed ATEX, the conditions of the leak are first found, and in particular the leak flow rate, representative of the rate with which the gas will be discharged into the locals.

For a given gas, the leak flow rate itself only depends on the following parameters :

- the section of the leak orifice,
- the orifice coefficient,
- the internal pressure of the network.

- **Assumption on the section of the leak orifice**

The leak rate first depends on the section of the leak orifice: considering that the question is to evaluate the risks of explosion of an ATEX following a safety approach, EDF-SEPTEN has defined the section of the leak orifice as the one which corresponds to the guillotine rupture of the duct installed in the locals.

Let us note that, for nominal diameters of more than 50 mm, EDF-SEPTEN considers under certain assumptions leak diameters equal to the nominal diameter divided by $\sqrt{2}$, which amounts to stating that the leak section is half of the one which corresponds to the nominal diameter.

In certain locals, it is not only a duct which is installed, but also the capacity which is connected to this duct ; as each capacity is equipped with a manhole, EDF-SEPTEN has contemplated the guillotine rupture of tapping of the manhole and the consecutive discharge of the gas through the totality of the section of the manhole. This is why the diameter of the tapping of the manhole, the thickness of the metal sheet of this tapping and the diameter of the leak orifice appear for these leak conditions.

- **Influence of the network pressure on the leak rate**

For a leak orifice with a given section, the leak rate depends on the value of the pressure of the network relatively to the critical pressure which has the value of 1.89 bars absolute :

- if the pressure of the network is greater than the critical pressure, the leak occurs under supercritical flow conditions,
- if the pressure of the network is less than the critical pressure, the leak occurs under subsonic flow conditions.

INERIS has several leak calculation models for characterizing explosive atmosphere. Taking into account the leak conditions retained in the scenarios considered here, INERIS has envisioned to perform calculations with the computational code of INERIS EXPLOJET 4.0 [6] derived from fluid mechanics and the commercial software package PHAST in its version 6.4 [7].

➢ ***Case of supercritical flow***

In order to deal with the cases corresponding to supercritical flow, we used the EXPLOJET software package [6] which was developed by INERIS during the nineties, with which supercritical gas jets with densities different from that of air may be simulated.

Developed within the scope of European experimental projects and compared with tests, it is more particularly suitable for simulating supercritical jets. The output data (concentration, rate, flammable mass, distance to LEL) are defined by the dynamics of the jet.

With this software, air overpressures likely to be generated consecutively to the inflammation of free jets of flammable gases may be calculated (pseudo-similarity theory applicable to turbulent jets with variable density).

In its version 4.0, this code consists of three computational modules which allow :

– calculation of the concentration, rate and turbulence fields prevailing in the formed explosible atmosphere (based on the applications of similarity laws, the existence of which has been proved in certain cases during experiments conducted by INERIS),
– calculation of the flame velocities likely to be observed, as a function of the position of the inflammation point (a module designed to be applied to the case of supersonic jets with variable density, however, its design is such that this module may also be applied to the case of subsonic jets with variable density),
– calculation of the air overpressure field generated by the propagation of flames [8], essentially, it should be recalled that this is not strictly applicable if the flame velocities exceed 250 m/s.

Assuming that the gas is discharged in the form of a free leak jet and without any obstacle, and also supposing that the jet is stationary, i.e. fed under the initial pressure of the network, assumed to be constant, with EXPLOJET, it is possible to calculate the following quantities :

- the volume flow rate of the discharged gas (in Nm^3/h),
- the distance from the point of the axis of the jet where the discharged gas content exactly reaches the LEL, (x_{LEL} (m)),
- the volume of the formed ATEX, V_{ATEX} (m^3),
- the mass of hydrogen contained in the formed ATEX (m_{H2} (kg)).

In order to reach these results, the EXPLOJET software mainly needs the following data :

- Data relating to the ambient air :
- its temperature,
- the atmospheric pressure.

- Data relating to the discharge gas :
  - its pressure in the network,
  - the kinematics viscosity and the γ ratio of the specific heats,
  - its temperature,
  - its LEL in air.
- Data relating to the leak orifice supposed to be circular :
  - its diameter,
  - the orifice coefficient C_D (for all the cases, the assumption of a guillotine duct rupture has led us to consider a single value equal to 0.77). Generally, the influence of the shape of the orifice in a capacity is expressed by a discharge coefficient ($0 < C_D < 1$). In order to provide assistance in determining the value of the discharge coefficient C_D, an INERIS report [9] specifies values of C_D, depending on the shape of the orifice for turbulent flows of homogeneous fluids (either totally liquid or totally gaseous).

The assumption of a stationary jet is of course not verified in the investigated cases here and it amounts to increasing the maximum volume of the ATEX likely to be formed; indeed the diameters of the leak orifices are often large and would allow rapid decrease of the network pressure, so that the maximum volume of the formed ATEX would necessarily be less than the volume calculated by EXPLOJET.

Further, in order to express the final result, in reality, the maximum amount of H_2-N_2 mixture likely to be contained in the formed ATEX should be taken into account. Of course, we must ensured that the volume provided by EXPLOJET does not exceed the amount of H_2-N_2 mixture initially contained in the network.

➢ ***Case of subsonic flow***

In order to deal with the cases corresponding to subsonic flow, we used the software package PHAST [7] version 6.4 developed by DNV Software. The versions 6.0 and 6.1 of the software package were subject to full evaluation by INERIS, the version 6.4 being placed in the continuity of the evaluated versions.

The PHAST software is based on the UDM (Unified Dispersion Model) method which combines models of jets, of dispersion of dense or lightweight gases and of passive dispersion. PHAST in particular allows simulation of gas jets, with variable density in their inertial phase and beyond in the entrainment phase in which inertial forces are dominated by atmospheric dispersion and gravity forces.

The complexity of this type of software is intermediate between Gaussian type dispersion calculations and three-dimensional software packages proceeding with finite volumes or elements.

The version 6.1 of the PHAST software was the subject of an evaluation of its capability of determining the consequences of accidents according to a strict methodology. An evaluation report has been established [10] and is available on the INERIS internet site (http://www.ineris.fr).

Like EXPLOJET, this software leads to the mass leakage flow rate and to the distance to the LEL on the axis of the jet, but it does not allow direct access to the volume of the ATEX or to the hydrogen mass contained in the ATEX.

It essentially requires the same data as EXPLOJET; let us note however that it has several limitations as to the amounts of discharged gas; moreover, it is basically reserved for calculations of atmospheric dispersion of gases, i.e. outdoors (in particular, stability characteristics of the atmosphere and wind speed are required), and dispersion in confined locals is treated with significant uncertainty. To be more specific, in order to limit this uncertainty, the software should be used with the maximum stability index of the atmosphere.

2.2.2 Parameters related to the amount of discharged hydrogen-nitrogen mixture

The maximum volume of ATEX which would be formed in the case of a leak, first depends on the total amount of H_2-N_2 mixture (insofar that it is limited) which is contained in the network and which is therefore capable of being discharged into the locals through the leak orifice; this amount itself depends on the following parameters :

- the proportion of H_2 in the H_2-N_2 mixture contained in the network, as well as the explosivity limits in the air of this mixture,
- the internal volume of the network,
- the pressure of the H_2-N_2 mixture contained in the network.

- ***Flammability characteristics and H_2 content of the H_2-N_2 mixture contained in the network***

In the case when the gas of the network is not pure hydrogen, this is a hydrogen-nitrogen mixture, the hydrogen volume content of which is comprised between 3.2 and 43.1 %.

For having a hydrogen-nitrogen mixture form an ATEX in the presence of air, it is required that the content in the hydrogen-nitrogen mixture belong to the explosivity domain of this mixture, comprised between the lower and upper explosivity limits [LEL and UEL (Upper Explosive limit) respectively] :

- in the case of a mixture without any nitrogen (pure hydrogen case), the LEL of hydrogen is equal to 4 %,
- in the case of ternary air-hydrogen-nitrogen mixtures, the LEL of each mixture may be obtained from a ternary diagram such as the one in the following Fig. 1.

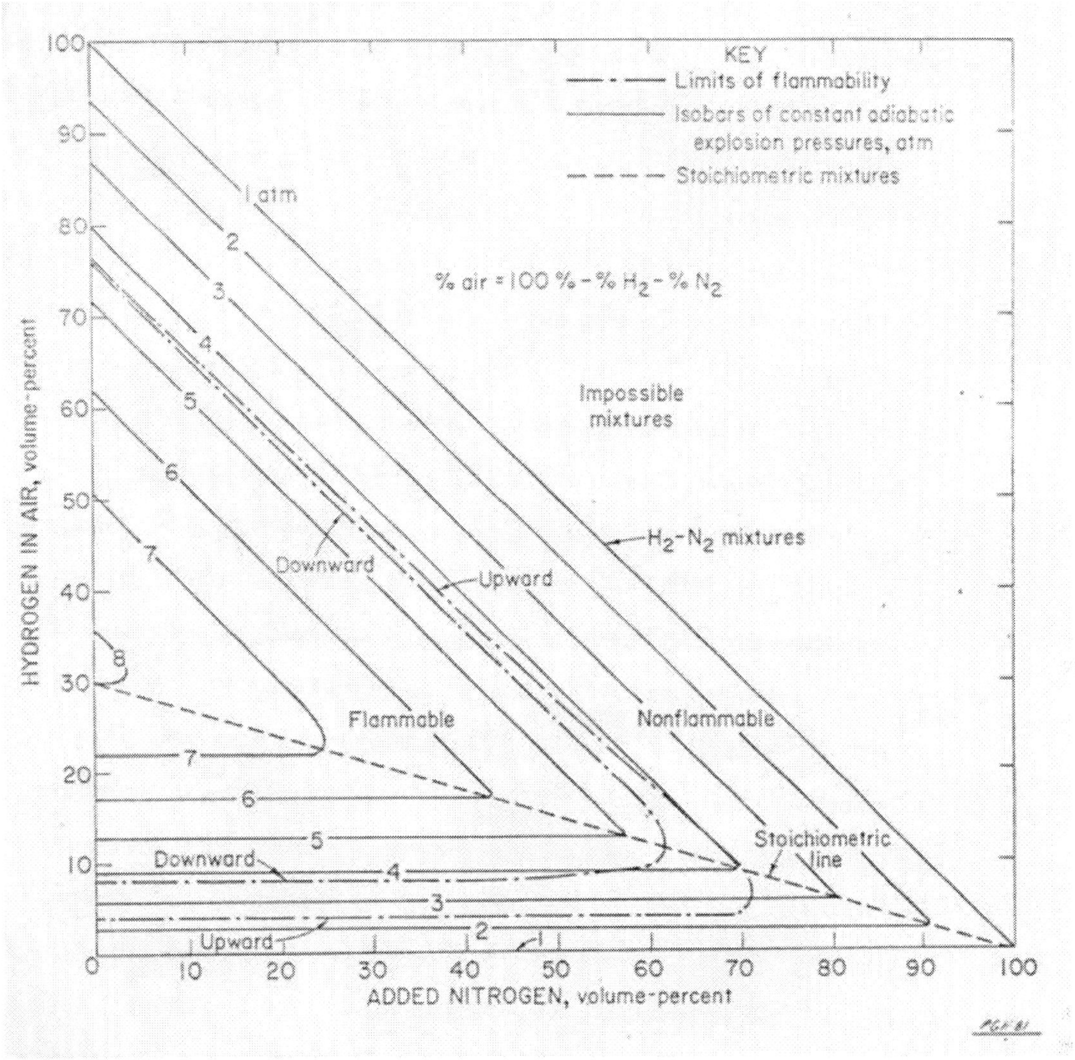

Fig. 1 : Ternary hydrogen-nitrogen-air diagram

Source : Flammability limits and pressure development in H_2-air mixtures, PRC Report No. 4305, M. Hertzberg, US Bureau of Mines, 1981.

The result of this diagram is that a hydrogen-nitrogen mixture, for which the hydrogen content is less than or equal to 5.2 % by Vol. is not flammable in air at ambient pressure and temperature and therefore cannot form any ATEX.

Table 1 hereafter indicates the LEL value of the other hydrogen-nitrogen mixtures considered here.

Hydrogen content of the H_2-N_2 mixture (% vol.)	LEL (% vol.)
43.1	9.3
42.6	9.3
23.5	16.4
18.8	20.0

Table 1: Values of LEL for the applied hydrogen-nitrogen mixtures

The curve of Fig. 2 allows determination of the value of the LEL of an H_2-N_2 mixture depending on its H_2 content $C_{H2\text{-}N2}$.

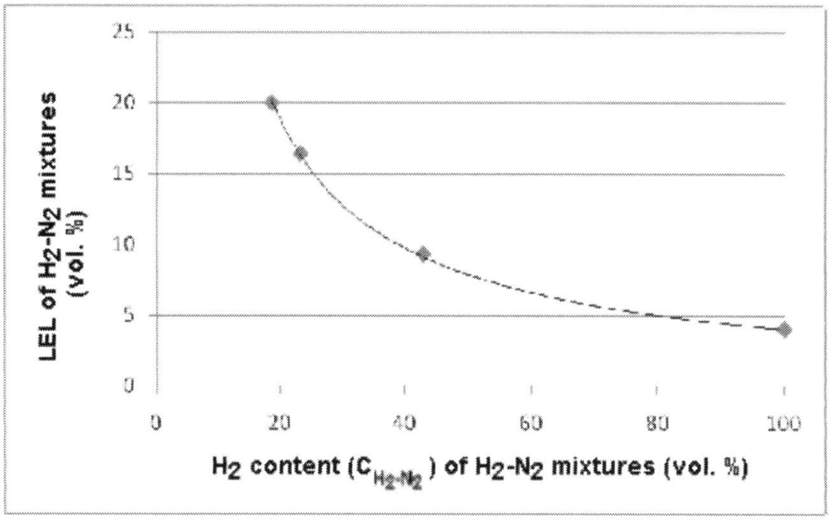

Fig. 2 : LEL variation of H_2-N_2 mixtures versus their H_2 content $C_{H2\text{-}N2}$

- ***Internal volume of the networks***

Each network consists of several duct sections defined by their inner diameter and their length, and the internal volume of which may be calculated; however this volume is always much smaller than the internal volume of the capacity(ies) to which the relevant networks here are connected, so that the internal volume of the network may be considered as equal to the sole volume of this or these capacities.

The value of the internal volume of the networks considered here (inclusive capacity(ies)) is comprised between 1 m^3 and 60 m^3. The networks connected to the compressed gas supply frames are generally assumed to have infinite capacity (permanent operating conditions) in the case of the absence of any cut-off system.

- ***Internal pressure of the gas of the networks***

In the present report, all the internal pressure values are expressed in bars absolute. For a few locals, this pressure has a value of 155 bars absolute; for the whole of the other locals, it is comprised between 1.3 and 7 bars absolute.

- ***Calculation of the maximum amount of discharged hydrogen***

The amount of H_2-N_2 mixture which would be discharged may be calculated by performing the product of the 2 following parameters :

- the volume of the capacity connected to the faulty member of the network,
- the relative gas pressure in this network,

As for the amount of hydrogen contained in the network, it may be obtained by multiplying the amount of H_2-N_2 mixture contained in the network by the proportion of H_2 in this mixture.

- ***Parameters related to the ventilation conditions of the locals***

In the case of leakage on a network, it is conceivable that the maximum volume of the formed ATEX as well as the period during which this ATEX would persist, depend on the relative extent of the ventilation flow rate of the locals (i.e. the rate with which the atmosphere of the locals is renewed with fresh air) with respect to the leak rate (i.e. relatively to the rate with which the air of the locals is loaded with hydrogen).

3. Results and interpretation

We indicated earlier that the case of leaks under supercritical flow conditions which correspond to P > 1.9 bars absolute, and the case of leaks under subsonic flow conditions, which correspond to P ≤ 1.9 bars absolute, should be considered separately.

3.1 Case of leaks under supercritical flow conditions

Establishing the simplified model consisted of clarifying the functions which each relate the following quantities :

- x_{LEL},
- V_{ATEX},

as a function of the following parameters :

- P,
- d,
- C_{H2-N2}.

In order to specify the quantities x_{LEL} and V_{ATEX}, with the EXPLOJET model, we calculated the values of each quantity for various pairs of d and P values.

For each pair, we imposed fictitious values of the LEL of hydrogen, which amounts to studying the influence of the H_2 content of the H_2-N_2 mixtures, designated as C_{H2-N2}.

The values of the LEL of the H_2-N_2 mixture versus their H_2 content (C_{H2-N2}) are grouped in Table 1 and shown in Fig. 2.

Let us add that in order to take into account the assumption of a guillotine rupture of the duct, the orifice coefficient C_d was taken in every case to be equal to the value proposed by the PHAST model, i.e. 0.77, as well as for the calculations performed with EXPLOJET.

The calculations led to the following relationships :

$$x_{LEL} = P^{0.5} \times d \times C^{1.3}{}_{H2-N2} \qquad (1)$$

and

$$V_{ATEX} = P^{1.5} \times d^{3} \times C^{2.68}{}_{H2-N2} \qquad (2)$$

wherein the quantities and parameters have the definitions and units as indicated above.

As regards the leaks under supercritical flow conditions, the relationships (1) and (2) show :

- the exponent 0.5 in how the quantity x_{LEL} varies with P,
- the exponent 1 in how the quantity x_{LEL} varies with d,
- the exponent 1.5 in how the quantity V_{ATEX} varies with P,
- the exponent 3 in how the quantity V_{ATEX} varies with d.

These correlations were established in the domain of validity of EXPLOJET and are therefore reliable.

As regards how the quantities x_{LEL} and V_{ATEX} vary with C_{H2-N2}, the values of the exponents [1.3 for relationship (1) and 2.68 for relationship (2)] are approximate values.

3.2 Case of leaks under subsonic flow conditions

As in the case of leaks under supercritical flow conditions, establishing the simplified model consisted of specifying the functions which relate x_{LEL} and V_{ATEX} to P, d and C_{H2-N2}.

But this time, we calculated the values of the first two quantities for different pairs of d and P values with the PHAST model.

In order to study the influence of the H_2 content of H_2-N_2 mixtures (C_{H2-N2}), we again imposed here fictitious LEL values for hydrogen, by using the values of Table 1 and as shown in Fig. 2.

The calculations led to the following relationships :

$$x_{LEL} = P^{0.38} \times d \times C_{H2-N2} \qquad (3)$$

and

$$V_{ATEX} = 0.004 \times x^3_{LEL} = 0.004 \times P^{1.14} \times d^3 \times C^3_{H2-N2} \qquad (4)$$

wherein the quantities and parameters have the definitions and units as indicated above.

As regards the leaks under subsonic flow conditions, the relationship (3) shows :

- the exponent 0.38 in how the quantity x_{LEL} varies with P,
- the exponent 1 in how the quantity x_{LEL} varies with d,
- the exponent 1 in how the quantity V_{ATEX} varies with C_{H2-N2}.

Unlike the EXPLOJET model which directly provides the volume of the ATEX by integrating the function (integral of (c(x, y, z)dxdydz), the PHAST model only considers the extension of the ATEX on the axis of the jet, i.e. x_{LEL}.

Consequently, the volume of the ATEX is calculated in a relatively inaccurate way by assimilating it to that of an ellipsoid for which the major axis and the minor axis are estimated from the value of x_{LEL}. This is why the relationship (4) shows that the quantity V_{ATEX} varies as to the power 3 of x_{LEL}.

4. Conclusion

The INERIS study for EDF-SEPTEN basically consisted of establishing a simplified computational model for determining the parameters V_{ATEX} and x_{LEL}, for defined H_2-N_2 mixture leaks.

From the moment that a flammable H_2-N_2 mixture is discharged into the air of locals, through a leak orifice, it is established that :

- an ATEX is always present at the leak orifice,
- the volume of this ATEX depends on the section of the leak orifice and on the network pressure upstream from the leak,
- depending on whether this pressure is greater or less than the critical pressure, i.e. a little less than 2 bars absolute, the leak will occur under the conditions of a respectively supercritical or subsonic jet,
- regardless of the flow conditions, the leak is characterized by its flow rate,
- depending on the values of this flow rate and of the ventilation flow rate of the locals, an ATEX may form in the locals and may accumulate there, further away from the leak orifice,
- the dangerousness of each of the ATEX which form in the locals is all the more significant since their volume is greater.

Let us add that as the present evaluation is conducted within the scope of a safety step, the dangerousness of an ATEX is appreciated with regard to the effects which its explosion would produce on sensitive pieces of equipment. The tolerance on this account is greater than in terms of protecting workers for example.

In the case of a leak under supercritical flow conditions, the formed volume of ATEX at the leak orifice may be calculated by means of the EXPLOJET software developed by INERIS, assuming the jet to be stationary, free and without any obstacle, which amounts to obtaining an upper bound value of the actual volume.

In the case of a leak under subsonic flow conditions, we evaluated the volume of the ATEX formed at the leak orifice by using the PHAST software but the obtained results are marred with significant uncertainty, because above all this is an atmospheric dispersion software which we therefore used at the limits of its capabilities.

Finally, we compared the results obtained with those resulting from the application of the homogeneous approach, a method which consists of :

- assuming that the discharged H_2-N_2 mixture homogeneously mixes with the atmosphere of the locals,
- comparing the content of this mixture with its LEL,
- deducing the possibility of formation of a dangerous ATEX in the locals.

In the case of freely developing jets, the evaluation of the ATEX volumes and distances on the jet is a useful addition to the approach by homogeneous dilution. In the first case, the instantaneous ATEX is characterized, in the second case the ATEX is characterized by accumulation. In the investigations conducted by EDF-SEPTEN, they were able to ascertain that the first type may become sensitive in the case of large size locals where accumulation does not necessarily lead to a notable rise in the average values.

References

1. B. Gautier EDF-SEPTEN "Reference Base for safety requirements concerning protection against the risk of internal explosion for the EPR" ENGSIN080356 A 2009.
2. European Directive ATEX 1999/92/EC relating to minimum prescriptions directed to improving protection as regards the safety and health of workers likely to be exposed to the risk of explosive atmospheres (Official Journal of the European Communities).
3. Studies of hydrogen leaks in large locals (INERIS – DRA -06 – P71922).
4. Study of dimensioning of the formed ATEX in the case of H_2-N_2 mixture leaks from pressurized ducts (INERIS DRA-07-89258-08949A).
5. Guide of good practices with no restriction for implementing the ATEX 1999/92/EC Directive (January 2003).
6. Explojet: Code for calculating explosions in turbulent flammable gas jets. Summary of research work and evaluation of version 4.0 (INERIS EMA-YMo-ERu/YMo-98-16FA26).
7. Evaluation of PHAST versions 6.0 and 6.1, summary report (INERIS DRA- 2002-29576 - SDu/Jco/Ebe, Nov 2002).
8. Deshaies B., Clavin P. (1979). Dynamic effects generated by a spherical flame at constant velocity (Journal de mécanique, Vol 18, n°2, pp 213-223).
   Deshaies B., Leyer J.C. (1981). Flow field induced by unconfined spherical accelerating flames (Combustion and Flame, Vol 40, pp 145-153).
   Cleaver R.P, Robinson C.G. (1996). An analysis of the mechanisms of overpressures generation in vapour cloud explosions (Journal of Hazardous Materials, Vol 45, pp 27-44).
9. INERIS Ω–19 Report « Determination of characteristic quantities of the source term required for using a model for atmospheric dispersion of accidental emissions » (INERIS-DRA-2005-P46055-C51076).
10. Duplantier et al., 2002: Evaluation of PHAST versions 6.0 and 6.1 (INERIS-DRA-2002-29576 report).

(30d) Desulfurization of Refinery off-Gases for Hydrogen Production

Ambalavanan Jayaraman, TDA Research Inc., Wheat Ridge, CO

Gökhan Alptekin, TDA Research Inc., Wheat Ridge, CO

Margarita Dubovik, TDA Research Inc., Wheat Ridge, CO

Michael Ware, TDA Research Inc., Wheat Ridge, CO

Amanda Parker, TDA Research Inc., Wheat Ridge, CO

Petroleum refining has become the second largest industrial consumer of energy in the U.S, accounting for approximately 10% of the total manufacturing energy use. The petroleum industry was also responsible for 3% of the toxic waste emissions and 17% of the greenhouse gas emissions from manufacturing section. A large fraction of these emissions come from the combustion (in some cases flaring) of various by-product fuel gases. In a typical refinery, these off-gases are generated from various separation units (crude distillation, cracking, hydro-treating) generate off-gasses. Because of the lack of the technologies that could convert these refinery off-gases into valuable products, many of these gases are currently flared. The off-gases from various refinery unit operations could be converted into valuable chemicals such as hydrogen instead of being sent to flare which contributes to energy loses and greenhouse gas emissions. However, the refinery off-gases contain large concentrations of sulfur and other impurities such as arsenic that must be removed to prevent the poisoning catalysts used in hydrogen production. Traditionally, deep desulfurization can be achieved by a two-step process consisting of hydrodesulfurization (HDS) and subsequent removal of H_2S with an expendable chemical absorbent. Although this approach has long been used for desulfurization of natural gas feedstocks (where the sulfur level does not exceed 5-10 ppmv), the one-time use of expendable metal oxide type sorbents for cleanup of the refinery off-gas is not practical due to the high concentration of sulfur and the presence of organic sulfur compounds which are not absorbed at near ambient temperatures.

TDA Research, Inc. (TDA) is developing a novel physical adsorbent, which will be operated in a regenerable manner to desulfurize refinery off-gases. The sorbent operates at near ambient temperatures (20-60°C) while reducing the sulfur content of the gas stream to less than ppmv levels (and can reduce the sulfur content to ppb levels if needed) to protect the catalyst used in the downstream conversion process. Unlike conventional sorbents, in addition to H_2S, our material removes organic sulfur compounds (such as mercaptans, sulfides and thiophenes) with high capacity and higher removal efficiency. The sorbent can be regenerated by applying a mild temperature swing. In this paper we will demonstrate the performance of the regenerable sorbent under multiple adsorption/regeneration cycles and the economic viability of our desulfurization process for producing hydrogen from sulfur contaminated refinery off-gas streams.

See more of this Session: Advancements in Hydrogen Technology II
See more of this Group/Topical: Topical 6: 10th Topical Conference on Gas Utilization

ITM Syngas: Ceramic Membrane Technology for Lower Cost Conversion of Natural Gas

John M. Repasky[1*], William E. Waldron[1], Christopher F. Miller[1], Michael F. Carolan[1*], David W. Studer[1*], Clifford M. Lowe[2], Karl F. Gerdes[2], Joris Smit[3]

[1.] *Air Products and Chemicals, Inc., 7201 Hamilton Blvd, Allentown, PA, 18195 USA*
[2.] *Chevron Energy Technology Company, 100 Chevron Way, Richmond, CA 94802 USA*
[3.] *Sasol Technology (Pty) Ltd, Vlierstraat 111, 7544 GG, Enschede, The Netherlands*
** Correspondence should be addressed to:*
J.M. Repasky, email: repaskjm@airproducts.com, tel: +1 (610) 481-5653
M.F. Carolan, email: carolamf@airproducts.com, tel: +1 (610) 481-2185
D.W. Studer, email: studerdw@airproducts.com, tel: +1 (610) 481-8333

Prepared for Presentation at the AIChE Spring National Meeting
San Antonio, TX March 21-25, 2010

Introduction

The ITM Syngas Team, comprising Air Products, Chevron, Sasol, Ceramatec, and other partners, in collaboration with the U.S. Department of Energy, is developing Ion Transport Membrane (ITM) technology for the production of synthesis gas. ITM ceramic membranes are fabricated from non-porous, multi-component metallic oxides and operate at high temperatures (typically over 700°C) with exceptionally high oxygen flux and selectivity. A conceptualization of the ITM Syngas technology is shown in Figure 1. Oxygen from low-pressure air permeates as oxygen ions through the ceramic membrane and is consumed through chemical reactions, thus creating a chemical driving force that pulls oxygen ions across the membrane at high rates. The oxygen reacts with natural gas in a partial oxidation process to produce a hydrogen and carbon monoxide mixture – synthesis gas.

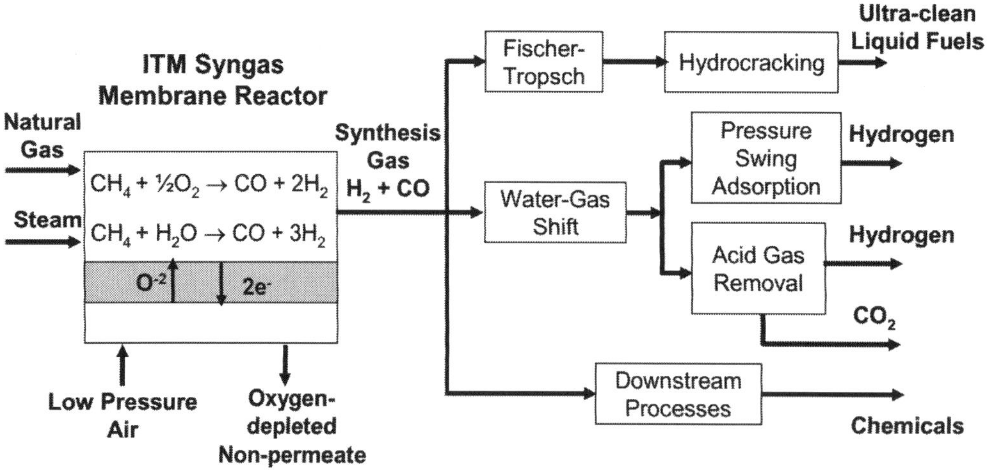

Figure 1. Conceptual ITM Syngas process showing multiple synthesis gas applications.

Applications

The ITM Syngas process is a breakthrough technology that combines air separation and high-temperature synthesis gas generation processes into a single ceramic membrane reactor, with potential for significant savings in the cost of synthesis gas production. Because synthesis gas is a feedstock for a range of different processes, ITM Syngas represents a technology platform that has numerous applications, such as Gas-to-Liquids; hydrogen; ultra-clean fuels, including liquid transportation fuels; and chemicals such as methanol. Pathways to these market applications are illustrated in Figure 1.

The ITM Syngas process is also readily configured for carbon capture from its high pressure syngas product (shown in Figure 1 using Water Gas Shift and Acid Gas Removal), for example toward the production of fuel-grade hydrogen. This inherent feature makes ITM Syngas an attractive technology toward emerging climate-friendly energy applications, including decarburization of refinery fuels, and natural gas fired steam boilers and natural gas combined cycle (NGCC) power generation with carbon capture and storage (CCS).

Membrane Design and Manufacture

The economics of the ITM Syngas process are most attractive when a high-pressure natural gas feed and a low-pressure air feed are used to produce a synthesis gas product at a high pressure that matches downstream process pressures. This arrangement avoids the expenses associated with compressing the air or the synthesis gas. Ceramic membranes for this application must withstand the stresses generated by the large pressure difference between the high-pressure natural gas/synthesis gas on one side of the membrane and the low-pressure air on the opposite side.

Within the ITM Syngas development program, a planar ceramic membrane structure was designed and fabricated that is able to achieve high oxygen fluxes while withstanding the thermo-mechanical stresses encountered during operation [1, 2]. This membrane structure, or wafer, design incorporates micro-channel features to reduce the thermo-mechanical stresses that occur within the wafer assembly due to the pressure loads and the exothermic oxidation reactions. The micro-channel features also provide for efficient distribution of the gases to the membrane surfaces: high-pressure natural gas/synthesis gas flowing outside the membrane wafer and low-pressure air flowing through the micro-channels within the membrane wafer. Figure 2 illustrates the structure of the ITM Syngas micro-channeled planar membrane wafer. Figure 3 further compares a cross-sectional photograph of the wafer internals (end view, on left) against a schematic diagram (side view, on right).

The planar wafer design enables very efficient packing of membrane surfaces within the reactor volume. An ITM Syngas module is assembled by stacking wafers vertically. Figure 4 shows a sub-module built from five full-size commercial-scale ceramic wafers. Ceramic spacers are placed between the individual wafers to create channels for the flow of natural gas and synthesis gas between wafers. Various sizes of modules consisting of pilot-scale wafers and full-size wafers have been joined and sealed using a ceramic sealant suitable for operating temperatures up to 1100°C (2010°F) [3].

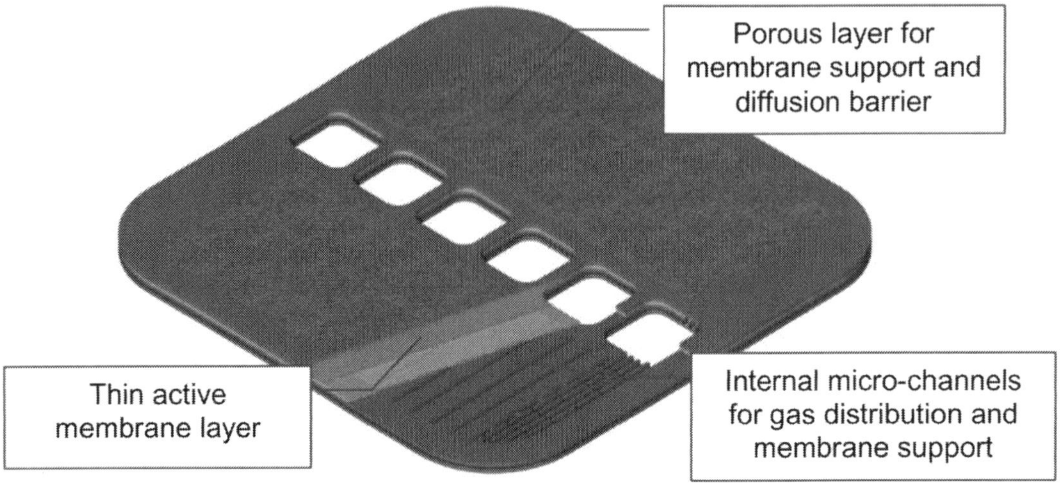

Figure 2. Cut-away view of the ITM Syngas ceramic wafer design showing the multi-layer structure.

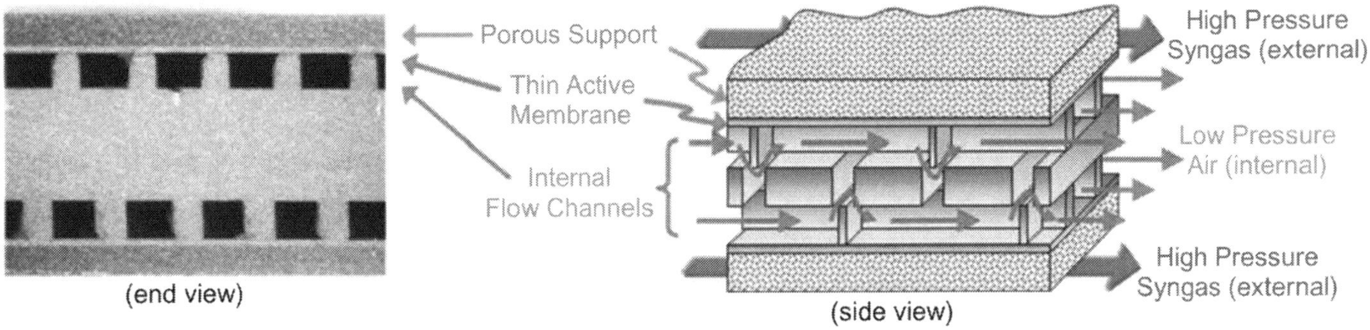

Figure 3. Cross-sectional photograph of wafer internals (end view, on left) and schematic diagram (side view, on right).

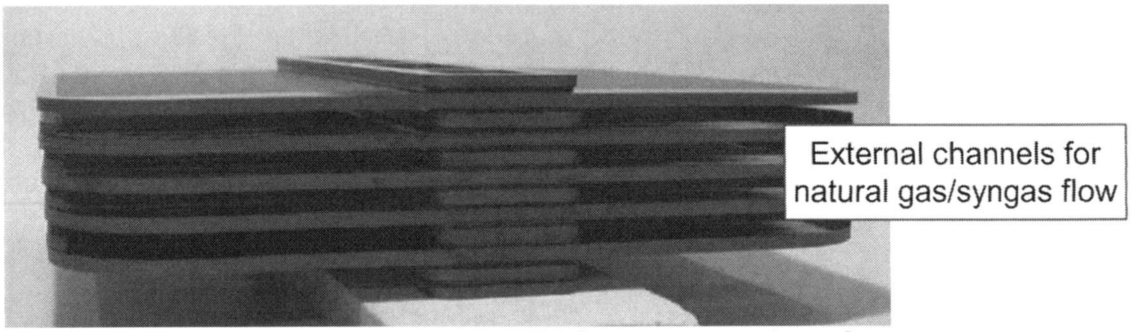

Figure 4. Full-size commercial-scale ITM Syngas ceramic wafers shown as part of a five-wafer sub-module.

A primary challenge in commercializing micro-channel devices has been the ability to mass produce the micro-channel components in the preferred materials at a commercially viable cost. The ITM Syngas planar wafers were designed with a multi-layer structure that can be manufactured using commercial-scale production and automation equipment.

In order to attain dimensional tolerances, a Laminated Object Manufacturing (LOM) method was developed. This LOM process begins with ceramic powders within a specified process window for surface area and particle size. These powders are blended with a mix of organic binders, plasticizers, and solvents to form a flowable ceramic slip. This slip is then cast onto a polyester film that dries and cures into a flexible, ceramic-filled plastic-like tape. This process is called tape casting. Once cured, gas channel features are added to the tapes for the various layers of the wafer using machining technologies, such as laser machining and punching. The featured, plastic layers are then pressure-laminated and bonded together. Once bonded, the wafers are heated to remove the organics and to densify the ceramic powders. The densification step is called sintering. Illustration of the LOM manufacturing method is shown in Figure 5.

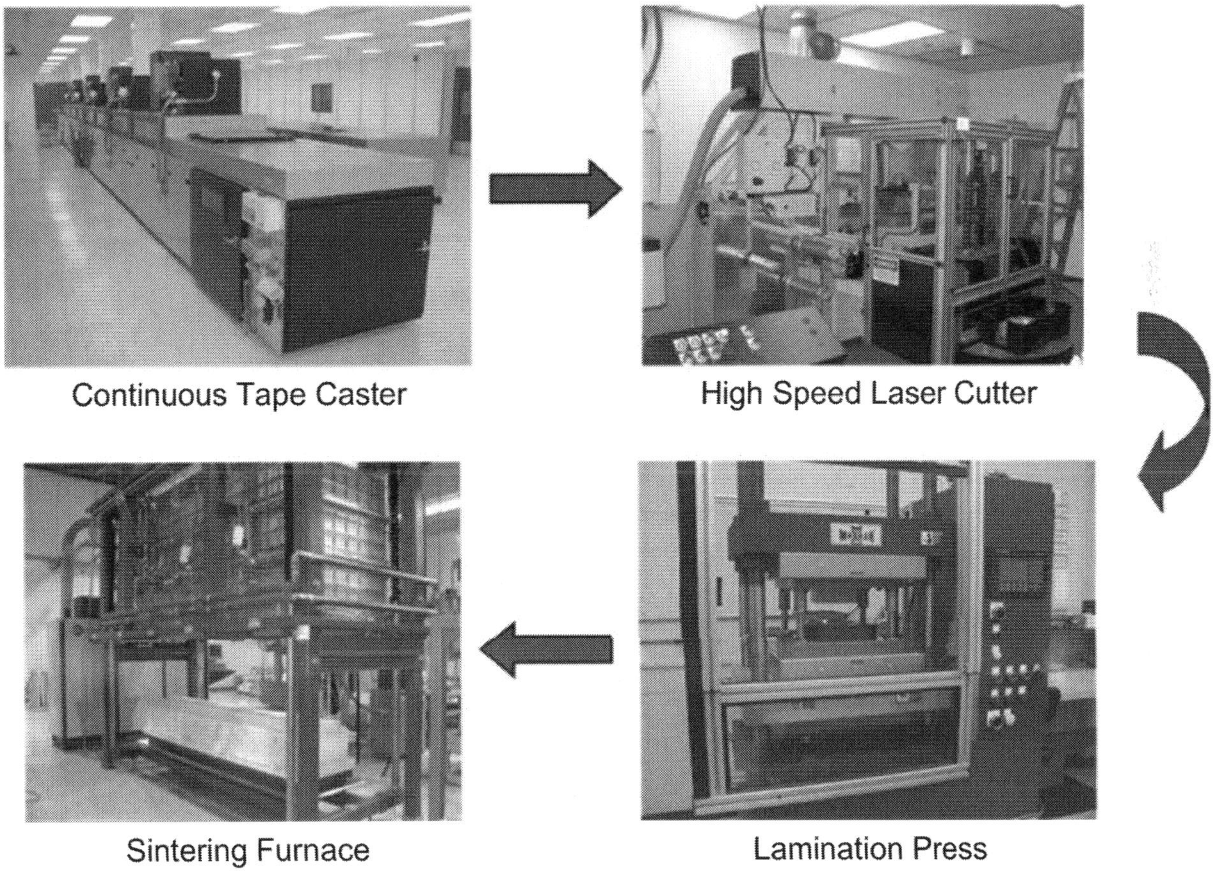

Figure 5. Laminated Object Manufacturing (LOM) method for manufacture of ITM Syngas planar membranes.

Test Results

A laboratory Process Development Unit (PDU) was constructed at Air Products and is used to test the planar ceramic membranes. The PDU operates at nominally 27 Nm^3/hr (24 thousand standard cubic feet per day) of synthesis gas capacity, and is used to evaluate membrane performance under full process operating conditions.

PDU tests with pilot-size wafers have included operation under synthesis gas simulating the conditions of a commercial ITM Syngas reactor, including temperatures up to 900°C (1650°F) and syngas pressures up to 29 barg (425 psig). These wafers demonstrated the ability to operate through a range of temperature, pressure, and composition conditions. Figure 6 shows the progression of oxygen flux demonstration, including the achievement of fluxes at and above the commercial flux target under simulated commercial conditions. Ceramic-to-ceramic seals [3, 4] are used to join the sintered ceramic components together to form a single module and ceramic-to-metal seals [5, 6] are used to connect the ceramic module to reactor internal piping. Seals have also been demonstrated and cycled at full process conditions against full pressure differential up to 28 bar (410 psid) and temperatures up to 900°C (1650°F).

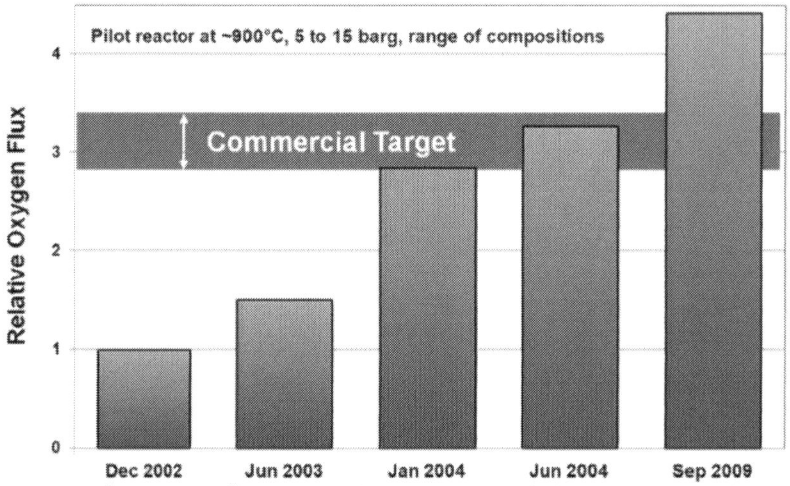

Figure 6. Achievement of oxygen fluxes at the commercial target was demonstrated in 2004. Later tests demonstrated fluxes above the commercial target.

ITM Syngas process equipment may include component materials that contain silicon or chromium. When exposed to the feed streams of the ITM Syngas process, such as air or partially reformed methane, volatile species of silicon or chromium may form as described in reactions (1) and (2):

(1) $SiO_2 + 2\ H_2O\ (g) = Si(OH)_4\ (g)$
(2) $Cr_2O_3 + 3/2\ O_2\ (g) = 2\ CrO_3\ (g)$

These volatile oxides or oxy-hydroxides, if present in high enough concentrations, can react with the membrane material and degrade the flux performance of the membranes.

To prevent contamination of the membranes by Si or Cr, guard beds containing an MgO getter can be used to remove the contaminants from the feed gases [7, 8, 9]. An MgO getter can reduce the partial pressure of the contaminant species by several orders of magnitude and eliminate flux degradation due to contamination as described in reactions (3) and (4):

(3) $2\ MgO + Si(OH)_4\ (g) = Mg_2SiO_4 + 2\ H_2O\ (g)$
(4) $MgO + 2\ CrO_3\ (g) = MgCr_2O_4 + 3/2\ O_2\ (g)$

In addition, non-reactive refractory or liners can be used in ITM Syngas process equipment to prevent potential contaminants from entering the process streams [10, 11].

Guard beds and liners have been used with great success to prevent contamination of ITM Syngas membranes and to produce stable and high fluxes. For example, during testing of pilot-sized wafers in the PDU, liners were used on the syngas side piping in conjunction with guard beds upstream of the wafers. Figure 7 shows the contamination on a membrane following testing along with the flux degradation observed when liners and guard beds were not used. Figure 8 shows a membrane tested under the same conditions, but using liners and guard beds. Not only was the contamination eliminated, but the flux was also significantly increased and stable. [12]

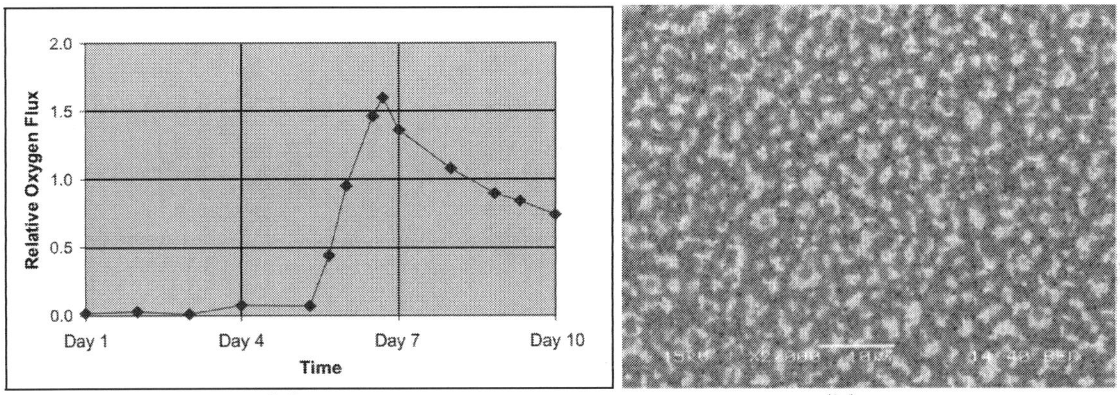

(a) (b)

Figure 7. (a) Oxygen flux decay was observed during initial startup of the PDU when not using guard beds and liners. (b) Scanning electron microscopy of the airside surface of the membrane revealed contamination (*i.e.*, the dark phases in the image are chromium-rich secondary phases that formed during operation due to contamination reactions).

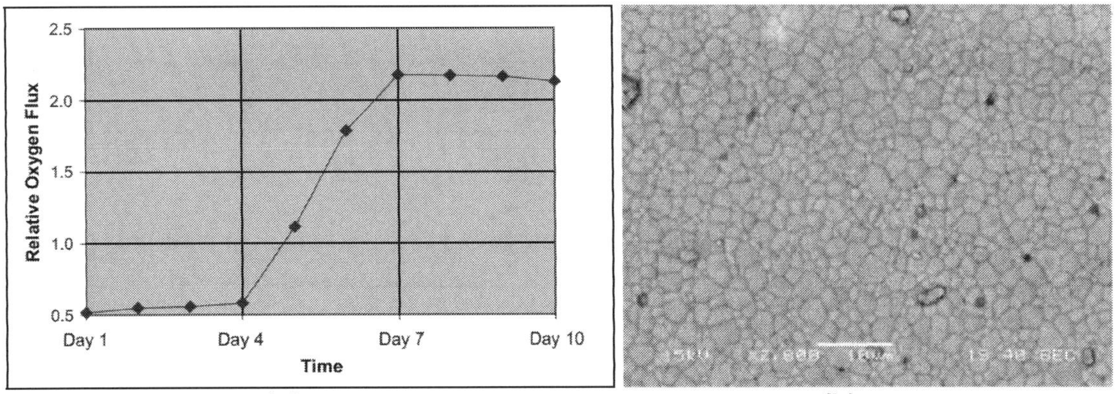

(a) (b)

Figure 8. (a) Higher and more stable oxygen fluxes were observed during initial startup of the PDU when using guard beds and liners. (b) Scanning electron microscopy of the airside surface of the membrane revealed a contaminant-free surface.

Full-size commercial-scale wafers have also been tested in the PDU reactor. Full-size wafers have more membrane area than pilot-size wafers, and thus produce more oxygen, require higher air feed flows, and use a larger ceramic-to-metal seal design for the air manifold connections. Photographs of a full-size commercial-scale wafer assembled in the PDU flow containment housing are shown in Figure 9. Full-size wafers have been operated for over 150 days, including at temperatures over 900°C (1650°F) and pressures up to 15 barg (215 psig), with excellent seal performance.

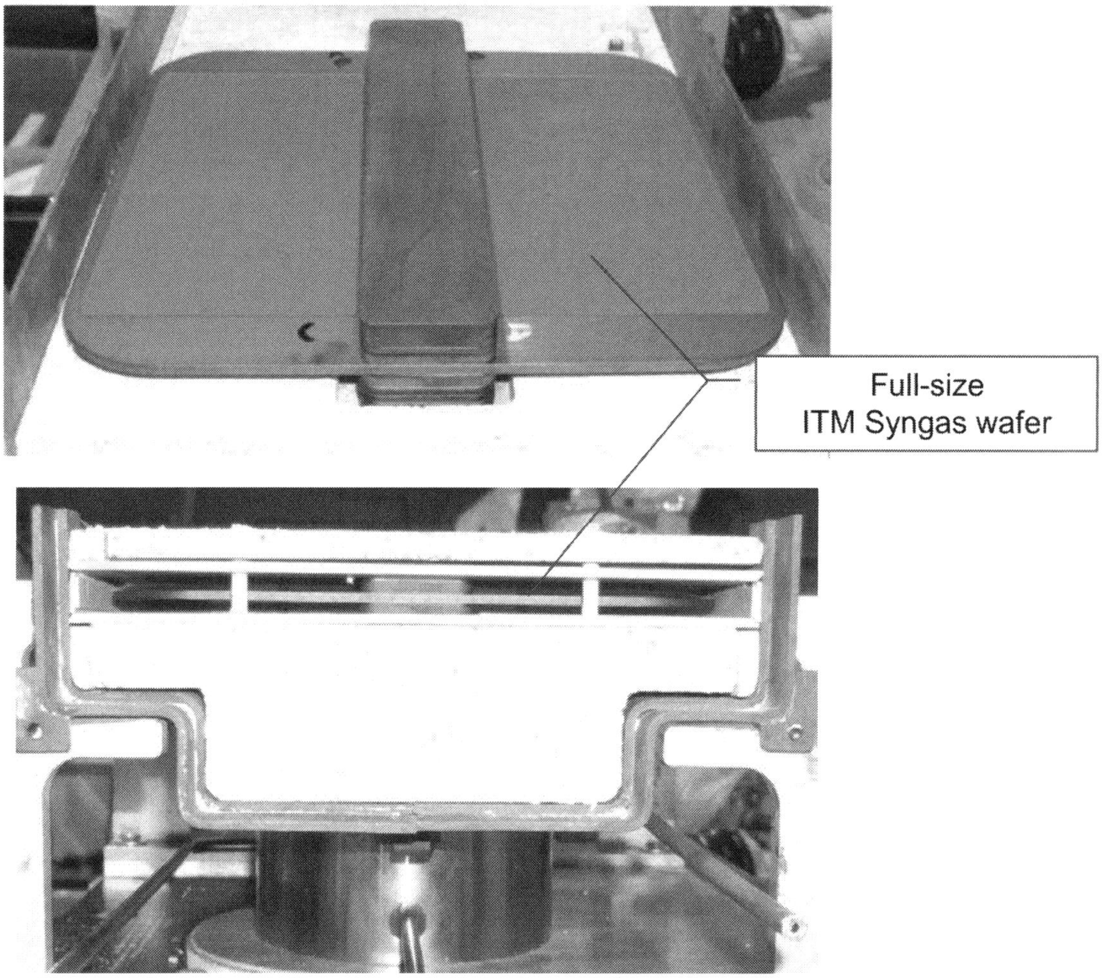

Figure 9. Full-size commercial-scale ITM Syngas wafer assembled in the Process Development Unit flow containment housing. Top and end views.

Process Development

Process development has been conducted on the ITM Syngas process and its integration into a Fischer-Tropsch Gas-to-Liquids (GTL) site [13]. A process flow sheet and control scheme have been developed to provide an efficient and operable ITM Syngas process [14, 15].

Figure 10 illustrates an optimized ITM Syngas process flow sheet. Natural gas feed is preheated against syngas product and sulfur is removed in a typical NiMo/CoMo/ZnO hydrodesulfurization (HDS) unit. High pressure steam is superheated and mixed with the hydrodesulfurization unit effluent. The mixed steam and natural gas feed is further preheated and pre-reformed over nickel catalyst. The pre-reformed natural gas, steam, and recycle gas from Fischer-Tropsch synthesis are heated and enter the ITM Syngas reactor on the permeate side of the membranes. Feed air to the ITM Syngas reactor is increased in pressure using a blower, heated against effluent non-permeate, further heated by direct combustion with natural gas, and enters the ITM Syngas reactor on the air feed/non-permeate side of the membranes. Oxygen permeates through the Ion Transport Membranes in the ITM Syngas reactor and reacts with the natural gas, steam, and recycle gas to form syngas. Syngas exiting the ITM Syngas reactor is cooled in a steam boiler (which generates steam for the process) to below temperatures associated with metal dusting before providing heat to the natural gas feed and recycle gas(es). Further heat is extracted from the syngas to preheat boiler feed water and make-up water to the deaerator; final syngas cooling is provided by air coolers; water is knocked-out; and the syngas product proceeds to the Fischer-Tropsch synthesis section of the facility (not shown). [15]

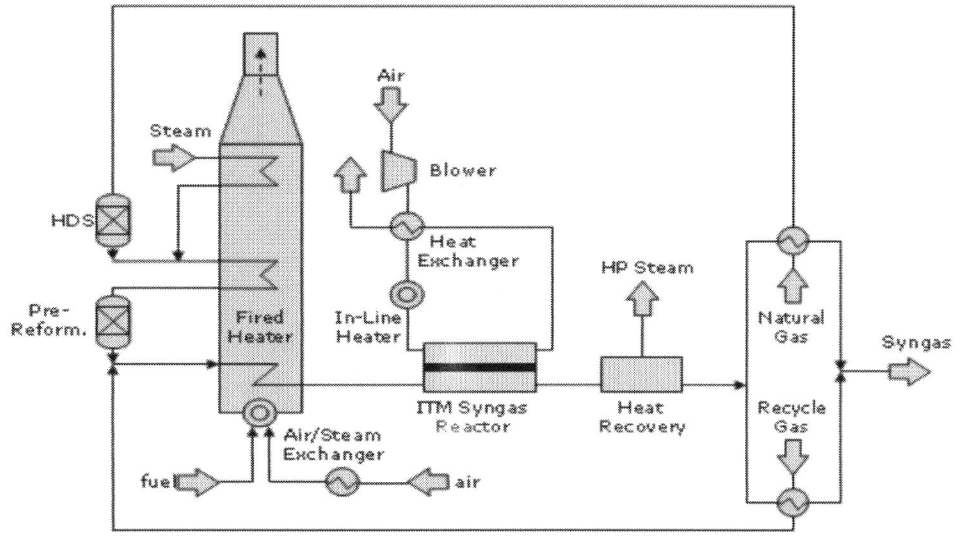

Figure 10. An efficient and operable ITM Syngas process has been developed.

Economic Analysis

Equipment designs and economic analyses were developed for a fully integrated two-train 31,200 bpd Fischer-Tropsch Gas-to-Liquids site, in order to compare ITM Syngas with oxygen-blown Autothermal Reforming (ATR) with a cryogenic Air Separation Unit (ASU). The evaluation included a matrix of nine design cases over a wide range of operating conditions and syngas H2:CO ratios. Cases were also included to understand the potential benefit of Gas Heated Reforming (GHR) technology relative to both ITM Syngas and ATR/ASU. Detailed analysis of Fischer-Tropsch synthesis and integration, as well as all power generation and utility requirements, were included in the analyses.

An optimized ITM Syngas process, proposed by Air Products and simulated by Sasol, with excess steam being condensed at no credit, showed potential for a percentage point advantage in GTL plant thermal efficiency over the optimized ATR/ASU process. As seen in previous analyses [15], ITM Syngas also provides significant excess steam which may be exported or converted to power for site-specific integration, e.g. LNG, if available. The addition of GHR technology [16, 17], as proposed by Air Products and simulated by Sasol, showed additional potential thermal efficiency improvement on both processes: almost 1.5 percentage points on the ATR/ASU process and greater than 3 percentage points on the ITM Syngas process. Carbon efficiencies showed a similar trend.

Equipment sizing and capital investment analysis was completed by Air Products for each of the nine design cases. Comparison of the ITM Syngas technology base case against the state-of-the-art ATR/ASU technology base case showed the potential for significant reduction in capital investment for synthesis gas generation. Economic sensitivity analysis was conducted to understand the impact of major factors affecting the economics, including steam-to-carbon ratio, key operating temperatures, syngas H2:CO ratio, ATR reactor cost, ASU cost, ITM Syngas reactor cost, and GHR reactor cost. ITM Syngas showed attractive economics over all sensitivity cases.

The current design and economic analysis confirmed the results of previous analyses developed for single-train Fischer-Tropsch Gas-to-Liquids plants producing 6,000 to 18,400 bpd, in which the capital cost of the synthesis gas generation process was projected to be significantly lower for ITM Syngas compared to conventional technology employing oxygen-blown Autothermal Reforming (ATR) with a cryogenic Air Separation Unit (ASU). The 6,000 bpd GTL plant was evaluated for a Floating Production Storage and Offloading (FPSO) application, and footprint reductions of up to 40% were identified compared to ATR/ASU. [15, 18, 19]

Applications to co-produce hydrogen and carbon dioxide, such as the use of hydrogen as a decarbonized fuel for centralized or distributed power applications, were also previously evaluated. ITM Syngas processes producing 170 to 850 KNm^3/hr (150 to 760 million standard cubic feet per day) of fuel-grade hydrogen were compared to conventional processes based on ATR/ASU. Economic evaluation of the ITM Syngas process showed the potential for 30% capital cost savings in the synthesis gas generation process area and over 20% capital cost savings for the overall synthesis gas/H_2/CO_2 production process. The ITM Syngas process also had a higher thermal efficiency of 74% compared to 71% for the oxygen-blown ATR process. [18, 19]

The potential economic benefits of ITM Syngas technology continue to be very promising. The capital cost and efficiency advantages for ITM Syngas are due to the combination of oxygen separation and high temperature synthesis gas production into a single unit operation and the ability to use low-pressure air as the oxidant source.

Conclusions

Significant progress has been made in designing and manufacturing planar ceramic membranes, and in demonstrating targeted membrane oxygen fluxes, which are necessary steps toward commercializing ITM Syngas technology. Full-size commercial-scale planar membrane wafers have been fabricated and successfully tested in the Process Development Unit. A process flow sheet has been developed to provide an efficient ITM Syngas process, and the potential economic benefits of the technology continue to be very promising.

Acknowledgement

The authors gratefully acknowledge the contributions of ITM Syngas team members at Air Products and Chemicals Inc., Ceramatec Inc., Chevron Energy Technology Co., Sasol Technology (Pty) Ltd, Norsk Hydro, Eltron Research Inc., SOFCo-EFS Holdings LLC, Penn State University, University of Pennsylvania, University of Alaska Fairbanks, and Pacific Northwest National Lab. This technology development has been supported in part by the U.S. Department of Energy (DOE) under contract No. DE-FC26-97FT96052.

References

[1] U.S. Patent US 7279027 B2, M.F. Carolan et al., "Planar ceramic membrane assembly and oxidation reactor system."

[2] U.S. Patent US 7513932 B2, M.F. Carolan et al., "Planar ceramic membrane assembly and oxidation reactor system."

[3] U.S. Patent US 7094301 B2, D.P. Butt et al., "Method of forming a joint."

[4] U.S. Patent US 7011898 B2, D.P. Butt et al., "Method of joining ITM materials using a partially or fully-transient liquid phase."

[5] U.S. Patent US 6302402 B1, S.W. Rynders et al., "Compliant high temperature seals for dissimilar materials."

[6] U.S. Patent Application US 2008/0260455 A1, M. F. Carolan et al., "Composite seal."

[7] U.S. Patent US 7179323 B2, V.E. Stein et al., "Ion Transport Membrane module and vessel system."

[8] U.S. Patent US 7335247 B2, V.E. Stein et al., "Ion Transport Membrane module and vessel system."

[9] U.S. Patent US 7425231 B2, M.F. Carolan et al., "Feed gas contaminant removal in Ion Transport Membrane systems."

[10] U.S. Patent Application US 2008/0168901 A1, M.F Carolan et al., "Liners for Ion Transport Membrane systems."

[11] U.S. Patent Application US 7556675 B2, M.F. Carolan et al., "Feed gas contaminant control in Ion Transport Membrane systems."

[12] C.F. Miller et al. "ITM ceramic membrane technology to produce synthesis gas," 214th Electrochemical Society & PRiME 2008 Meeting, Honolulu, HI, 15 October 2008.

[13] U.S. Patent US 6114400, S. Nataraj et al., "Synthesis gas production by mixed conducting membranes with integrated conversion into liquid products."

[14] U.S. Patent Application US 2008/0302013 A1, J.M. Repasky et al., "Staged membrane oxidation reactor system."

[15] J.M. Repasky et al. "ITM Syngas: ceramic membrane technology for lower cost conversion of natural gas," AIChE Fall National Meeting, 4 November 2007.

[16] U.S. Patent US 6048472, S. Nataraj et al., "Production of synthesis gas by mixed conducting membranes."

[17] U.S. Patent US 6110979, S. Nataraj and S.L. Russek, "Utilization of synthesis gas produced by mixed conducting membranes."

[18] C.M. Chen et al. "ITM Syngas: ceramic membrane technology for lower cost conversion of natural gas," AIChE Spring National Meeting, 25 April 2006.

[19] D.L. Bennett et al. "ITM Syngas: ceramic membrane technology for lower cost conversion of natural gas," AIChE Spring National Meeting, 12 April 2005.

(33b) Diesel Length Aldehydes From Iron-Based Fischer Tropsch Synthesis

Ed Durham, Department of Chemical Engineering, Auburn University, Auburn, AL

Sihe Zhang, Department of Chemical Engineering, Auburn University, Auburn, AL

Christopher B. Roberts, Auburn University, Auburn University, AL

Carrying out Fischer Tropsch Synthesis on an iron catalyst promoted with zinc, copper, and potassium under supercritical conditions resulted in a significant selectivity to diesel-length aldehydes. Performing the reaction under traditional gas phase and slurry phase conditions gave a similar overall product distribution, but did not give detectable quantities of these aldehydes. Additionally, studies in which the residence time was varied indicated that aldehydes are primary products that are, through secondary reactions, converted to olefins. Paraffins also are primary products, with the supercritical media effectively supressing their formation from olefins. These results carry implications for the steps through which the different types of product are formed and the mechanism of chain propagation.

See more of this Session: Coal, Biomass and Natural Gas to Liquids II

See more of this Group/Topical: Topical 6: 10th Topical Conference on Gas Utilization

(33c) Olefin Pseudo-Equilibrium in the Fischer-Tropsch Reaction

Cornelius Mduduzi Masuku, Centre of Material and Process Synthesis (COMPS), School of Chemical and Metallurgical Engineering, University of the Witwatersrand, Johannesburg, South Africa

Diane. Hildebrandt, Centre of Material and Process Synthesis, University of the Witwatersrand, Johannesburg, South Africa

David Glasser, Chemical and Metallurgical Engineering, University of the Witwatersrand, Johannesburg, Johannesburg, South Africa

It is generally accepted that the product distribution obtained in the Fischer-Tropsch (FT) synthesis over various catalysts can be described by the Anderson-Schulz-Flory (ASF) product distribution or some slight variation of it. Most attempts to describe the FT product distribution have a kinetic basis. If it did have a kinetic basis one would expect that for very low conversions one would find that the ASF distribution would not hold (fewer than predicted higher members of the homologous series) as it would take longer to make the later products in the distribution. This does not seem to be the case.

Few attempts have been made to derive the FT product distribution from thermodynamic considerations. Overall or global thermodynamic equilibrium predicts a predominately methane product for the FT synthesis reaction. However, although the FT product distribution may not be described by overall thermodynamic equilibrium, there may be aspects of it that can be described by equilibrium.

We postulate that an equilibrium is set up in the FT reaction between species, i.e. between hydrocarbon of chain length n and that of chain length n+1, and thus although we do not have global equilibrium, we have an approach to equilibrium between the product species.

We initially postulate that an equilibrium distribution is set up between the alpha olefins or a species that leads to alpha-olefins. Thermodynamic equilibrium and reaction reversibility often are different manifestations of the same phenomena. As olefins are reactive under FT reaction conditions, these are obvious candidates to postulate that equilibrium could exist between the species. We tested this hypothesis by theoretically calculating what the product distribution would be. We also checked what the theoretical predictions would give if we assumed that equilibrium was set up between the members of the paraffin series as well as the alcohol species.

The calculated results predict that equilibrium between the species of any homologous series would results in an ASF type distribution, where the ratio of the moles of species n to that of the moles of species n+1 would be constant. However when the value of this ratio α is calculated, it is found that the experimentally measured values match those calculated and thus the distribution of the olefins. As mentioned above paraffins and alcohols would also follow a single α product distribution at Fischer-Tropsch reaction conditions However, assuming that the alcohols or paraffin distribution approaches equilibrium, leads to calculated values of α that are not consistent with the experimental results. Thus we postulate that there is a pseudo-equilibrium set up among the alpha olefins.

Keywords: Fischer-Tropsch product distribution, FT thermodynamics, olefin reactivity.

See more of this Session: Coal, Biomass and Natural Gas to Liquids II
See more of this Group/Topical: Topical 6: 10th Topical Conference on Gas Utilization

(33d) Low Temperature Conversion of Bio-Mass to Ultra-Clean Gasoline

Amit Goyal, Exelus Inc., Livingston, NJ

Kelly Coley, Exelus Inc., Livingston, NJ

The need for renewable fuels to reduce carbon emissions, improve economic security, and achieve energy independence is widely acknowledged and understood. Displacing large quantities of imported oil with domestic biofuels requires certain enabling technologies that are equivalent to those in an oil refinery.

Two of the most prominent technologies under consideration for biomass conversion, pyrolysis and gasification, have significant intrinsic limitations. Both gasification and pyrolysis involve very high temperature (500-1000°C) endothermic reactions and generally exhibit poor yields and problematic solid waste streams. Gasification is also very inefficient when the feedstock material contains more than 10% water. Enzyme or microbial routes to cellulosic ethanol are also popular. Even if the cost of the requisite enzymes or microbes can be greatly reduced, these technologies will still fail due to low product concentrations, high energy costs and long reaction times.

Exelus proposes a first-of-a-kind process to deconstruct biomass into a stable liquid feedstock and convert it to liquid transportation fuels. The final product is a mixture of alcohols that serve as a direct replacement for gasoline. This process finds the "middle ground" between the slow, low temperature reactions catalyzed by enzymes and the fast, high-temperature, and un-selective reactions occurring in pyrolysis or gasification.

The Exelus concept overcomes the issue of highly reactive intermediates (sugars) reacting to form un-selective products like polymers and acids through the use of environmentally benign chemistry and innovative reactor designs. Liquid-phase aqueous processing is used to minimize the need to dry biomass and to provide both a solvent and reactant in the hydrolysis of biomass. A combination of low capital investment, high biomass utilization and lower processing costs translates into a cost of fuel produced via the new route of about $2.00/gal.

See more of this Session: Coal, Biomass and Natural Gas to Liquids II
See more of this Group/Topical: Topical 6: 10th Topical Conference on Gas Utilization

Role of cobalt carbide in high alcohols (C_1-C_{18}) synthesis from syngas over activated carbon supported cobalt catalysts

*Yunjie Ding**, *Yanpeng Pei*, *Hejun Zhu*, Xiangen Song, Juan Zang, Wenda Dong and Yuan Lu

Dalian National Laboratory for Clean Energy, Dalian Institute of Chemical physics, Chinese Academy of Sciences, 116023 Dalian China.Email: dyj@dicp.ac.cn

1. Introduction

The mixed linear α-alcohols of C_6-C_{18} are used as intermediates of plasticizers, detergents and lubricants[1], and their synthesis from natural gas, coal, extra heavy oil or biomass resources via syngas (H_2/CO) is of high interests in both industry and academia [2,3]. It has long been recognized that the mixed alcohols of C_1-C_6 could be successfully produced from syngas. Extensive studies have been carried out in an attempt to improve the selectivity toward C_1-C_{18}, especially C_6-C_{18} alcohols, in the FT process, particularly under the conditions with low selectivity to methane and CO_2. Exxon reported a Th-promoted Fe catalyst, on which about 8 wt% C_6-C_{12} alcohols were produced in the total organic products [4]. Recently, Buess et al. prepared the CoCuMn and CoCuMg catalysts via co-precipitated for the synthesis of C_1-C_{12} alcohol. They found that the catalysts exhibited 2-7 % selectivity to C_6^+ alcohols on the basis of converted CO. Thus, the direct synthesis of C_6-C_{16} alcohols from syngas with high yield is highly desirable.

2. Experimental

Activated carbon is made from coconut shell and it was crushed into 20-40 mesh and washed with de-ionized water before use. The activated carbon supported Co catalysts were prepared by incipient wetness impregnation method according to the following procedure: the activated carbon was vacuated for 30 minutes and then impregnated in a rotary evaporator at room temperature with an aqueous solution of $Co(NO_3)_2 \cdot 6H_2O$. After evaporating and drying in air at 363 K for 8.0 h, the sample was calcined at 623 K for 3.0 h in the argon atmosphere. The Zr and La promoted Co catalysts were prepared by impregnating the activated carbon with the mixture of $Co(NO_3)_2$, $ZrO(NO_3)_2$ and $La(NO_3)_3$, drying and calcinating under the same conditions as above. The Co catalysts were characterized using nitrogen sorption, X-ray diffraction, TG-DTA, and TEM techniques, detailed information was described previously.

3. Result and discussion

The XRD patterns of various reduced and used Co-based catalysts demonstrated that diffraction peak at 2θ of 44.3° for Co(0) (JCPDS 15-0806) and those at 2θ of 36.8° and 44.8° for Co_3O_4 (JCPDS 09-0418) were observed on the reduced and passivated samples (Figure 1). These peaks were broad and weak due to the high dispersion of cobalt on the support. XRD patterns of used catalysts (Figure 2) revealed the new peaks due to Co_2C species (2θ = 37.0°, 41.3°, 42.6°, 45.7° and 56.6°, JCPDS 72-1369). Apparently, the Co_2C species was formed during the F-T reaction. In addition, the XRD results clearly indicated that the peaks of Co_2C in the Co-La/AC and Co-Zr-La/AC were much stronger than those in the Co/AC and Co-Zr/AC. This seems to suggest that the presence of La favored the formation of Co_2C species while Zr did not seem to have such effect during the F-T reaction.

Fig. 1 shows the HRTEM images of Co-Zr-La/AC before (Fig. 1a) and after the FT reaction (Fig. 1b, 1c and 1d). Many particles were clearly seen with the core-shell structure for the catalysts after the FT reaction. The average particle size was found to be about 15 nm with a shell thickness of about 4.2

nm. Fig. 1c depicts a typical HRTEM image of such core-shell structure. The core contained $Co_2C(011)$ and $Co_2C(111)$ with crystal lattices of 3.117 Å and 2.122 Å, respectively, while the shell contained mainly CoO(110) and metallic Co(111) with the crystal lattice of 2.452 Å and 2.040 Å, respectively. Only Co(111) particles with a crystal lattice distance of 2.040 Å (Fig. 1a) were observed for the sample before the FT reaction, while metallic Co (Fig. 1b and 1d), cobalt oxide (Fig. 1d) and cobalt carbide (Fig. 1c and 1d) were found in the catalysts after the FT reaction.

Mixed linear α-alcohols (C_1-C_{18}), naphtha distillates and diesel fuels were directly synthesized from syngas over the activated carbon supported Co catalysts through the one-step synthesis technique. The conversion of CO was low (14 %), but the selectivity to hydrocarbons was high, over the Co/AC. The addition of Zr greatly increased the conversion of CO and decreased the selectivity to methane and CO_2. Mixed linear C_2-C_{18} α-alcohols were the main products with methanol lower than 2% based on the total organic products. The addition of La also greatly increased the conversion of CO, with a significantly enhanced selectivity to α-alcohols (from 21 to 38 %) at the same time. The formation of CH_4 and CO_2 was further suppressed with the addition of La. It seems that the addition of Zr in Co/AC increased the conversion of CO only while the addition of La in Co/AC enhanced not only the conversion of CO, but also the selectivity to α-alcohols. This result was further confirmed by the reactivity of Co-Zr-La/AC for the FT reaction. The conversion of CO could be as high as 71 % with a low selectivity to methane (11 %) over the Co-Zr-La/AC reduced in H_2 at 693 K. At the same time, the concentration of C_6-C_{18} α-alcohols among the total alcohols remained high (45 %).

4. Conclusion

In summary, the promotion effects of Zr and La for the carbon supported cobalt catalysts were clearly observed for the F-T reaction. While the addition of Zr increased the conversion of CO, the addition of La enhanced both the conversion of CO and the selectivity to high alcohols. The catalyst Co-Zr-La/AC exhibited high conversion of CO (71.4%) and high yield of $C_{6\text{-}18}$ alcohols (39.1%) during the F-T synthesis. A linear correlation was found between the Co2C/Co(0) in the catalysts and the selectivity to α-alcohols, suggesting that the Co_2C might play an important role for the production of α-alcohols during the F-T synthesis.

5. References

[1] Scheibel, J. J., J. Surf. Deter. 7(2004) 319-327.

[2] Vosloo, A.C., Fuel Process Technol. 71(2001)149-155.

[3] Dry, M.E., Appl. Catal. 276(2004)1-3.

[4] Wright, F. J.; Richard, M. A.; Pirkle, J. C., US Pat.t 4504600.(1985).

[5] Buess, P.; Caers, R.F.I.; Frennet, A.; Ghenne, E.; Hubert, C.; Kruse, N. US Pat., 6,362,239.(2002).

(39a) Cogeneration Options in LNG Liquefaction

Cyrus B. Meher-Homji, LNG Technology- Turbomachinery Group, Bechtel Corporation, Houston, TX

Karl Masani, ConocoPhillips Company, Houston, TX

Neeraj Bhatia, LNG Engineering, Bechtel Corporation, Houston, TX

Ram Tekumalla, Advanced Simulation Group, Bechtel Oil, Gas & Chemicals Inc., Houston, TX

High thermal efficiency for LNG liquefaction plants is of importance due to the need to minimize feed usage and to reduce CO2 emissions. The need for high efficiency becomes very important in gas constrained situations where any savings in the fuel auto consumption of the plant for liquefaction chilling and power generation can be converted into LNG production. The application of aeroderivative engines allows a significantly lower CO2 footprint of about 30% compared to the use of simple cycle heavy duty engines. Aeroderivative engines offer very attractive efficiencies where steam systems are not viable or desired by the end customer. When steam systems are acceptable, a cogeneration type liquefaction facility can be attractive. This paper will cover concepts relating to cogeneration options for LNG facilities using both aeroderivative engines and heavy duty engines. Parametric analysis relating to different cogeneration options and schemes will be presented with varying levels of heat integration between the liquefaction block and also the powergeneration block. The paper will discuss the interaction of gas turbine driver parameters with cogen / combined cycles.

See more of this Session: Professor Cedomir M. Sliepcevich Memorial Session One - LNG Technology
See more of this Group/Topical: Topical 6: 10th Topical Conference on Gas Utilization

IMPACT OF PRESSURE RECOVERY ON C2 RECOVERY PROCESSES – COMPARING PERFORMANCES BETWEEN EXPANDER AND JT VALVES

Vedran Makitan, Chevron Energy Technology Pty Ltd, Perth, WA, Australia
Stanley Huang, Chevron Energy Technology Company, Houston, TX, USA
Karl Gerdes, Chevron Energy Technology Company, Richmond, CA, USA

ABSTRACT

Natural Gas Liquid (NGL) separation is an important link in any LNG plant. Recently, expander-based process has received increased attention, thanks to better appreciation of the refrigeration integration between NGL and LNG sections. SIMAR, which stands for System Intrinsic Maximum Recovery, is a systematic approach in process design. The main objective of this paper is to illuminate the advantages of turbo-expanders over Joule-Thomson (JT) valves. The second objective is to demonstrate that SIMAR has adequate resolutions for such tasks. In view of recent development of new equipment, for example, TWISTER^{TM}, SIMAR may find use in process screening and comparison.

1.0 INTRODUCTION

There are several approaches to recover intermediate (C_2+) components from natural gas (NG), such as light-oil absorption, refrigeration-based process, and absorption. Since the early 1980's, improvements in machinery manufacturing have made expander-based processes the mainstay technology in this field. Quantum progresses and incremental improvements have pushed this art to an unprecedented level in process efficiencies. As a witness of this trend, there are hundreds of patents filed in the U. S. alone for process configurations.

SIMAR, which stands for System Intrinsic Maximum Recovery, is a methodology developed in recent years to provide a systematic approach in process design. Simply stated, SIMAR reduces a two-dimensional search for optimal process parameters to a one-dimensional search. Detailed descriptions have been reported previously [Huang et al., 2006; McMullen and Huang, 2007; Yengle and Huang, 2008 & 2009].

Sections 2 to 3 of this paper provide a brief summary of SIMAR. The focus of this paper is on the impacts of replacing turbo-expanders by different mechanical items, for example, Joule-Thomson (JT) valves. Accordingly, Section 4 provides the study basis. Section 5 applies the SIMAR procedure to identify the impacts of replacing the turbo-expander with JT valve. This work is further extended to include the results of applying SIMAR to both lean and rich gases in gas processing. Finally, Section 6 lists the conclusions of this paper.

2.0 TECHNICAL BACKGROUND

Figure 1 contains two illustrations: 1a and 1b. Figure 1a shows a generalized scheme for C_2 recovery based on expander technology for natural gas systems. Treated NG goes through an inlet chilling section, where the gas is chilled to a suitable level before entering the heart of the plant: an expander-based C_2 recovery scheme. The refrigeration of the inlet chilling section is mainly provided by the returning residue gas and by side-draws from the de-methanizer (DeC1, side-draws not shown), or supplemented by an external refrigeration (not shown). The residue gas is recompressed and returned to the pipeline. The main components of the expander section include a separator, expander, and DeC1. A dominant consideration in the process design is the refrigeration supply. Often times, the needs to balance refrigeration complicate the entire design considerations.

Figure 1b is created with an intention to isolate refrigeration provisions from design considerations. The method is to replace NG feed by Liquefied Natural Gas (LNG) of the same compositions. All other process configurations remain unchanged except that the residue stream from the DeC1 top goes directly to the re-compressors without refrigeration recovery. This simplification of process flow is feasible because LNG contains unlimited refrigeration for practical purposes. As a result, the design works can be focused on recovery levels and process efficiencies. SIMAR was initially developed based on Figure 1b, then extended to 1a for real applications.

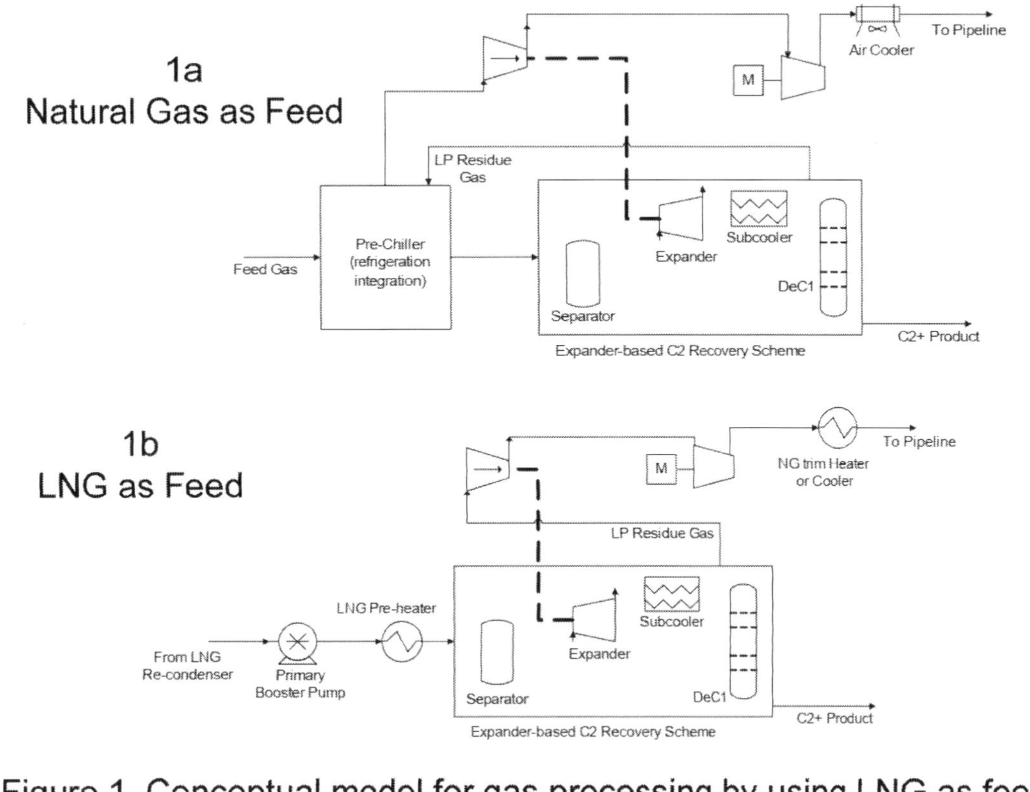

Figure 1 Conceptual model for gas processing by using LNG as feed

Existing process configurations for expander based C_2 recovery schemes can be classified to facilitate discussions. Figure 2 shows four categories of expander-based schemes. The classification is based on the handling of the vapor stream of the separator, namely, separator top not rectified (XPDR 1), separator top fully rectified (XPDR 2), and separator top self-rectified (XPDR 3). The last category includes a base scheme with embedded heat pumps. A heat pump is recognized by the use of a compressor, a cooler for rejecting heat to a high-temperature sink, a JT valve or a second expander and optionally, a second exchanger to take heat from the low-temperature source. The impact of heat pumps was covered in the references and not to be discussed in the paper.

3.0 SUMMARY OF SIMAR METHODOLOGY

There are two subsections. The first one explains the basic principles based on Figure 1b, which uses LNG as feed. The second subsection explains how to apply the SIMAR methodology to natural gas systems.

3.1 Basic principles

Refer to Figure 2 for illustrations. For a given C_2+ recovery configuration, the separator temperature and the DeC1 pressure are the two independent parameters which sensitively affect the C_2 recovery levels. This two-dimensional search presents a challenge in designing a new process. Traditional procedures demand extensive accumulated experiences in performing this art. SIMAR was created to simplify the search for optimal process conditions.

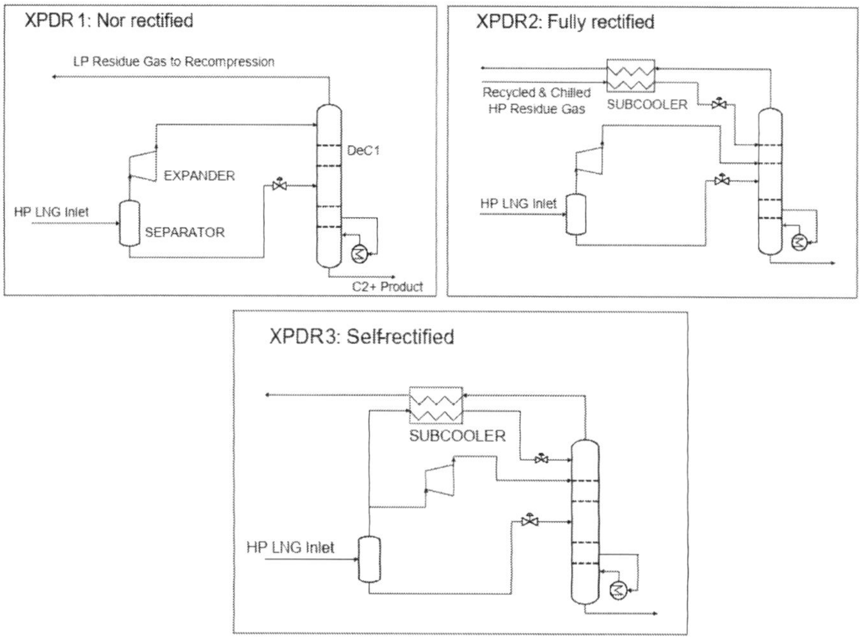

Figure 2 Categorizing expander based schemes

Figure 3 shows two phase envelopes corresponding to two natural gas compositions. The two gas compositions are designated as medium and rich. The rich composition has an enlarged two-phase region toward higher temperatures and pressures.

A SIMAR curve can be constructed following a few simple steps. The procedure starts from a relatively high temperature at a reasonable pressure level. The track of testing temperatures penetrates through the two phase region and ends at an arbitrarily chosen level of -100°C. The fluid would remain liquid at and below this temperature level.

Figure 4 plots the C_2 recovery level when the separator temperature decreases systematically from a relatively high level. Both XPDR 1 and XPDR 2 show monotonic trends of improvement as the temperature decreases. The trend continues even after the inlet gas is totally liquefied, when the expander is replaced by a JT valve. In contrast, for XPDR 3 as the separator temperature decreases, the C_2 recovery goes through a maximum. In other words, blindly increasing the refrigeration to the inlet chilling section of a configuration may not always improve the C_2 recovery.

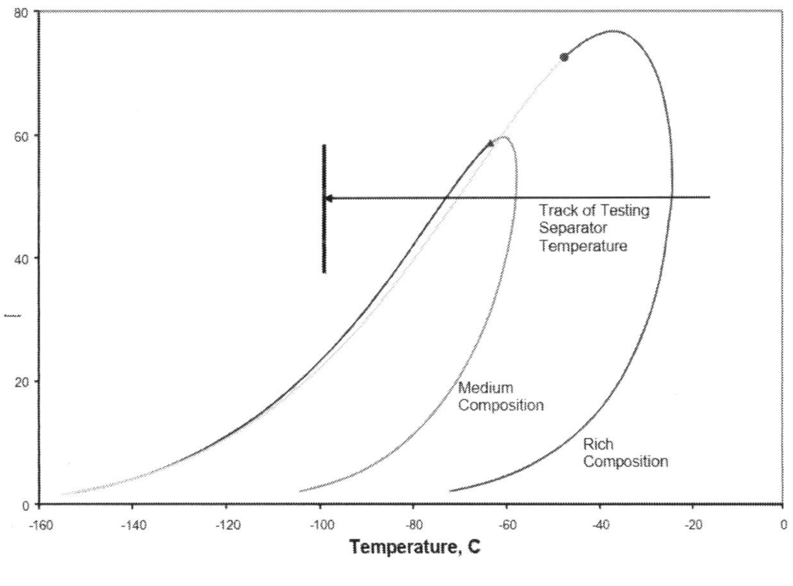

Figure 3 Phase envelope of inlet gases

For the cases showing maxima, such as XPDR 3, the SIMAR point is defined as the maximum of the trend. For the cases showing monotonic trends, such as XPDR 1 and XPDR 2, the SIMAR point is defined at the temperature where the separator fluid has 30% of vapor fraction (VF) as no expanders would be installed below this level. The collection of all SIMAR points, at different DeC1 operating pressures, forms a SIMAR curve for the process configuration.

Figure 5a depicts the SIMAR curves for the three types of process schemes. Some observations of the three categories can be made. (1) The reflux stream makes XPDR 2 more efficient than XPDR 1 throughout the entire range, which corresponds to the operating pressures of DeC1 from 17 to 42 bar. (2) The efficiencies of XPDR 2 and XPDR 3 are comparable.

Similarly, figure 5b shows the SIMAR curves for selected patented processes. The identification of individual processes is intentionally left out. The figure indicates that SIMAR procedure provides adequate resolutions to distinguish different process configurations. Furthermore, some of the patented processes improve the C_2 recovery from the generic XPDR3; however, there are exceptions. Users' discretion is required in the process selection. The impacts of inlet pressure, inlet gas compositions, and heat pumps have been reported elsewhere.

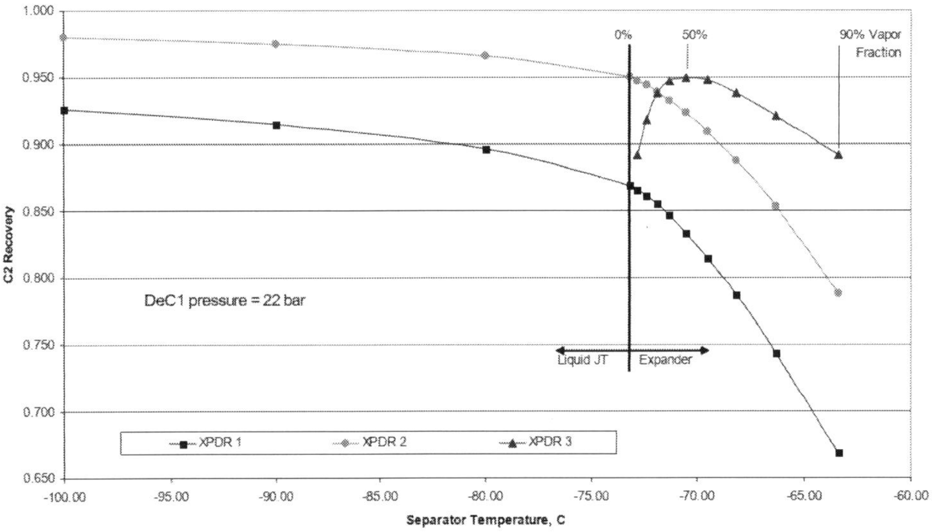

Figure 4 Two types of behavior for different categories.
VF stands for Vapor Fraction.

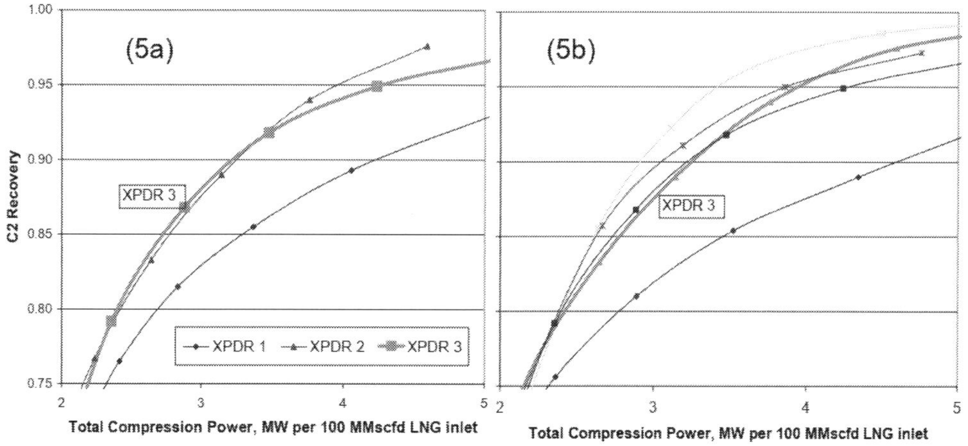

Figure 5 Comparing three categories (5a) and different patented designs (5b)

3.2 SIMAR applied to gas processing

When the feed is reverted to natural gas, as shown in Figure 1a, the refrigeration in the returning residue gas has to be recovered to support the operation of the inlet chiller. As a result, the recompression work increases. Figure 6 shows the impacts. The first curve from the left is the SIMAR curve with LNG feed. The second curve is the case with natural gas feed. The two curves are roughly parallel to each other and the difference is about 2 MW per 100 MMscfd. This gap is the impact of refrigeration recovery.

In addition to recompression, the total compression should also include the power needed in the supply of external refrigeration. The two thick curves on the right side of Figure 6 present the total compression work by using one or two side-boilers (SB). The case of two SB requires a lower total compression work because the provided refrigeration from two SB matches better with the chilling requirements. This closer match improves the overall system efficiency.

Moving along the curves in Figure 6 from high to low C_2 recovery implies increasing DeC1 operating pressures. When the DeC1 operating pressure increases, the recompression work decreases. Meanwhile, the refrigeration generated by the expander also decreases. For the system to operate at SIMAR conditions, the demand for external refrigeration increases. This explains the widening gap between recompression and total compression at lower C_2 recovery levels.

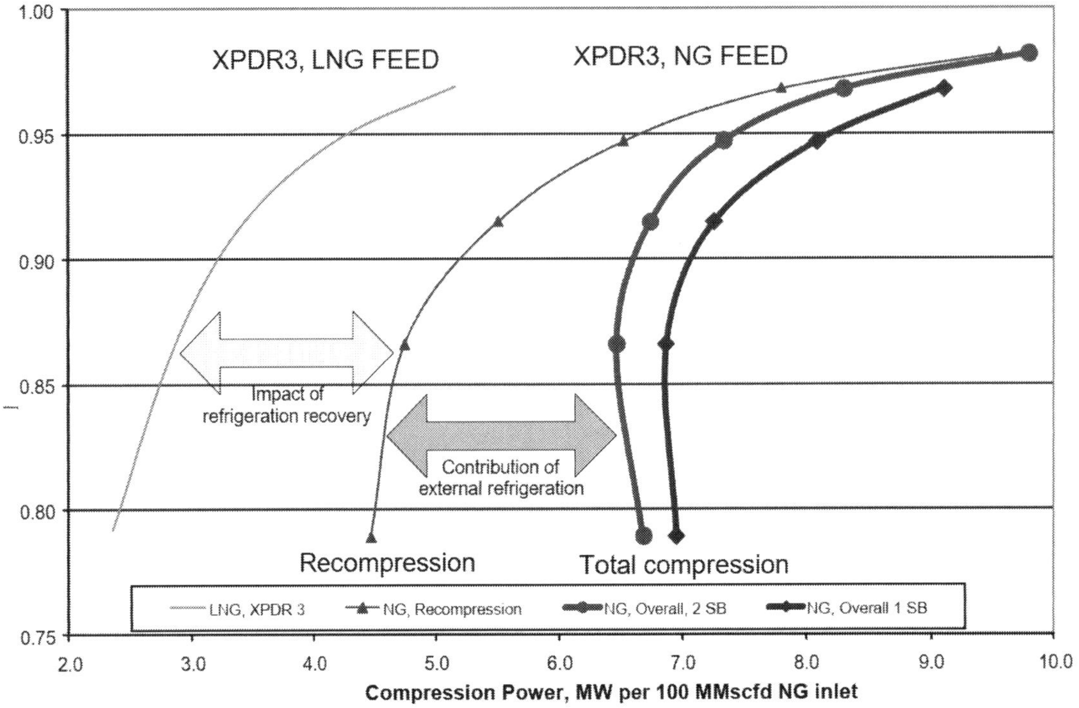

Figure 6 Compression power and impact of side – boilers (SB)

It is possible to eliminate the external refrigeration by increasing the expander-generated refrigeration at high C_2 recovery levels. This is demonstrated by the convergence of the curves of recompression and total compression. The final choice is an economic decision taking into account the availability of both the feed pressure and external refrigeration.

Not all gas processing facilities are designed to operate at SIMAR conditions. Provision of external refrigeration may become a limiting factor for achieving those conditions. The designer may want to sacrifice the DeC1 utilization efficiency in exchange for savings in the refrigeration systems. The discussion of sub-SIMAR conditions are provided in the references.

4.0 STUDY BASIS

Unless stated otherwise, all results presented are based on HYSYS simulations. Typical industrial parameters are used in simulations, as shown in Table 1 below. Two compositions of different richness are included. The inlet pressure of the rich gas is chosen so that the reduced pressures (pressure over critical pressure) of both cases remain at the same level.

Table 1. Simulation parameters used in this work

Natural Gas inlet	Medium	Rich
Feed to gas processing section	50 bar	65 bar
Plant capacity, inlet LNG	100 MMscfd	100 MMscfd
Send-out gas pressure	100 bar	100 bar
Components	Mole Fraction	Mole Fraction
Nitrogen	0.00127	0.0000
CO2	0.00008	0.0000
Methane	0.90692	0.8695
Ethane	0.08617	0.0792
Propane	0.00552	0.0332
i-Butane	0.00003	0.0074
n-Butane	0.00001	0.0055
i-Pentane	0.00000	0.0037
n-Pentane	0.00000	0.0015
TOTAL	1.00000	1.0000
HHV, Btu/SCF	1,082	1,165
Critical pressure, bar	58.63	75.69
Expander polytropic efficiency	80%	80%
Compressor polytropic efficiency	80%	80%
Pressure drop through plate-fin exchanger	0.2-0.3 bar	0.2-0.3 bar
Pressure drop through air coolers	0.6 bar	0.6 bar
Minimum delta T in plate-fin exchanger	2.5 ± 0.2 °C	2.5 ± 0.2 °C
DeC1 operating pressure	17 to 42 bar	17 to 42 bar
Number of trays in DeC1 column	15	15
C1 composition in DeC1 bottom product	1%	1%

In the following work, only the results of two side-boilers will be presented. The relative distribution of loads on the two side-boilers is mainly determined by the requirement of the pre-chiller. The C_2 recovery level is insensitive to the distributions.

Additionally, conceptual thermodynamic heat pump model is adopted to account for external refrigeration so as to simplify the discussion. Thermodynamic heat pumps are assumed to be 50% of Carnot efficiency. For example, to move 1 MW of heat from -60 °C (213 K) to 40 °C (313 K), a Carnot engine requires 0.47 (= 313/213 - 1) MW of compression power. In this work, twice the amount is required due to the assumed 50% efficiency.

5.0 REPLACING EXPANDER BY JOULE-THOMSOM VALVE

There are three subsections. The first one uses LNG as feed. The purpose is to illustrate the major points by keeping the system simple. The second subsection uses medium NG as feed to demonstrate behavior of real applications. The third section uses rich NG as feed to show the impact of richness in the feed.

5.1 LNG as feed

When LNG is used as feed, refrigeration recovery is unnecessary from the return residue. Therefore, recompression power is very much reduced. At cryogenic conditions, isentropic lines have very sharp slopes on P-H diagrams. Consequently, expansions by expander or JT valve are not anticipated to have big differences. Figure 7 confirms this anticipation. In the range of our interest, the difference is only a fraction of one percentage.

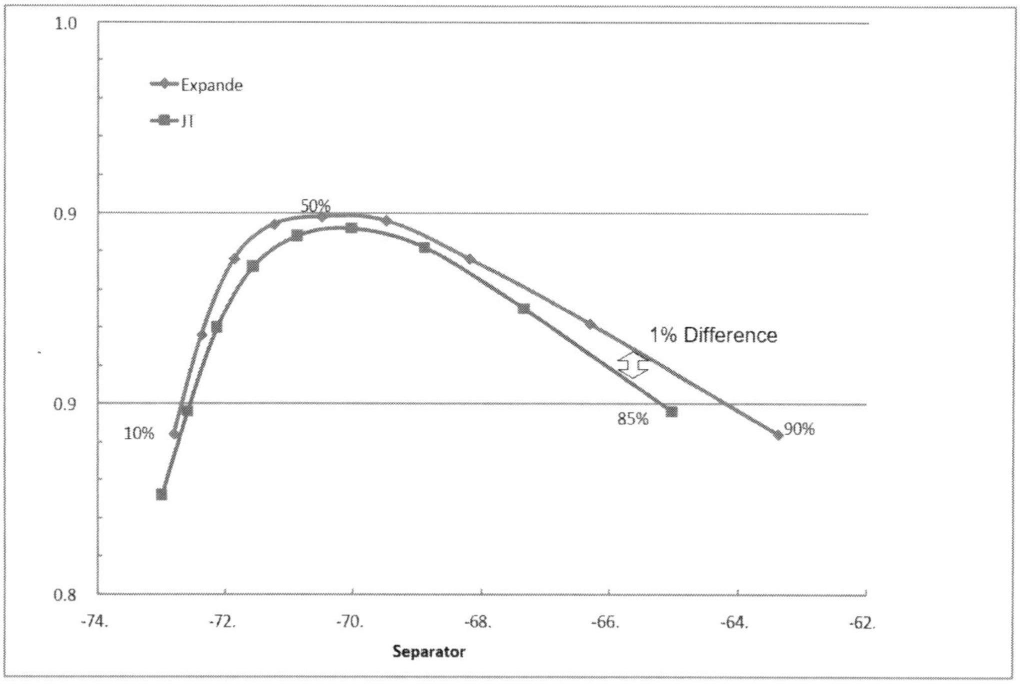

Figure 7 Expander slightly outperforms JT valve in recovery

However, using JT valves would miss the opportunity of recovering the letdown pressure. Consequently, the re-compressor would have to make up the difference. Figure 8 shows the impact. There is a step increase in compression power for the JT case. The gap is roughly 0.3 MW per 100 MMscfd LNG inlet.

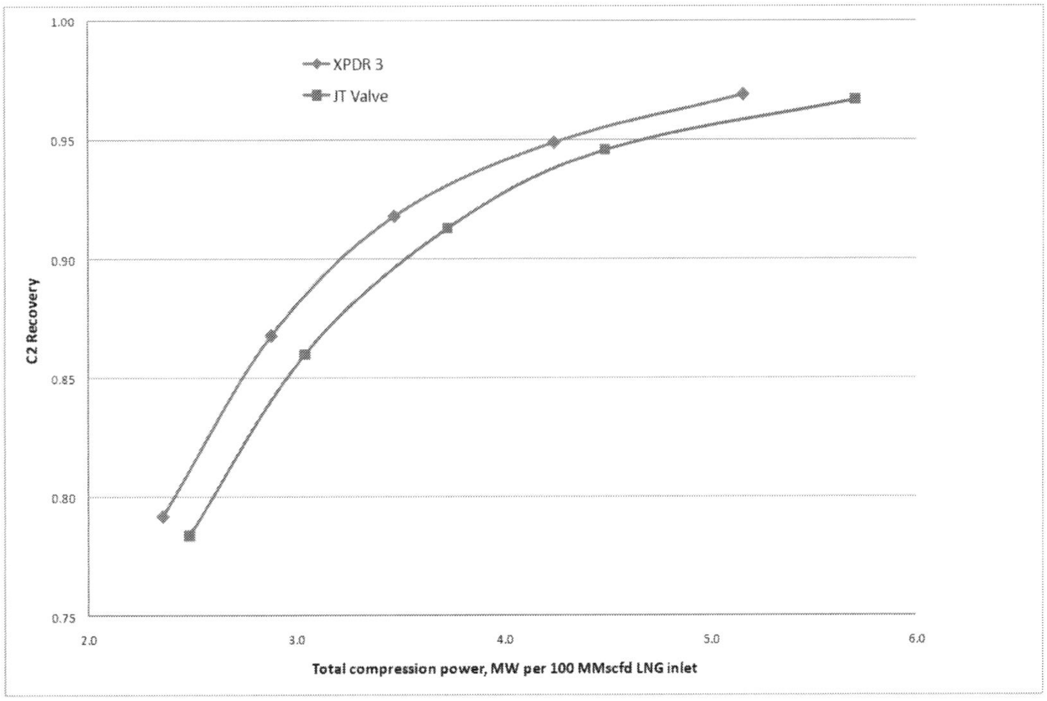

Figure 8 Expander outperforms JT valve in compression power

5.2 Applied to lean gas in gas processing

Figure 9 superimposes the results of JT valve case on Figure 6. Only the results of double side-boiler case are presented. As anticipated, the general trend of using JT valve is similar to that of using expander, which was elaborated in Figure 6.

For each of the three sets of curves, there is a gap between the two curves. For the two sets of recompression data, the gaps are small and consistent. They are attributable to the lost opportunity of pressure recovery.

For the total compression curves, the bigger gap in the higher recovery region is attributable to the fact that JT valve generates less refrigeration for the lower section of the DeC1. Consequently, less refrigeration can be recovered from the two SB and additional external refrigeration has to be supplied. Since external refrigeration is generated based on JT valves in this work, increased demand for external refrigeration enhances the difference.

As the DeC1 operating pressure increases and C_2 recovery decrease, the advantages of expander over JT valve are obscured. Elevated levels of external refrigeration should be supplied in both cases. Hence, the two cases merge with each other.

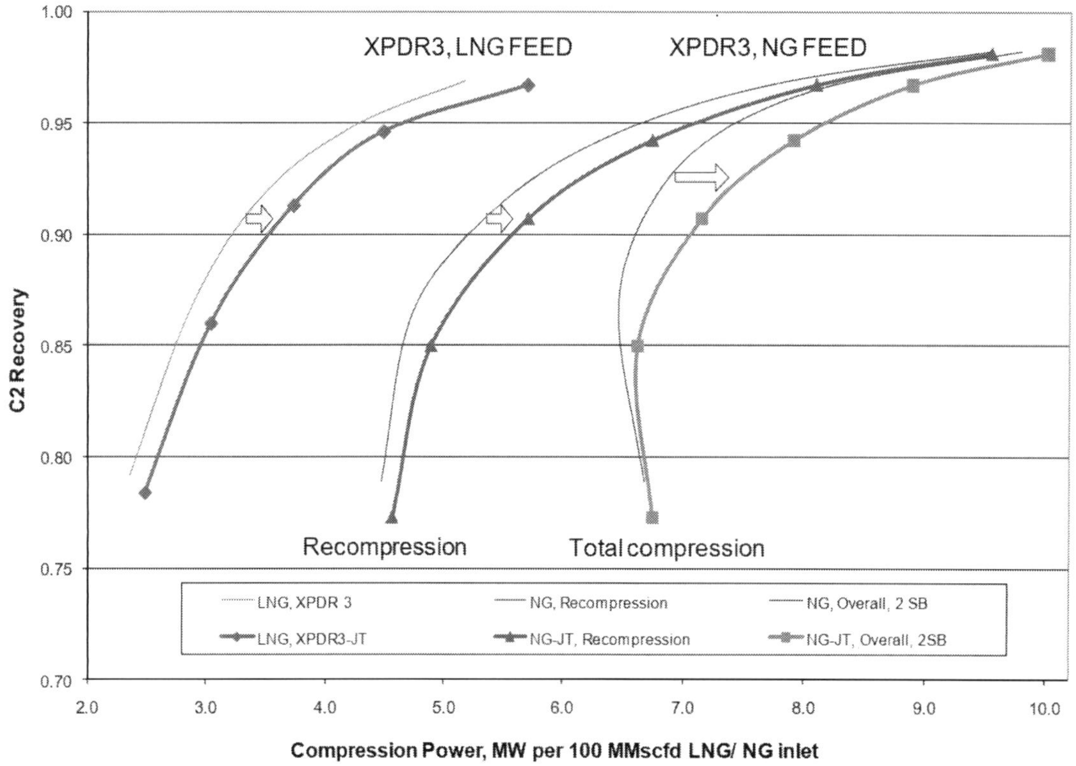

Figure 9 Expander outperforms JT valve in compression power

5.3 Applied to rich gas in gas processing

Figures 10 and 11 show the results of applying SIMAR to the rich case, a conventional expander in XPDR3 configuration is used in both figures. The minor reductions in recompression power are caused by reduced flow rates in residue gases as heavier components are removed from DeC1 columns.

The external refrigeration requirements are elaborated below. In Figure 10, the rich gas requires significantly reduced external refrigeration than the medium gas. This is contrary to the conventional wisdom that it takes more external refrigeration to process rich gas. The reason is attributable to the inlet gas pressures. Table 1 indicates that rich gas has higher inlet pressure than the medium gas in order to maintain a relatively constant reduced pressure (ratio between actual pressure and critical pressure of gas). The increased inlet pressure is converted to refrigeration in the process configurations.

When the inlet pressure of the rich gas is reduced to the same level (50 bar) as the medium gas, the rich gas requires more external refrigeration. This is demonstrated in Figure 11.

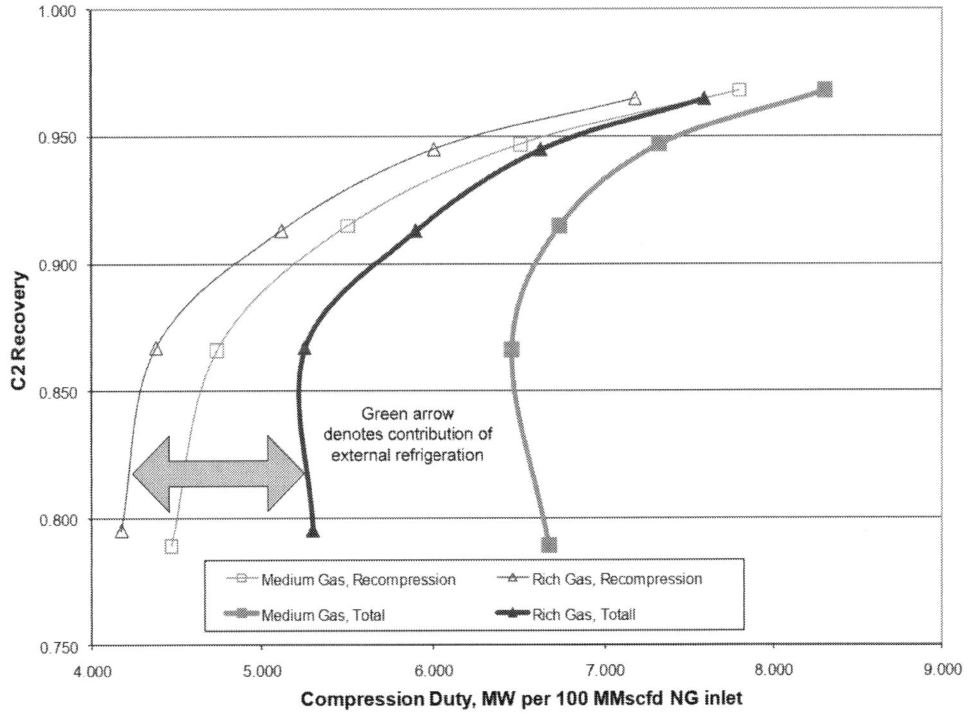

Figure 10 Improvements in Rich Gas case attributable to increased inlet pressure

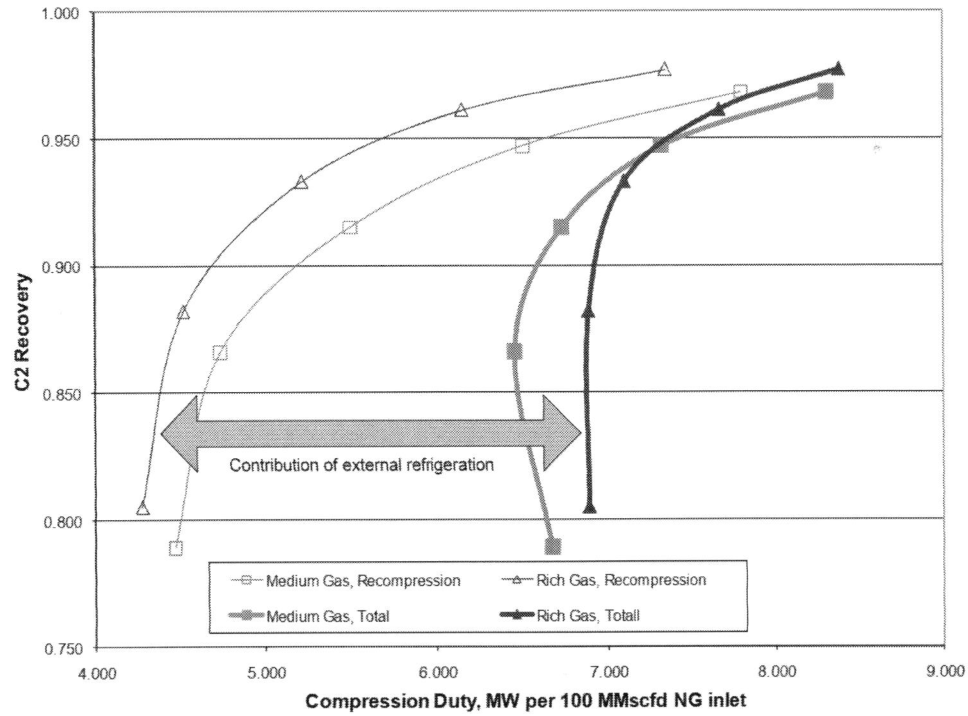

Figure 11 Rich Gas case with 50 bar inlet pressure.
Notice the increased demand for external refrigeration.

Figure 12 compares the performance of the expander against JT valve for the rich gas case. Similar comparison was made for medium gas in section 4.2. For the overall compression curves, the large gap between the expander and JT valve performance is due to the fact that JT valve generates less refrigeration in the lower section of DeC1. If compared to the medium case than it is noticeable that this gap is much bigger. The reason for this is increased pressure drop due to higher inlet pressure of rich gas which further amplifies the difference between the refrigeration achieved by expander and JT valve. This has direct effect on the external refrigeration requirements and it the main reason why rich gas requires relatively little external refrigeration, as shown in Figure 10.

As a result, the rich gas shift from expander to JT cases in Figure 12 should be a parallel one and therefore no convergence is observed as in Figure 9.

The two sets of recompression data exhibit very similar characteristics of lean case, throughout the whole range of recovery levels. As expected, the expander clearly outperforms the JT valve.

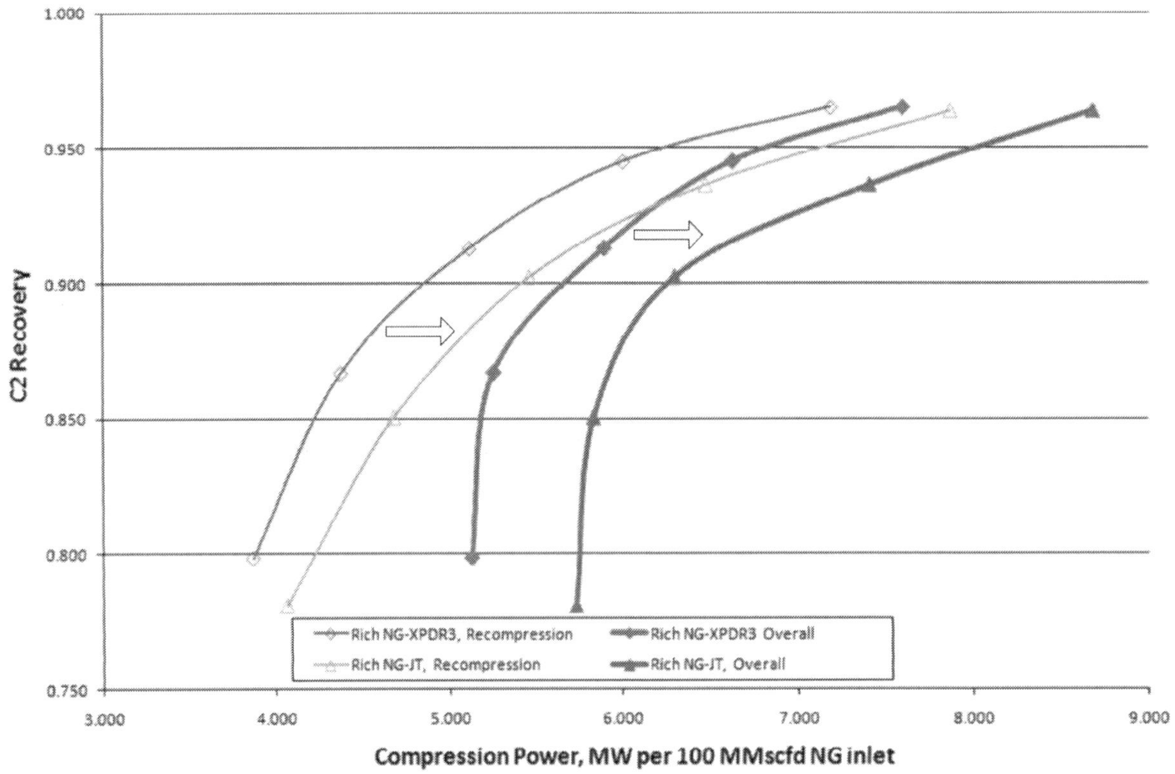

Figure 12 Rich Case Expander outperforms JT valve in overall compression power, however at high and low DeC1 pressures this is not the case for recompression power alone.

6.0 CONCLUSIONS

The conclusions of this work are listed as follows:

- A brief summary of SIMAR is provided, which is a methodology developed to facilitate process designs for gas processing facilities.
- Processes using expander outperform counterparts with JT valves. The reasons are attributable to lost opportunities in recovering letdown pressures and/or reduced refrigeration generation for the same level of pressure letdown.
- When the inlet pressure remains the same, the conventional wisdom holds that richer gases require more external refrigeration to process.
- When the inlet pressure of the rich gas is held at the same reduced pressure as the medium gas, the behavior of the two systems is similar.

REFERENCES

- Huang, S., Chen, R. J. J., Yao, J., D., Elliot, D.G., "Processes for High C2 Recovery from LNG: Part III: SIMAR Applied to Gas Processing", Paper presented at AIChE Spring Meeting, Orlando, Florida (2006).
- McMullen, C., Huang, S., "SIMAR Applications 2: Optimal Design of Expander-Based C2+ Recovery in Gas Processing", paper presented at AIChE Spring Meeting, Houston, April 22-26 (2007).
- Yengle, C. M., Huang, S., "SIMAR Application in Gas Processing: Challenges in Supercritical Regions", paper presented at AIChE Spring Meeting, New Orleans, April 6-10 (2008).
- Yengle, C., Huang, S., "A Universal Methodology Based on SIMAR for Composing and Evaluating Expander-Based Processes", paper presented at 1st Annual Gas Processing Symposium, Doha, Qatar (2009).

Biographies

Vedran Makitan is a process engineer for Chevron Energy Technology Pty Ltd. He graduated in 2006 from the Curtin University of Technology in Perth with Bachelors Degree in Chemical Engineering and Commerce and received his Master's degree in Chemical Engineering in 2009. Vedran has been with Chevron for just over a year focusing on technical development and strategic research within gas processing and conversion.
E-mail: VMakitan@chevron.com

Dr. Stanley Huang has a specialty area in cryogenic applications, particularly in LNG and gas processing. He graduated from National Taiwan University with a B.S. degree and attended Purdue University in 1981. He earned his Master and Ph.D. degrees there, all in Chemical Engineering. Additionally, he also acquired a Master of Science in physics. Stanley has 20 years of industrial experiences and is a Registered Professional Engineer in Texas. Currently he is a Staff LNG Process Engineer of Chevron.

Dr. Karl Gerdes has been engaged in the development of new technology for gas processing for more than 25 years. He graduated from the South Dakota School of Mines & Technology and attended UC Davis, where he earned MS and PhD degrees in Chemical Engineering. Karl joined Chevron in 1973 and has been with the company in a variety of technical assignments for his entire career. He is chairman of the research committee of the Gas Processors Association.

(39c) Optimal NGL Recovery in Natural Gas Liquefaction Plants

Frank L. Del Nogal, KBR, Houston, TX

Heinz Kotzot, KBR, Houston, TX

Refrigerant make-up, product specification constraints, avoiding hydrocarbon freezing and improving revenue are reasons why most LNG production facilities are designed with the capability of extracting some amount of ethane and/or heavier hydrocarbons from the feed gas. The capital cost and potential revenue from NGL extraction can have a significant impact on the overall economics of an LNG project. Existing and planned liquefaction facilities feature various methods to extract NGL for LPG recovery and heating value adjustment as well as varying degrees of NGL extraction. This paper explores the performance implications and cost-effectiveness of using typical turbo-expander and scrub column configurations at different degrees of LPG recovery when developing a grassroots LNG facility.
See more of this Session: Professor Cedomir M. Sliepcevich Memorial Session One - LNG Technology
See more of this Group/Topical: Topical 6: 10th Topical Conference on Gas Utilization

ConocoPhillips
The Optimized Cascade® Process –
Not Just for Frame 5 Any More
AIChE Spring Meeting
San Antonio, Texas
March 22, 2010

Trends in LNG Liquefaction

- Lower fuel consumption
  - Price of fuel
  - Greenhouse gas emissions
- Larger train capacities
- Smaller train capacities
- Floating
- Higher capital costs
- Challenged resources and locations

Importance of Refrigeration Driver/Compressors in LNG

- Capacity is proportional to power
- Availability
- Reliability
- Largest fuel consumer
- Significant capital cost
- Significant operating costs

Refrigeration compressor selection is among the most critical process decision affecting project economics

Desirable Features of Turbines

- Proven
- Sized to match the desired plant capacity
- Efficient
- Flexible (variable speed/multi shaft)
- Quick startup
- Identical equipment across services
- High availability (little downtime for maintenance)
- High reliability (little downtime for unplanned events)
- Low maintenance cost
- Low capital cost
- Power intensive
- Operating speed to match compressor
- Capable of future upgrading

Options for LNG Drivers

- Steam Turbines
- Gas Turbines
  - Industrial
  - Aeroderivative
- Electric Motors

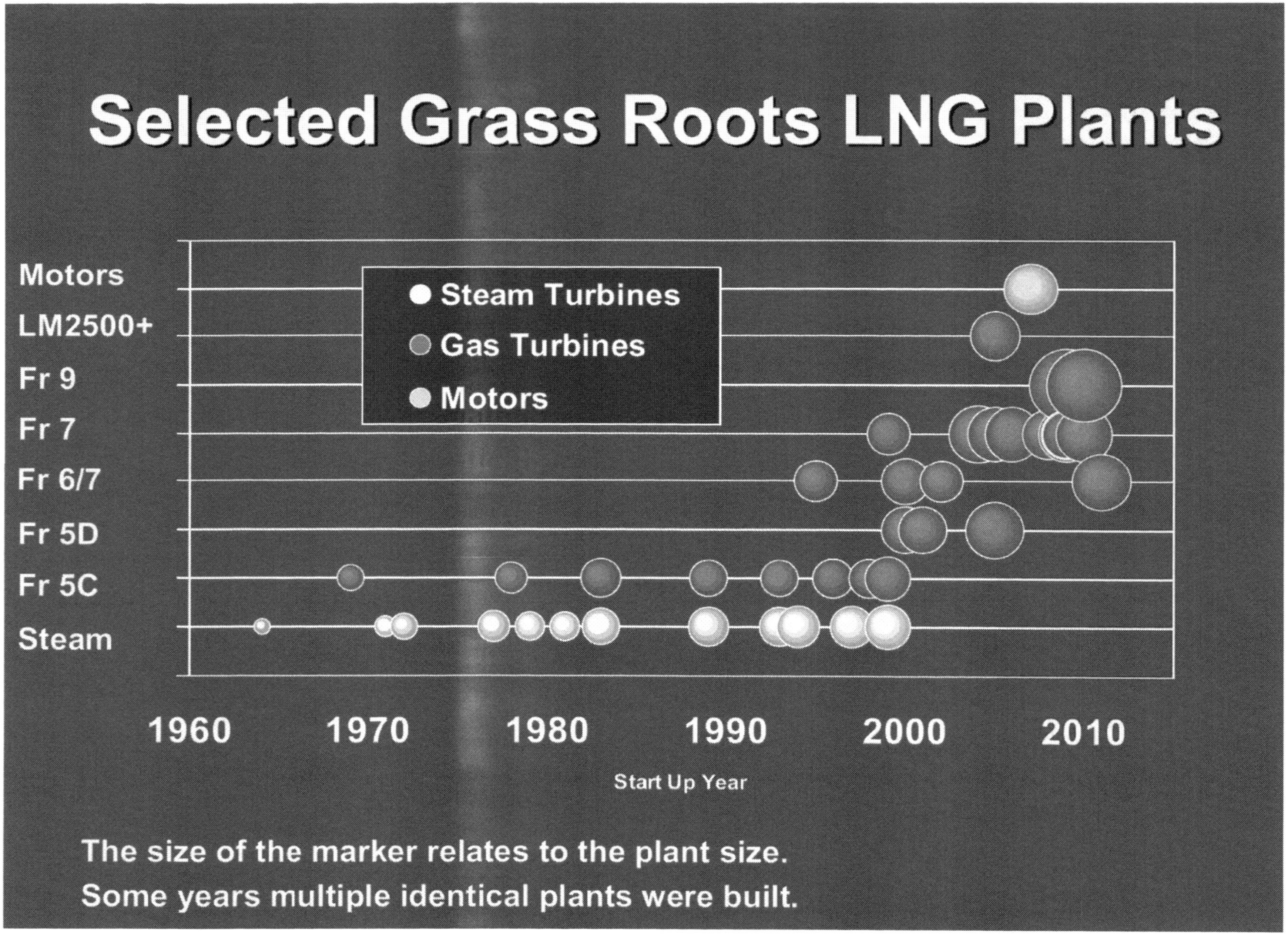
Selected Grass Roots LNG Plants
Steam Turbines
Gas Turbines
Motors
Motors
LM2500+
Fr 9
Fr 7
Fr 6/7
Fr 5D
Fr 5C
Steam
1960
1970
1980
1990
2000
2010
Start Up Year
The size of the marker relates to the plant size.
Some years multiple identical plants were built.

Selected LNG Turbines

Turbine	Shaft	Power (kW)	Efficiency	Fuel Consumption (Indexed)	Scheduled Downtime
Frame 5D	Dual	32,580	29.4%	100	2.6%
LM2500+	Dual	31,364	41.1%	72	1.6%
LM2500+G4	Dual	34,000	41.1%	72	1.6%
LM6000	Dual	44,740	42.6%	69	1.6%
Frame 6B	Single	43,530	33.3%	88	4.4%
Frame 7E	Single	87,300	33.0%	89	4.4%
Frame 9E	Single	130,100	34.6%	85	4.6%

Values depend on emissions standards, options, and maintenance contracts and philosophy.

Availability does not include unscheduled turbine downtime and is NOT plant downtime.

Dual shafts enhance plant operability and start up

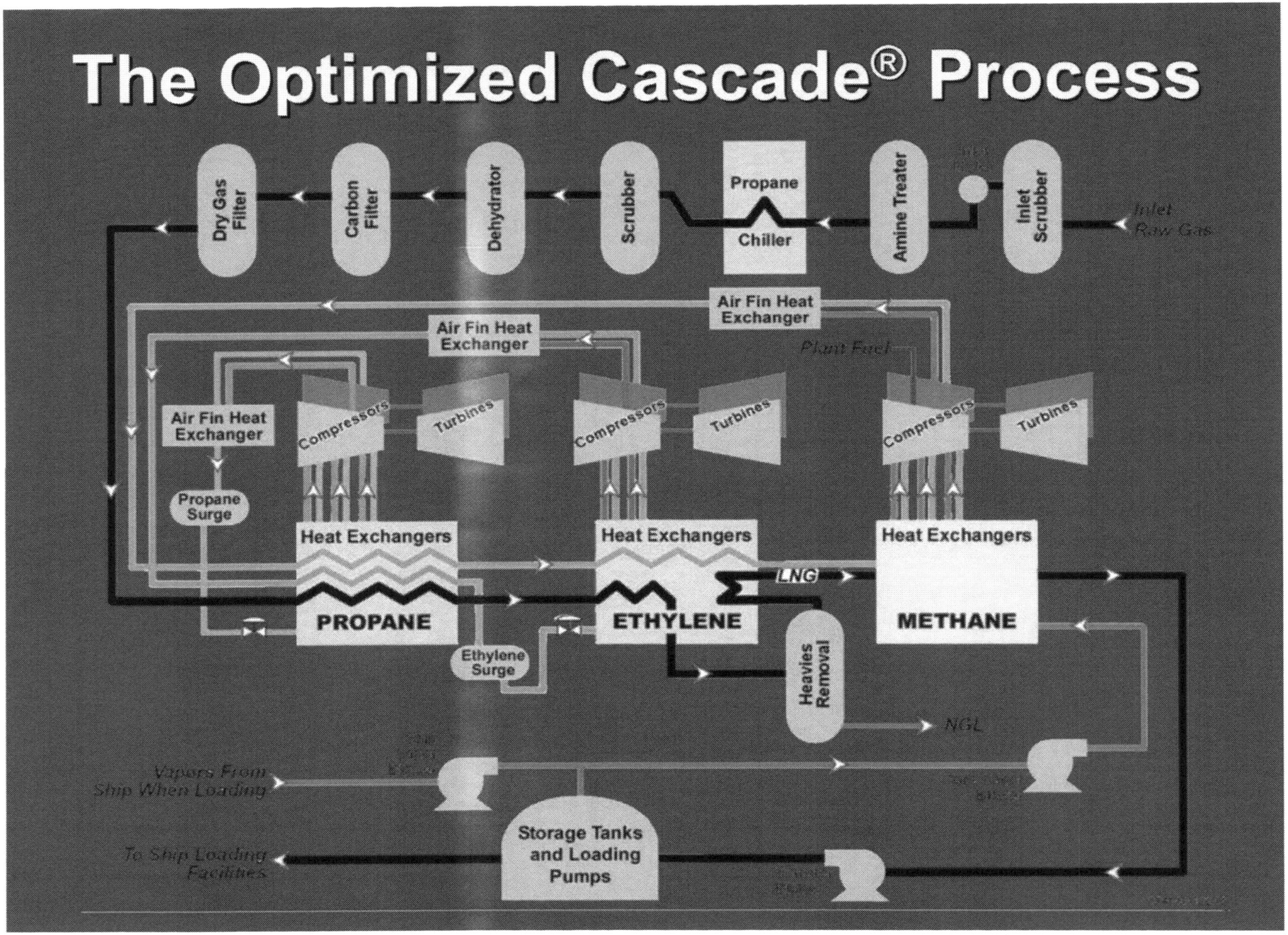
The Optimized Cascade® Process
Dry Gas Filter
Carbon Filter
Dehydrator
Scrubber
Propane
Chiller
Amine Treater
Inlet Scrubber
Inlet Raw Gas
Air Fin Heat Exchanger
Air Fin Heat Exchanger
Plant Fuel
Air Fin Heat Exchanger
Compressors
Turbines
Compressors
Turbines
Compressors
Turbines
Propane Surge
Heat Exchangers
PROPANE
Heat Exchangers
ETHYLENE
LNG
Heat Exchangers
METHANE
Ethylene Surge
Heavies Removal
NGL
Vapors From Ship When Loading
Storage Tanks and Loading Pumps
To Ship Loading Facilities

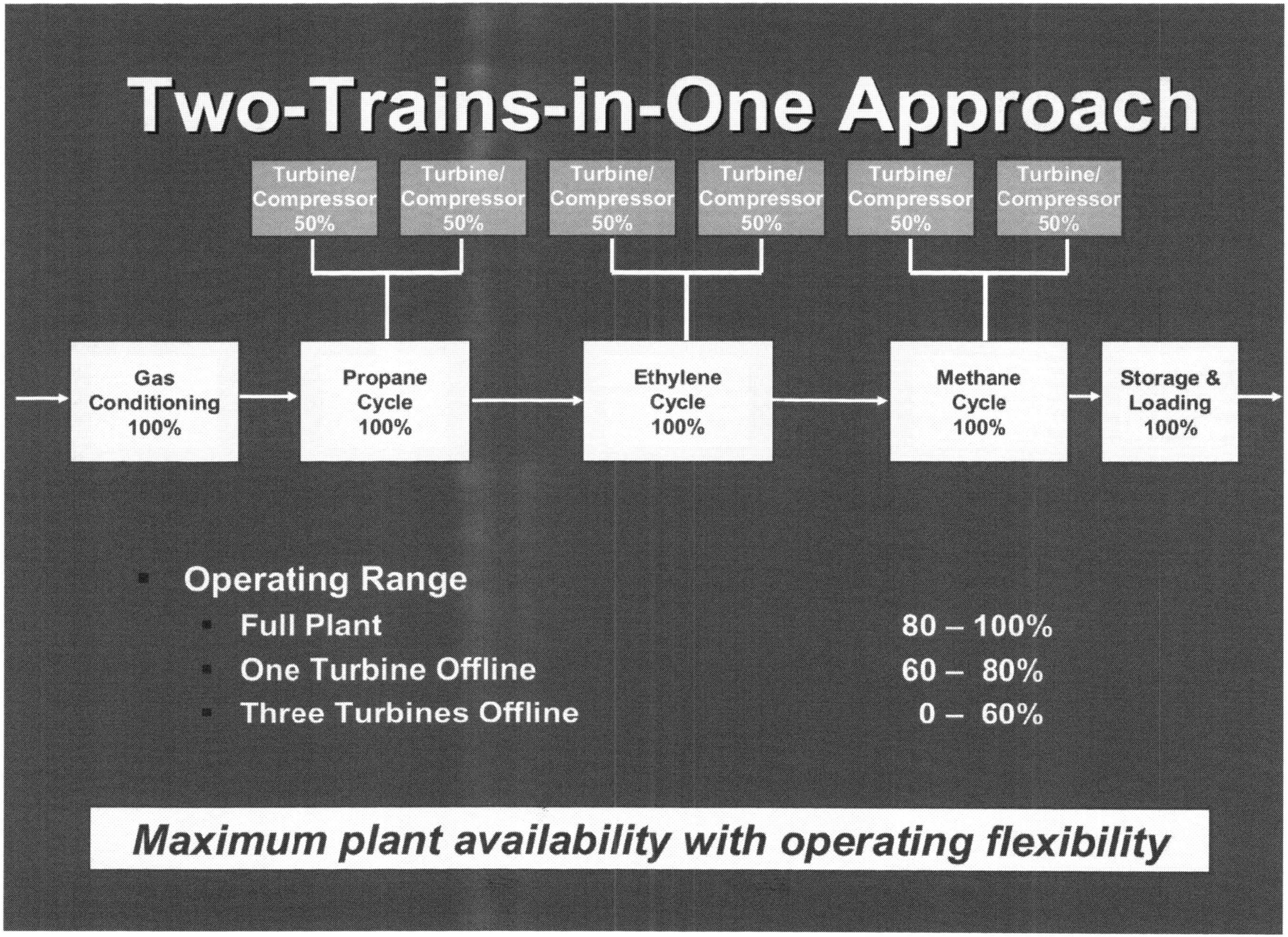
Two-Trains-in-One Approach
Turbine/ Compressor 50%
Turbine/ Compressor 50%
Turbine/ Compressor 50%
Turbine/ Compressor 50%
Turbine/ Compressor 50%
Turbine/ Compressor 50%
Gas Conditioning 100%
Propane Cycle 100%
Ethylene Cycle 100%
Methane Cycle 100%
Storage & Loading 100%
Operating Range
Full Plant 80 – 100%
One Turbine Offline 60 – 80%
Three Turbines Offline 0 – 60%
Maximum plant availability with operating flexibility

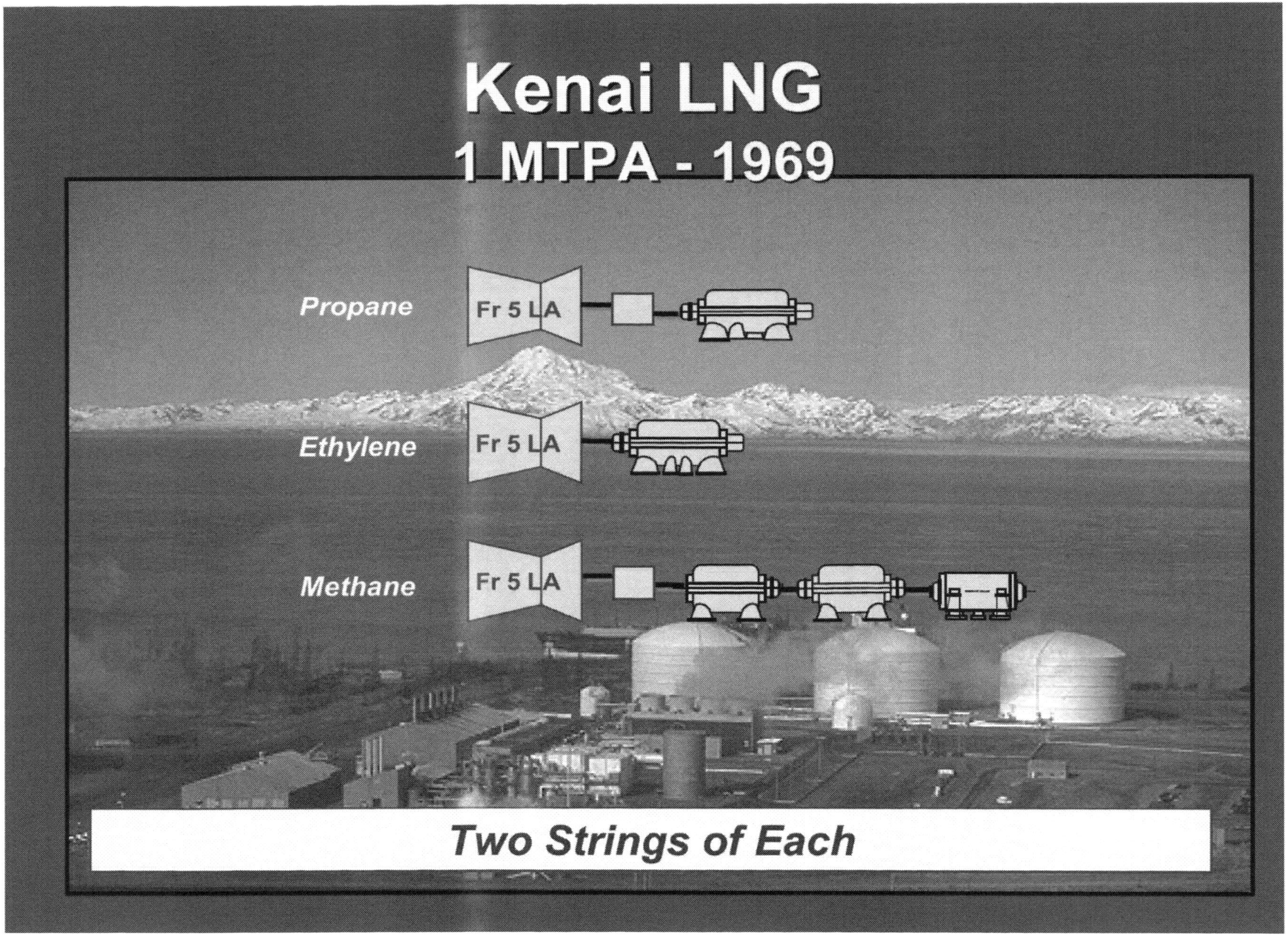
Kenai LNG
1 MTPA - 1969
Propane
Ethylene
Methane
Fr 5 LA
Fr 5 LA
Fr 5 LA
Two Strings of Each

Atlantic LNG, Train 1
3 MTPA - 1999

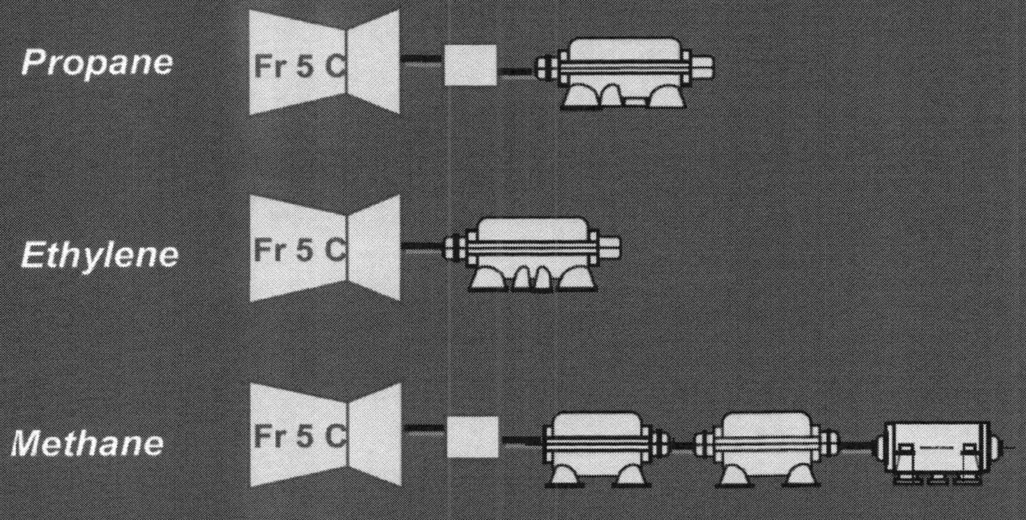

Two Strings of Each

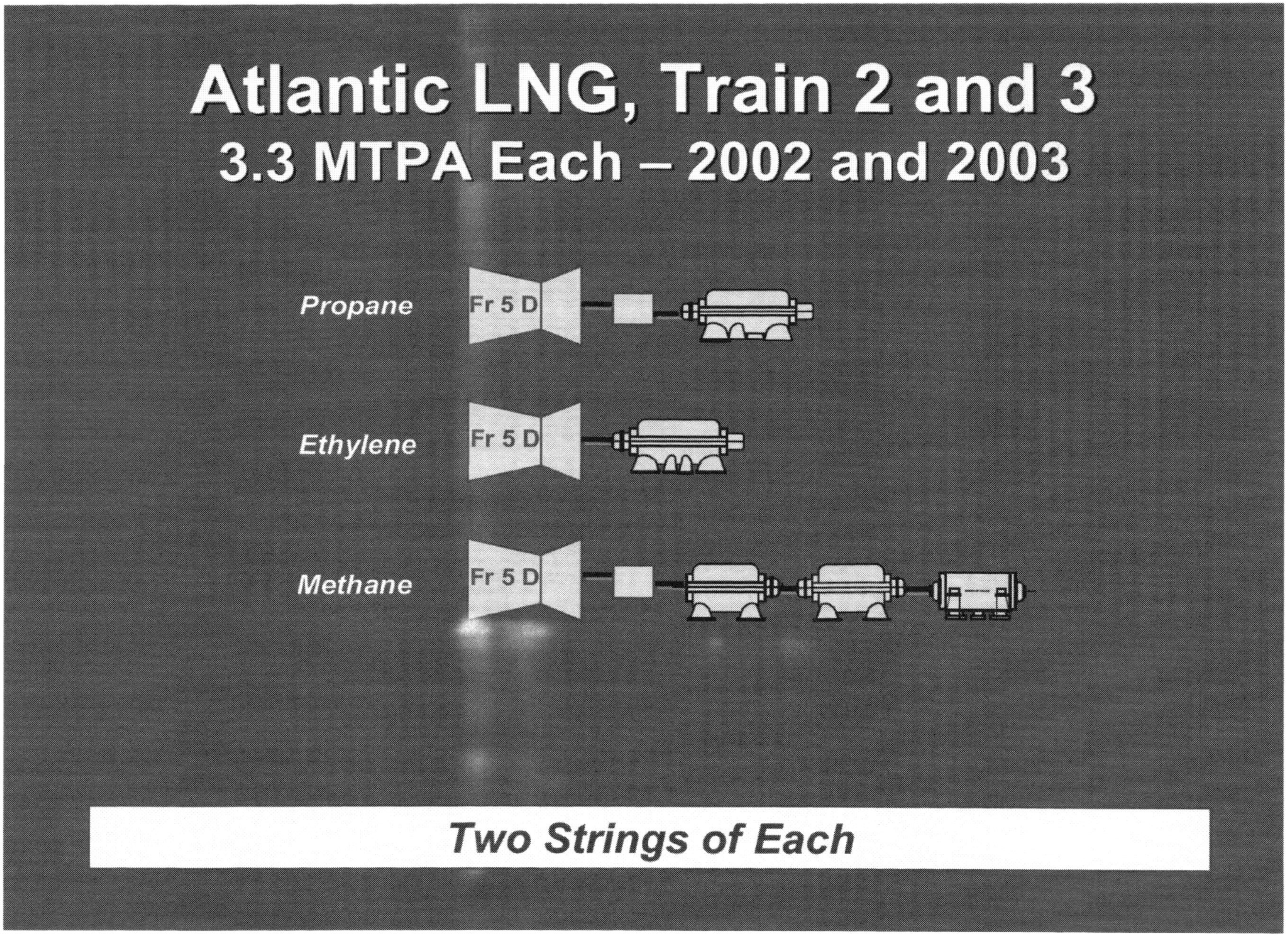
Atlantic LNG, Train 2 and 3
3.3 MTPA Each – 2002 and 2003
Propane
Fr 5 D
Ethylene
Fr 5 D
Methane
Fr 5 D
Two Strings of Each

Egypt LNG and Equatorial Guinea LNG

3.6 and 3.4 MTPA

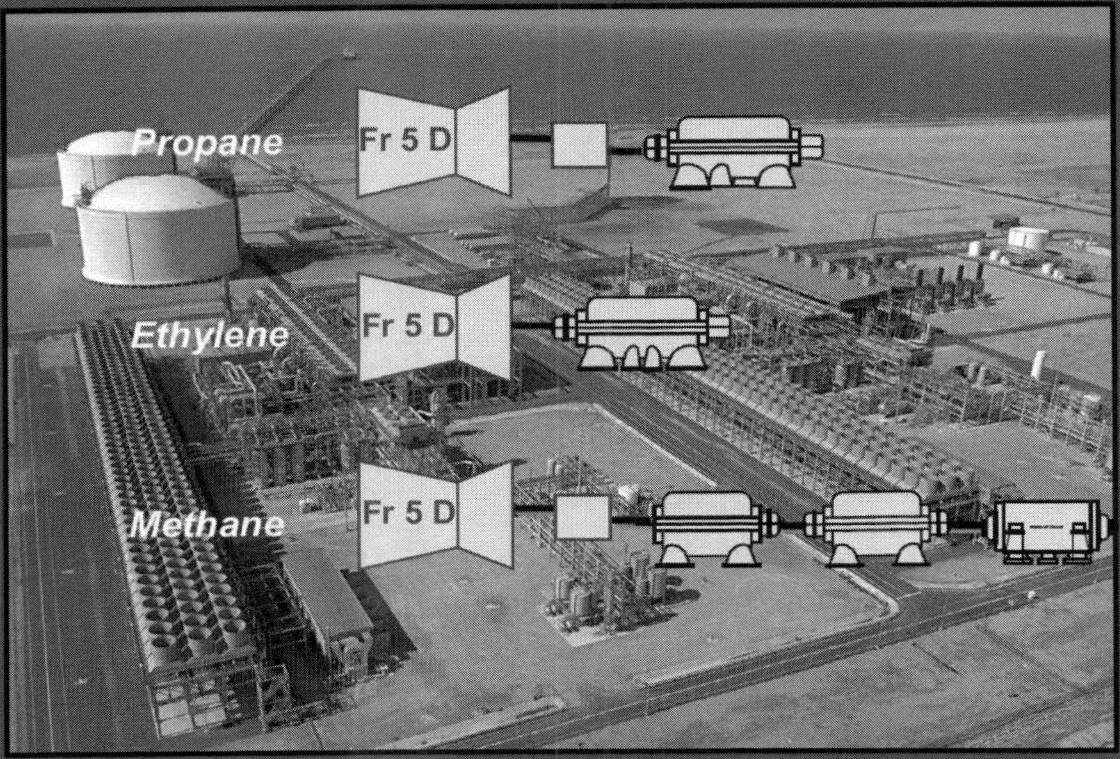

Two Strings of Each

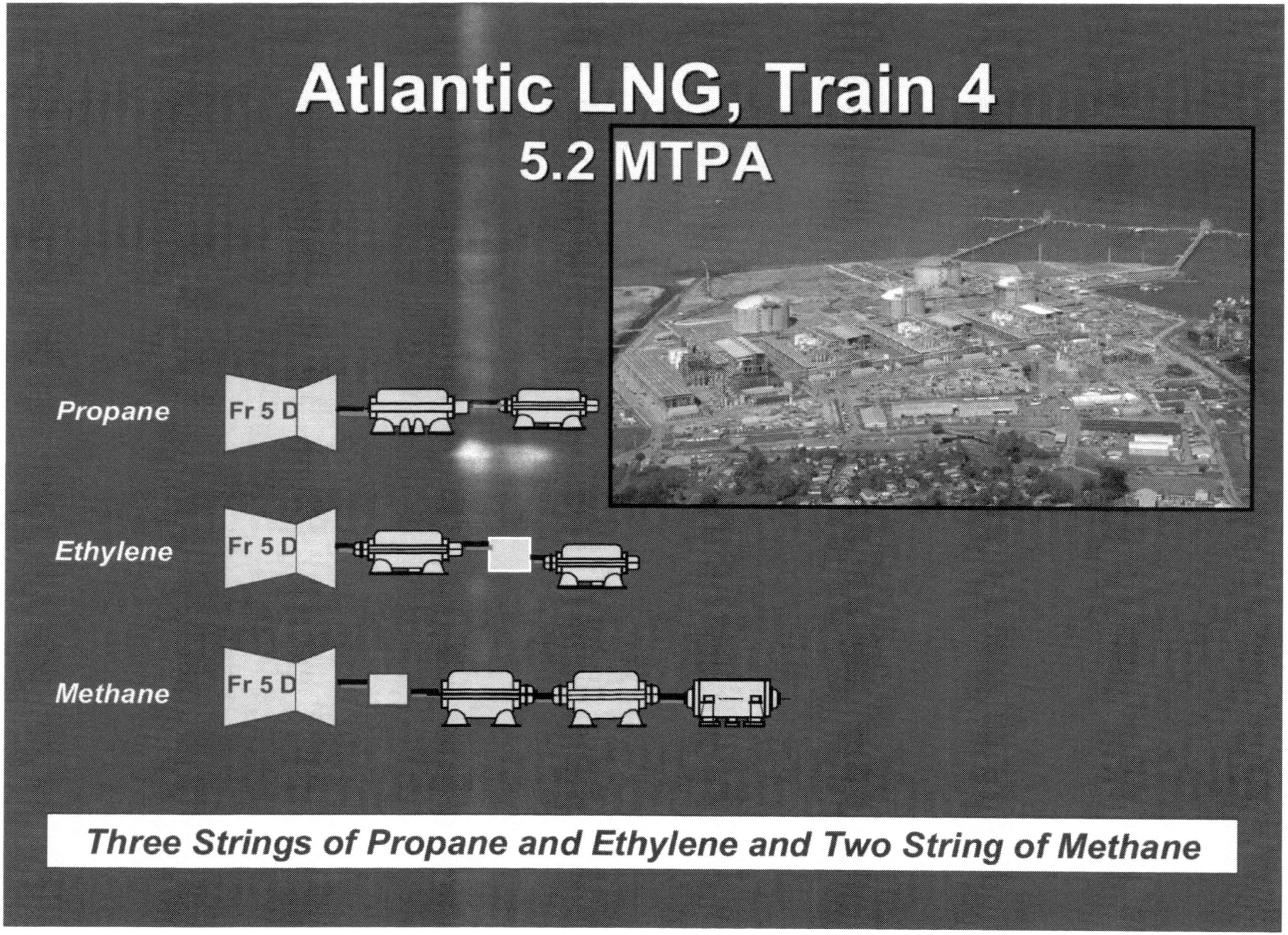
Atlantic LNG, Train 4
5.2 MTPA
Propane
Fr 5 D
Ethylene
Fr 5 D
Methane
Fr 5 D
Three Strings of Propane and Ethylene and Two String of Methane

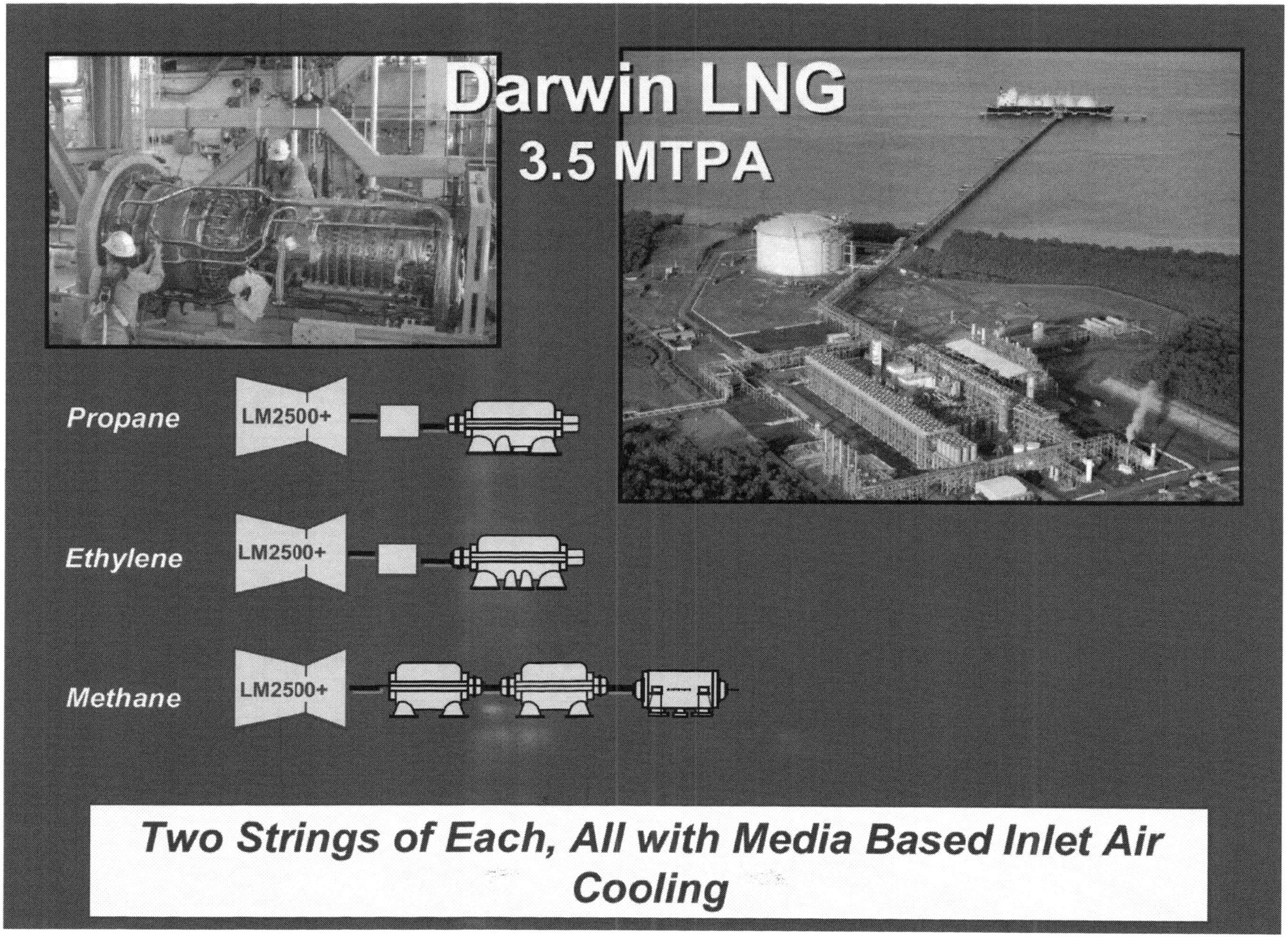
Darwin LNG
3.5 MTPA
Propane
LM2500+
Ethylene
LM2500+
Methane
LM2500+
Two Strings of Each, All with Media Based Inlet Air Cooling

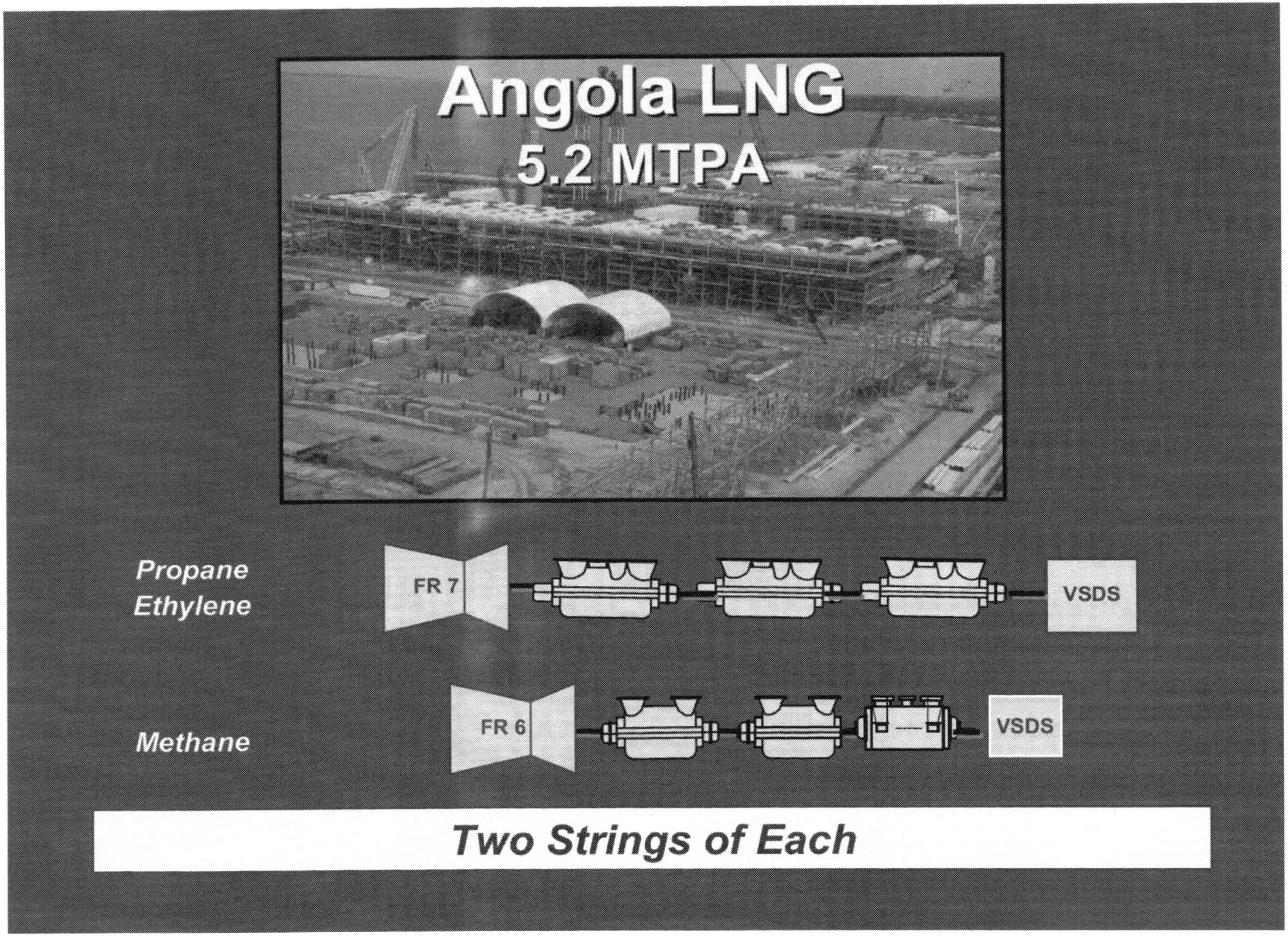
Angola LNG
5.2 MTPA
Propane
Ethylene
FR 7
VSDS
Methane
FR 6
VSDS
Two Strings of Each

Angola LNG

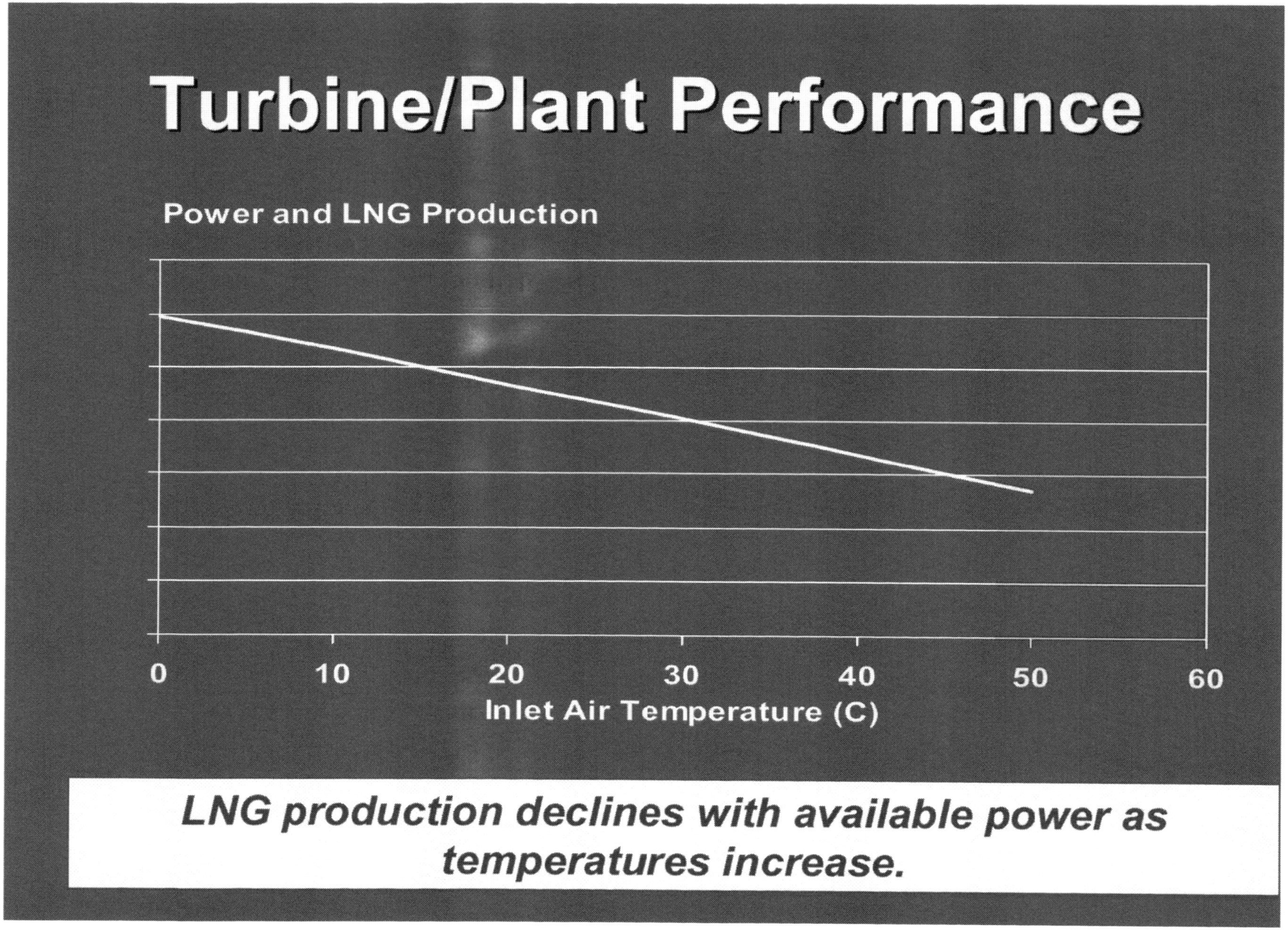
Turbine/Plant Performance
Power and LNG Production
0
10
20
30
40
50
60
Inlet Air Temperature (C)
LNG production declines with available power as temperatures increase.

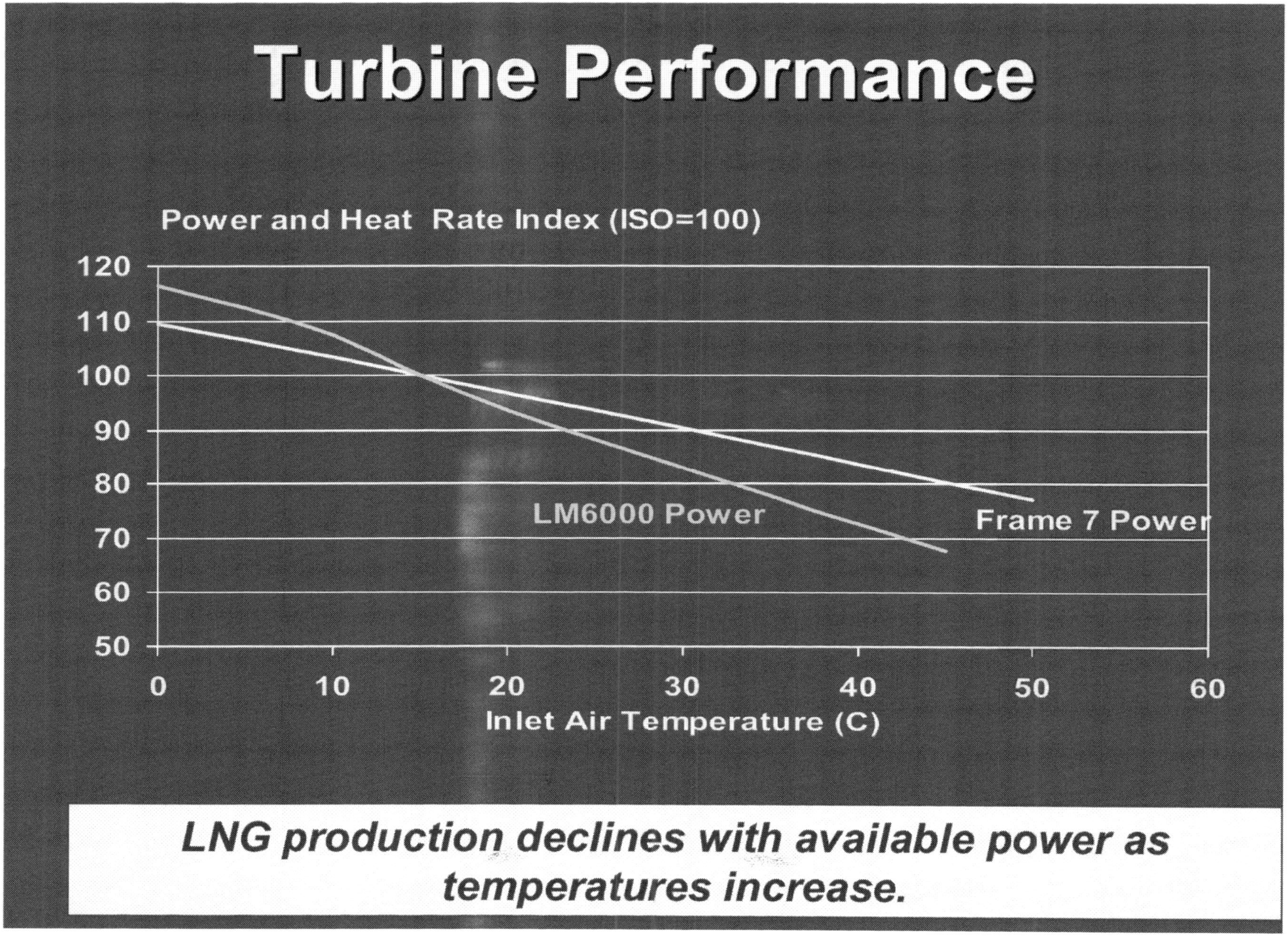
Turbine Performance
Power and Heat Rate Index (ISO=100)
120
110
100
90
80
70
60
50
0
10
20
30
40
50
60
Inlet Air Temperature (C)
LM6000 Power
Frame 7 Power
LNG production declines with available power as temperatures increase.

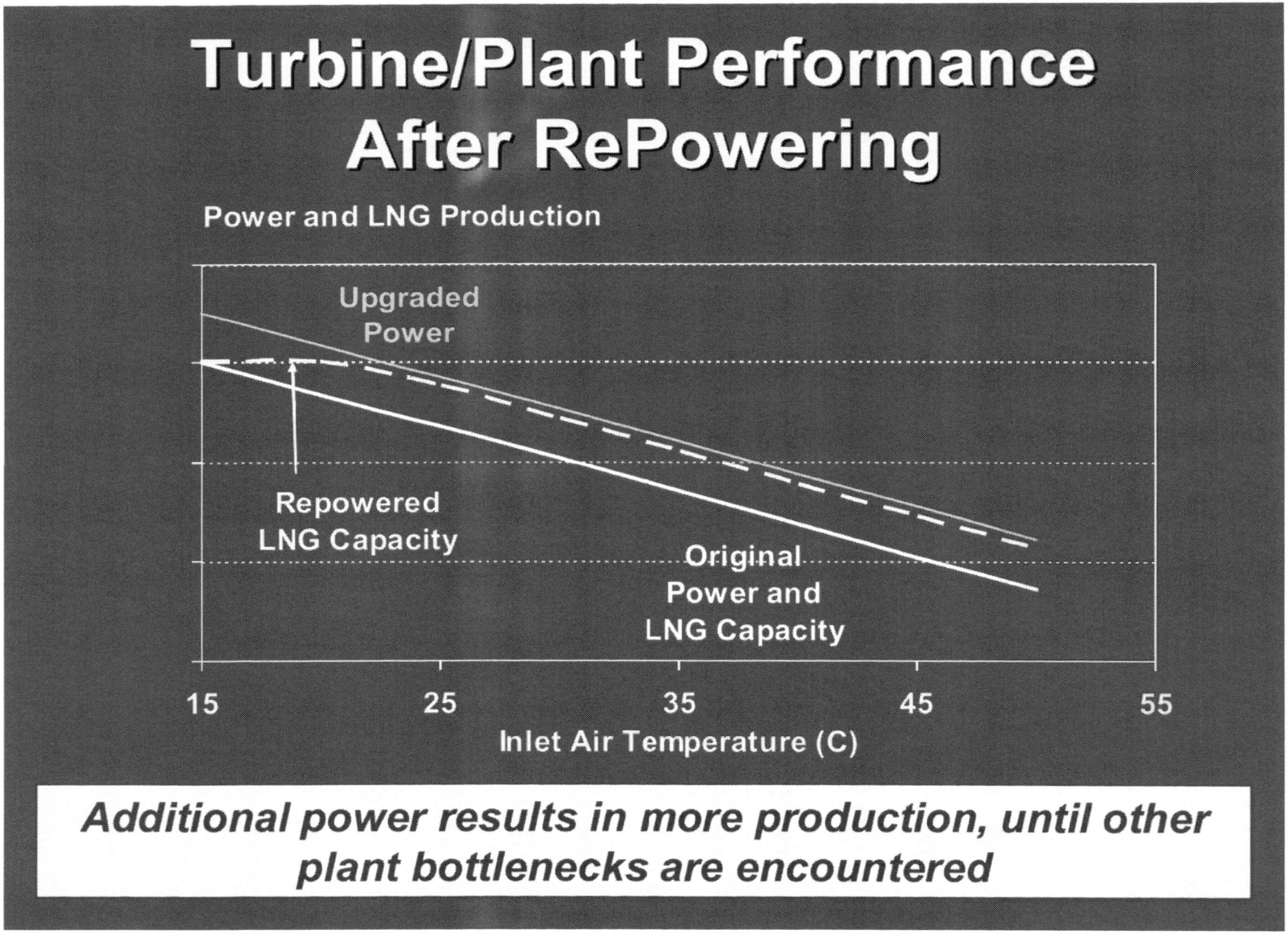
Turbine/Plant Performance
After RePowering
Power and LNG Production
Upgraded
Power
Repowered
LNG Capacity
Original
Power and
LNG Capacity
15
25
35
45
55
Inlet Air Temperature (C)
Additional power results in more production, until other plant bottlenecks are encountered

Options for Increasing Power

- Upgrade turbines
- Media based evaporative cooling
- Inlet air chilling
- Compressor modifications
- Power crystals

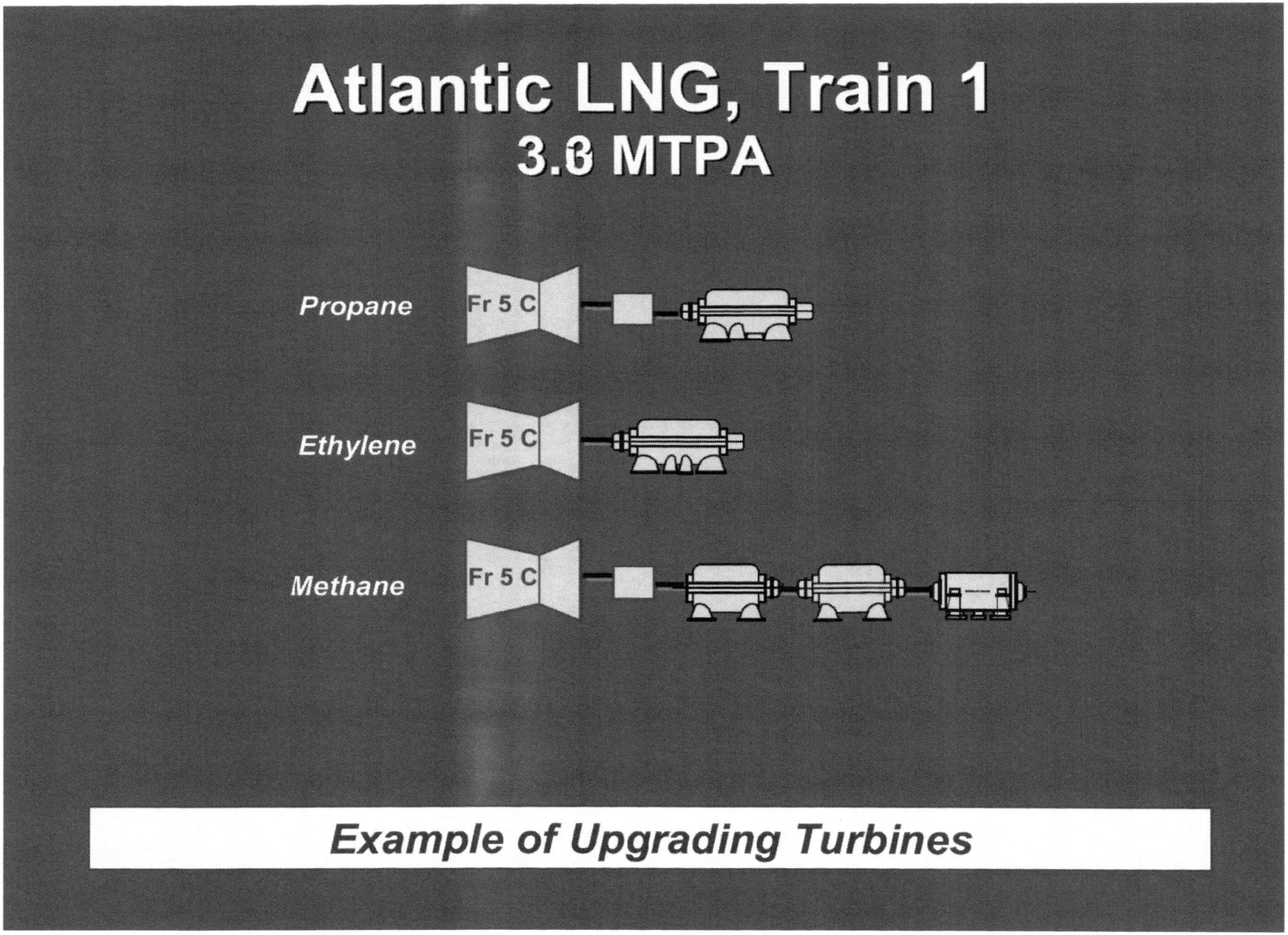
Atlantic LNG, Train 1
3.0 MTPA
Propane
Fr 5 C
Ethylene
Fr 5 C
Methane
Fr 5 C
Example of Upgrading Turbines

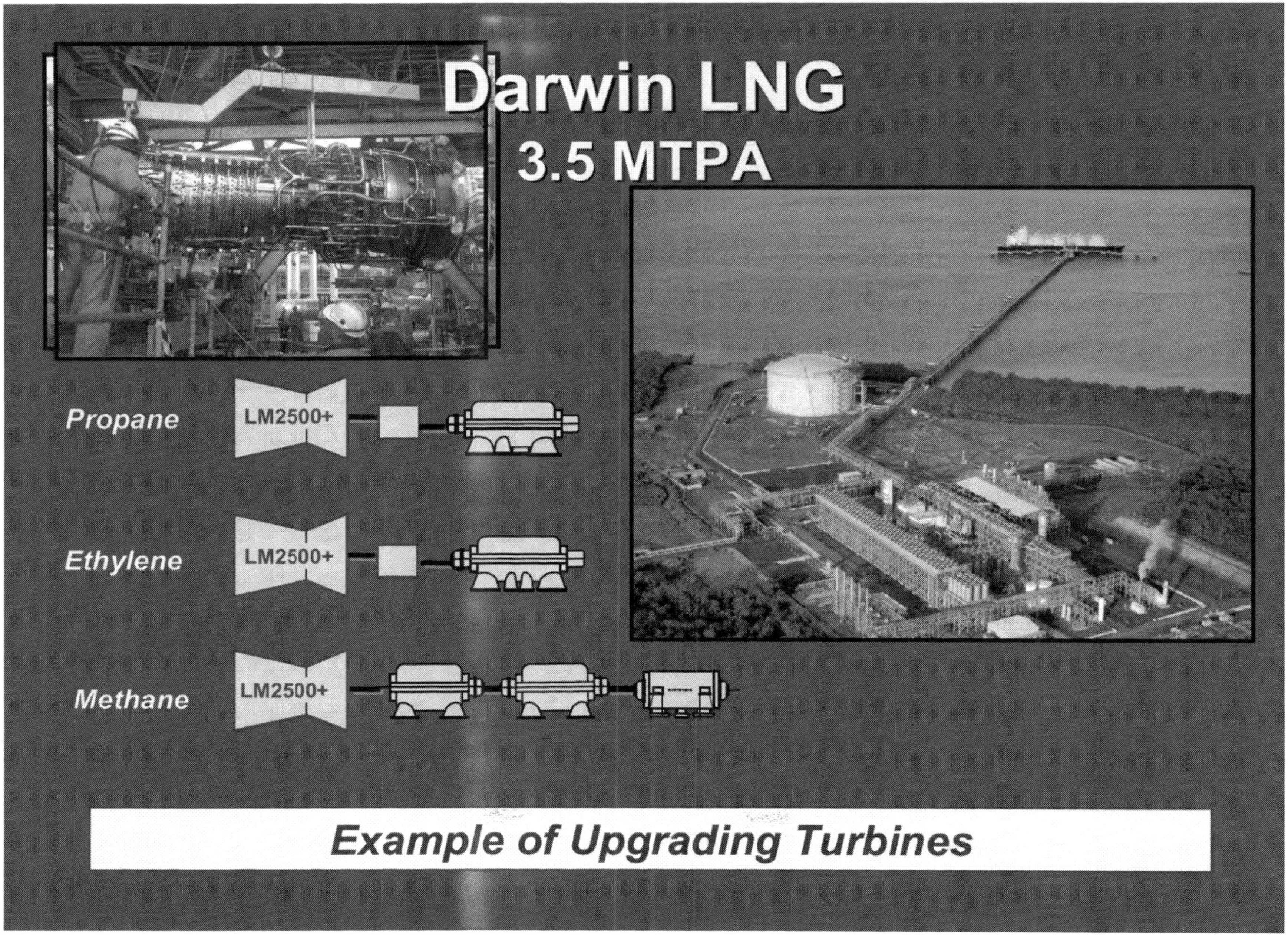
Darwin LNG
3.5 MTPA
Propane
LM2500+
Ethylene
LM2500+
Methane
LM2500+
Example of Upgrading Turbines

Current Plant Designs
Front End Engineering & Design

- LM 6000
  - Larger plant capacity with six turbines
- Inlet Air Chilling
  - More LNG at design
  - Flatter production curve
- Liquid Expanders
  - More LNG
  - Higher Efficiency

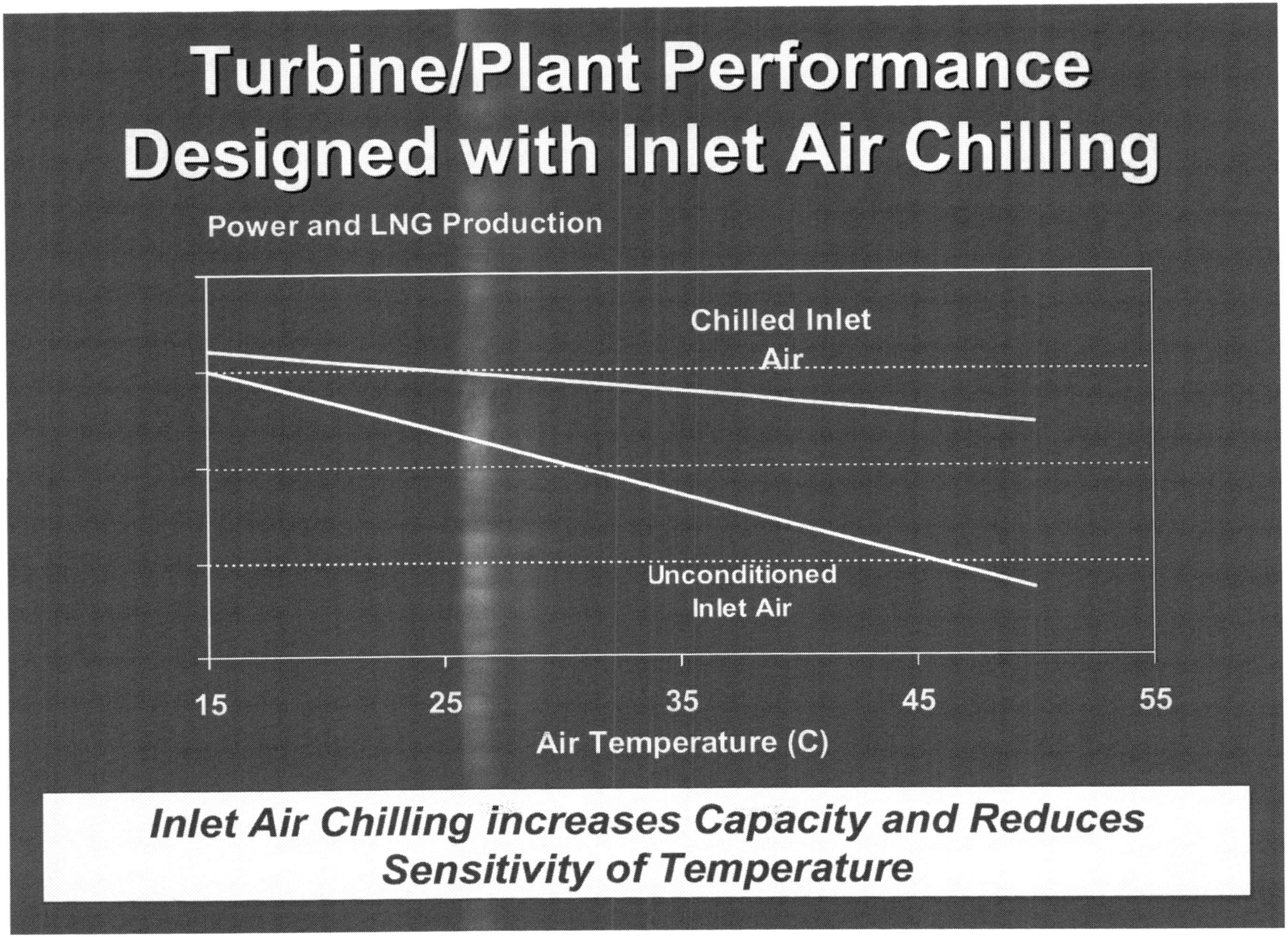
Turbine/Plant Performance
Designed with Inlet Air Chilling
Power and LNG Production
Chilled Inlet
Air
Unconditioned
Inlet Air
15
25
35
45
55
Air Temperature (C)
Inlet Air Chilling increases Capacity and Reduces
Sensitivity of Temperature

Further Enhancements

- LMS 100 / Frame 9
  - One option similar to Angola LNG design
- Gas Turbine Driver with waste heat steam generation and steam turbines
  - Concept presented LNG 6
  - Example Frame 9 drivers for C3/C2= and steam turbines for C1
- Electric Motor Option Powered by Gas Turbine Combined Cycle

Floating LNG

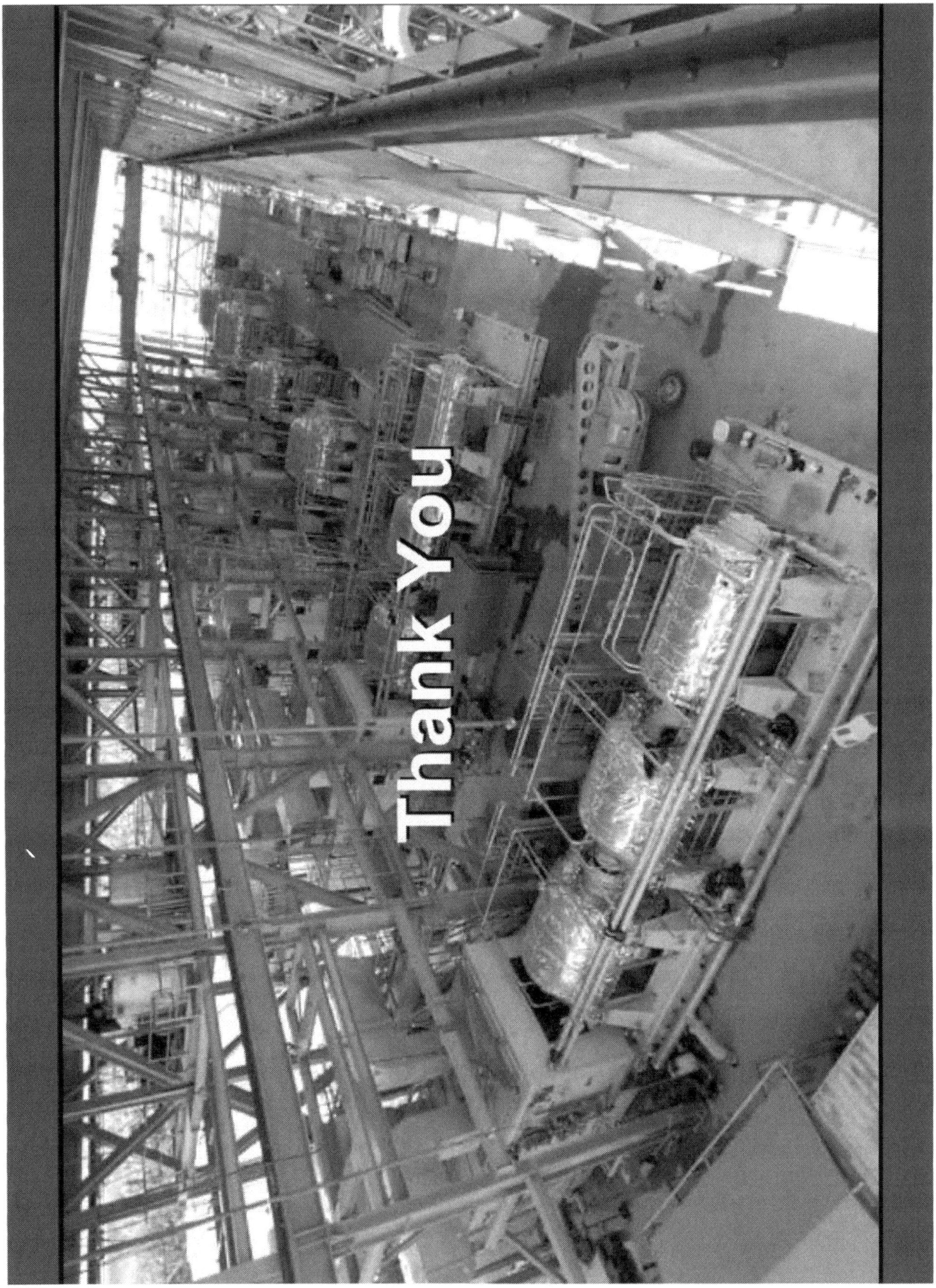
Thank You

(39e) Anticipating Future Lean Feed Gas for the LNG Plant: Badak LNG Plant Case

Agung Eka Purnawan, Technical Department, Process Engineering Section, PT Badak NGL Plant Site , Timur, Indonesia

Johan Anindito Indriawan, PT Badak NGL Plant Site, Timur, Indonesia

Introduction

Badak LNG plant consists of 8 LNG Trains having a total production capacity of 22.5 mtpa LNG and 1.0 mtpa LPG. The feed gas delivered from the existing gas field will be replaced gradually by another source i.e.: deep water gas and even coal bed methane. The challenge in the Badak LNG Plant now is change of the feed gas compositions. The future new feed gas sources are lean which having methane content of 93 – 97 %-vol, compare to the current methane content of 83-85 %-vol.

Conclusions

Assessment of the existing plant facilities is required for anticipating the future feed gas composition change. Lean gas will not cause remarkable changes at the gas treating unit. Some modifications or operational changes of the fractionation, liquefaction and refrigerant unit might be required.

References Cited

1. Chatterjee et al, "Operational Flexibility of LNG Plants using the Propane Precooled Multicomponent Refrigerant MCR Process", 5th International Conference on Liquefied Natural Gas, 1977 ---

2. Spilsbury, Cris. et al, "Optimizing the LNG Liquefaction Process", LNG Journal January/February, Page 43, 2005.---

See more of this Session: Professor Cedomir M. Sliepcevich Memorial Session One - LNG Technology
See more of this Group/Topical: Topical 6: 10th Topical Conference on Gas Utilization

RESOLVING CATALYTIC SOOT AND COKE FORMATIONS CHALLENGES IN A LIQUEFIED NATURAL GAS PLANT: OPERATIONAL EXPERIENCES

Siti Rafidah Moslim
PETRONAS LNG Complex, Bintulu, Sarawak, Malaysia,
sitirafidah_moslim@petronas.com.my

ABSTRACT

In an extremely competitive Oil and Gas industry, heat and energy optimisation is a crucial design consideration in plant designs to ensure optimum business performance. One common approach is via the employment of Waste Heat Recovery Unit (WHRU) at the flue gas path of gas turbines. Incompatible WHRU materials and process conditions could potentially lead to conducive environment that promotes unexpected catalytic soot and coke formations that result in severe equipment fouling and integrity concerns, which in turn could adversely impact sustainable plant production. With understanding of factors that promotes conducive environment for catalytic soot and coke formations, these can be avoided and sufficient designs considerations can be made from Project phase in ensuring long term plant availability and utilization. PETRONAS Liquefied Natural Gas (LNG) Complex; nestled in a small town of Bintulu, in the coastline of Sarawak, Malaysia, consists of three LNG Processing Plants operated as one integrated site. The three plants namely Malaysia LNG (MLNG), Malaysia LNG Dua (MLNG DUA) and Malaysia LNG Tiga (MLNG TIGA) with total LNG deliverable capacity of about 23 million tonnes per annum (mtpa). The first plant, MLNG with three modules began operation in 1983. This was followed by the second plant, MLNG DUA, which started its operations in 1994 also with three modules. The third plant, MLNG TIGA with two modules was later integrated with the existing site and came into operations in 2003. This paper enlightens the accomplishment of the implementation of a redesign of the Regeneration Gas Heater (RGH) in the WHRU in MLNG TIGA to fully resolve catalytic soot and coke formations. Most importantly, this paper shares the lessons learnt from the Root Cause Analysis (RCA) study and the mitigations actions carried out in order to manage the tremendous operational challenges faced by MLNG TIGA team to maximise plant availability, all while working for the long term implementation of the RGH redesign.

Keywords: Catalytic Soot, Coke, Liquefied Natural Gas (LNG), Fouling, Mercury Adsorber, Waste Heat Recovery

1. INTRODUCTION

Malaysia LNG Group of Companies (a subsidiary of PETRONAS) has three LNG Processing Plants, collectively known as the PETRONAS LNG Complex, located in a small town of Bintulu, Sarawak, Malaysia. The three LNG Processing plants, Malaysia LNG (MLNG), Malaysia LNG Dua (MLNG DUA) and Malaysia LNG Tiga (MLNG TIGA) are operated as an integrated site with total LNG deliverable capacity of about 23 million tonnes per annum (mtpa). The first plant, MLNG with three modules began operation in 1983. This was followed by the second plant, MLNG DUA, which started its operations in 1994 also with three modules. The third plant, MLNG TIGA with two modules was later integrated with the existing site and came into operations in 2003.

Overview of LNG Process

The three LNG Processing plants employ a typical LNG Process, which consists of four main steps, namely gas receiving, gas treating, liquefaction; and storage and terminal.

In the first step, raw natural gas is transported via pipeline from offshore gas fields from Central Luconia basin. Once the raw gas lands at the onshore facility, the condensate is separated from the gas phase. The raw natural gas is then metered and sent to the gas treating section. The raw feed gas consists of unwanted components namely carbon dioxide (CO_2), hydrogen sulphide (H_2S), water (H_2O) and mercury (Hg), which needs to be removed to meet product specification and to prevent freezing in the liquefaction section. Hg also needs to be removed to prevent corrosion to the aluminum alloys in the Main Cryogenic Heat Exchanger in the liquefaction section. In the gas treating section, acid gases (CO_2 and H_2S) are removed using an amine-based solvent and later incinerated before being discharged safely to the environment. After the removal of the acid gasses, the gas then enters the dehydration and mercury removal units where the H_2O content is reduced to less than 1 ppmv and the Hg content is reduced to less than 10 ng/Sm^3. The clean and dry natural gas is later liquefied to -161 degC using Air Products and Chemicals Inc. (APCI)'s Propane Pre-Cooled Mixed Component Refrigerant process. The LNG is then piped to the storage area where it is kept at atmospheric pressure. With the arrival of an LNG tanker, LNG is then pumped to the ship via loading arms at the jetty head before being sent to the customers.

The LNG process in the Complex is supported by key utilities systems including heat supply (steam and heat transfer fluid), cooling system (via sea water and air cooling), instrument air, nitrogen and power systems.

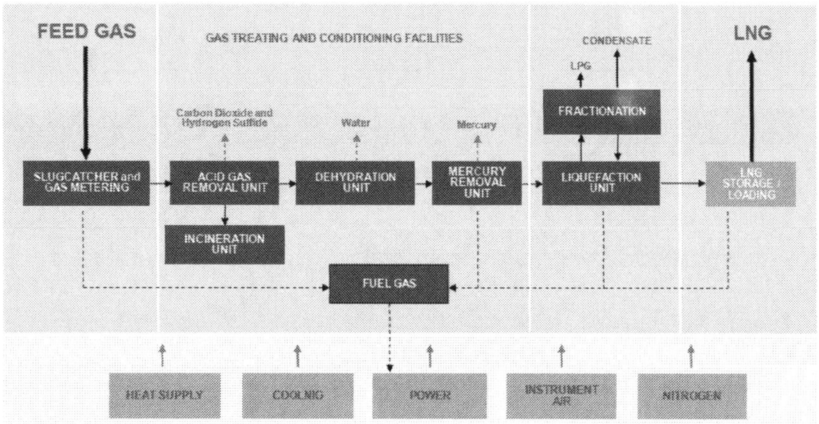

Figure 1. Simplified block diagram of LNG process

Overview of MLNG TIGA Dehydration and Hg Removal Units

MLNG TIGA has two identical modules. Each of the modules has a Dehydration and Hg Removal Units. Natural gas is dehydrated to prevent plugging in the downstream cryogenic equipment. Treated gas from the Acid Gas Removal Unit is pre-cooled to 21 degC by propane chilling to condense the bulk water, to reduce the load on the molecular sieve dryer beds. The bulk water is removed in a separator vessel. Natural gas from the separator vessel then flows downward through two molecular sieve drier beds removing the water to below 1 ppmv, while the third drier bed is being regenerated. Regeneration of the dryer beds is achieved by means of a slip stream of dehydrated gas, which is heated in the Regeneration Gas Heater (RGH) using waste heat from the gas turbine driving the propane compressor. RGH is located at the Waste Heat Recovery Unit (WHRU). The WHRU consists of two parallel convection banks, which is installed with RGH coils and Heat Transfer Fluid (HTF) coils per convection bank. RGH and HTF banks are installed to recover and utilize the heat available from the gas turbine driving the propane compressor. After cooling and water removal, the regeneration gas is compressed back to the main natural gas feed stream, upstream of the propane chiller. The dehydrated gas later goes through a sulphur impregnated activated carbon bed (Hg adsorber bed) where Hg content is reduced to less than 10ng/Sm^3. A mechanical dust filter is installed downstream of the Hg bed to capture any dust or unwanted solid particles from blocking downstream equipments.

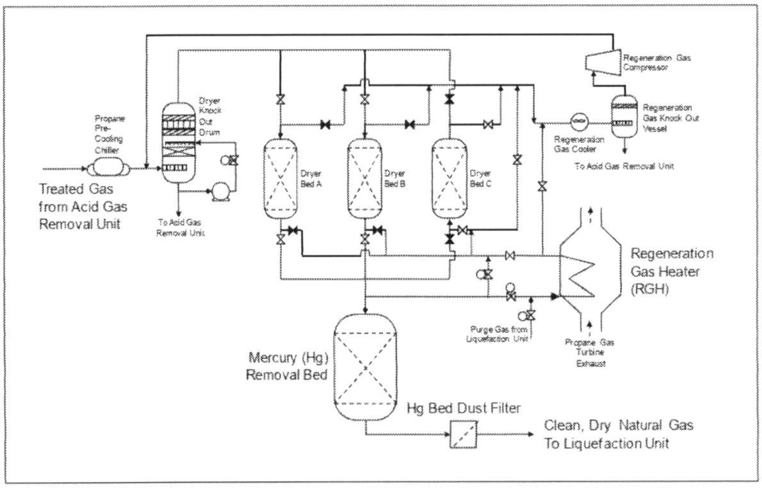

Figure 2. Simplified flow diagram of Dehydration and Hg Removal Units

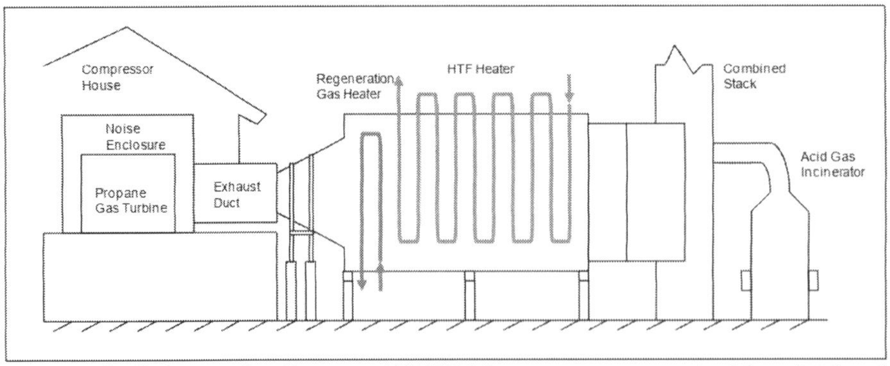

Figure 3. Schematic diagram WHRU for propane gas turbine (side view)

Operational Challenges post Start Up

In the early operations of MLNG TIGA, severe fouling of Hg bed as well as its downstream dust filter was experienced in both modules impacting sustainable LNG plant operations. Fouling was also experienced in the RGH, which resulted with mechanical integrity concerns. Equipment inspection consistently confirmed the presence of unexpected black solids, which predominantly appeared fine, soft and fluffy; and some hard and brittle.

Fouling occurs in an exponential manner, which reduces the life span of the Hg bed from typically between three to five years (design) to less than six months (actual). After the maximum allowable pressure drop (dP) across the Hg bed is reached, the LNG module has to be shutdown for duration of seven days for the removal and replacement of the fouled Hg bed. The downstream dust filter also experienced very frequent filter element change-out, which is approximately three times per year (actual) as compared to typically between every three to five years (design).

RGH in both modules also suffered coils deformation; and tube thinning failure. This was due to uneven blockage of the RGH coils, whereby some tubes were fully blocked, some partially blocked while some were not blocked at all. Tubes blockages resulted with uneven and restrained thermal expansion in the tubes, which in turn caused irregular coils movements resulting in coils rubbing and deformation as well as thinning failure. This was first found in the second module's statutory shutdown. Ultrasonic (UT) examination in the second module's RGH demonstrated that two tubes had significantly lost its original installed tube thickness and considered to have less than the minimum thickness allowed by design code requirements, while two other tubes had lost its fins and still have wall thickness within design allowable limit. A decision was made to shutdown the first module for RGH inspection and consistent findings were found. In both modules, immediate mechanical design improvements were made to ensure that integrity of the RGH coils is not jeopardized.

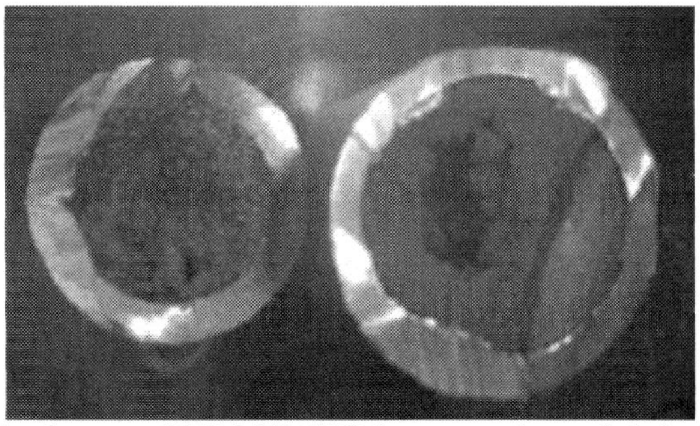

Figure 4. Blockage of MLNG TIGA second module's RGH coils

2. ROOT CAUSE ANALYSIS

The root cause of the severe fouling in Hg bed, its downstream dust filter and the RGH coils is due to continuous soot and coke formations in the RGH coils. Soot is the results of cracking of gaseous hydrocarbon, which is typically black in colour, fine and fluffy in nature. While coke is the results of cracking of liquid hydrocarbon, which is typically black in colour, globular, hard and brittle in nature. The unexpected continuous soot and coke formations are attributed due to the exposure of the regeneration gas to high temperature and the presence of catalyst in the RGH, predominantly nickel (Ni) from the RGH's Incoloy 800 coils material. This was confirmed though extensive process data analysis, actual equipment inspection and micro structural analysis of black solids (soot and coke) samples.

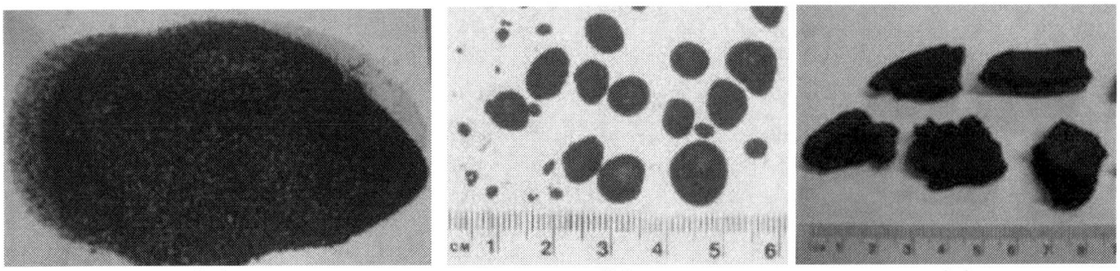

(a) (b) (c)

Figure 5. (a) Soot collected from Hg bed downstream dust filter; (b) Coke collected from Hg bed inlet distributor; (c) Coke collected from RGH

Feed to the RGH

Regeneration gas (in the range of methane (C_1) to decane (C_{10})) enters the RGH coils predominantly in gaseous phase and a small amount in liquid phase. The formation of liquid hydrocarbon in the regeneration gas is believed to take place mostly during the heating cycle, particularly at the RGH inlet flow control valve as this valve has a significant pressure drop and the hydrocarbon gas is at or near its dew point. In addition, during the stand by cycle, the regeneration gas is isolated from the RGH and purge gas (predominantly C_1 and ethane (C_2)) from the liquefaction unit, which is at -50 degC will be introduced into RGH to displace the regeneration gas. This results in the cooling of RGH inlet header and potentially contributes to the formation of liquid hydrocarbons at the inlet header.

Exposure to High Temperature

During actual operation at design regeneration flow of 980t/d, 670t/d of the flow enters the RGH coils while the remaining 32% of the flow by-passes the RGH coils. This resulted with RGH outlet bulk temperature during heating steady of approximately 500 degC, which is 180 degC higher compared to the design figure of 320degC. During heating ramp-up and ramp-down, the actual RGH outlet film temperature is estimated to increase and peaked to an even higher temperature of approximately 523 degC. It is also found that based on actual operations, the propane gas turbine flue gas temperature is in the range of 550 degC to 575 degC, much higher than what was assumed in design of 515 degC.

Presence of Catalyst

Different type of hydrocarbon decomposes and cracks at different temperature levels. The heavier the hydrocarbon, the lower the temperature that cracking starts to occur, regardless whether a catalyst is present or not. Transition metals, especially metals from group VIII (iron (Fe), cobalt (Co) and Ni) in the periodic table, are extensively used in the industry in catalytic decomposition of C_1. With regards to the catalysts for the decomposition of C_1, it is well known that Ni is the most effective catalyst where it shows highly catalytic activity at temperatures of between 450 degC to 550 degC (Chai et al. 2004). It is noted that RGH Incoloy 800 coils consists of about 30 to 35% of Ni and the regeneration gas (in the range of C_1

to C_{10}) is exposed to temperature as high as 523 degC (estimated RGH outlet film temperature).

In order to confirm the presence of catalytic soot formation in the RGH, carbonaceous samples of black solids from both modules were analysed by transmission electron microscopy (TEM) to provide structural and compositional information; and scanning transmission electron microscopy (STEM) with associated elemental analysis by energy dispersive X-ray analysis (EDX) together with conventional scanning electron microscopy (SEM) to determine bulk elemental composition. All samples confirmed the presence of high proportions of filamentous soot, indicating high degree of soot production through catalytic process. The soot particles were characterised by the presence of metal-containing tips and carbon-rich filaments, with Ni, Fe and Cr detected at the tips. The filaments were also found hollow and lacy in appearance, consistent with the characteristics of carbon nanotubes. The levels of Ni, Fe and Cr levels found were at similar relative proportions in all samples analysed, suggesting a common origin from RGH's Incoloy 800. The only source of Ni and Cr in the Dehydration unit that is exposed under high temperature (within the temperature range at which decomposition and cracking of hydrocarbon is possible) is the RGH coils. Catalytic soot formation mechanism involves the formation of soot at the active Incoloy 800 metals (predominantly Ni) through chemisorptions of hydrocarbon species from the gas phase that subsequently dehydrogenate to form carbonaceous deposits. Soon after that, the active metal underneath the carbonaceous deposits diffuses through the deposit, to present a fresh active site at which further deposition occurs. The action by the active metal sustains catalytic soot production.

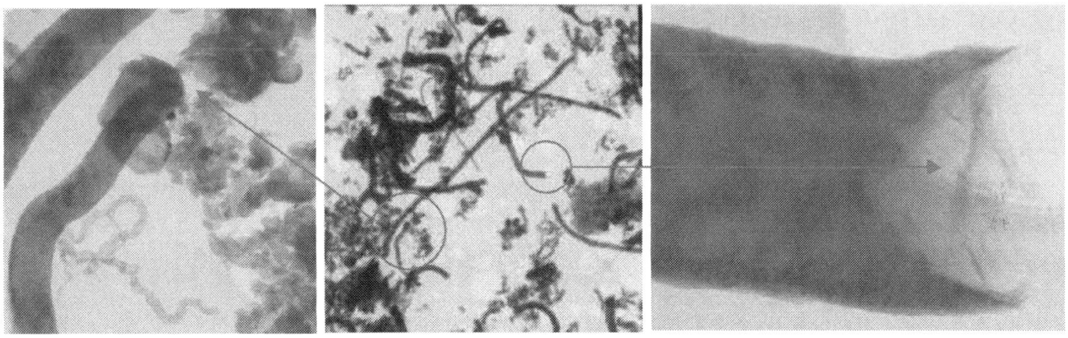

(a) (b) (c)

Figure 6. Soot formed by catalytic mechanism (a) 50000x; (b) 5000x; (c) 100000x magnification views (Titchener and Smith, 2004)

The exposure of gaseous and liquid hydrocarbon to the actual high temperature levels had resulted with the formation of soot (via catalytic mechanism) and coke (via condensation mechanism) in the RGH. Catalytic soot formation is considered more dominant compared to condensation mechanism. This process occurs preliminary during dryer stand-by and secondary during dryer heating ramp-up.

Soot and Coke Movements resulting in Hg Bed and Downstream Dust Filter

During Dryer Bed Regeneration, soot and coke from the RGH are carried together with the regeneration gas, which flows upwards through the dryer bed. The soot and coke will then be deposited at the bottom of the dryer bed. During Dryer Beds Adsorption, when the dryer bed is switched from regeneration to adsorption mode, the flow direction through the bed is reversed. Soot and coke are consequently pushed to the Hg bed together with the main natural gas flow. Coke particles are mostly captured by the Hg bed inlet distributor. The finer soot particles foul the Hg bed.

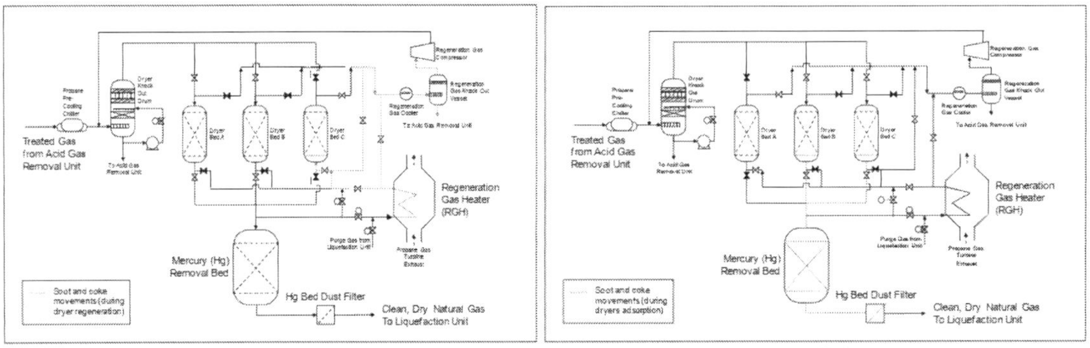

(a) (b)

Figure 7. Soot and coke movement during dryer bed (a) regeneration; and (b) adsorption

3. CORRECTIVE MEASURES TAKEN

Short Term

RGH Process Condition Changes

With the understanding of the root cause, immediate process changes were made to minimise further soot and coke formation rate in both modules. The regeneration gas flow through the RGH was increased by 43%, within the maximum capacity of the regeneration gas compressor. This was achieved by increasing the regeneration temperature set point higher by 10 degC, resulting in the closing the by-pass valve across the RGH. This significantly reduced the bulk temperature of the regeneration gas from 500 degC to 340 degC; and reduction of RGH coil residence time from 6.6 seconds to 5.1 seconds. Reduction of bulk and film temperatures; and residence time results in lower soot and coke formation rate, which is translated to reduction in Hg bed dP build-up from more than 0.1bar/day to 0.055bar/day (on average).

Hg Bed Re-Rating

Fouling in the Hg bed results in rapid dP build up, which dictates the run length of the modules. Low maximum allowable dP results in shorter run length and higher number of unscheduled module shutdowns. The true maximum allowable dP for the whole Hg bed was evaluated via re-rating calculation, focusing on the weakest point of the Hg bed, which is the outlet distributor, originally designed for maximum allowable dP of 4.05 bar. The outcome of the re-rated calculation confirmed that the true maximum allowable dP for the bed is 10.02 bar, which allowed an additional two months run length, before Hg bed replacement. Nevertheless, as the Hg bed is operated with a higher dP, this results in a lower natural gas pressure going to the downstream liquefaction unit. Reduction in the natural gas pressure increases its entropy. This result with a higher liquefaction power required to liquefy the same mass of natural gas. As the maximum available liquefaction power is fixed, higher Hg bed dP results with LNG production impact.

Restriction Orifice at RGH outlet

A restriction orifice, similar dP as per the RGH inlet control valve was installed at the downstream of the RGH coils. This was done in order to minimise the presence of liquid in the regeneration gas entering the RGH coils, which in turn minimises the formation of coke via condensation mechanism.

Coke Catcher in Hg Bed

In August 2004, an innovative trial was done in the first module's Hg bed, where the Hg bed configuration was modified to include the installation of a coke catcher protective layer at the top of the Hg bed. This protective layer is aimed to collect most of the soot in the coke catcher's open ring structures. In selecting the new Hg bed scheme, while working with the available room in the Hg bed, care was taken to ensure that minimum sulphur impregnated mass transfer zone (MTZ) height is available to guarantee that the Hg bed is able to remove the designed feed gas Hg content of 1510 ng/Sm^3 and reduced the value to less than 10 ng/Sm^3. In addition, some margin was made available on top of the MTZ to cater for the possibility of retrograde condensation due to the expected high dP in the bed; and sufficient bed height is also made available to ensure even flow distribution through the bed to prevent bed channelling. The installed coke catcher comprised of 8 x 8 x 1.5 mm Raschig rings, 6 x 6 x 1.5 mm Raschig rings and 3/16" Halder Topsoe TK30 rings, which have higher void fractions than ceramic balls that is placed on top of the sulphur impregnated activated carbon bed.

With the coke catcher protective layer, the actual Hg bed dP build-up in the first module's Hg bed was observed to be 10 times slower compared to the build-up without the coke catcher materials. This innovative installation resulted with higher run length of the Hg bed and avoided seven days unscheduled shutdown particularly during the peak LNG demand period.

Usage of Alternative Hg Bed Adsorbent Type

In September 2004, a trial was carried out in the second module where the Hg bed sulphur impregnated activated carbon type, originally designed with naturally broken adsorbent type, was replaced with extrudates adsorbent type. The former is typically not so uniform in sized, wider particle size distribution, and smaller average particle size, which is more prone to fouling due to the small soot particles entering the bed.

With the extrudates adsorber type, the second module's Hg bed dP build up was observed to be very low and showed very minute build-up. However, the Hg bed dust filter dP seemed to build-up much faster, whereby the maximum allowable dP is reached after two months of operations, resulting in higher change out frequency. It is believed that the usage of extrudates activated carbon had allowed soot to pass though the Hg bed, which ends up in the downstream dust filter.

Long Term

RGH Redesign

With the understanding of the root cause of soot and coke formations, a decision was made to redesign the RGH to eliminate soot and coke formations. The RGH was redesigned; new RGH coils were fabricated and installed at a lower propane gas turbine flue gas path, located at the existing location of the third HTF coils bank. The implementation was successfully completed in the first and second modules in July 2005 and March 2006 respectively. After the installation of the new RGH coils, the old RGH coils were blank-off. The simplified sketches of the original RGH design as well as the new redesigned RGH are shown below.

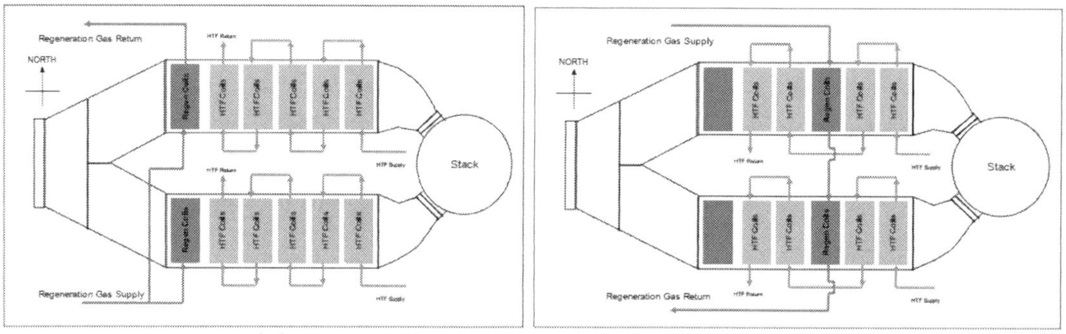

(a) (b)

Figure 8. Simplified sketch showing top view of WHRU with RGH and HTF coils (a) Original RGH design configuration; (b) New RGH design configuration

The new RGH is designed to avoid exposure of the regeneration gas to propane gas turbine flue gas temperature of above 400 degC. The maximum film temperature in the new coils is less than 400 degC at all times. The regeneration gas bulk outlet temperature is designed to be below 350 degC at all times. The coils location ensures that the HTF and RGH design duties are not compromised. The new design eliminates the purge cycle to mitigate the possibility of coke formation. The new design also minimizes the presence of catalyst materials in the coils. The selected optimum material is carbon steel with stainless steel at the first pass only to prevent any potential of cold end corrosion from CO_2 and H_2S condensation. The new design ensures that the mechanical support system is sufficient to prevent tube-thinning or rubbing damage to coils. The design also ensures that only minimum changes are required at the existing WHRU structure and support.

Since the implementation of the new RGH coils redesign, the RGH coils in both modules have been operating within the intended design limits. After close to three to four years of new RGH redesign implementation in both modules, no re-occurrence of soot and coke formations in the new RGH has been observed. No fouling is observed at the Hg bed and its downstream dust filter as well. MLNG TIGA Dehydration and Mercury Removal Units have been restored to its original design intent, which is to support sustainable LNG productions. Unscheduled downtime, LNG production impact as well as high maintenance costs due to soot and coke formations have been fully eliminated. Figure 9 shows the trends of both modules' RGH regeneration gas outlet temperature before and after RGH redesign implementation together with the Hg bed and Hg dust filter normalised dP.

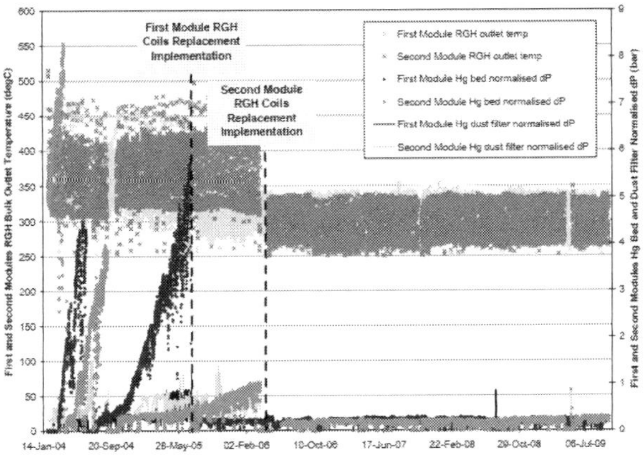

Figure 9. RGH regeneration gas bulk outlet temperature before and after RGH re-design implementation together with the Hg bed and Hg dust filter normalised dP

4. CONCLUDING REMARKS

The challenges faced in MLNG TIGA during its initial operations had served as valuable lessons learnt, particularly in the areas of heat and energy optimisation in WHRU design. The experiences in TIGA RGH have proven the importance of combination of gaseous and liquid hydrocarbon, high temperature and coils material selection to ensure that it would not result with a conducive environment for soot and coke formations that eventually would cause severe fouling of the heater coils and downstream equipment.

The immediate measures taken to manage the tremendous operational challenges faced by MLNG TIGA team to maximise plant availability, all while working for the long term implementation of the RGH redesign were proved to be effective.

MLNG TIGA RGH coils had been redesigned by taking into considerations lessons learnt on key root causes of soot and coke formations. The redesign efforts have been successfully implemented in both modules. After close to 3 to 4 years of implementation, the outcomes have been proven effective. No re-occurrence of soot and coke formations have been observed, which directly supports sustainable LNG productions in MLNG TIGA.

REFERENCES

1. Chai S.P., Zein S.H.S. and Mohamed A.R. (2004), 'A Review on Carbon Nanotubes Production via Catalytic Methane Decomposition'.
2. Titchener K.J. and Smith G.C. (2004), 'Analysis Report: Micro Structural Analysis of Activated Carbon and Soot Samples ex. Petronas/MLNG'.

(63a) Gas Hydrates in Porous Media

Yuanyuan He, Technology University of Delft, Delft, Netherlands

E. Susanne J. Rudolph, Technology University of Delft, Delft, Netherlands

Pacelli. L. J. Zitha, Technology University of Delft, Delft, Netherlands

Description: Natural gas hydrates occur in large quantities below the seafloor and in permafrost regions. The amount of hydrocarbon-hydrates deposits worldwide is enormous: this makes them one of the most important future energy resources. In this paper, we present an experimental study on the formation of gas hydrates in porous media. Non-destructive X-ray Computed Tomography techniques are used for visualization of the gas hydrate formation in porous media.

Applications: The data of this study give insight into the gas hydrate formation in porous medium. The data are compared to experimental data on hydrate formation in the bulk phase. With this knowledge differences in the physical-chemical behavior in the bulk and the porous medium can be identified and analyzed. This forms the base for modification of existing models so that they are applicable for porous medium. Additionally, it will give insight into the possibility to sequestrate CO2 in the form of gas hydrate.

Observations: The gas hydrate formation in porous medium is studied on a well defined model porous medium made of uniformly sized glass beads. The core is first filled with water up to a representative water saturation of 70%. Then gas, either methane or carbon dioxide, is injected into the core at a constant flow rate. Due to gravity segregation the gas mainly occupies the upper part of the core. The core is then shut in at the desired pressure. After shutting-in the core the gas hydrates starts to form which is visualized with the help of CT-scan images. CT images show that gas hydrates are mainly formed in the upper part of the core where initially gas was present. In the lower part hydrate could not be detected. The experiments show that formation of gas hydrate in porous media is slower than in the bulk. The retarding of the process can be explained by the fact that in porous medium the transfer of the gas to the water to allow the formation of gas hydrates is dominated by diffusion. Further on, in porous medium the formation of CH4 hydrates is order of magnitude slower than the formation of carbon dioxide; just as it was observed during bulk phase experiments.

Conclusion: The process of gas hydrate formation in a water-saturated core is a diffusion limited crystal-growth process. It was found out that a fast pressure build up in the core by injecting the gas without controlling the flow rate affects the hydrate formation kinetics as well as the rate of formation. Fast and uncontrolled injection appears to shorten the formation time before a constant pressure is established (equilibrium). However, the data are not very reproducible. If gas is injected slowly and with a constant flow rate the data are reproducible. The final (equilibrium) pressure is the same as when injecting uncontrolled but the time to reach the equilibrium pressure is by a factor of 3 shorter. It can be concluded from the experiments in the core and in the bulk phase that CO2 hydrate formation is faster and at lower pressures (at the same temperature) than methane hydrate formation. This supports the idea of sequestrating CO2 into the reservoir after producing methane from the hydrates.

See more of this Session: Gas Separation Technologies
See more of this Group/Topical: Topical 6: 10th Topical Conference on Gas Utilization

New Natural Gas Storage and Transportation Capabilities Utilizing Rapid Methane Hydrate Formation Techniques

Thomas D. Brown, Charles E. Taylor, and Mark Bernardo***
U.S. Department of Energy
National Energy Technology Laboratory
Pittsburgh, PA, USA

Introduction

Natural gas (methane as the major component) is a vital fossil fuel for the United States and around the world. One of the problems with some of this natural gas is that it is in remote areas where there is little or no local use for the gas. Nearly 50 percent worldwide natural gas reserves of ~6,254.4 trillion ft^3 (tcf) is considered as stranded gas, with 36 percent or ~86 tcf of the U.S natural gas reserves totaling ~239 tcf, as stranded gas [1] [2]. The worldwide total does not include the new estimates by U.S. Geological Survey of 1,669 tcf of natural gas north of the Arctic Circle, [3] and the U.S. ~200,000 tcf of natural gas or methane hydrates, most of which are stranded gas reserves. Domestically and globally there is a need for newer and more economic storage, transportation and processing capabilities to deliver the natural gas to markets.

In order to bring this resource to market, one of several expensive methods must be used:

1. Construction and operation of a natural gas pipeline
2. Construction of a storage and compression facility to compress the natural gas (CNG) at 3,000 to 3,600 psi, increasing its energy density to a point where it is more economical to ship, or
3. Construction of a cryogenic liquefaction facility to produce LNG, (requiring cryogenic temperatures at <-161 °C) and construction of a cryogenic receiving port.

Each of these options for the transport requires large capital investment along with elaborate safety systems.

* Author to contact, Thomas D. Brown, 412-386-4691, thomas.brown@netl.doe.gov .
** Undergraduate Researcher through DOE's Oak Ridge Institute for Science and Education (ORISE) Program.

Innovative Approach(s)

The Department of Energy's Office of Research and Development Laboratories at the National Energy Technology Laboratory (NETL) is investigating new and novel approaches for rapid and continuous formation and production of synthetic NGHs. These synthetic hydrates can store up to 164 times their volume in gas while being maintained at 1 atmosphere and between -10 to -20°C for several weeks. Owing to these properties, a new process for the economic storage and transportation of these synthetic hydrates could be envisioned for stranded gas reserves.

The recent experiments and their results from the testing within NETL's 15-Liter Hydrate Cell Facility exhibit promising results. Introduction of water at the desired temperature and pressure through an NETL designed nozzle into a temperature controlled methane environment within the 15-Liter Hydrate Cell (Figure No. 1) allowed for instantaneous formation of methane hydrates. The instantaneous and continuous hydrate formation process was repeated over several days while varying the flow rate of water, its' temperature, and the overall temperature of the methane environment. These results clearly indicated that hydrates formed immediately after the methane and water left the nozzle at temperatures above the freezing point of water throughout the range of operating conditions. Figure No. 2a is a still photo from a video showing the hydrate spray and subsequent hydrate formation away from the NETL designed nozzle. Furthermore, Roman Spectroscopy Analysis on samples collected from the Cell verify formation of Type I methane hydrate (Figure No. 2b).

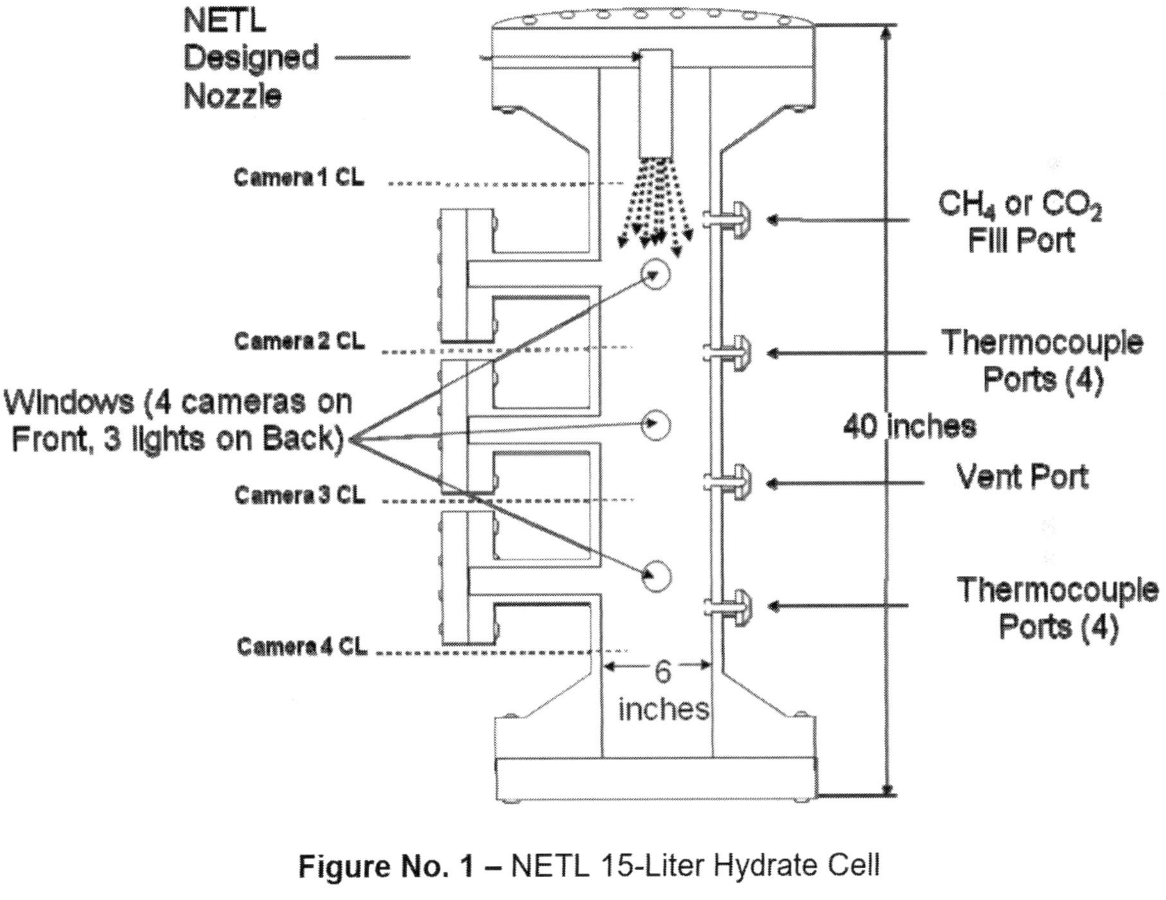

Figure No. 1 – NETL 15-Liter Hydrate Cell

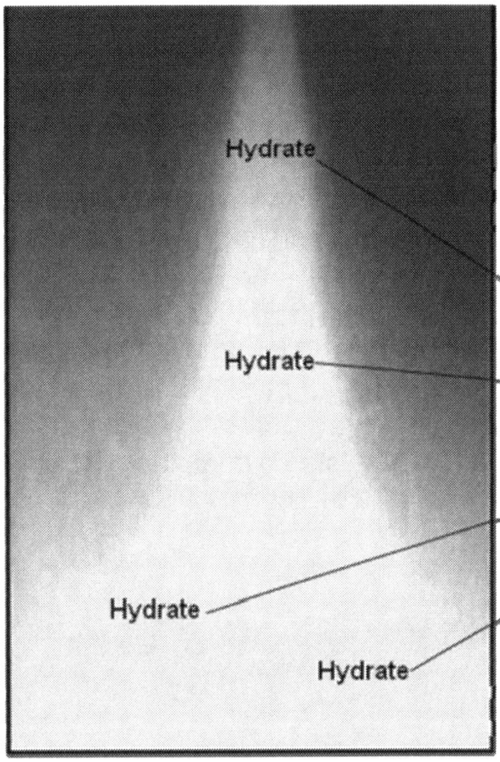

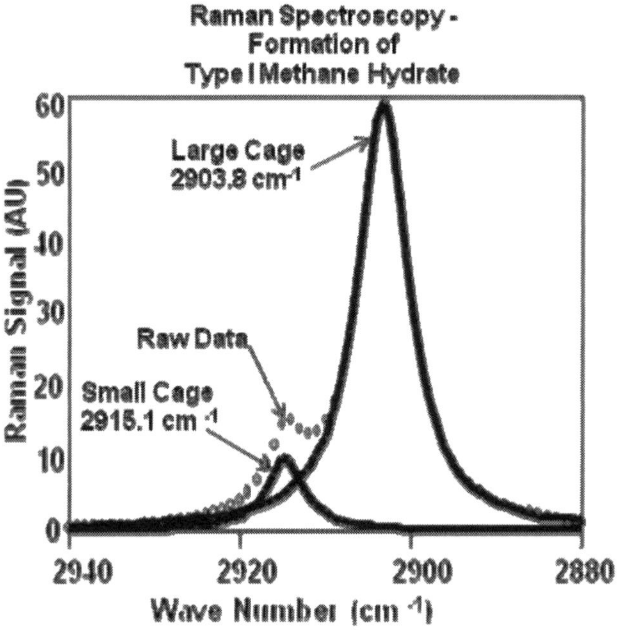

Figure No. 2a – Typical Hydrate spray and subsequent hydrate formation downstream from NETL designed nozzle, Camera 1 CL location within Figure No. 1.

Figure No. 2b - Raman Spectrum of rapidly formed hydrate in the 15L Cell utilizing NETL designed nozzle (11/25/09).

Figures 3a through 3d are still photos from videos starting at the first few minutes of the experiment showing the hydrate spray and subsequent deposition over a twenty-five minute period within the 15-Liter Hydrate Cell at Camera 3 CL location, as referenced in Figure No. 1.

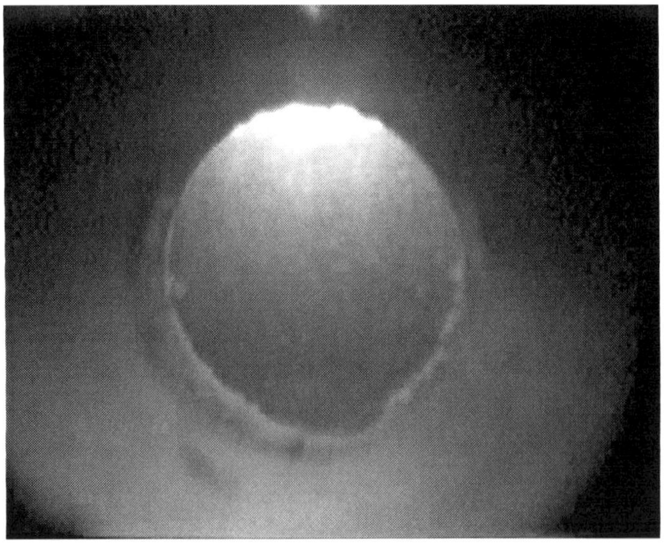

Figure No. 3a – Typical hydrate spray and subsequent hydrate deposition within NETL 15-Liter Hydrate Cell two (2) minutes into experiment – Camera 3 CL in Figure No. 1.

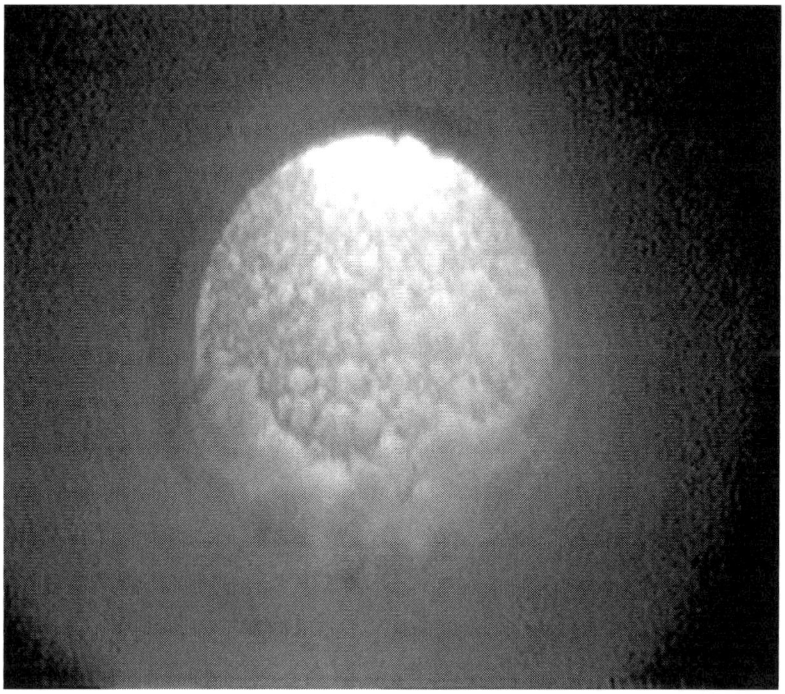

Figure No. 3b – Typical hydrate spray and subsequent hydrate deposition within NETL 15-Liter Hydrate Cell ten (10) minutes into experiment – Camera 3 CL in Figure No. 1.

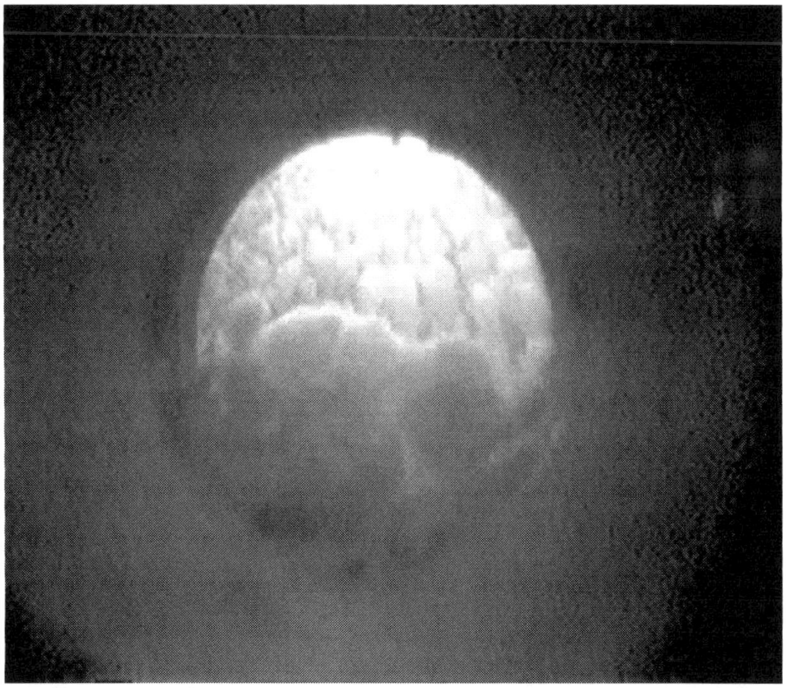

Figure No. 3c – Typical hydrate spray and subsequent hydrate deposition within NETL 15-Liter Hydrate Cell fifteen (15) minutes into experiment – Camera 3 CL Figure No. 1.

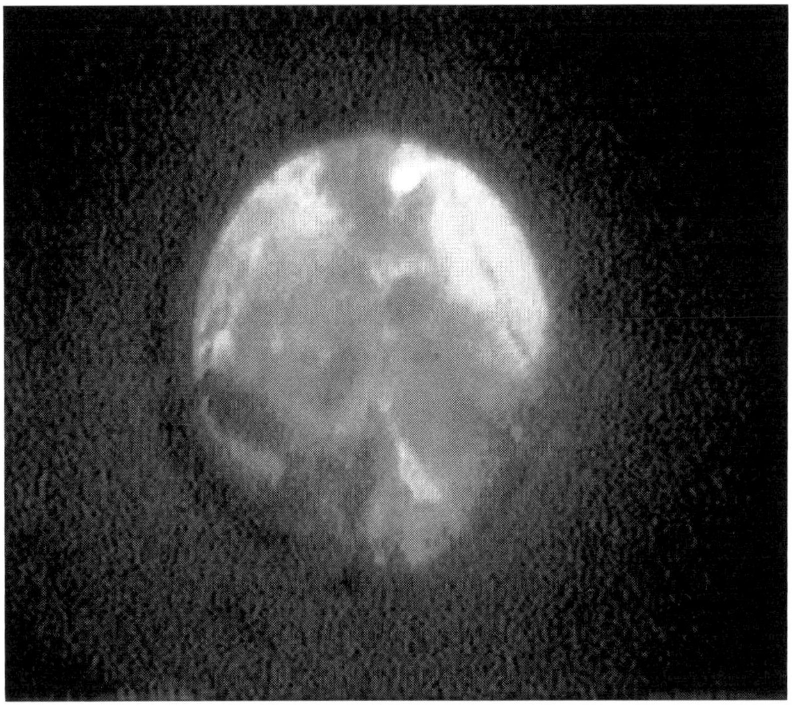

Figure No. 3d – Typical hydrate spray and subsequent hydrate deposition within NETL 15-Liter Hydrate Cell twenty-five (25) minutes into experiment – Camera 3 CL in Figure No. 1.

Owing to the success of the NETL rapid and continuous methane hydrate formation technique [4], investigations have begun on the separation of carbon dioxide (CO_2) from natural gas. Natural gas at the well head can contain up to 8% CO_2 [5] (as shown in Table No. 1) with higher values associated with foreign natural gas reserves. Figure No. 4 also indicates "other" gases contained in natural gas and their respective hydrate stability curves [6]. These curves and their respective concentrations shown in Table No. 1 provide for possible separation as individual gas hydrates to increase the overall methane content and heating value of production natural gas.

Table No. 1 - Composition of US Natural Gas [5]

Methane	CH_4	70-90%
Ethane	C_2H_6	0-20%
Propane	C_3H_8	
Butane	C_4H_{10}	
Carbon Dioxide	CO_2	0-8%
Oxygen	O_2	0-0.2%
Nitrogen	N_2	0-5%
Hydrogen sulphide	H_2S	0-5%
Rare gases	A, He, Ne, Xe	Trace

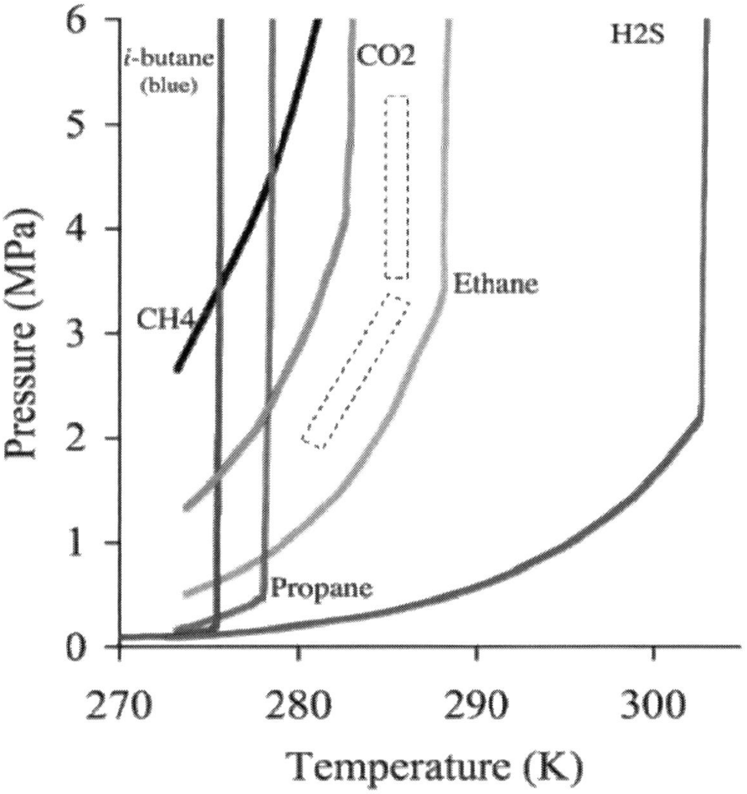

Figure No. 4 - Hydrate Stability Curves of Natural Gas Constituents [6]

In theory, operating within each of the individual hydrate curves (to the left of each curve as shown in Figure No. 4) and outside (to the right of the next gas hydrate stability curve as shown in Figure No. 4) of the temperatures and pressures of other gas hydrate stability curve, would allow for possible separation of each gas as a hydrate. For example, when operating within a defined range of temperature(s) and pressure(s) as shown in the dashed-lined rectangles (left) or inside the ethane hydrate stability curve and outside (right) of the CO_2 hydrate stability curve to form ethane hydrate but not CO_2 hydrate. Utilizing NETL rapid hydrate formation techniques could enhance the separation of these gases via hydrate since the operations are within a tight range of temperatures and pressures conducive to individual gas hydrate formation. Owing to the lower concentrations and subsequent lower partial pressures of each gas contained in natural gas, experiments will be designed to form methane hydrate first from a mixture of methane and CO_2. After which, a concentrated stream, in this case, CO_2 with a lower concentration of methane (lower partial pressure than CO_2) could be drawn from one hydrate cell and fed into another hydrate cell to form only CO_2 hydrate.

The presentation at the 2010 AIChE Spring Meeting will detail the experiments to separate CO_2 from methane in a variety of temperature and pressure controlled hydrate cells.

References

[1] Oil and Gas Journal Vol. 160.48, Dec 22, 2008.

[2] http://www.eia.doe.gov/oiaf/servicerpt/natgas/chapter3.html and http://www.eia.doe.gov/oiaf/servicerpt/natgas/pdf/tbl7.pdf
[3] U.S. Geological Survey, "Circum-Arctic Resource Appraisal: Estimates of Undiscovered Oil and Gas North of the Arctic Circle," May 2008.
[4] Brown, T. D., and Taylor, C. E., "A Rapid Hydrate Formation Process for New Storage and Transportation Capabilities of Gas Phase Energy Sources," Energies (ISSN 1996-1073), Special Issue, "Natural Gas Hydrate"; spring 2010.
[5] http://www.naturalgas.org/overview/background.asp.
[6] http://knol.google.com/k/gas-hydrates

(63c) Development of Hollow Fiber Sorbent for Gas Separation

Naoki Bessho, School of Chemical & Biomolecular Engineering, Georgia Institute of Technology, Atlanta, GA

William J. Koros, School of Chemical & Biomolecular Engineering, Georgia Institute of Technology, Atlanta, GA

A new concept of "fiber sorbent" is investigated. The fiber sorbent is created as a pseudo-monolithic material comprising polymer and zeolite by applying hollow fiber spinning technology. Phase separation of the polymer solution provides appropriately porous structure through the material. In addition, the zeolite crystals are homogeneously dispersed in the polymer matrix with high loading. The zeolite mainly contributes to sorption capacity of the fiber sorbent. Mass transfer processes in the fiber sorbent module for gas separation is analyzed theoretically and compared with results for a packed bed. The model indicates advantageous points of the fiber sorbent module over the packed bed technology by changing the outer and bore diameters to maintain the same pressure drop or to reduce the pressure drop. The fiber sorbent was spun successfully and showed high sorption capacity and highly porous structure. This new material has been proven in the application to pressure swing adsorption process as an alternative to traditional zeolite sorbent pellets.

See more of this Session: Gas Separation Technologies

See more of this Group/Topical: Topical 6: 10th Topical Conference on Gas Utilization

(63d) Polyimide Membranes with Internal Free Volume for Gas Separation

Yoon Jin Cho, School of Chemical Engineering and WCU Department of Energy Engineering, Hanyang University, Seoul, South Korea

Ho Bum Park, School of Chemical Engineering and WCU Department of Energy Engineering, Hanyang University, Seoul, South Korea

These days, polymer membranes have served as a key element in many useful scientific and technological applications such as sensors, fuel cells, water purification and desalination and gas and vapor separation. With the advent of a new era, demanding energy-saving and environment-friendly innovative processes, however, conventional polymer membranes for gas separation are confronted with some drawbacks – poor chemical and thermal stability with relatively low mass transport rates and low separation efficiency. To compete effectively with other materials, it is necessary for polymer membranes to achieve both high permeability and selectivity as well as to impart strong environmental adaptability. Here we show new polyimide membranes having high internal free volume for gas separation. High internal free volume in polyimide membranes was achieved by incorporating triptycene-based diamine into polyimide structure. Triptycene is the simplest member of iptycenes, and in triptycene the three bulky phenyl groups are joined together by a single hinge as the bridgehead carbons. An unusual combination of properties of triptycene makes it useful in polymer design with high internal free volume, expecting high diffusion coefficients of gases and increased sorption capability. As such, triptycene-based polyimide membranes show relatively higher gas permeability and selectivity as compared to other polyimide membranes with rigid or semi-rigid structures. In addition, they exhibit good solubility to common organic solvents, which means they could be processed as useful membrane forms such as thin-film composites or asymmetric membranes by solution-casting or phase inversion methods.

See more of this Session: Gas Separation Technologies

See more of this Group/Topical: Topical 6: 10th Topical Conference on Gas Utilization

(63e) Polysulfone-Co-Poly(ethylene oxide) Random Copolymer Membranes for CO_2 Separation

Hyo Won Kim, School of Chemical Engineering and WCU Department of Energy Engineering, Hanyang University, Seoul, South Korea

Ho Bum Park, School of Chemical Engineering and WCU Department of Energy Engineering, Hanyang University, Seoul, South Korea

Membrane-based gas separation is actually developing industrial membrane separation technologies, and the market size and number of applications served are expanding because it offers a number of advantages in terms of energy and capital cost. Current membrane-based gas processes and potential applications include hydrogen separation and recovery from ammonia purge gas, refinery, and syngas stream in the petrochemical industry. Other applications include carbon dioxide separation from flue gas in power plants, natural gas processing, and landfill gas upgrading. Here we have developed new polysulfone membranes including poly(ethylene oxide) (PEO) segments for selective CO_2 separation. Generally, polysulfone (PSf) is a glassy polymer that has been extensively used for membrane applications such as gas separation, ultrafiltration and support membranes for nanofiltration and reverse osmosis. It is well-known that PEO with ether oxygen units appears to provide excellent CO_2 separation and permeation properties, by enhancing CO_2 sorption capability in PEO segments relative to other small gas molecules such as hydrogen and nitrogen. In this work, polysulfone-co-poly(ethylene oxide) random copolymer membranes were prepared by varying molar ratio of PEO segments and molecular weight of PEG segments, and their gas separation performances were deeply discussed in terms of correlation between CO_2 permeability and CO_2/H_2, CO_2/CH_4, and CO_2/N_2 selectivity, respectively.

See more of this Session: Gas Separation Technologies

See more of this Group/Topical: Topical 6: 10th Topical Conference on Gas Utilization

Fischer–Tropsch Synthesis over Rhodium-promoted Co/SiO_2 Catalysts

Zhen Yan[1], Dragomir B. Bukur[2] and D. Wayne Goodman[1]

[1]*Department of Chemistry, Texas A&M University, College Station, TX, USA*
[2]*Department of Chemical Engineering, Texas A&M University at Qatar, Doha, Qatar and Texas A&M University, College Station, TX, USA*

Introduction

Supported cobalt catalysts have been extensively studied for Fischer-Tropsch synthesis (FTS) because of their high activity and selectivity for linear hydrocarbons [1-6]. The promoting effects of the addition of a second metal to supported cobalt catalysts, including increased reducibility of cobalt oxides, hydrogen spillover, and formation of bimetallic particles [7], have been reported in numerous studies. Ru [8-9], Re [10], Pt [11], and Pd [12] are among the most studied promoters; on the other hand, there have been very limited studies of Rh-Co bimetallic systems [13-15]. In this work we report the promoting effect of Rh on Co/SiO_2 FTS catalysts, as well as the results on Ag-Co and Au-Co bimetallic catalysts.

Experimental

Cobalt-based bimetallic catalysts were prepared by incipient wetness co-impregnation of silica (Davison 952) with aqueous solutions containing cobalt and a second metal. The metal salts used were $Co(NO_3)_2{\cdot}6H_2O$, $AgNO_3$, $HAuCl_4{\cdot}3H_2O$, and $RhCl_3{\cdot}3H_2O$. After impregnation, the samples were dried at 353 K overnight and calcined in air at 673 K for three hours.

Hydrogen chemisorption and temperature programmed experiments were carried out using a Micromeritics AutoChem II 2920 system, which incorporates a thermal conductivity detector. After reduction, catalysts were cooled to 220 K in flowing hydrogen before purged with argon carrier gas. The temperature was then increased to 673 K at 10 K/min and the hydrogen desorption spectrum was recorded. The reducibility of cobalt oxides was measured by injecting oxygen pulses at 573 K after H_2 TPD. TPR experiments were carried out using 10% H_2/Ar after the calcined catalyst was pretreated in helium at 473 K and cooled to room temperature.

FTS was carried out at 493 K and 5 bar in a plug flow reactor after the catalyst (40 – 60 mesh, 0.5 g) was reduced in flowing hydrogen at 673 K for 12 hours. CO/N_2 (3:1) mixture (99.99%) was passed through a hot trap kept at 513 K to remove metal carbonyls, and hydrogen (99.995%) was used without further purification.

Results and Discussion

The catalysts were characterized by elemental analysis, XRD, TEM, and H_2 chemisorption. The average diameter of cobalt particles was 18 ± 2 nm. Both TPR and oxygen pulse experiments indicated that the cobalt oxide in Co/SiO_2 and Rh-Co/SiO_2 catalysts was completely reduced after the reduction step. TPR profiles showed that the addition of Rh lowered the reduction temperatures of cobalt oxides.

The CO conversions as a function of time on Co and bimetallic Co catalysts are shown in Figures 1 and 2. Both Au-Co and Rh-Co catalysts showed an induction period of about 40 hours, during which the CO conversions increased significantly. It is believed that the initial low activities were caused by residual chloride from the precursors, which poisons the active metal sites. It has been reported that the chloride can be removed by water vapor during FTS reaction. This is consistent with the gradually increased activities during the first runs and the disappearance of the induction period in the second runs. The activity of the Ag-Co catalyst was lower than that of Co/SiO_2, while the steady-state activity of the Au-Co catalyst was slightly higher than Co/SiO_2. The addition of a small amount of Rh (0.1–0.5 wt%) to Co/SiO_2 increased the FTS activity by ~50%. The deactivated Rh-Co catalysts can be completely regenerated by reduction at 673 K as shown in Figure 2, or by oxidation at 773 K followed by reduction. No sintering of cobalt particles was observed after repeated FTS reaction and regeneration cycles.

Both the addition of Ag and Au resulted in a slightly lower selectivity to C_{5+} products. The selectivity to C_{5+} products on the Rh-Co catalysts increased with time during the induction period, and at the steady-state conditions, the product distributions were the same as on Co/SiO_2.

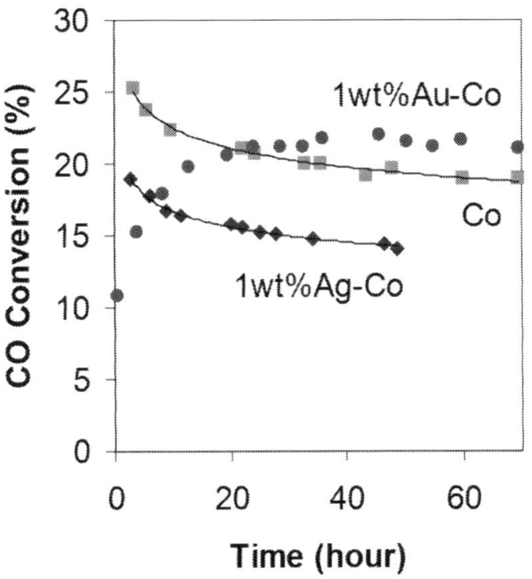

Figure 1. CO conversion as a function of time on Ag and Au promoted 15 wt% Co/SiO_2.

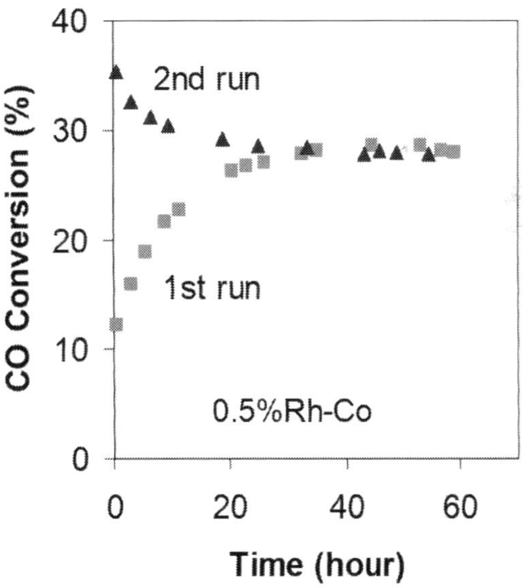

Figure 2. CO conversion as a function of time on Rh promoted 15 wt% Co/SiO_2 (Regeneration conditions: H_2, 673 K).

The results in this study suggest that, although the addition of Rh catalyzes the reduction of cobalt oxides, it does not explain why Rh-promoted cobalt catalysts are more active than unpromoted Co/SiO_2, because cobalt oxides in both Co and Rh-Co catalysts were completely reduced, and the number density of surface cobalt sites was the same for Co and Rh-Co catalysts. Therefore, the higher activity of Rh-Co catalysts is caused by some promoting effects of Rh during FTS reaction, most likely the spillover of hydrogen from Rh [16] to metallic cobalt surfaces.

Conclusions

The FTS activity on the cobalt-based bimetallic catalysts follows an order of Rh-Co > Au-Co > Co > Ag-Co. The addition of a very small amount of Rh (0.1–0.5 wt%) to Co/SiO_2 increased the CO conversion by 50%, while the selectivity was not affected. It is proposed that hydrogen spillover from rhodium to cobalt during FTS reaction is responsible for the high activity of Rh-promoted Co/SiO_2 catalysts.

References

[1] M.E. Dry, Catal. Today 71 (2002) 227.
[2] G. Van Der Laan, A. Beenackers, Catal. Rev. -Sci. Eng. 41 (1999) 255.
[3] E. Iglesia, Appl. Catal. A 161 (1997) 59.
[4] A.Y. Khodakov, W. Chu, P. Fongarland, Chem. Rev. 107 (2007) 1692.
[5] A.A. Adesina, Appl. Catal. A 138 (1996) 345.
[6] B.H. Davis, Fuel Process. Technol. 71 (2001) 157.
[7] F. Diehl, A.Y. Khodakov, Oil Gas Sci. Technol. 64 (2009) 11.
[8] E. Iglesia, S.L. Soled, R.A. Fiato, G.H. Via, J. Catal. 143 (1993) 345.
[9] A. Kogelbauer, J.J.G. Goodwin, R. Oukaci, J. Catal. 160 (1996) 125.
[10] A.M. Hilmen, D. Schanke, A. Holmen, Catal. Lett. 38 (1996) 143.
[11] S. Vada, A. Hoff, E. Ådnanes, D. Schanke, A. Holmen, Top. Catal. 2 (1995) 155.
[12] N. Tsubaki, S. Sun, K. Fujimoto, J. Catal. 199 (2001) 236.
[13] H.F.J. van 't Blik, D.C. Koningsberger, R. Prins, J. Catal. 97 (1986) 210.
[14] J. Kiviaho, M. Reinikainen, M.K. Niemelä, K. Kataja, S. Jääskeläinen, J. Mol. Catal. A: Chem. 106 (1996) 187.
[15] F.B. Noronha, A. Frydman, D.A.G. Aranda, C. Perez, R.R. Soares, B. Morawek, D. Castner, C.T. Campbell, R. Frety, M. Schmal, Catal. Today 28 (1996) 147.
[16] S. Bernal, J.J. Calvino, G.A. Cifredo, A. Laachir, V. Perrichon, J.M. Herrmann, Langmuir 10 (1994) 717.

DEACTIVATION OF IRON BASED FISCHER-TROPSCH CATALYST-A CRITICAL PROBLEM

D. Ferdous
B. Demirel

Rentech Inc. Commerce City, CO 80022

Biography-Deena Ferdous

Deena Ferdous obtained her bachelor degree in Chemical Engineering in 1993 from Bangladesh University of Engineering and Technology. She worked as a Research Engineer for three years at Bangladesh Council of Scientific and Industrial Research, Dhaka. She moved to Canada in 1998 to pursue higher studies and completed her masters and PhD degrees from the University of Saskatchewan in 2000 and 2003, respectively. She is currently working as a Research and Development Engineer for Rentech, Inc. Rentech is involved in the development of process and catalyst for the production of liquid fuel from synthesis gas. Her main responsibility at Rentech is the development of highly active and attrition resistant catalyst for the commercial Fischer-Tropsch plant.

Abstract

Fischer-Tropsch synthesis (FTS) catalytically converts syngas, a mixture of CO and H_2 to hydrocarbons through a surface polymerization reaction. Iron-based FTS catalyst precursors consist of Fe_2O_3 nanocrystals. Promoters are often added to these nanocrystals in order to improve catalyst performance. However, the exact structural composition of the active sites and the deactivation mechanism of this catalyst are still not clearly understood. Iron based Fischer-Tropsch catalysts undergo two types of deactivation during use. One is due to the physical degradation of the catalyst because it destroys the integrity of the catalyst. This is considered as a serious problem because it forms fine particles that cannot be easily separated from the heavy wax products. It also increases the viscosity of the slurry which may cause extreme pressure drops. The other deactivation pathway can be classified as being due to chemical or phase change. Moreover, carbon deposition, pore blocking, sintering and deposition of other chemicals such as sulfur on the catalyst are also considered as the possible reasons of Fe based Fischer-Tropsch catalyst deactivation.

1.0 Introduction

FTS is a strongly exothermic reaction. Often the catalysts and reaction conditions are tuned in such a way to obtain a wax consisting of long hydrocarbon chain products (C_{20+}). A typical catalyst contains promoters like copper to enhance catalyst reducibility, potassium to improve CO dissociation and silica or zinc oxide to improve the attrition resistance of the catalyst which in turn also improves the amount of iron atoms interacting with the gas phase. Catalysts are normally treated in H_2, CO or syngas to convert Fe_2O_3 to its active form. It is generally recognized that a complex mixture of iron phases such as metallic iron, carbide iron and iron oxide is formed after activation and during FTS reaction. The iron based catalysts lose activity with time on stream and the rate of deactivation depends on the presence/absence of promoters.

It is believed that carburized iron is an active catalyst. However, it is still disputed in the literature whether or not bulk carbide phases themselves play an active role during FT synthesis. During pretreatment and FT synthesis of iron based FT catalyst bulk iron carbide is formed because the activation energy for carbon diffusion (43.9-69.0 kJ/mol)[1] is lower than for the FT reaction (89.1 ± 3.8 kJ/mol). Many researchers claim that iron carbides are absolutely necessary for a FTS catalyst to be active. On the other hand other claims that metallic iron[2], Fe_2O_3[3] and Fe_3O_4[4] are the active phases in FTS.

The iron based FT catalyst system is one of the oldest and perhaps most studied systems known in the field of heterogeneous catalysis. However, even after more than of 80 years of research many important questions remain unanswered. This purpose of this paper is review the deactivation mechanisms of iron based FT catalyst.

2.0. Iron Carbide Formation and FT Reactions

For iron based FT catalyst, iron carbides are normally formed during activation treatment or under FT reaction conditions. Hexagonal ε-Fe_2C can be formed by carburizing finely divided iron or iron oxide powder at low temperature (typically 170°C). Hexagonal ε-Fe_2C is transformed into χ-Fe_5C_2 Hägg carbide above 250 °C, which is eventually decomposed into θ-Fe_3C cementite above 450°C.[5]

Dependent on the gas composition and temperature, the following reactions takes place during catalyst pretreatment.[5]

$$Fe_xO_y + yH_2 \longleftrightarrow xFe + yH_2O \quad (1)$$

$$Fe_xO_y + yCO \longleftrightarrow xFe + y\,CO_2 \quad (2)$$

CO can adsorb on the metallic iron active surface sites (represented by *) and subsequently gets dissociated. The carbon species can then react with iron species and form bulk iron carbide.

$$* + CO^* \longrightarrow C^* + O^* \quad (3)$$

$$yC^* + xFe \longrightarrow Fe_xC_y \quad (4)$$

Equation (4) takes place readily on iron at typical FT reaction temperatures because of the low apparent activation energy of this reaction. Dissociated carbon can also take part in two other reactions. First, in the presence of dissociated hydrogen, surface carbide species can be hydrogenated according to the following reaction.

$$C^* + xH^* \longrightarrow CH^*_x + x^* \quad (5)$$

The CH_x species can then further react with each other on the surface and form longer chain products.

Secondly, when more and more carbon species are available on the surface and hydrogen adsorbed atoms are sparse, a side reaction can take place.

$$C^* + C^* \longrightarrow C_i^* \qquad (6)$$

Where, C_i is the inactive surface carbon species. The build-up of these species on the catalyst surface can lead to the blocking of the active surface and lowering the overall catalyst activity.

3.0. Iron Catalyst Deactivation

During FTS, Fischer-Tropsch catalysts undergo deactivation due to the physical degradation and also due to the chemical change. In the FTS, inherent catalyst deactivation and operational catalyst deactivation will occur. Inherent catalyst deactivation occurs due the exposure of the catalyst to the realistic (clean) Fischer-Tropsch conditions such as temperature, partial pressures of H_2, CO, CO_2 and H_2O and the possible product compounds whereas additional deactivation routes may play a role in the operation of the catalyst. Four main mechanisms of deactivation have been described in the literature.[5]

1. Transformation of active iron phases (χ-carbide, ε-carbide, ε'-carbide, more generally Fe_xC of metallic α-Fe) to less active or inactive phases. Most groups report that the active phase is gradually oxidized to magnetite (Fe_3O_4). On the other hand, the other groups believe that catalyst deactivation occurs due to the interconversion of one kind of iron carbide to other kind of iron carbide.
2. In the second mechanism, it is proposed that the deactivation occurs due to the pore plugging and deposition of inactive carbonaceous compounds (graphite carbon, amorphous carbon coke) on the catalyst surface which in turn limiting the contact between reactant gases and catalytically active phase.
3. Some groups claim that sintering, the loss of catalytic surface area due to reopening or migration and coalescence of the iron phase is a main cause for the loss of catalyst activity.
4. Finally, iron based catalyst could be poisoned and deactivated by different chemical compounds present in syngas such as sulfur, carbonyle, ammonia, chlorine etc.

The first three mechanisms are known as inherent catalyst deactivation mechanism whereas the 4th mechanism is considered to be as operational deactivation.[6] Some researchers claim that a single mechanism is responsible for catalyst deactivation; on the other hand, most groups propose that catalyst deactivation occurs due to the contributions of more than one mechanism.

3.1. Deactivation through phase change

For iron based catalyst whether the catalyst exists as carbides or as oxides depends on the reducing behavior of the atmosphere present in the reactor.[7] Catalyst is exposed to an atmosphere of hydrogen, carbon monoxide, and the reaction products such as water, carbon dioxide etc. It is believed that the oxidation of iron carbide and metallic iron occurs in the presence of water and carbon dioxide. O'Brien et al.[7] from the thermodynamic data analysis on the iron/iron oxide system, reported that oxidation conditions exist at a H_2O/H_2 ratio>0.03 or a CO_2/CO ratio>2.8, and the Fe_3O_4 is the stable phase compared to metallic iron. However, they have not reported whether these conditions are applicable for high temperature or low temperature FTS. During FTS, iron carbide can be oxidized to Fe_3O_4 according to the following equations if H_2O/H_2 or CO_2/CO ratios are high enough.[8]

$$Fe_xC_y + yH_2O \longleftrightarrow yC^* + yH_2 + Fe_xO_y \qquad (7)$$

$$Fe_xC_y + yCO_2 \longleftrightarrow yC^* + yCO + Fe_xO_y \qquad (8)$$

The above reactions sum up is considered as the one of the main proposed mechanisms of deactivation of iron based catalyst during FTS.

It has been reported that iron carbides are considerably more resistant to oxidation by water than metallic iron.[5] It has also been reported that smaller catalyst particles carburize more rapidly and are not oxidized during reaction whether larger particles showed higher oxide : carbide ratio.[8] In contrast, Rustad et al.[9] reported that oxidation of nano-sized iron carbide thermodynamically more feasible than the oxidation of bulk iron carbide, due to higher surface energy of iron carbide compared to magnetite. Therefore, the oxidation of small crystallites is possible to occur at much lower water partial pressure.

It is observed that the relationship between the catalyst activity and catalyst time on stream strongly depends on the composition of the catalyst, the activation conditions and the reactor configuration. For example, Bukur et al.[10] reported a continuous decrease in catalyst activity when the catalyst was pretreated by carbon monoxide or syngas whereas better catalyst stability was observed when it was pretreated by hydrogen. Their hypothesis was that the higher surface concentration of hydrogen on these catalysts somehow stabilized catalyst activity. The hydrogen reduced catalyst showed higher selectivity toward methane, lower hydrocarbons, isomers and olefins as compared to carbon monoxide or syngas pretreated samples. Deactivation of the carbon monoxide or syngas pretreated catalyst was attributed to the conversion of the χ-Fe_5C_2 to less active oxide phases and the build-up of carbonaceous material.

From literature it is clearly observed that the oxidation of iron carbide and the catalyst activity solely depends on the catalyst composition.[11] Davis[11] studied time on stream phase composition of 100Fe/3.6Si/0.71K catalyst under FT reaction conditions (270°C, 175 psig, H_2/CO= 0.7 at a flow rate of 3.4nL/h-gFe) using Mossbauer. This catalyst was pretreated at the temperature, pressure and the flow rate of 270°C, 175 psig and 2.0 nL/h-gFe, respectively. The holding time was 24 hours. This catalyst was quite stable for more than 450 hours when the conversion was maintained ~82% and the oxides on the catalyst surface was more than 80% after 350 hours of time on stream. The total carbide and super paramagnetic at this time on stream was 0 and 18%, respectively which clearly shows that for this catalyst there is no correlation between catalyst activity and the iron carbide. These authors claimed that the activity of the catalyst is coming from the super paramagnetic materials and minimum amount of iron carbide is required to maintain stable catalytic activity. However, Raje et al.[12], Zhang et al.[13] and Hayakawa et al.[14] supported that deactivation of iron based catalyst occurs by the oxidation of iron carbide.

3.2. Deactivation due to pore plugging and the deposition of carbonaceous material on the catalyst

Deactivation of iron based catalyst may also occur due to the deposition of carbon on the catalyst surface. This is also known as the inherent cause of iron based catalyst deactivation.[6,15] It has been reported that under Fischer-Tropsch reaction conditions, carbon deposition is thermodynamically favorable and the rate of carbon deposition increases with the increase in crystalline size.[6] Two types of carbon species are mainly formed on the catalyst surface (1) amorphous; (2) graphite. Amorphous carbons are mainly formed during low temperature (<280°C) Fischer-Tropsch reaction. On the other hand, the graphite (insoluble) carbon formation is favorable at the temperature >280°C which leads to coke deposition.

During FT synthesis, dissociation of CO causes the formation of carbon atom which in turn forms surface carbon species due to the direct coupling according to Eq. (6) which is known as amorphous carbon. These carbon species reduce the catalyst activity by blocking the active catalyst sites. However, these carbon species can be removed by hydrogen treatment at elevated temperature (>350 °C).

Eliason and Bartholomew[16] reported that under mild steady state reaction conditions, Eq. (6) is not a serious competitor for carbon atoms. It starts to play an important role as the reaction temperatures increase. In the beginning of FT synthesis, the effect of significantly less active carbon species becomes evident immediately with the increase in the formation of hydrocarbon products and the catalyst activity decreases. They have reported to decrease in 40% activity within 2 hours at 280 °C for unpromoted catalyst. It was proposed that at higher temperature the deactivation of catalysts occur due to the following reactions.

$$C^* + yC^* \longrightarrow C^*_{y+1} \text{ (polymeric } \beta\text{-carbon)} \longrightarrow [-C^{*y+1}-] \text{ (graphite } \delta\text{-carbon)} \quad (9)$$

$$Fe_{2.2}C(\varepsilon') \longrightarrow Fe_5C_2\ (\chi) \longrightarrow Fe_3C\ (\theta) \quad (10)$$

Carbon deposition may also occur due to the Boudouard reaction which causes the formation of insoluble carbon on the catalyst surface

$$C^* + 2CO \longrightarrow C\text{-}C^* + CO_2 \quad (11)$$

$$C^* + CO + H_2 \longrightarrow C\text{-}C^* + H_2O \quad (12)$$

Jung and Thomson[17] speculated that carbon atoms precipitated by the ε' to χ transformation may serve as nucleation sites for the Boudouard reactions (Equations 11 and 12). However, Eliason and Bartholomew[16] indicated that these carbon atoms formed in carbide transformations may serve as nucleation sites for polymerization to β-carbon and/or crystallization to graphitic δ-carbon.

Hao et al.[18] found that the carburization is enhanced with increasing reduction temperature and further carburization takes place during FT reaction. They have reported that the deposition of free carbon might partition the catalyst surface into some "island", restricting the diffusion of monomers for chain growth. Therefore, the increase in deposited carbon might result in the decline of the selectivity of the heavier products. This group found that the catalyst activity decreased with the increase in the reduction temperature from 240 to 280 °C. However, the stability of the catalyst markedly improved with the increase in reduction temperature. They interpreted that the decrease in activity could be due to the content of working catalyst loss in the heavier wax. Davis[11] reported to increase in carbon deposition tenfold with the increase in temperature from 285 to 338 °C indicates the strong temperature sensitivity for deactivation by carbon deposition. However, from their work, no correlation was observed between catalyst activity and carbon deposition.

3.3. Deactivation by poisoning

Poisoning is referred as the strong chemisorption of reactants, products or impurities on the catalytic sites. Poisoning can be slow or fast depending on the poison's concentration and it can be either reversible or irreversible depending on the strength of the poison. Poisoning is a complex process and occurs due to the some or all of the following reasons [15]: (1) physical blockage of one or more catalytic sites by strongly adsorbed poison; (2) electronic modification of nearest neighbor atoms and sometimes even next-nearest neighbor atoms; (3) restructuring of the adsorbent surface, and (4) hindering surface diffusion of adsorbed reactants, thereby preventing reaction.

Sulfur poisoning of iron catalysts is considered as the one of the most serious problem in FT circles. A small concentration of sulfur can deactivate the catalyst very quickly. Typical sulfur levels currently allowed for Fe-based FT are <5 ppb in syngas feed to the reactor.[19] It has been indicated that the low level of sulfur poisoning through the pretreatment of reduced iron catalysts have been shown to enhance the activity of the catalyst, whereas high levels led to deactivation.[6] Poisoning by H_2S is slower on iron carbide than on metallic iron[20] and the transformation of iron to iron sulfide which is difficult to reduce is thermodynamically feasible. During poisoning by sulfur, H_2S adsorbs on the catalyst surface and deactivates the catalyst by accumulating it on the catalyst. Literature shows that sulfur not only deactivates the catalyst but also the selectivity shifts toward less olefinic material.[19] Shultz et al.[21] reported that alkali promoters, particularly potassium provide resistance to sulfur deactivation. Manganese and boron were reported as sulfur sinks by several authors.[22]. However, the effect of sulfur on the catalyst surface depends on the nature of sulfur species.[23]. Several researchers indicated that the sulfur enhanced catalyst life and olefin selectivity[21,26], on both iron[24] and iron carbide.[25] For example, Bromfield and Coville[26] in their work prepared a series of catalyst by adding small amounts Na_2S to precipitated iron. Na_2S concentrations were varied from 500-20000 ppm. Catalyst with low sulfide concentrations were up to four time more active in FT reaction than a sulfur free catalyst, while catalyst with high sulfide loadings were apparently poisoned.

Different compounds such as ammonia, HCN and carbonyl mainly act as a kinetic inhibitor during FTS. Rausch[27] indicted that co-feeding of ammonia to FTS did not result in decrease in conversion for iron catalyst. It has been reported to decrease in CO conversion over cobalt catalyst in the presence of HCN.[28] No information is available on the impact of HCN on iron catalyst. Deactivation of cobalt catalyst by carbonyl material and chlorine was reported but no data is available on the impact of these compounds on iron catalysts.

3.4. Deactivation by sintering

Sintering of catalyst occurs because of the loss of catalytic surface area due to crystallite growth of the catalytic phase, and loss of support area due to support collapse and of catalytic surface area due to pore collapse on crystallites of the active phase. Smit and Weckhuysen[5] stated that all iron based FTS catalysts to some extent deactivate by sintering of the active phase. In contrast, other group explicitly concluded that sintering is not playing a significant role in the deactivation of FT catalyst.[16] Many groups reported that sintering is one of the reasons for FT catalyst deactivation. Mansker et al.[29] reported that the change in carbide phase composition from Fe_7C_3 to χ-Fe_5C_2 was obtained by a progressive growth in crystallite size after ~350 hours on stream.

Sintering is driven by the thermodynamic force to minimize the exposure of highly energetic surfaces, which proceeds either through atomic migration and/or crystallite migration and this process is generally accelerated by the presence of water vapor.[15] Sintering rates depend on the temperature, atmosphere, metal type, metal dispersion, promoters/impurities and support area, texture and porosity.[15] Sintering mainly occurred when the temperature is above the Tamman temperature. When the temperature is above the Tamman temperature, atoms on the surface of crystallites are expected to become mobile and as a result crystallites can sinter.[29] The Tamman temperature for iron is 633 °C which is much higher than the typical FT reaction temperature (200-400 °C). As the FT reaction is very exothermic, there is a possibility that the local temperature of crystallites could be very high. Van Steen and Claeys[6] reported that sintering can be reduced by improving the distribution of catalytically active component. On the other hand, Rao et al.[30] indicated that sintering can be prevented by stabilizing the active phase using the binders.

4.0. Conclusion

The main challenge in the design of novel iron-based FTS catalysts is to overcome their notoriously high deactivation rates. Even after more than 80 years of research on iron-based FTS catalysts, no straight forward relationship between catalyst structure and performance is available. In-situ time on stream characterization could be useful to find the possible reason of FT catalyst deactivation.

5.0. References

1. Baker, R. T. K. (1979), "In-Situ Electron Microscopy Studies of Catalyst Particle Behavior", Catal. Review, 19, pp. 161-209.
2. Shashkin, D. P., Shiryaev, P. A., Chichagov, A. V., and Morozova, O. S. (2004), "Structural Transformation of Iron Oxides during the Formation of Fischer-Tropsch Catalysts", Kinet. Catal., 33(2), pp. 744-749.

3. Grzybek, T., Papp, H., and Baerns, M. (1987), "Fe/Mn Oxide Catalysts for Fischer-Tropsch Synthesis - Part V. XPS Surface Characterization of Calcined and Reduced Samples", Appl. Catal., 29(2), pp. 335-350.
4. Mahajan, D., Gütlich, P., Ensling, J., Pandya, K., Stumm, U., and Vijayaraghavan, P. (2003), "Evaluation of Nanosized Iron in Slurry-Phase Fischer-Tropsch Synthesis", Energy & Fuels, 17(5), pp. 1210-1221.
5. de Smit, E. and Weckhuysen, M. B. (2008), "The Resistance of Iron-Based Fischer-Tropsch Synthesis: on the Multifaceted Catalyst Deactivation Behavior", Chem. Soc. Rev. 37, pp. 2758-2781.
6. Van Steen, E. and Claeys, M., (2008), "Fischer-Tropsch Catalysts for the Biomass to Liquid (BTL)-Process, Chem. Engg, Tech. 31(5), pp. 655-666.
7. O'Brien, R. J., Xu, L., Spicer, R. L., and Davis, B. H. (1996), "Activation Study of Precipitated Iron Fischer-Tropsch Catalysts," Energy & Fuels, 10(4), pp. 921-926.
8. Dry, M. E. (1981), "The Fischer-Tropsch Synthesis," in "Catalysis, Science and Technology", (J.R. Anderson, M. Boudart, eds.), Springer-Verlag, 1, pp. 159-256.
9. Rustad, R. J., Wasserman, E., and Flemy, R. A.(1999), "A Molecular Dynamics Investigation of Surface Reconstruction on Magnetite (001)", *Surface Science,* **432** , pp. 583-588.
10. Bukur, D. B., Koranne, M., Lang, X., Rao, K. R. P. M., and Huffman, G. P. (1995), "Pretreatment Effect Studies with a Precipitated Iron Fischer-Tropsch Catalyst", Appl. Catal., 126, pp. 85-113.
11. Davis, H. B., (1999), "Technology Development for Iron-Fischer Tropsch Catalysis,", DOE Contract DE-AC22-94PC94055, Final Report, March.
12. Raje, A., O'Brien, R. J., Xu, L., and Davis, B. H. (1997), "Deactivation of Iron-Based Catalysts for Slurry Phase Fischer-Tropsch Synthesis," in "Catalyst Deactivation", Stud. Surf. Sci. Catal., 111, pp. 527-533.
13. Zhang, Y., Sirimanothan, N., O'Brien, J. R., Hamdeh, H. H., and Davis, H. B. (2001), "Study of Deactivation of Iron-Based Fischer-Tropsch Synthesis Catalysts", Studies in Surface Science and Catalysis, 139, pp. 125-132.
14. Hayakawa, H., Tanaka, H., and Fujimoto, K. (2006), "Studies on Precipitated Iron Catalysts for Fischer–Tropsch Synthesis", App. Catal., 310, pp. 24-30.
15. Bartholomew, C. H. (2001), "Mechanisms of Catalyst Deactivation", Appl. Catal. A, 212(1-2), pp. 17-60.
16. Eliason, S. A., and Bartholomew, C. H. (1999), "Reaction and Deactivation Kinetics for Fischer-Tropsch Synthesis on Unpromoted and Potassium-Promoted Iron Catalysts," Appl. Catal. A, 186, pp. 229-243.
17. Jung, H., and Thomson, W. J., (1992), "Dynamic X-ray Diffraction Studies of an Unsupported Iron Catalyst ij Fischer-Tropsch Synthesis", J. Catal., 134, pp. 654-667.
18. Qing-Lan Hao, Q.-L., Liu, F.-X., Wang, H., Chang, J., Zhang, C.-H., Bai, L., Xiang, H. W., Li, Y.-W., Yic, F., and Xu, B.-F., (2007), Effect of Reduction Temperature on a Spray-Dried Iron-Based Catalyst for Slurry Fischer–Tropsch Synthesis", J. Mol. Catal. 261, pp. 104-111.

19. Kritzinger, J. A. (2002), "The Role of Sulfur in Commercial Iron-Based Fischer-Tropsch Catalysis with Focus on C_2-Product Selectivity and Yield", Catal. Today, 71(3-4), pp. 307-318.
20. Karn, F. S., Shultz, J. F., Kelly, R. E., Anderson, R. B. (1964), "Hydrogen Sulfide Poisoning of Nitrided and Carbided Iron Catalysts in the Fischer-Tropsch Synthesis", Ind. Eng. Chem. Prod. Res. Dev., 3(1), pp. 33-38.
21. Shultz, J. R., Hofer, L. J. E., Karn, F. S., and Anderson, R. B. (1962), "Studies of the Fischer-Tropsch synthesis", J. Phys. Chem., 6(6), pp. 501-506.
22. Bartholomew, C. H., and Bowman, R. M. (1985), "Sulfur Poisoning of Cobalt and Iron Fischer-Tropsch Catalysts," Appl. Catal., 15, pp. 59-67.
23. van Dijk, W. L., Niemantsverdriet, J. W., van dar Kraan,A. M., and van der Baan, H. S. (1982), "Effects of Manganese Oxide and Sulphate on the Olefin Selectivity of Iron Catalysts in the Fischer Tropsch Reaction, Appl. Catal. 2(4-5), pp. 273-288.
24. Chaffee, A. L., Campbell, I., and Valentine, N. (1989), "Sulfur Poisoning of Fischer-Tropsch Synthesis Catalysts in a Fixed-Bed Reactor", Appl. Catal., 47(2), pp. 253-276.
25. Madon, R. J., and Taylor, W. F., (1979), "Effect of Sulfur on the Fischer-Tropsch Synthesis: Alkali-Promoted Precipitated Cobalt-Based Catalysts", chapter 8 in "Hydrocarbon Synthesis from Carbon Monoxide and Hydrogen" (E. L. Kugler, F. W. Steffgen, eds.), ACS Adv. Chem. Ser. 178, pp. 93-112.
26. Bromfield, T. C., and Coville, N. J. (1999), "The Effect of Sulfide Ions on a Precipitated Iron Fischer-Tropsch Catalyst", Appl. Catal., 186(1-2), pp. 297-307.
27. Rausch, A. K. (2008), "C-N-Kopplung and Heterogenen Kobalthaltigen Katalysatoren", PhD thesis Carl von Ossietzky Universität Oldenburg.
28. Oliva, C., van den Berg, C., Niemantsverdriet. J. W., and Curulla-Ferre, D. (2007) "A Density Functional Theory of HCN Hydrogenation to Methylamine on Co(111)", J. *Catal.*, **248, pp.** 38-45.
29. Mansker, L. D., Jin, Y., Bukur, D. B., and Datye, A. K. (1999), "Characterization of Slurry Phase Iron Catalysts for Fischer-Tropsch Synthesis", Appl. Catal., 186(1-2), pp. 277-296.
30. Rao, V. U. S, Stiegel, G. J., Cinquegrane, G. J., and Srivastave, R. D., (1992), "Iron Based Catalysts for Slurry Phase Fischer-Tropsch Process: Technology Review", Fuel Proc. Tech., 30, pp. 83-107.

FISCHER-TROPSCHSYNTHESIS: KINETICS ON 25%Co/Al_2O_3 CATALYSTS CALCINED IN AIR OR NITRIC OXIDE

Wenping Ma, Gary Jacobs, Keogh Robert, and Burtron H. Davis
Center for Applied Energy Research, University of Kentucky, 2540 Research Park Drive, Lexington, Kentucky 40511, USA

Introduction

Although Fischer-Tropsch synthesis (FTS) has been widely studied over conventional iron, cobalt, nickel, and ruthenium based catalysts in the past 90 years or so, the impact of water produced by the FTS on catalyst performance parameters still remains unclear. Several groups have investigated the water effect on FTS performance using the water co-feeding technique [1-6]. It was found that support type, which primarily governs the interaction between support and metals, plays an important role in determining the water effect. Recently, CAER researchers developed a kinetic approach to study the water effect during FTS. An empirical model to define the water effect on cobalt based catalysts has been successfully developed and used[1,7,12]:

$$r_{CO} = k\, P_{CO}^{a} P_{H2}^{b} / (1 + m\, P_{H2O}/P_{H2}) \qquad (1)$$

where $k = k_0 \exp(-E_a/RT)$, is the kinetic rate constant, a and b are the reaction orders for CO and H_2, respectively, and m is a water effect constant.

However, the water effect during FTS may also be linked with Co particle size, even though the turnover frequencies of CO over the surface Co metal atoms on supported Co catalysts has been reported to be constant at a given temperature [8]. The traditional impregnation plus air calcination usually led to larger metal particles over Co/SiO_2 catalysts. Recently, a novel calcination procedure using nitric oxide (NO) gas to create smaller cobalt crystallites over silica was patented [9] and was confirmed in our recent study [10,11]. Continuing to investigate the sensitivity of Co particle size to FTS activity, selectivity and the water effect, we employed the conventional air calcination and calcination with nitric oxide gas to prepare two Al_2O_3 supported cobalt catalysts (25%Co/Al_2O_3), which were expected to display differences in Co size as well. The two air and NO calcined 25%/Al_2O_3 catalysts were characterized by means of BET, TPR, hydrogen chemisorption pulse reoxidation and extended X-ray absorption fine structure (EXAFS), and kinetic behavior was tested using a 1-L stirred tank slurry reactor. The effect of Co cluster size on catalyst behavior including activity, selectivity and water effect is reported. A possible deactivation mechanism on the cobalt catalysts is proposed based on the EXAFS results.

Experimental

Catalyst preparation In this study, a medium pore puralox HP14-Al_2O_3 (SA ~150 m^2/g), was used as catalyst support and cobalt nitrate was used as the precursor. Details of procedures for the preparation of Co/Al_2O_3 catalysts are reported elsewhere [10,11]. In brief, a slurry impregnation of Al_2O_3 was employed. The volume of the impregnated nitrate solution was controlled by adjusting the ratio of the volume of solution to the weight of alumina to 2.5. Two impregnation steps were used, with interval drying at 273 K, prior to air and nitric oxide (5% NO in N_2) calcination at 673K.

Catalyst characterization BET measurements for the catalysts were conducted using a Micromeritics Tri-Star system. Prior to the measurement, samples were slowly ramped to 160 °C and evacuated for 24 hours to approximately 50 mTorr. Temperature programmed reduction (TPR) profiles and hydrogen chemisorption measurements of fresh catalyst samples were performed using a Zeton Altamira AMI-200 unit, which incorporates a thermal conductivity detector (TCD). TPR was performed using 30 cc/min 10%H_2/Ar mixture referenced to argon. The ramp rate was 5 °C/min from 50 °C to 1100 °C, and the sample was held at 1100 °C for 30 min. For the hydrogen chemisorption with pulse reoxidation, the catalyst was activated at 350°C for 10 hrs using a flow of pure hydrogen and then cooled under flowing hydrogen to 100°C. The sample was held at 100°C under flowing argon to prevent physisorption of weakly bound species prior to increasing the temperature slowly to the activation temperature. After TPD of hydrogen, the sample was reoxidized at the activation temperature by injecting pulses of pure oxygen in helium referenced to helium gas. The measurement results of BET, TPR and H_2 chemisorption pulse oxidation was summarized in Table 1.

Table 1. Results of BET and hydrogen Chemisorption with pulse reoxidation

Description	BET			Chemisorption						
	BET Area	Pore Volume	Average Pore	H_2 desorbed	O_2 pulsed	Reduction [a]	Uncorrected	Uncorrected Co	Corrected	Corrected Co
	m²/g	cm³/g	Diameter, nm	μmol/gcat	μmol/gcat	Degree, %	Disp., %	Cluster size, nm	Disp., %	Cluster size, nm
Medium Pore Puralox HP14 Al_2O_3 (Suppo	152.6	0.92	20.65							
25%Co/Al_2O_3, NO calcined	152.06	0.308	6.763	97.4	1526.5	54	4.95	22.5	9.02	11.4
25%Co/Al_2O_3, Air calcined	127.49	0.3009	8.044	98.4	1695	60	4.64	22.2	7.74	13.3

Extended X-ray Absorption Fine Structure (EXAFS) The catalysts after FTS runs (in slurry form) and Co references were evaluated by EXAFS spectroscopy at Brookhaven National Laboratory. EXAFS spectra were recorded in transmission mode at the National Synchrotron Light Source (NSLS) at Brookhaven National Laboratory, Upton, New York, Beamline X18-b. The EXAFS spectra were recorded near the Co K edge. The spectra were recorded near the boiling temperature of liquid nitrogen to minimize contributions to the dynamic Debye-Waller factor. Sample thickness was determined by calculating the amount in grams per square centimeter of sample, w_D, by utilizing the thickness equation:

$$w_D = \ln(I_0/I_t)/\Sigma\{(m/r)_j w_j\}$$

where m/r is the total cross section (absorption coefficient/density) of element "j" in the sample at the absorption edge of the EXAFS element under study in cm²/g, w_j is the weight fraction of element j in the sample, and $\ln(I_0/I_t)$ was taken over a typical range of 1 – 2.5. EXAFS data reduction and fitting were carried out using the WinXAS, Atoms, FEFF, and FEFFIT programs. The k- and r-ranges were chosen to be 3-15 Å-1 and 1.0 – 3.0 Å, respectively.

Catalyst pretreatment The air and NO calcined 25%Co/Al_2O_3 (~10 g) were ground and sieved to 80-325 mesh before loading into a fixed-bed reactor for 10 h of *ex-situ* reduction at 350°C and atmospheric pressure using a gas mixture of H_2/He with a molar ratio of 1:3. The reduced catalyst was then transferred into a 1-L continuously stirred tank reactor (CSTR), under the protection of N_2 inert gas, which was previously charged with 315 g of melted Polywax 3000. The transferred catalyst was further reduced *in-situ* at 230 °C at atmospheric pressure using pure hydrogen for another 10 h before starting the FTS reaction.

Kinetic design The kinetic experiment was carried out using a wide range of conditions: 205-220 °C, 280 psig, H_2/CO ratio of 2.5-1.0 and space velocity of 3-22 Nl/g-cat/h. Details of experimental procedures can be found elsewhere [12, 13].

Results and discussion

BET, TPR and chemisorptions results

From table 1, the NO and air calcined 25%Co/Al_2O_3 have a BET surface area of about 127-150 m^2/g and a average pore size of 6-8 nm, which falls into a mesopore size range. The NO calcined 25%Co/Al_2O_3 catalyst showed relatively low reduction degree, but Co dispersion is relatively higher compared to the air calcined one. The average Co cluster sizes for the NO and air calcined 25%Co/Al_2O_3 catalysts determined by hydrogen chemisorption/pulse oxidation were 11.4 and 13.3 nm, respectively.

FTS reaction over air and NO calcined 25%Co/Al_2O_3

Regarding the stability of the air and NO calcined 25%Co/Al_2O_3 catalysts, Figure 1 displays changes in CO conversion with time on stream for the two catalysts over a 200 h testing period. Under the same reaction conditions (i.e., 220 °C, 2.03 MPa, 10 Nl/g-cat/h, H_2/CO=2.5), the NO calcined 25%Co/Al_2O_3 catalyst displayed much higher CO conversion relative to the air calcined one. During 200 h testing, the average CO conversion was about 49.6% for the NO calcined 25%Co/Al_2O_3 and 28.2% for the air calcined catalyst. This demonstrates that employing the NO calcination pretreatment prior to activation significantly improved catalyst activity. The most probable reason for the observed higher activity is that the NO calcination created a smaller average Co cluster size (table 1), thus resulting in a higher Co active site density on the catalyst surface. Unfortunately, NO calcination also led to a higher deactivation rate relative to the air calcination method (1.87%/day versus 0.29%/day).

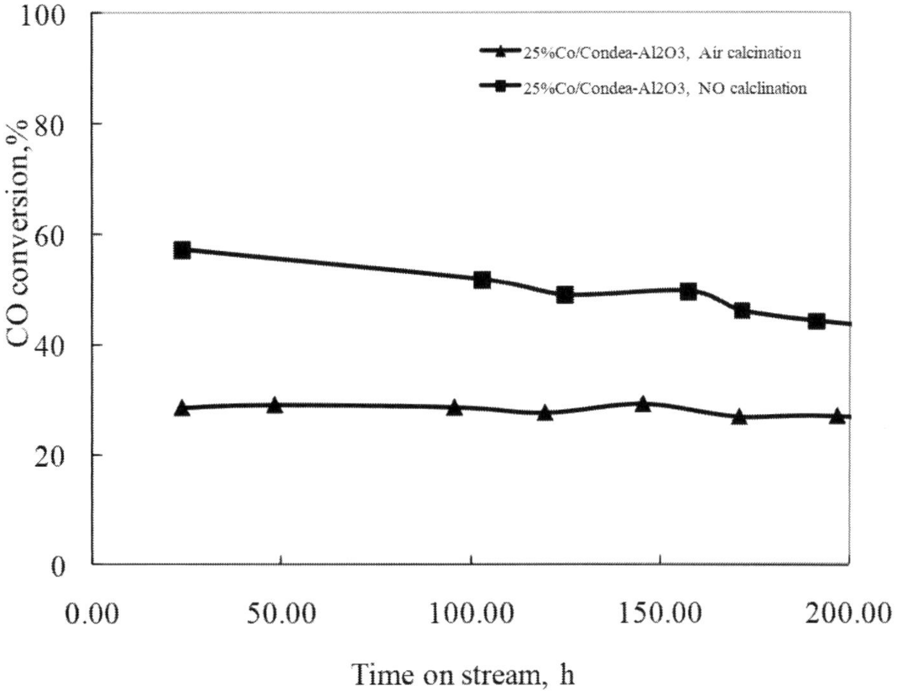

Figure 1. CO conversion as a function of time on air and NO calcined 25%Co/γ-Al_2O_3 (220 °C, 2.03 MPa, 10 Nl/g-cat/h, H_2/CO=2.5, N_2 = 12.5%).

The mean hydrocarbon selectivities toward FTS products including CH_4, C_{5+}, olefins and inter-olefins for the two catalysts are summarized in Table 2. Clearly, NO calcination resulted in relatively less CH_4 and higher C_{5+} selectivities. The difference in Co cluster size on the two catalysts is assumed to be responsible for the observed selectivity difference, the smaller Co cluster size exhibiting the higher C_{5+} selectivity. In comparison with the olefin content over the air calcined 25%Co/Al_2O_3 catalyst, the NO calcined one displayed lower olefin content (34 vs 40%), reflecting higher activity of the secondary reaction of olefins at higher conversion level for the NO calcined catalyst. This assumption is consistent with the higher 2-olefin selectivity observed for the NO calcined catalyst, as shown in Table 2.

Table 2. Summary of selectivities to CH_4, C_{5+}, olefins and inter-olefins(d)

Catalyst	Hydrocarbon selectivity, mol% (a)		Lump olefin fraction, % (b)	Lump 2-olefin sel, % (c)
	CH_4	C_{5+}	C_2-C_{10}	C_2-C_{10}
25%Co/Al_2O_3, NO calcined	10.3	80.1	33.8	27.9
25%Co/Al_2O_3, Air calcined	10.8	79.6	40.1	10.1

(a) Based on carbon atom (b) Fraction of C_2-C_{10} olefins in C_1-C_{10} hydrocarbons (c) Fraction of internal C_2-C_{10} olefins in C_2-C_{10} olefins (d) 220 °C, 2.03 MPa, 10 Nl/g-cat/h, H_2/CO=2.5, N_2 = 12.5%, 24-200 h.

Kinetic parameter values calculated based on kinetic data at 220 °C are summarized in Table 3. On both catalysts, the order of partial pressure of CO is negative, and the order of partial pressure of H_2 and the water effect term, m, are positive, indicating much stronger CO adsorption relative to hydrogen for the two catalysts, and a negative water effect for the two catalysts during FTS. As a result, CO adversely affects the FT reaction rate, while the observed negative water effect occurring over both catalysts is consistent with the results observed over other Co/Al_2O_3 catalysts [13]. The value of *b* for NASA 2B is close to 0.5, suggesting adsorbed H_2 was dissociated on the catalyst surface before taking part in the FT reaction. Comparing *k* values for the NO and air calcined 25%Co/γ-Al_2O_3, the *k* value on the NO calcined one is about 40% higher, defining the particle size effect on the reaction rate constant. Furthermore, higher *m* values were obtained for the NO calcined 25%Co/γ-Al_2O_3 catalyst relative to the air calcined one, evidence that the sensitivity of the water effect depends on Co particle size, with the smaller Co particle size exhibiting a more pronounced sensitivity to the water effect during FTS.

Table 3. Summary of kinetic parameter values at 220 °C

$$r_{CO} = k\, P_{CO}^{a} P_{H2}^{b} / (1 + m\, P_{H2O}/P_{H2})$$

Catalyst Description	Reaction order		Water effect	Reaction rate constant
	a	b	constant, *m*	*k*, mol/g-cat/h/MPa$^{(a+b)}$
25%Co/Al_2O_3, NO calcin	-0.85	0.71	0.79	0.0253
25%Co/Al_2O_3, Air calcine	-0.73	0.56	0.28	0.0182

EXAFS study of spent air and NO calcined 25%Co/Al_2O_3 catalysts

To understand if Co cluster size increased (e.g., agglomeration or ripening) and if oxidation of cobalt by water was taking place during FTS, EXAFS and XANES experiments of used air and NO calcined 25%Co/Al_2O_3 catalysts and reduced catalysts without FTS reaction were performed. EXAFS fitting results are summarized in Figure 2 and Table 4, where a simultaneous fitting procedure was used to constrain the fitting. An excellent fitting was obtained, as characterized by the low r-factor. Results reveal that the fresh catalyst could be fitted using a combination of only Co-Co of the Co metal and Co-O and Co-Co of the CoO oxide. However, Co-O, Co-Co, and Co-Al parameters for $CoAl_2O_4$ had to be added in order to provide a good fitting of the catalysts after long-term usage, indicating that a fraction of Co did oxidize to cobalt aluminate during FTS. In agreement with this, growth of a feature in the XANES derivative spectra at 7.17 keV was recorded, typical of cobalt aluminate formation (not shown).

Conclusions

Both NO calcination also can produce smaller Co cluster size on Al_2O_3 support. NO calcination resulted in a more active catalyst, but the catalyst deactivation was faster. Both air and NO calcinated 25%Co/ Al_2O_3 showed negative water effect. Smaller cobalt clusters on Al_2O_3 support were more sensitive to deactivation induced by water.

In EXAFS and XANES results, catalyst deactivation was accompanied by structural changes in the catalyst. Cobalt aluminate was a characteristic component in the catalysts after long-term usage and there was loss of Co-Co coordination in the metal; the fresh catalysts were more reduced and contained virtually only Co metal and CoO oxide.

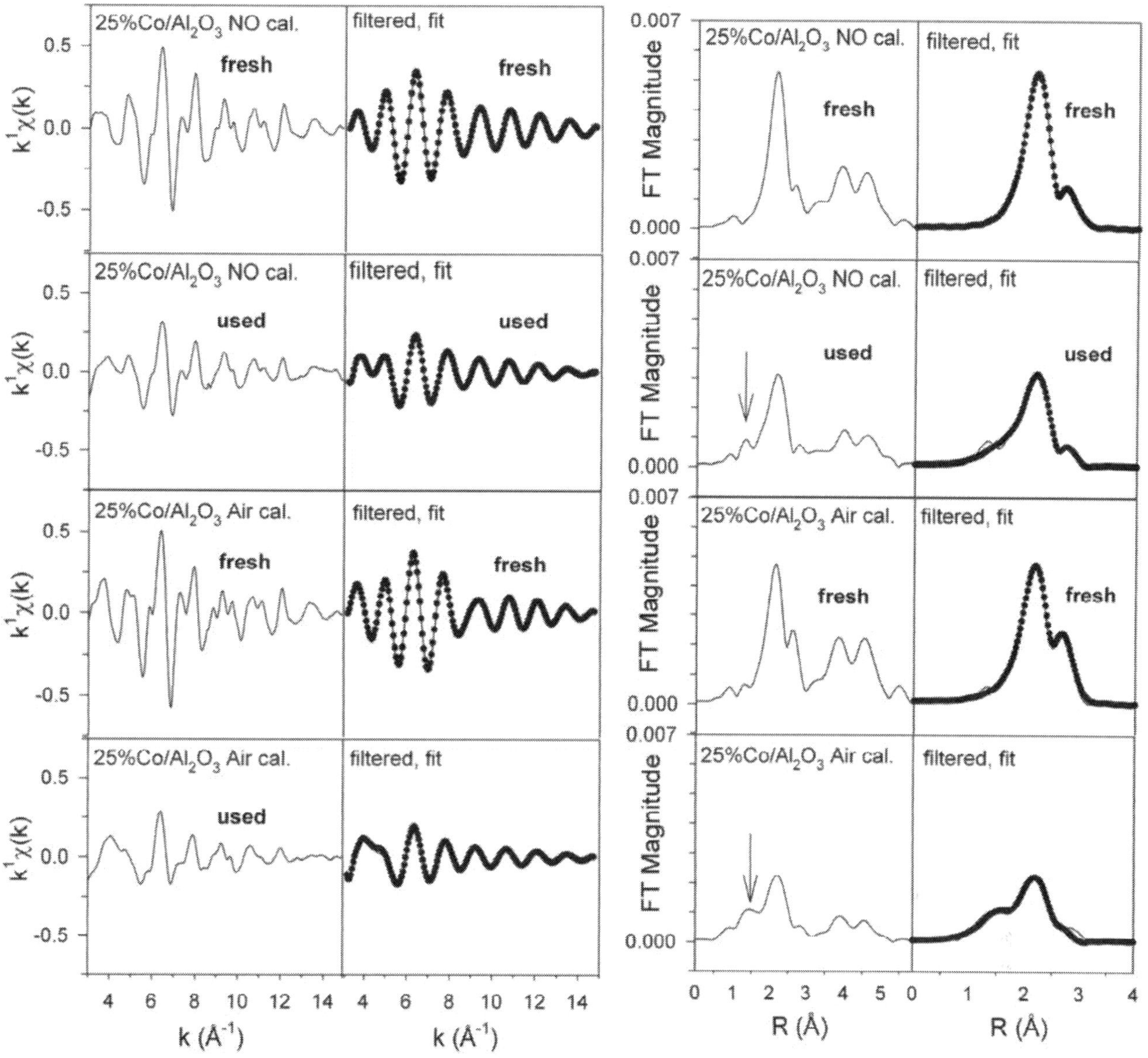

Figure 2. Results of EXAFS fitting (arrow denotes growth of Co-O in $CoAl_2O_4$).

Table 4. Results of EXAFS fitting parameters for data acquired near the Co K edge for fresh and used catalysts, cooled and solidified in the wax in the in-situ state. The fitting ranges were approximately $\Delta k = 3 - 15$ Å^{-1} and $\Delta R = 1.0 - 3.0$ Å. $S_0^2 = 0.9$.

Sample Description	Calcin.	N Co-O for CoO and CoAl2O4	R Co-O (Å) for CoO and CoAl2O4	N Co-Co metal	R Co-Co (Å) metal	N Co-Co & Co-Al for CoO & CoAl2O4	R Co-Co & Co-Al (Å) for CoO & $CoAl_2O_4$	e_0 (eV)	σ^2 (Å^2)	r-factor
25%Co/Al_2O_3	Nitric oxide	**1.58**	2.0702	**4.55**	2.5011	**3.17**	3.032	-1.348	0.00236	0.00971
start of run		-0.662	-0.0831	-1.764	-0.00735	-1.323	-0.0949	-1.135	-0.0011	
								-0.5059	0.00624	
								-1.7985	-0.00455	
25%Co/Al_2O_3	Nitric oxide	**0.19**	2.0702	**3.35**	2.5011	Co-Co		-4.05	0.00366	
end of run		-0.347	-0.0831	-0.822	-0.00735	**0.37**	3.032	-3.079	-0.00151	
		2.59	1.9844			-0.693	-0.0949		0.000102	
		-1.484	-0.0376			**1.3**	2.7675		-0.0108	
						-0.742	-0.136		0.0109	
						Co-Al			-0.00869	
						1.3	2.7675		0.00832	
						-0.742	-0.136		-0.0102	
25%Co/Al_2O_3	Air	**2.37**	2.0982	**4.22**	2.5061	**4.74**	3.0197		0.00287	
start of run		-0.372	-0.02516	-0.544	-0.00535	-0.744	-0.0146		-0.000929	
									0.00516	
									-0.0016	
25%Co/Al_2O_3	Air	**0.66**	2.0982	**2.69**	2.5061	Co-Co			0.00326	
end of run		-1.494	-0.02516	-0.367	-0.00535	**1.31**	3.0197		-0.000772	
		2.72	1.9465			-2.988	-0.0146		0.0147	
		-1.057	-0.01826			**1.36**	2.6853		-0.0209	
						-0.528	-0.04154		0.009	
						Co-Al			-0.0038	
						1.36	2.6853		0.00986	
						-0.528	-0.04154		-0.00612	

Acknowledgments

This work was supported by NASA contract, #NNX07AB93A and the Commonwealth of Kentucky.

References

1. J.L. Li, X.D. Zhan, G. Jacobs, B. H. Davis, Appl. Catal. 208 (2002) 203.
2. G. Jacobs, T.K. Das, P.M. Patterson, J.L. Li, L. Sanchez, B.H. Davis, Appl. Catal. 247 (2003) 335.
3. M. Claeys, E. van Steen, Catal. Today 71 (2002) 419.
4. H. Schulz, M. Claeys, S. Harms, Stud. Surf. Sci. Catal. 107 (1997) 193.
5. S. Storsæter, S., Borg, Ø., Blekkan, E.A., Holmen, A., J. Catal. 2005, 231, 405.
6. O. Borg, S. Storsaeter, S. Eri, H. Wigum, E. Rytter, A. Holmen, Catal. Lett. 107 (2006) 95.

7. T. Das, X.D. Zhan, J.L. Li, G. Jacobs, M. E. Dry and B. Davis, Stud. Surf. Sci. Catal. 163 (2007) 1430.
8. E. Iglesia, Appl. Catal. 161 (1997) 59.
9. J.R.A. Sietsma, A.J. van Dillen, P.E. de Jongh, and K.P. de Jong, PCT Int. Appl. (2008) WO 2008029177.
10. W.P. Ma, G. Jacob, M.K.Gnanamani, U.M. Graham, B.H. Davis, in Advances in Fischer Tropsch Synthesis, Catalysts, and Catalysis, eds. Burt H. Davis and Mario L. Occelli. Taylor & Francis Group, 2009, p31.
11. G. Jacobs, W.P. Ma, Y.Y.Ji, S. Khalid, B.H. Davis, in Advances in Fischer Tropsch Synthesis, Catalysts, and Catalysis, eds. Burt H. Davis and Mario L. Occelli. Taylor & Francis Group, 2009, p147.
12. W.P. Ma, Janet C. Mohandas, G. Jacob, B.H. Davis, NASA annual report, 2009.
13. W.P. Ma, G. Jacob, B.H. Davis et al. paper submitted to Fuel for publication, December, 2009.

(66d) Fischer Tropsch Synthesis in Supercritical Carbon Dioxide

Derek M. Perry, Department of Mechanical Engineering and Energy Processes, Southern Illinois University, Carbondale, IL

Jitendra Gautam, Department of Mechanical Engineering and Energy Processes, Southern Illinois University, Carbondale, IL

Kanchan Mondal, Department of Mechanical Engineering and Energy Processes, Southern Illinois University, Carbondale, IL

The results from studies on Fischer Tropsch synthesis wherein syngas was dissolved in CO2 are presented. The syngas used was typical of that obtained from coal gasification, i.e. CO:H2 of one. Under these conditions Co-based catalysts without any water gas shift catalyst does not perform well while Fe – based catalysts have been found to be useful. However, the Fe based catalysts have a propensity towards CO2 selectivity via the primary FT reaction, Boudouard reaction and the water gas shift reaction. The use of CO2 as a solvent was found to suppress the CO2 and CH4 selectivity while enhancing the hydrocarbon selectivity and CO conversion when FT synthesis was conducted using coal derived syngas on Fe-Zn-K catalysts. The effects were found to be significantly pronounced at pressures higher than or equal to 1200 psi. It should be noted that CO2 is supercritical at pressures higher than 1050 psi and 31.4 oC. The effect of CO2 partial pressure, reactor pressure, reaction temperature, catalyst loading and H2:CO ratio in syngas on the liquid product distribution will be presented. Some of the notable findings include negative CO2 selectivities while obtaining nearly 98 % CO conversion. Increasing the reactor pressure was found to favor longer chain growth. In addition, it is noted that the ratio between CO2:H2 in the reaction mixture is an important factor in the liquid product distribution. A higher value of the ratio is seen to favor hydrocarbon synthesis, while a lower value of the ratio favors oxygenate production particularly pentanols and butanols. In addition, the data on the once through fractionation of the products utilizing the solubilities in supercritical CO2 and pressure tuning will be presented. It will be shown that the products can be easily fractionated into narrow carbon chain length distributions downstream of the reactor by simply reducing the pressures in each collection vessel. It was generally observed that the oxygenates and higher n-alkanes were collected in the higher pressure trap and lower hydrocarbons in the subsequent lower pressure traps. This research was made possible with support, in part, by the Illinois Department of Commerce and Economic Opportunity through the Office of Coal Development and the Illinois Clean Coal Institute and the Office of Vice Chancellor for Research at Southern Illinois University, Carbondale.

See more of this Session: Kinetics, Design and Modeling of Fischer-Tropsch Catalysts and Reactions I

See more of this Group/Topical: Topical 6: 10th Topical Conference on Gas Utilization

Microemulsion synthesis of Co nanoparticles supported on CNT: Effects of Nanoparticles Size on reducibility, activity and selectivity in FTS.

Mariane Trépanier[a], Ajay K. Dalai[a], Nicolas Abatzoglou[b]

a- Catalysis & Chemical Reaction Engineering Laboratories, Department of Chemical Engineering., University of Saskatchewan, Saskatoon, SK, S7N5C5 Canada
b- Department of Chemical Engineering, Université de Sherbrooke, Sherbrooke, Qc, J1K

Abstract

The influence of cobalt particle size on catalyst performance in Fischer-Tropsch synthesis (FTS) has been investigated using inert carbon nanotube (CNT)-supported catalysts. The catalysts were produced by the core reverse micelle reactions with cobalt particles of various sizes (3 to 10 nm). It has been shown that particle size is proportional to the water-to-surfactant ratio (3 to 10) used for the catalyst preparation. Very narrow particle size distributions have been produced by the microemulsion technique and at relatively high loading (Co 10 wt %). Selectivity and activity were found to be dependant on cobalt particle size. The FTS rate increases from 0.36 to 0.44 gHC/gcat./h and the C_{5+} selectivity increases from 89 to 92.5 wt% with increasing the average cobalt particle size from 2-3 nm to 9-10nm, respectively. According to TEM analysis, small Co particles (2-6nm) are mostly confined inside the CNTs where influence of its electron deficiency in the inside surface has changed the commonly expected catalyst's structure-sensitive results. Finally, the CNT-supported cobalt nanoparticles synthesized by the proposed microemulsion technique increased the CO conversion by 15% compared to those prepared by incipient wetness impregnation.

Keywords: Microemulsion; cobalt catalysts; carbon nanotubes; Fischer-Tropsch synthesis; nanoparticles

1. Introduction

There is a renewed interest in Fischer-Tropsch synthesis (FTS) in both academia and industry, largely as a result of the demand for clean and renewable transportation [1]. In the Fischer-Tropsch (FT) reaction, Syngas (a mixture of CO and H_2), is converted into liquid fuel via catalytic surface polymerization which leads to a large variety of products such as paraffins, olefins, alcohols and aldehydes [1-3]. Supported cobalt catalysts are well-known for their activity and selectivity towards FTS. High chain growth probability, lower deactivation rates, low water-gas shift activity, and low costs make cobalt catalysts the best candidates for converting syngas to clean liquid fuels [4, 5]. Many investigations have been carried out to study the influence of catalyst properties on Fischer Tropsch synthesis, including the influence of catalyst preparation techniques (such as microemulsion), to have better understanding of the structure sensitive effects in FT catalysis [1, 6-7]. A sub-category of structure-sensitive reactions regards the dependence of both catalytic activity and selectivity on catalytic metal particle size [8, 10]. It is well documented that the metal particle size of the catalyst is a parameter of importance for the CO hydrogenation mechanism [1, 8, 10]. Microemulsion, a novel technique for catalyst preparation, enables the control of metal particle size with a narrow particle size distribution, regardless of metal content [6, 8, 9, 11]. Briefly, a microemulsion consists of nano-sized water droplets surrounded by an oil phase, stabilized by a surfactant [11]. The size of the cobalt particles formed in water-in-oil (w/o) microemulsions is controlled by changing the micelle size (the water-to-surfactant ratio) [11]. Hanaoka et al. and Hayashi et al. have shown that the catalysts prepared by water-in-oil microemulsion increases the CO hydrogenation rate, the H_2 chemisorption rate and the C_2^+ selectivity [7-12]. Also, Hayashi et al. found that the activity of a Fe/SiO_2 catalyst prepared by microemulsion was higher than that of the same average particle size catalyst prepared by incipient wetness impregnation [7]. However, it has been also shown that cobalt nanoparticles obtained by microemulsion techniques interact strongly with oxygen-carrying ceramic supports (such as Al_2O_3, SiO_2 and TiO_2). Such interactions lead to a decrease of the catalyst reduction efficiency [6,13,14]. Therefore, oxidic carriers impose serious limitations to the investigation of structure-sensitive effects in FT catalysis because of the co-existence of incompletely reduced cobalt phase, caused by strong interaction with the oxidic carrier [2, 6, 13, 15]. Specifically, nanoparticles synthesized by microemulsion are known to be more difficult to reduce on oxidic carriers. Although, Martinèz et al., has shown that with modifying the catalyst carrier (silylated ITQ-2), high reducibility can be obtained using microemulsion as the catalyst preparation route [6]. Moreover, our previous work has shown that carbon nanotubes (CNT), when used as a cobalt catalyst support, allow a better metal dispersion control and minimize the metal phase interaction (formation of mixed compounds) with the support [2, 16,17]. Chen et al. observed that the confinement of the Fe particles within the CNT enables a better reducibility and leads to higher rates of the CO dissociative adsorption on the metal surface [18]. Since the chemisorption of reactants is the rate determining step on FTS reaction, a cobalt particle located inside the tubes must be more active than one on the outer surface of the CNT [15-18]. Although, it is not clear until now, if this phenomenon is due only because of the metal-CNT walls particular interactions.

This present work compares the proposed microemulsion technique for catalyst preparation with the incipient wetness impregnation method for the control of cobalt metal particle size using CNTs as a catalyst carrier. The influences of cobalt particle size on the FTS Co/CNT catalysts on their activity and selectivity as well as the reducibility were evaluated and reported. A critical discussion allows drawing useful scientific and technical conclusions.

2. Experimental

2.1. Catalyst preparation

Prior to catalyst preparation, the Mknano-MWCNT (>95%) support was treated with 30 wt.% HNO_3 at 100°C overnight, washed with distilled water, and dried at 120°C for 6h. Cobalt particles were synthesized in a reverse microemulsion using a nonionic surfactant Triton X-100 (Aldrich) and cyclohexane (C_6H_{12}) as the oil phase. The concentration of cobalt was adjusted using aqueous cobalt nitrate ($Co(NO_3)_2.6H_2O$ 99.0%, (Merck) up to 10 wt.%. The water-to-surfactant molar ratio (W/S) was varied from 3 to 10. After vigorous stirring, a microemulsion was obtained (15 min). Hydrazine was added in excess (Hydrazine/Co = 10) to improve cobalt nanoparticle formation in the core of the micelles by reducing the cobalt oxide [6, 11, 19]. Then, the appropriate weight of purified carbon nanotubes was added under stirring. During the 3h of stirring, tetrahydrofurane (THF), an emulsion destabilizing agent, was added dropwise (1ml/min). A fast addition could lead to fast particle agglomeration and uncontrolled particle deposition on the support (CNT) [6, 11, 19]. The mixture was left to mature and settle slowly overnight and then decanted. The solid sample was recovered by vacuum filtration using ash less 90mm Ø x 100 circle filtration paper (Whatman ®) and washed several times with ethanol. In order to remove the remaining traces of surfactant and nitrates, the catalysts were calcined under argon (Ar) flow at 450°C for 3h and slowly exposed to an oxygen atmosphere during the cooling step. The catalysts prepared by means of this protocol were denoted as $MECo_a$ for a W/S ratio of 3, $MECo_b$ for a W/S ratio of 5 and $MECo_c$ for a W/S ratio of 10.The control catalyst was prepared using the incipient wetness impregnation method ($IWCo_d$). Cobalt nitrate ($Co(NO_3)_2.6H_2O$ 99.0%, Merck) solution was added to the treated CNT as the support up to 10 wt.% Co. After the impregnation step, the catalyst was dried at 120°C and calcined at 450°C under Ar flow for 3h and slowly exposed to an oxygen atmosphere during the cooling step. The cobalt loading of 10%.wt in all the calcined catalysts was verified by an Inductively Coupled Plasma Atomic Emission Spectroscopy (ICP-AES) system.

2.2. Catalysts characterization

The treated CNTs and the catalysts were characterized by Transmission Electron Microscopy (TEM), X-ray diffraction (XRD), temperature programmed reduction (TPR), H_2 chemisorption (TPD) and BET (measurement of surface area).

2.3. Reaction testing

The catalysts were evaluated in terms of their FTS activity (g HC produced/ g cat./ h) and selectivity (the percentage of the converted CO that appears as a hydrocarbon product) in a fixed bed micro-reactor (Fig.1). Catalyst activation was conducted first in-situ under pure hydrogen at a flow rate of 60 ml/min and a temperature of 380 °C for 20h. After the catalyst reduction period, the mixed gases (CO, H_2 and Ar) were fed at a flow of 30ml/min, a temperature of 220°C, a H_2/CO =2 and a pressure of 2 MPa. The products were continuously removed from the reactor and passed through two traps, one maintained at 100°C (hot trap) and the other at 0°C (cold trap). The uncondensed vapor stream was depressurized to atmospheric pressure through a back pressure regulator. The composition of the outlet gas stream was determined using an on-line GC-2014 Shimadzu gas chromatograph. The contents of hot and cold traps were removed every 24 h. The hydrocarbon and water fractions were separated, and then analyzed by Varian 3400 GC liquids analyzer.

3. Results and discussion

3.1. Catalysts characterization

The TEM images of the catalysts made by the proposed microemulsion preparation route are shown in Figure 1 a to e. The $MECo_a$ and $MECo_b$ catalyst particles are both dispersed mostly inside the tubes and on the outer surface of the CNT walls. For the $MECo_c$ (9-10 nm) catalyst, the percentage of the particles lying at the outer surface of the CNT walls is higher compared to the other catalysts prepared by microemulsion (Fig.1d). Indeed, the narrow inner diameter of the CNT channels (8-10 nm) restricted the insertion of particles in sizes close to the channel diameter (10 nm). The average particle sizes for $MECo_a$, $MECo_b$ and $MECo_c$ are linearly depending upon their respective water-to-surfactant ratio (3, 5 and 10). The most abundant particle sizes for the $MECo_a$ catalyst are within the range of 2-3 nm (Fig.1 a and b). $MECo_b$ and $MECo_c$ catalysts have Co particle sizes in the range of 5-6nm and 9-10nm, respectively (see Fig.1 c, d and e). Figures 1 f and g show the TEM images of the $IWCo_d$ catalyst made by the incipient wetness impregnation route. Similar to our previous works, the smaller cobalt particles are lying inside the CNT channels and the larger particles on the outside. The average particle size of the $IWCo_d$ catalyst is within the range of 6 to 10nm [16-17]. In this case, the cobalt particles are formed within the CNT during the impregnation and calcination steps. CNT channels have restricted the growth of the particles inside the tubes from 4 to 9nm. Almost all particles of sizes 10 nm and over are lying on the outer surface of the CNT walls.

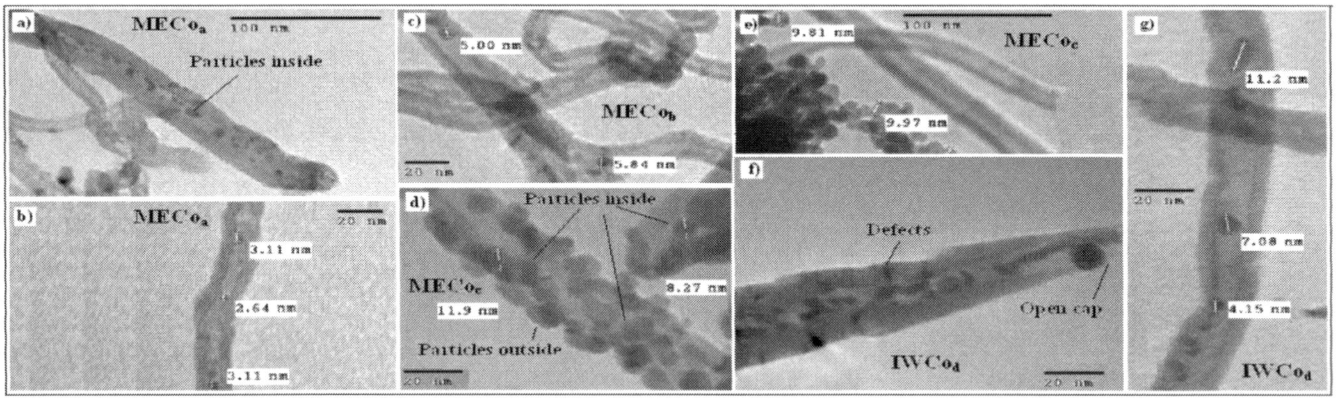

Figure 1: TEM pictures of the calcined catalysts showing (a) the particles inside the CNT for the $MECo_a$ catalyst and the particles size of the (b) $MECo_a$ catalyst (3nm), (c) $MECo_b$ catalyst (5nm), (d) the particle outside and inside the $MECo_c$ catalyst, (e) the particle size of the $MECo_c$ catalyst (10 nm), (f) the defects and the open cap onto the CNT after acid treatment and (g) the particles size of the $IWCo_d$ catalyst.

Figure 2 depicts the size distribution of the cobalt particles, which is determined using the population of the total cobalt particles of each $MECo_a$, $MECo_b$, $MECo_c$ and $IWCo_d$ catalysts based on data taken from 10 TEM pictures. This figure, as well as the TEM pictures, shows that microemulsion preparation route enables to control a narrow nanoparticle size distribution with different water to surfactant ratios. For example, Figure 4 shows that for the W/S ratio of 3, 5 and 10, the average particle sizes are 2-3, 5-6 and 8-10 nm, respectively. According to Figure 4, the average particle size for the catalyst prepared by incipient wetness impregnation method ($IWCo_d$) is 6-10nm and over. This results is in accordance with our previous work and Tavasoli et al., observation for 10 wt.% Co/CNT [15-17].

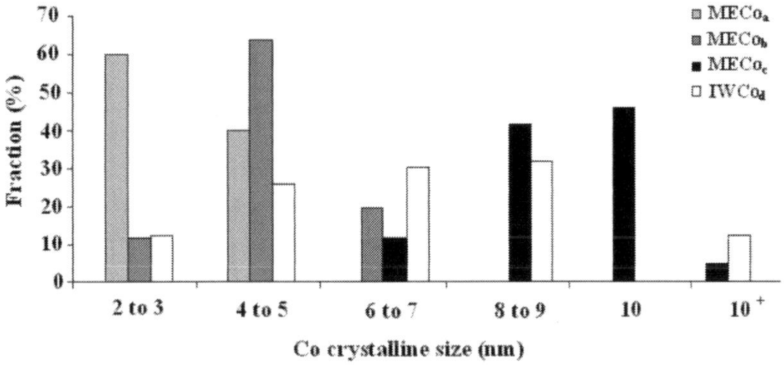

Figure 2: A bar graph depicting the particles size distributions for the calcined $MECo_a$, $MECo_b$, $MECo_c$, $IWCo_d$ catalysts.

Table 1 shows the average Co_3O_4 particle size of the catalysts calculated from XRD spectra using the Scherer equation at 2θ value of 36.8° [21]. The average Co_3O_4 cluster size was determined after calcinations for the $MECo_a$, $MECo_b$, $MECo_c$ and $IWCo_d$ as approximately 5.6, 7.1, 10.5 and 11.2 nm, corresponding to 4.2, 5.3, 7.6 and 8.5 nm when reduced to metal, respectively. This agrees reasonably well with the cobalt particle diameter obtained with the H_2 chemisorption results (Table 1). As shown in Table 1, the average particle sizes of Co_3O_4 are linearly correlated with the water-to-surfactant ratio used during the microemulsion catalyst preparation route; the latter is also confirmed by TEM pictures of the $MECo_a$, $MECo_b$, and $MECo_c$ (Figure 1).

Table 1: Selected catalysts properties

Catalysts	Co (wt. %)	Preparation route	BET (m^2/g)	Pore volume (cm^3/g)	Pore radius (nm)	XRD d (Co_3O_4) nm	Co^o dp (nm)
CNT	-	-	210	0.63	6.1	-	-
$MECo_a$	10	Microemulsion	185	0.54	5.5	5.6	3.6
$MECo_b$	10	Microemulsion	177	0.54	5.7	7.1	5.9
$MECo_c$	10	Microemulsion	173	0.53	6.6	10.5	7.8
$IWCo_d$	10	Incipient Wetness	191	0.60	6.8	11.2	7.9

The reducibility of the catalysts in H_2 atmosphere was determined by TPR experiments. The TPR spectra of the calcined $MECo_a$, $MECo_b$, $MECo_c$ and $IWCo_d$ are shown in Figure 3. The low temperature peak (300-400°C) is typically assigned to reduction of Co_3O_4 to CoO, although a fraction of the peak likely comprises the reduction of the larger, bulk-like CoO species to Co^o [22, 23]. The second broad peak is assigned to reduction of small CoO to Co^o species, which also includes the reduction of cobalt species that interact with the support. The small peak at about 600°C in the TPR spectra of the catalysts can be assigned to the gasification of support as indicated by TPR of pure CNT support at about 600°C (Fig.3).

According to Figure 3, the deposition of cobalt nanoparticles synthesized by microemulsion on the CNTs shift the reduction step of the Co_3O_4 to CoO species (1st peak) to a lower temperature compared to the catalyst impregnated by the common method; indicating higher reducibility for uniform particles ($MECo_c$) . Interestingly, the reduction temperature of the first peak also decreases with decreasing cobalt particle size from 10 to 3 nm. According to the TEM pictures of the catalysts prepared by microemulsion technique, most of the small particles (< 9nm) are inside the tubes, especially for the $MECo_a$ catalyst, which in turn leads to a better interaction with the intern electron deficient walls of the CNTs and favors the reduction of Co_3O_4 species [18]. The tubular morphology of the graphene layers make CNTs different compared to other carbonaceous supports. Thus, it is well known that the exterior surfaces of the CNTs are electron-rich, whereas the interior ones are electron-deficient, which could influence metal and metal oxide particles in contact with either surface [15,18]. Chen et al. studies reveal that deviation of the graphene layers from planarity causes π-electron density to shift from the concave inner surface to the convex outer surface, leading to an interior electron-deficient surface and an exterior electron-enriched surface [18]. Moreover, Chen et al., have particularly found that particles inside the carbon nanotubes are more catalytically active than particles on the outside surface [18].

Although, the second reduction peak for the $MECo_a$, $MECo_b$ and $MECo_c$ catalysts shifts to higher temperatures such as 475, 470 and 465°C compared to 452°C for the $IWCo_d$ catalyst . As expected in case of larger cobalt particle sizes (>10nm), the catalyst prepared by impregnation ($IWCo_d$) displays better reducibility for the second reduction step. Indeed, the degree of interaction with the metal phase and the support varies with the cobalt particles size, larger CoO particles are reduced more easily than smaller CoO particles [1, 14,17]. Figure 3 also indicates that by decreasing the cobalt particle size, the broad shoulder of the second TPR peak becomes larger, suggesting more difficult reduction process for CoO species with 2-3 nm cobalt particle.

According to Figure 3, there is no significant evidence of formation of metal-support compounds on the catalyst surface due to the absence of significant reduction peaks above 500°C. Thus, Tavasoli et al., as well as Martinez et al. have noted that reduction peak present at temperatures above 530°C with oxidic carrier shows formation of cobalt species that are difficult to reduce (oxide compounds). CNT as an inert support for cobalt catalyst do not show any peak related to formation of metal-support compounds as compared to Co/ γ-Al_2O_3 catalysts suggesting an easier reduction process with CNT than with oxidic carriers [2, 15]. Martinez et al., also observed that microemulsion preparation technique with a less oxidic carrier almost suppresses the formation of cobalt species which are normally difficult to reduce with an oxidic support [6] .Indeed, Ernst et al., have observed that cobalt catalyst particles (1-5nm) prepared by microemulsion route are usually more difficult to reduce with oxidic support carriers than those produced by incipient wetness impregnation [24].

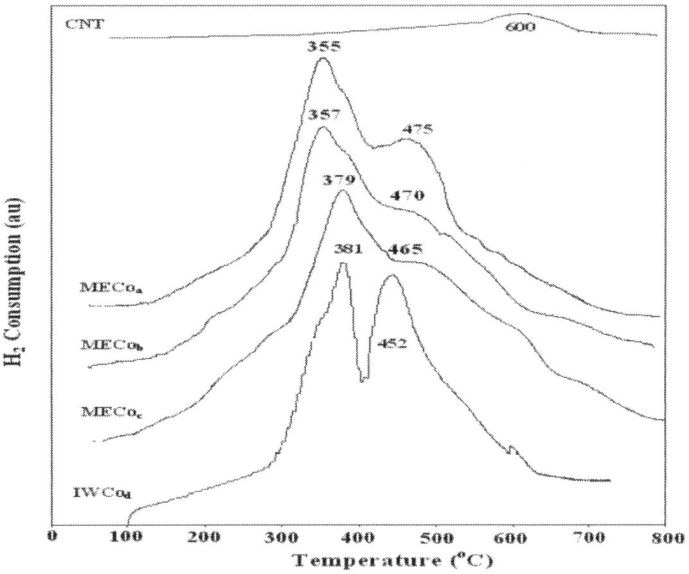

Figure 3: Temperature Programmed Reduction profiles for the calcined $MECo_a$, $MECo_b$, $MECo_c$ and $IWCo_d$ catalysts.

To conclude, uniformity of nanoparticles synthesized by core reverse micelle as well as particles confined inside the CNT improved the first reduction step of the catalysts, but increased the reduction temperature for the CoO to Co^o compounds. Thus, particles within the CNTs are easily reduced because of the confinement phenomenon for the first reduction step but the reducibility is still more influenced by the particle size for the second step of reduction. Therefore, the confinement of Co particles in the CNT channels allowed high catalyst reducibility which does not normally occurs with nanoparticle supported on other common carriers [6,18, 24, 25]. Consequently, no significant reducibility limitations were observed for any of the catalysts studied.

The H_2 uptake of the $MECo_c$ catalyst is higher than the $IWCo_d$ catalyst, inspite of similar Co particle size (7.8 and 7.9 nm). Martinez et al, have shown that narrow particle size distributions obtained by the microemulsion preparation route are more efficient for H_2 chemisorption than particle size distributions obtained with the impregnation method [6]. Moreover, Bezemer et al. observed that H_2 uptake is directly related to particle size for particles less than 10nm, a trend which levels-off for bigger particles. This can also explain the similarity of H_2 uptake between the $MECo_c$ and $IWCo_d$ catalysts [1].

3.2. Activity and product selectivity for FTS

The catalytic activity and product selectivity data have been calculated for runs showing catalyst stability within the first 24 h of operation and have been reproduced twice to confirm the results reproducibility. Figure 4 shows the influence of cobalt particles size on FTS rate (g HC produced/ g cat/ min), olefin/paraffin mass ratio, C_{5+} and methane (CH_4) selectivity. These preliminary results show that the FTS activity decreases as the size of the CNT-supported cobalt particles is decreased to < 10 nm. This is in accordance with Bezemer et al. observations showing lower activity for particles smaller than 10 nm [1,14]. Bezemer et al. found that the maximum concentration of surface Co^o sites and FTS activity are achieved with catalyst particle sizes less than 10 nm on carbon nanofiber supports [1]. In our case the optimum has not been evaluated due to the internal CNT diameter restriction and the particle confinement concern in our study. Indeed, Chen et al. observed that particle size is not the only property to have an effect on the FTS activity [18]. In our case, the particles are mostly inside the CNT walls for the $MECo_a$, $MECo_b$ and $IWCo_d$ (Figure 1). It appears that the particles located inside the CNT and the particle size itself influences the FTS activity [1, 14, 18]. However, the particle's confinement within CNTs may be more important than the particle size effect for the FTS activity. Figure 4 clearly shows that the CH_4 selectivity is reduced with increasing cobalt cluster size. Compared to $MECo_a$ and $MECo_b$ catalysts, the $MECo_c$ catalysts showed 1.2 % lower selectivity to methane. The systematic error of our test was ± 0.1% which indicates a significant change on the selectivity of methane. The lower production rate of CH_4 for the cobalt catalyst could be due to the effective participation of olefins in the carbon-carbon chain propagation. Moving upward with average particle sizes from 3 to 10 nm resulted in 4% improvement in the C_{5+} selectivity of the catalyst. Bezemer et al. reported that a minimum cobalt particle size of 8nm is necessary to give high C_{5+} selectivity [1]. According to Figure 4, at the range of 2-3 nm the C_{5+} selectivity is already 89 wt%. This result reveals that the dissociative adsorption of CO and chain growth probability (α–olefins re-adsorbtion) on the propagation step is efficient even with 2-3nm particles due to beneficial electronic properties of the internal CNT walls. The olefin/paraffin mass ratio decreases from 0.7 to 0.58 with increasing the particle size. Catalysts with high C_{5+} selectivity and cobalt particles smaller than 8nm usually show a substantial increase in the paraffin over olefin ratio [6]. It should be mentioned that the measurements for α-olefin/n-paraffin ratios were taken for C_2–C_8 hydrocarbons in this work. Therefore, the slight decreasing might be related to the lower amount of CH_4 for larger cobalt particles that can produce 1-alkene (olefins). 1-alkenes are regarded as primary products of the FTS propagation step and may be hydrogenated to alkanes or isomerized to 2-alkenes in the course of the reaction [26].

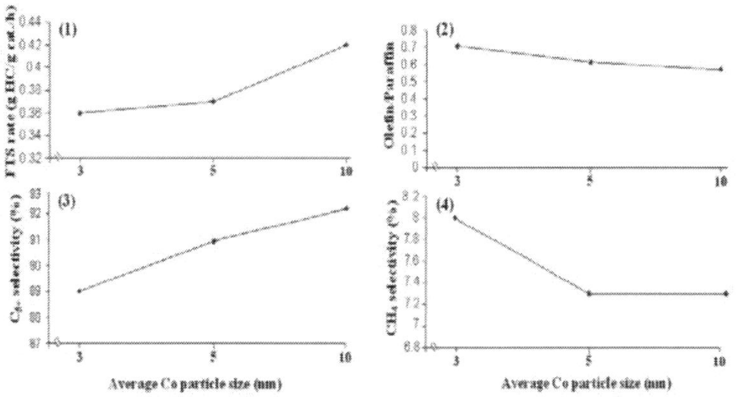

Figure 4: Influence of cobalt particles size on (1) FTS rate (gHC/gcat./h), (2) Olefin to paraffin ratio, (3) C_{5+} selectivity and (4) CH_4 selectivity.

The comparative C_{5+} selectivity, olefin/paraffin ratio and CH_4 selectivity for the $MECo_c$ and $IWCo_d$ catalysts have been studied. The catalyst prepared by the reverse micelle synthesis has higher selectivity for C_{5+} products than that of the $IWCo_d$ catalysts. The same results have been observed by Kim el al. with a zirconia supported palladium catalyst. The uniformity of the particle size improved the hydrogenation of carbon monoxide, which in turn increased the selectivity for C_{5+} products. Although, the catalyst prepared by the conventional impregnation method showed better selectivity for heavier hydrocarbon molecules (C_{10+}), which can be explained by the presence of larger cobalt particle sizes (10^+nm) than the catalyst prepared by the microemulsion route. Large particles (10^+nm) have high potential of re-adsorption and polymer chain initiation on the catalyst's surface [26]. Interestingly, the CH_4 selectivity is lower for the $MECo_c$. The decrease in the selectivity of CH_4 can be explained by the effective participation of olefins in the carbon-carbon chain propagation for the uniform cobalt clusters produced by microemulsion technique. The olefin/paraffin ratio is higher for the $MECo_c$ catalyst. On $MECo_c$ catalyst, α–olefins of the type R-CH=CH_2 can compete with carbon monoxide and heavier olefins for re-adsorption and chain initiation up to chain lengths of C_{11}. The higher H_2 chemisorption on the $MECo_c$ catalyst means higher probability of CHx + H reactions which consequently leads to a higher termination rate after C_{11} hydrocarbon chains. The shift towards olefin production could also indicate lower hydrogenation activities for particles outside the CNT tubes (larger particles) than the particles inside the tubes. As shown in TEM pictures, the $MECo_c$ catalyst has fewer particles inside the tube than $IWCo_d$ catalyst.

The study of FTS activity and selectivity reveals that catalysts prepared through the microemulsion synthesis technique show a higher FTS rate than those prepared by the incipient impregnation preparation route. %CO conversion increased from 54% in the case of $IWCo_d$ catalyst to 62% for the $MECo_c$ catalysts. $IWCo_d$ catalyst is characterized by different particle size distribution compared to $MECo_c$ catalyst showing a narrow particle size distribution. The cobalt sites produced by microemulsion are more stable by their uniformity than the ones produced by incipient wetness impregnation [7].Improvement of the uniformity of the catalyst particles leads to a better stability of the products and the FTS activity. These results reveal that FT activity of the catalysts is strongly dependent on the size distribution of the cobalt cluster.

4. Conclusions

This research has been carried out using inert carbon nanotubes (CNT)-supported cobalt catalysts to compare the effects of microemulsion technique for catalyst preparation with that of the incipient wetness impregnation method. Cobalt nanoparticles produced at relatively high loadings of 10 wt% with the microemulsion technique show a narrow particle size distribution. It was also found that FT activity and selectivity of the catalysts are dependent upon the size distribution of the cobalt cluster. According to TEM analysis, small Co particles (2-6nm) are mostly confined inside the CNTs where influence of its electron deficiency in the inside surface has changed the commonly expected results. Thus, CNTs as a catalyst carrier-support with Co nanoparticles maintained high reducibility of Co which will not normally occur with small particles supported on oxidic catalysts. The proposed microemulsion technique also increased the CO conversion by 15% compared to those prepared by incipient wetness impregnation. This new catalyst preparation method may offer an attractive alternative for nanoparticles synthesis by reverse microemulsion and for fundamental catalytic studies especially such for structure-sensitive FT catalysis.

5. Acknowledgements

The authors are indebted to Natural Science Engineering Research Council (NSERC) of Canada, Natural Resources Canada (CANMET) and Enerkem Technologies Inc. for funding this project.

6. References

[1] G.Leendert Bezemer, J.H.Bitter, H.P.C.E. Kuipers, H.Oosterbeek, J.E.Holewijin, X.Xu, F.Kapteijn, A.Jos van Dillen and K.P. de Jong, J.Am.Chem.Soc 128 (2006) 3956.
[2] A. Tavasoli, R.M. Malek Abbaslou, M. Trépanier and A.K. Dalai, Appl.Catal.A 345 (2008) 134.
[3] J.Li, G.Jacobs, Y.Zhang, T.Das, B.H.Davis, Appl..Catal.A 223 (2002) 195.
[4] M.E. Dry, J.Chem.Tech. and Biotec. 77 (2001) 43.
[5] G.Jacobs, T.K.Das, Y.Zhang, J.Li, G.Racoillet and B. H. Davis, Appl.Catal.A 233 (2002) 263.
[6] A.Martinèz, G.Prieto, J.Catal.245 (2007) 470.
[7] H.Hayashi, L.Zhe Chen, T.Tago, M.Kishiba, K. Wakabayashi, Appl.Catal. A 231 (2002) 81.
[8] W-Y.Kim, T.Haoka, M.Kishida, K. Wakabayashi, Appl.Catal. A 155 (1997) 283.
[9] T.Tago, T.Hanaoka, P.Dhupatemiya, H. Hayashi, M.Kishida and K.Wakabayashi, Catal.Lett. 64 (2000) 27.
[10] M.Ojeda, S.Rojas, m.boutonnet, F.J.Pérez-Alonso, F. Javier Garcia-Garcia and J.L.G.Fierro, Appl.Catal.A 274 (2004) 33.
[11] S. Eriksson, U. Nylén, S.Rojas and M.Boutonnet, Appl.Catal. A 265 (2004) 207.
[12] T.Hanaoka, W-Y.Kim, M.Kishida, H.Nagata and K. Wakabayashi, Chem.Lett. 17 (1997) 645.
[13] A.M.Saib, M,Claeys, E. van Steen, Catal.Today 71 (2002) 395.
[14] G.L Bezemer, P.B. Radstake, V.Koot, A.J. van Dillen, J.W.Geus and K.P de Jong, J.Catal 237 (2006) 291.

[15] A. Tavasoli, K. Sadagiani, F. Khorashe, A.A. Seifkordi, A.A. Rohani A. Nakhaeipour, Fuel.Process.Tech. 9 (2008) 491.
[16] M. Trépanier, A.Tavasoli, A.K. Dalai, N. Abatzoglou, Fuel Procc.Tech. 90 (2009) 367.
[17] M. Trépanier, A.Tavasoli, A.K.Dalai, N. Abatzoglou, Appl.Catal.A 353 (2009) 193.
[18] W. Chen, Z. Fan, X. Pan and X. Bao , J.Am.Chem.Soc 130 (2008) 9414.
[19] A. Dausher, R. Touroude, G.Maire, J.Kizling and M. Boutonnet-Kizling, J.Catal 143 (1993) 155.
[20] G. L. Bezemer, A. van laak, A. J. van Dillen, K. P. de Jong, Stud.Surf.Sci.Catal. 147 (2004) 259.
[21] R. Bechera, D.Balloy and D.Vanhove, Appl.Catal..A 207 (2001) 343.
[22] G.P.Huffman, N.Shah, J.Zhao, F.E.Huggins, T.E.Hoost, S.Halvorsen and J.G.Goodwin, J.Catal. 151 (1994) 17.
[23] H.-Y. Lin, Y.-W. Chen / Materials Chemistry and Physics 85 (2004) 171–175
[24] B.Ernst, S.Libs, P.Chaumette, A.Kiennermann, Appl.Catal.A 186 (1999)145.
[25] A.Martinèz, C.Lopèz, F.Marquèz, I.Diaz, J.Catal.220 (2003) 486.
[26] R.J. Madon, E. Iglesia, J.Catal. 139 (1993) 576–590.
[27] A. Tavasoli, Y. Mortazavi, A. Khodadadi and A. Karimi, I. J. Chem. Chem. Eng. 36 (2005) 25.

CHARACTERIZING THE INFLUENCE OF Mn AND K PROMOTERS ON THE CARBURIZATION RATES AND PERFORMANCE PARAMETERS OF Fe-BASED CATALYSTS BY TPR-EXAFS/TPR-XANES AND CATALYTIC TESTING

Mauro C. Ribeiro[1], Gary Jacobs[1], Ramana Pendyala[1], Burtron H. Davis[1], Donald C. Cronauer[2], A. Jeremy Kropf[2], and Christopher L. Marshall[2], (1) Center for Applied Energy Research, University of Kentucky, Lexington, KY, (2) Chemical Sciences and Engineering Division, Argonne National Laboratory, Argonne, IL.

Introduction

Fischer Tropsch synthesis (FTS) is the Fe or Co catalyzed hydrogenation of carbon monoxide to yield a mixture of hydrocarbons and oxygenates with molecular weights ranging from light gases (C_1-C_4) to heavy waxes (C_{30}+). Small quantities of Cu, Si and K are normally used as promoters in Fe catalysts to enhance, respectively, its surface area, stability, as well as the selectivity for high molecular weight product. Among the alternative promoters studied, Mn has also been studied, particularly after Kölbel et al. claimed in a patent [1] that higher selectivities for light products might be obtained on FeMn catalysts (promoted or not with Cu and K) catalysts. According to the authors, even if Fe is the minority component (the authors mention compositions with at least 50% wt. Mn in their catalysts) high activities with high selectivities for the light products (C_2-C_4) are obtained with low methane selectivity. Many authors have investigated this catalyst, but questions about why Mn would: (a) increase selectivity for low molecular weight hydrocarbons; or (b) why Mn, being present as an oxide, would not cause activity loss in the catalyst in the same manner as if Al or Si were present in high concentrations in the catalyst, remain to be answered. In order to investigate the effect of Mn in the structure/activity of FeMn catalysts, the carburization process of co-precipitated Fe catalysts containing different amounts of Mn, was studied by *in-situ* XAS from both Fe and Mn K-edges. Catalysts containing low (Fe:Mn 100:30) and high Mn (Fe:Mn 100:90) atomic ratios were tested for FTS and the results were compared to runs made with unpromoted Fe catalysts. It was found that Mn, present in the carburized catalyst basically in a MnO phase, decreases the amount of iron carbides in the catalyst carburized for 3 h at 300°C. However, the general activities (measured in CO conversion) are unchanged or even higher when Mn is present, even in high content. As far as the selectivities are concerned, higher methane and water gas shift activities were observed with Mn promoted catalysts. Differences in both catalyst activity and selectivity will be correlated with structural information.

Experimental

A series of FeSi based catalysts containing different amounts of Mn was prepared by coprecipitation with ammonia solution at (80±5) °C. After calcination at 350 °C for 5 h, the catalyst precursor samples were tested for FTS in a continuously stirred tank reactor, described elsewhere [2]. After *in-situ* activation under CO flow at 270 °C, 175 psi for 24 h, the catalyst samples were tested under syngas (H_2:CO = 0.7) at the same pressure and temperature.

The carburization of FeMnSi catalysts under CO/He flow was investigated by XAS at both the Fe and Mn K-edges (7.112 keV and 6.54 keV, respectively). The temperature was increased for 7 h from room temperature to 300°C and then held constant at that temperature for 3 h. Successive X-ray absorption spectra were recorded during sample carburization. Spectra were background subtracted and normalized using the WinXAS [3] software.

Results

XAS results

Figure 1 shows the iron carbide content in the catalyst samples as a function of the carburization time at 300 °C, calculated by linear combination of Fe K-edge XANES spectra. All the samples tend to the same carburization level in the end of the carburization treatment. However, the samples containing only K (100Fe:5.1Si:3K, atomic proportions) and a small amount of Mn (100Fe:5.1Si:5Mn:3K) reached higher carburization extents at shorter times than the unpromoted (100Fe:5.1Si) or any one of the heavily Mn promoted samples (100Fe:5.1Si:5Mn or 100Fe:5.1Si:5Mn:3K). In fact, the presence of Mn causes a slight hindering in the rate of the carburization process, when K is or is not present in the sample (e.g., compare samples 100Fe:5.1 and 100Fe:5.1Si:30Mn). From this figure, it can be concluded that, unlike K, Mn, especially in higher fractions in the catalyst, hinders Fe carburization.

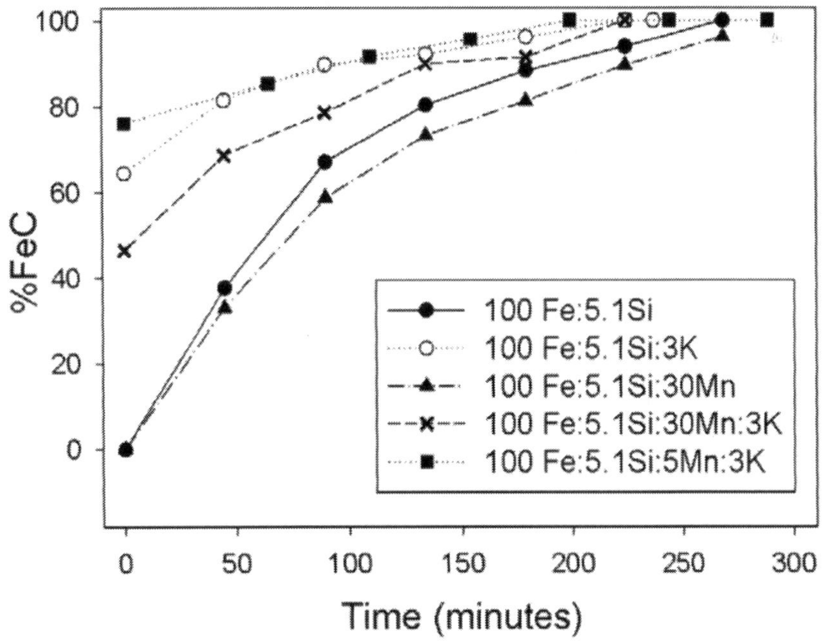

Figure 1. Concentration of iron carbides calculated from XANES spectra.

Figure 2 displays the TPR-XANES results from the FeMn samples. Regardless of the Mn content in the sample, Mn oxide is reduced to MnO.

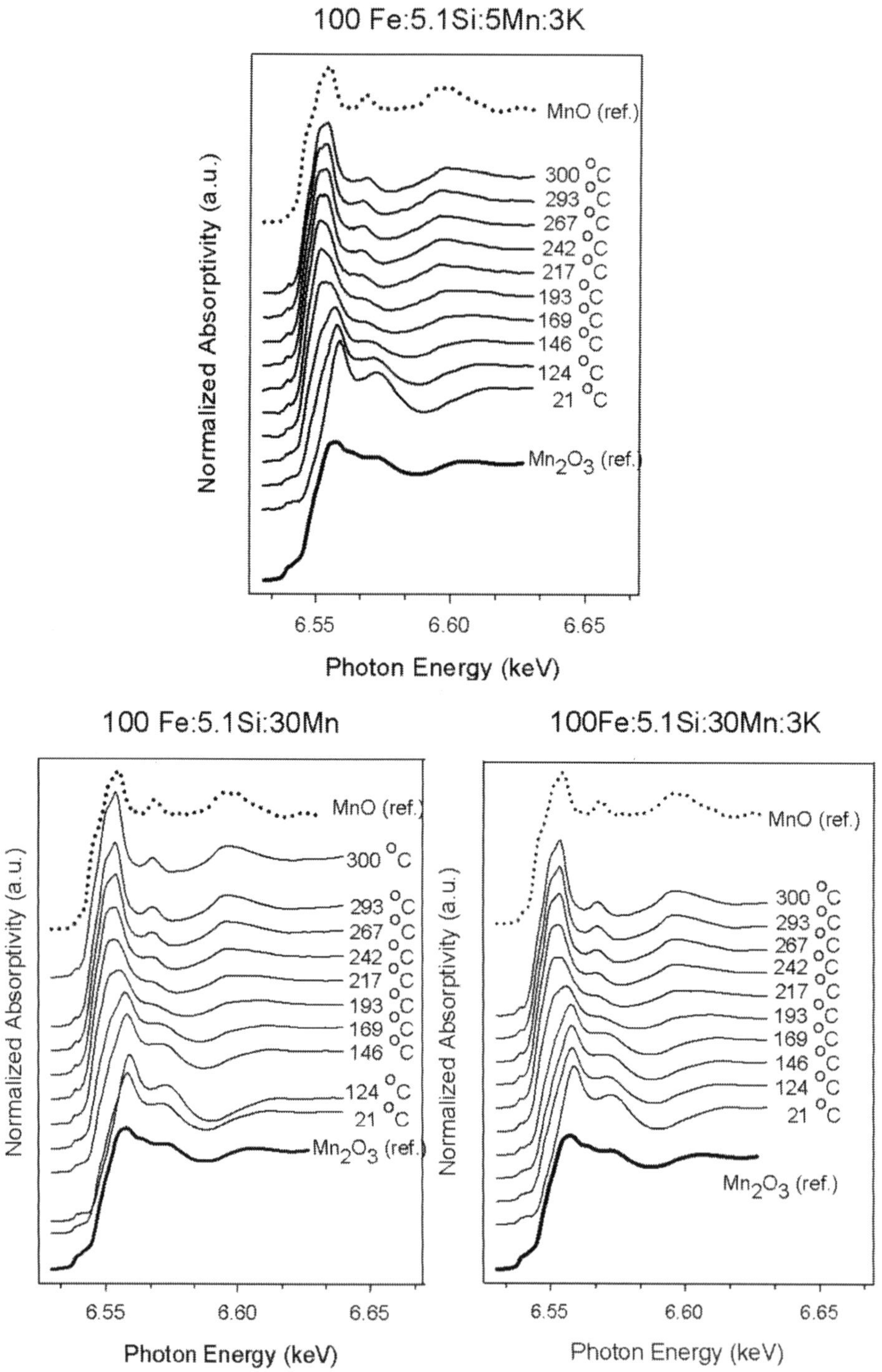

Figure 2. TPR-XANES of Mn K-edge from the FeMn samples.

FT testing

Figure 3 depicts the results of reactor testing. The baseline comparison is made with a catalyst containing a small quantity of Cu. As can be seen, all the catalysts present comparable CO conversion levels (figure 3a). Cu is known to enhance iron reducibility and therefore, to enhance the active surface area, with no significant modification in the selectivity. This fact, along with the fact that the Mn containing catalysts were tested at a considerably higher space velocity, leads to the conclusion that the Mn containing catalysts were highly active. After 150 h, the catalyst 100Fe:5.1Si:30Mn:3K deactivated quickly, unlike the catalyst 100Fe:5.1Si:90Mn:3K, which kept a steady CO conversion level during the entire test. The higher activity of the Mn containing catalyst can be explained by considering that MnO stabilized smaller Fe particles and prevented its sintering, in a comparable manner to SiO_2

Comparisons between the CH_4, CO_2 selectivities, as well as olefin fractions in the light (C_2-C_4) products can be made, as the conversion levels were comparable at testing times below 150 h. From this comparison, it can be seen that a) the presence of Mn increases the CH_4 and CO_2 selectivity (WGS activity). Likewise, the olefin fraction is decreased in the presence of Mn in the catalyst. As it could be seen in our XANES study, the presence of MnO itself leads to a decrease in the carburization rate. As the bulk carburization is preceded by CO dissociative adsorption on the Fe surface, MnO might influence the ability to adsorb/dissociate CO on the surface. If this is so, a smaller surface coverage of carbon species, necessary for the formation of larger molecular weight product, could be responsible for the higher CH_4 and lower chain probabilities (figure 4) obtained at high Mn content.

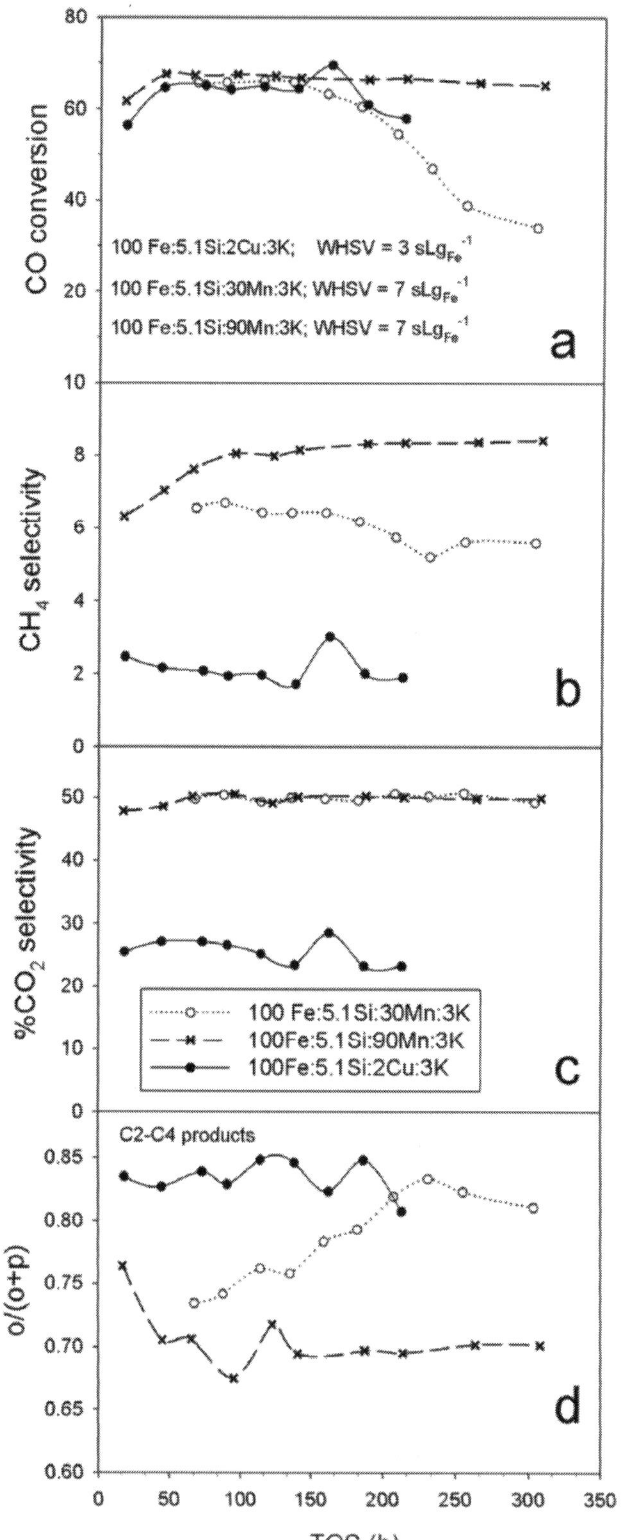

Figure 3. CO conversion (a), CH_4 and CO_2 selectivities (b,c); olefin fraction (d) in function of time on stream.

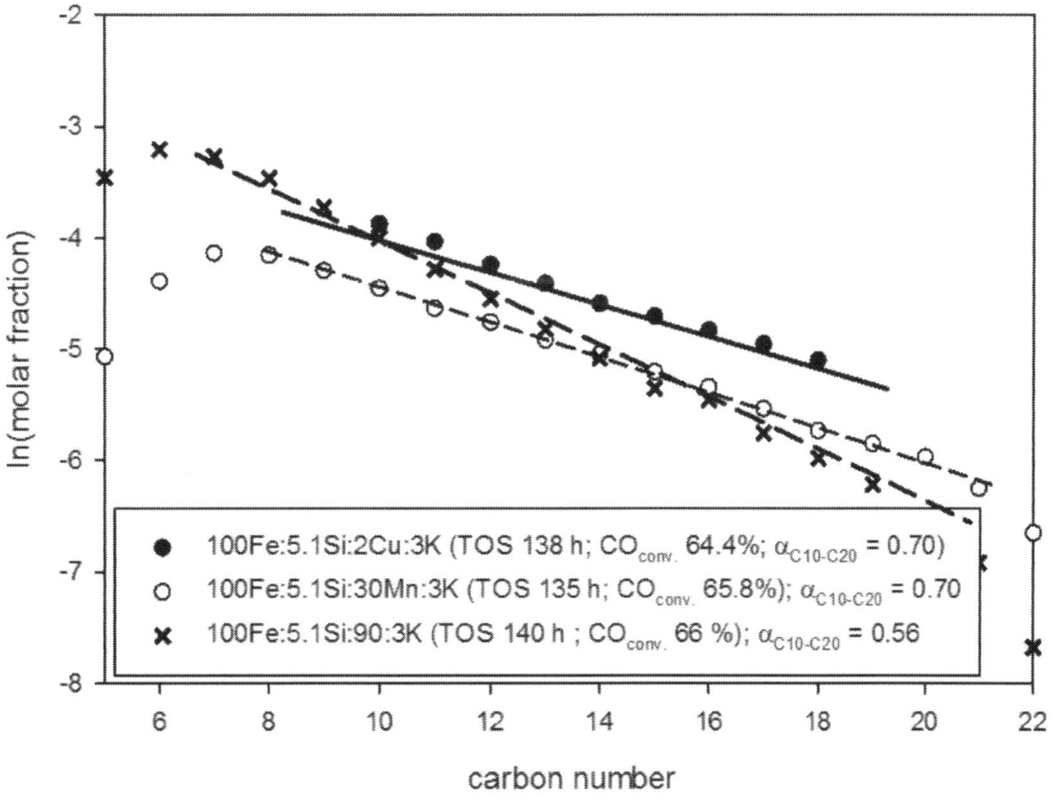

Figure 4. Alpha plots (C_{10}-C_{20}) from the liquid products obtained in the tests.

4. Conclusions

Mn, present in the carburized catalyst as MnO, increases the catalyst activity, yet it also increases CH_4 and WGS selectivities as much as it lowers chain growth probability. This trend in the selectivities may be explained by the proposal that the presence of MnO decreases the Fe carburization rate, which means a decrease in the surface coverage of carbon active species, necessary to the formation of long chain products. As for the increase in the activity, it is probably because MnO stabilizes smaller Fe particles, which leads to an increase in the Fe surface area.

References

1. Kolbel; Herbert, Tillmetz, Klaus D., US Patent (1979), US 19791204.

2.Dalai, A., Das, T., Chaudhari, K., Jacobs, G., Davis, B., *Applied Catalysis*, A 289 (2005) 135.

3. T. Ressler, *J. Synch. Rad.* **5** (1998) 118.

Effect of isomorphic substitution of zirconium on mesoporous silica as support for Cobalt Fischer-Tropsch synthesis catalysts

Congliang Tao[a], Jinlin Li[a]*, Yuhua Zhang[a], Kong Yong Liew[b]

[a] *Key Laboratory of Catalysis and Materials Science of the State Ethnic Affairs Commission & Ministry of Education, South-central University for Nationalities, Wuhan 430074, China*

[b] *Faculty of Industrial Science and Technology, University Malaysia Pahang, Gambang 26300, Kuantan, Pahang, Malaysia*

Abstract

Isomorphic substituted mesoporous SBA-15 with zirconium at Zr/Si atomic ratios of 1/100, 1/50, 1/20, 1/8 and 1/3 have been successfully synthesized. It was demonstrated that the Zr^{4+} ions were introduced into the framework of SBA-15 with the hexagonal array of uniform tubular channel configuration intact until Zr/Si ratio of 1/20. Another sample of zirconium incipient wetness impregnated SBA-15 with Zr/Si ratio of 1/20 was prepared for comparison. These mesoporous materials were loaded with 10 wt % cobalt by incipient wetness impregnation and characterized by XRD, TEM, N_2 absorption–desorption, UV-visible, H_2-TPR, H_2-TPD, O_2 titration as well as XPS. It was shown that introduction of zirconium increased the density of surface cobalt, improved the cobalt reducibility and generated more available active sites with low zirconium content (Zr/Si atomic ratio < 1/20). At high zirconium content, strong interaction between the cobalt species and the surface of the support was formed, to decrease cobalt reducibility detected by XPS and H_2-TPR. The catalytic performances of catalysts for FTS were tested in a fixed bed reactor, and the isomorphic substituted mesoporous SBA-15 with zirconium at Zr/Si ratio of 1/20 supported cobalt catalyst showed the best performance.

Keywords: Fischer–Tropsch synthesis; Cobalt catalyst; SBA-15; Zirconium; Isomorphic substitution

1. Introduction

Fischer–Tropsch synthesis (FTS) is attractive for production of high-quality liquid fuels from alternative sources, such as natural gas, coal and biomass via syngas conversion over cobalt or iron based catalysts [1-2]. The cobalt based catalysts have high activity and high selectivity to linear hydrocarbons and low activity for the water gas shift (WGS) reaction [3]. The catalytic activity depends mainly on the density of surface metallic cobalt, which is correlated to its dispersion and reducibility, which are significantly influenced by the texture and surface chemical properties of the supports [4–6].

SBA-15 is a silica-based mesoporous material with uniform hexagonal channels of 3 to 30 nm and has very narrow pore size distribution [7]. It is an attractive catalyst support with large surface area allowing the dispersion of a large number of catalytically active species. However, SBA-15, as a pure silica material, has low activity. Thus, incorporating SBA-15 with zirconium [8], titanium [9], aluminum [10], vanadium [11] has attracted much attention. For FTS research, the performance of SBA-15 supported cobalt catalysts has been reported by many researchers [12-15]. However, there are few reports on incorporation of metal on SBA-15 supported cobalt catalysts for Fischer-Tropsch synthesis.

Zirconia is a transition metal oxide with acid and alkali sites as well as redox ability. As a promoter, it enhances the activity and C_{5+} selectivity for FTS reaction [16-18]. Recently, mesoporous materials, such as Zr-MCM-41 and Zr-FSM-16 [19-20], incorporating zirconium have received considerable attention in FTS catalysis because of their high surface area and porosity. As supports for cobalt catalysts, these zirconium incorporated mesoporous materials have been shown to have good catalytic performance for FTS. Liu et al. [21] reported the incorporation of zirconium in SBA-15 using various methods, including wet impregnation, coprecipitation, isomorphic substitution and grafting. One of the interesting conclusions was that the zirconium incorporated SBA-15 prepared by isomorphic substitution as support for cobalt catalysts showed higher activity than the one prepared by incipient wetness impregnation, but lower activity than that prepared by grafting at the same cobalt loading. However, the effect of zirconium content and the interaction of metallic cobalt and the support were not elucidated. In the present work, a series of Co/SBA-15 catalysts with isomorphic substituted zirconium were synthesized, and another sample of Co/SBA-15 catalyst with zirconium impregnated by incipient wetness with Zr/Si ratio of 1/20 was prepared for comparison. These materials were characterized by XRD, TEM, N_2 absorption–desorption, UV-visible, H_2-TPR, H_2-TPD, O_2 titration as well as XPS, and their catalytic performances of catalysts for FTS were tested in a fixed bed reactor. The aim is to investigate systematically the effect of different zirconium contents on the structure of the mesoporous silica supports, the activity and selectivity of the cobalt catalysts. The incorporation of zirconium in SBA-15 by the two methods (isomorphic substitution and incipient wetness impregnation) at Zr/Si ratio of 1/20 of the supported cobalt catalysts were investigated in detail.

2. Experimental

2.1. Sample preparation

2.1.1. Materials

Tetraethyl orthosilicate (TEOS) and zirconium nitrate ($Zr(NO_3)_4 \cdot 5H_2O$) were used as silicon and zirconium sources respectively. Triblock poly (ethylene oxide) - poly (propylene oxide) – poly (ethylene oxide) (P123, EO_{20}-PO_{70}-EO_{20}, $M_{AV} = 5800$, Aldrich) was used as template.

2.1.2. Supports

Zirconium isomorphic substituted mesoporous SBA-15 were prepared following the procedure reported previously [8, 22]. The samples were labeled as ZS(x)-s with x standing for the Zr/Si atomic ratio. For comparison, another sample was prepared by incipient wetness impregnation of SBA-15 using an aqueous solution containing the desired amount of zirconium nitrate with Zr/Si atomic ratio of 1/20 labeled as ZS(1/20)-i.

SBA-15 and ZrO_2 were prepared following the reported procedures [7, 23]. The compositions of the supports are shown in Table 1.

Table 1 Supports composition

Sample	Zr/Si [a]	Zr [b] (wt %)	Zr [c] (wt %)
SBA-15	0	0	0
ZS(1/100)-s	1/100	1.41	1.27
ZS(1/50)-s	1/50	3.12	2.68
ZS(1/20)-s	1/20	7.07	6.43
ZS(1/20)-i	1/20	7.07	6.28
ZS(1/8)-s	1/8	14.47	14.09
ZS(1/3)-s	1/3	30.31	29.39
ZrO_2	-	-	-

[a] Atomic ratio in starting mixture
[b] Zr (wt %) in starting mixture
[c] Zr (wt %) obtained by ICP measurement

2.1.3. Catalysts

The catalysts of 10Co/SBA-15, 10Co/ZS(x)-s, 10Co/ZS(1/20)-i and 10Co/ZrO_2 with cobalt loading (10 wt %) were prepared by incipient wetness impregnation with the desired amount of aqueous cobalt nitrate ($Co(NO_3)_2{\cdot}6H_2O$). They were dried at 120 °C for 12 h and calcined at 450 °C in flowing air for 5 h.

2.2 Characterization

ICP element analysis; XRD; TEM; (DR) UV–vis [25]; N_2 absorption–desorption; XPS; H_2-TPR; H_2-TPD and O_2 titration[26].

2.3. Catalytic evaluation

Fischer–Tropsch synthesis was performed in a tubular fixed bed reactor (id = 12 mm). The catalyst (0.5 g) was mixed with 5 g carborundum and then reduced in H_2 (space velocity of 6 NL h^{-1} g^{-1}) at atmospheric pressure. Reaction conditions: 1.0MPa, 230 °C, CO/H_2=1:2, space velocity of syngas 8SLg^{-1} h^{-1}. The outlet gases were analyzed online by an Agilent 3000 GC, the oil (collected at −2 °C) was analyzed by an Agilent 6890 GC and the solid wax (collected at 130 °C) was analyzed by an Agilent 7890 GC.

3. Results and discussion

3.1. Characterization

3.1.1. X-ray powder diffraction.

The powder XRD patterns of the supports within the range of 0.5-5° are shown in Fig.1. When Zr/Si atomic ratio is lower than 1/20, three well-resolved characteristic diffraction peaks at 2θ values ranged from 0.9° to 2° can be clearly observed. These are indexed to the (100), (110) and (200) reflections of the hexagonal P6mm space group, similar to the SBA-15 reported by Zhao et al. [7]. It indicates that the hexagonal structure of SBA-15 had been preserved during the synthesis and subsequent calcinations of ZS(x)-s. When Zr/Si atomic ratio is higher than 1/20, the intensity of the peak near 2θ=0.9° decreases with increasing atomic ratio and the peaks indexed to (110) and (200) diffraction disappears, while the peak indexed to (100) remains clearly observable, indicating gradual lowering of the long rang order at higher content of zirconium. Comparing ZS(1/20)-s and ZS(1/20)-i, at the same content of zirconium, three characteristic peaks of SBA-15 materials can be observed from the former, while for the latter, two characteristic peaks of SBA-15 materials disappear, showing that the hexagonal structure of SBA-15 had been better preserved when the sample was prepared by isomorphic substitution.

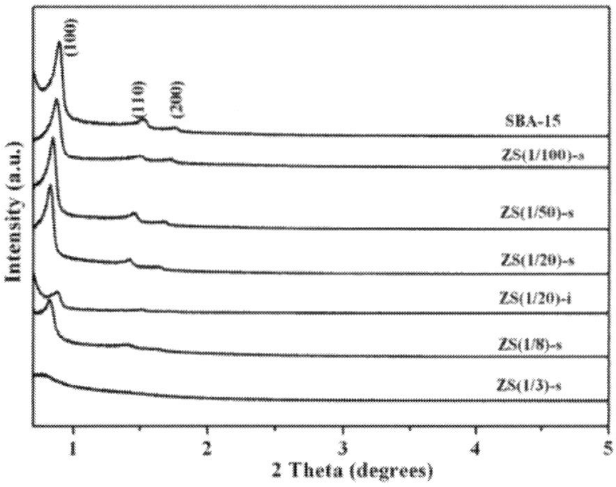

Fig.1. Small-angle XRD patterns of the supports

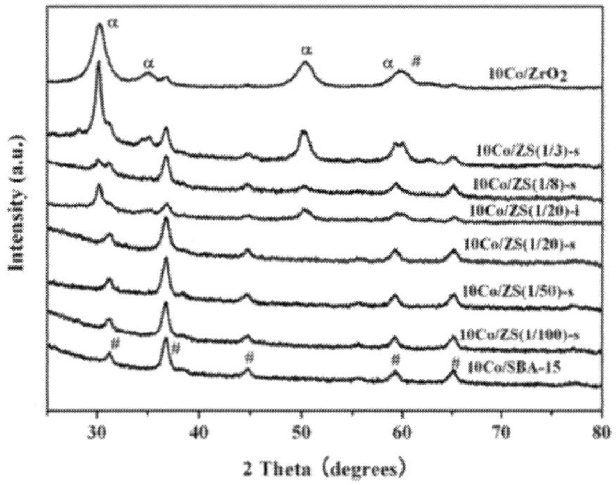

Fig.2. Wide-angle XRD patterns of the catalysts. (α: ZrO_2, #: Co_3O_4)

For catalysts with Zr/Si atomic ratio lower than 1/20, the diffraction peaks at 2θ of 31.3°, 36.9°, 45.1°, 59.4° and 65.4° (Fig. 2) indicate that cobalt species were primarily in the form of spinel Co_3O_4 on the catalysts after calcinations at 450 °C. The characteristic peaks of zirconium compounds are not observed, showing that nanoscale zirconium was well dispersed in the SBA-15 matrix. When the Zr/Si atomic ratio is higher than 1/20, the characteristic peaks (peak α) belonging to the tetragonal ZrO_2 phase are observed and the peak intensity of the tetragonal ZrO_2 phase increases with increasing Zr/Si ratio, showing that isolated particles of tetragonal ZrO_2 appeared on the surface of the catalysts. Comparing ZS(1/20)-s, for sample prepared by isomorphic substitution, and ZS(1/20)-i, for sample prepared by impregnation, isolated particles of tetragonal ZrO_2 phase appears in the latter, indicating that zirconium was not well dispersed in the structure of SBA-15 when prepared by impregnation.

3.1.2. Transmission electron microscopy

The transmission electron micrographs of the samples ZS (1/50)-s and ZS (1/8)-s are shown in Fig. 3. It can be observed that the sample ZS (1/50)-s still retains a regular hexagonal array of uniform channels, characteristic of mesoporous SBA-15 materials, while the meso-structure of sample ZS (1/8)-s is partially destroyed. Moreover, no particle of pure ZrO_2 can be observed in the nano-channels by TEM-mapping.

Fig. 3.TEM images of the supports: (a) ZS (1/50)-s; (b) ZS (1/8)-s.

3.1.3. Diffuse reflectance (DR) UV–vis spectroscopy

Diffuse reflectance (DR) UV–vis spectroscopy is a useful technique for studying the local coordination environment and electronic state of isolated transition metal ions or aggregated transition metal oxides [27]. It has been reported that characteristic absorption peaks in the 205 - 215 nm range were observed on zirconium modified mesoporous molecular sieves due to $O^{2-} \rightarrow Zr^{4+}$ of the p → d electronic transition in the framework of mesoporous molecular sieves [28-29]. Fig. 4 shows DR UV–vis spectra of the samples along with those of pure SBA-15 and ZrO_2. All of the samples containing zirconium exhibit a band around 210 nm attributable to $O^{2-} \rightarrow Zr^{4+}$ charge-transfer transition except ZS(1/20)-i and ZS(1/3)-s. These absorption peaks are obviously different from that of the pure ZrO_2 which has a characteristic absorption peak at 230 nm, showing that the Zr^{4+} ions had entered into the framework of SBA-15. The samples of ZS(1/20)-i and ZS(1/3)-s show a single band at about 230 nm, attributed to the characteristic absorption peak of the pure ZrO_2 sample, indicating that some Zr^{4+} ions isolated from the framework of the SBA-15 and the particles of tetragonal ZrO_2 phase appeared on the surface, revealed similarly by wide-angle XRD.

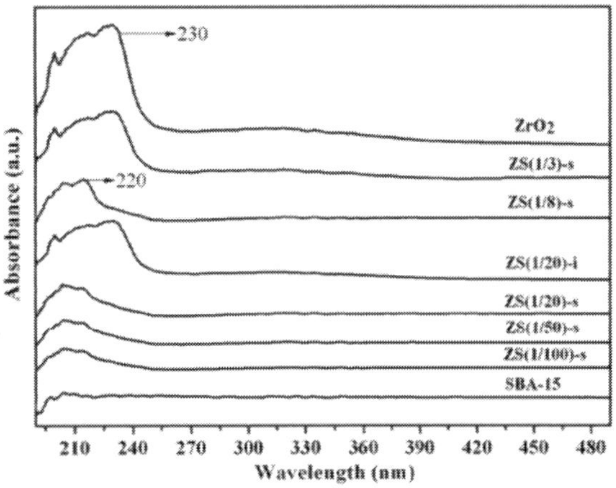

Fig.4. Diffuse reflectance UV–vis spectra of the supports

3.1.4. Nitrogen absorption-desorption

The nitrogen absorption-desorption isotherms of the catalysts are shown in Fig. 5. For the catalysts, except 10Co/ZS(1/20)-i and 10Co/ZS(1/3)-s, the nitrogen absorption-desorption isotherms show an obvious H1 hysteresis loop at high relative pressure, characteristic of ordered mesoporous materials, and the isotherms are similar to that of the SBA-15, indicating that the mesoporous structure did not change significantly during the synthesis and subsequent calcinations. However, for 10Co/ZS(1/20)-i and

10Co/ZS(1/3)-s, the nitrogen absorption-desorption isotherms are not very regular, suggesting that the mesoporous structure of SBA-15 had been partially destroyed.

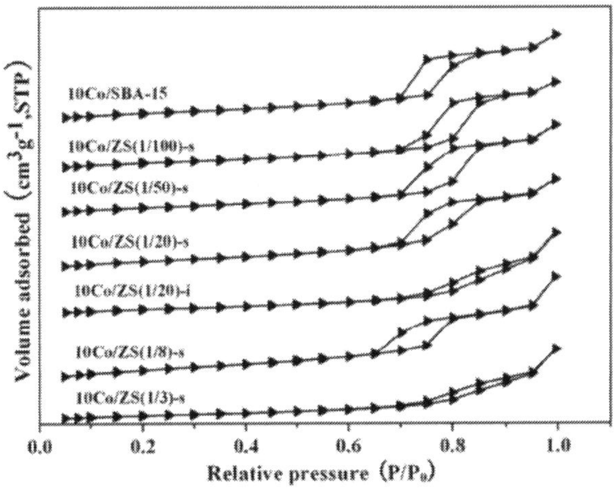

Fig.5. N_2 absorption–desorption isotherms of the catalysts.

As shown in Table 2, the BET surface area (S_{BET}), pore volume (V_{BJH}) and average pore diameter (d_{BJH}) of the catalysts decrease with increasing Zr/Si atomic ratio. The BET surface area and the pore volume decrease clearly with Zr/Si atomic ratio > 1/20, showing that the periodic mesoporous structure is partially collapsed during the synthesis and subsequent calcinations with high content of zirconium. Moreover, the pore size decreases gradually with increasing Zr/Si atomic ratio, perhaps due to the formation of single ZrO_2 particle outside the meso-structure blocking the mesoporous channels. Comparing 10Co/ZS(1/20)-s and 10Co/ZS(1/20)-i, with the same content of zirconium, the BET surface area, pore volume and average pore diameter of the former are larger than those of the latter, showing that the mesoporous structure of SBA-15 had been better preserved when the sample was prepared by isomorphic substitution, as indicated by XRD results.

Table 2 N_2 physisorption results

Sample	S_{BET} (m^2g^{-1})	V_{BJH} (cm^3g^{-1})	d_{BJH} (nm)
10Co/SBA-15	563.0	1.121	9.8
10CoZS(1/100)-s	515.9	1.097	8.5
10Co/ZS(1/50)-s	463.4	1.033	7.9
10Co/ZS(1/20)-s	419.9	1.028	7.7
10Co/ZS(1/20)-i	376.9	0.613	7.2
10Co/ZS(1/8)-s	291.2	0.969	6.5
10Co/ZS(1/3)-s	226.3	0.528	6.3
10Co/ZrO_2	113.2	0.256	5.8

3.1.5. X-ray photoelectron spectroscopy

X-ray photoelectron spectroscopy (XPS) is used to investigate the surface composition of catalysts. The XPS data are shown in Table 3 and Fig. 6. The binding energies (BEs) for $Co2p_{3/2}$ of the catalysts decrease apparently with increasing Zr/Si atomic ratio but remain in the region of 780.44 eV for 10Co/ZS-s(1/20) and 781.78 eV for 10Co/ZrO_2, This suggests that the main phase on the catalysts surface is Co_3O_4, in agreement with the XRD results. The BE of $Co2p_{3/2}$ in 10Co/SBA-15 is higher than that of pure Co_3O_4 (BE = 780.3 eV [30], showing that the cobalt-silica phase was formed on the catalyst surface, similar results reported by Zou et al. [31]. Moreover, the surface concentration of cobalt increases detected by XPS, showing that the density of surface cobalt increased with addition of zirconium.

An interesting feature in our XPS spectra of the Co 2p regions for catalysts is the shift to lower BE value of Co $2p_{3/2}$ with addition of zirconium, but still higher than that of pure Co_3O_4. For cobalt, the BE of Co-metal has the lowest BE for all cobalt chemical states. The binding energies of the Co $2p_{3/2}$ electrons increase from Co^0 to Co^{2+} in a predictable manner. However, Co_3O_4 has a lower BE than the CoO [32]. In the present study, the Co $2p_{3/2}$ matches neither with Co_3O_4 nor with CoO, and with increasing zirconium content the BE for Co $2p_{3/2}$ shifts to pure Co_3O_4, indicating that the addition of zirconium inhibited the formation of cobalt surface phase, thus, the strong interaction between the cobalt species and the support was perhaps suppressed. In the case of the catalysts with high zirconium content (Zr/Si atomic ratio > 1/20) the Co $2p_{3/2}$ peak shifts toward higher energies, this could be taken as evidence of the strong interaction of the cobalt species with the surface of the ZS(x) support at high zirconium content.

Table 3 Summaries of XPS characterization of the catalysts

Samples	$Co2p_{3/2}$ (eV)	$Si2p_{3/2}$ (eV)	$Zr3d_{3/5}$ (eV)	Concentration(atom%)			
				Si	O	Co	Zr
10Co/SBA-15	781.47	102.86	-	33.01	50.77	0.75	-
10Co/ZS(1/100)-s	781.08	102.06	182.42	28.05	51.01	1.49	0.37
10Co/ZS(1/50)-s	780.88	102.19	182.52	27.31	48.56	1.53	0.74
10Co/ZS(1/20)-s	780.44	102.05	182.77	26.99	46.45	1.62	1.34
10Co/ZS(1/20)-i	780.78	102.60	182.95	24.99	53.70	1.82	1.83
10Co/ZS(1/8)-s	780.68	102.37	182.83	25.63	46.11	1.95	1.78
10Co/ZS(1/3)-s	780.84	102.55	182.91	25.21	45.23	2.21	2.34
10Co/ZrO_2	781.78	-	183.17	-	50.65	1.86	9.96

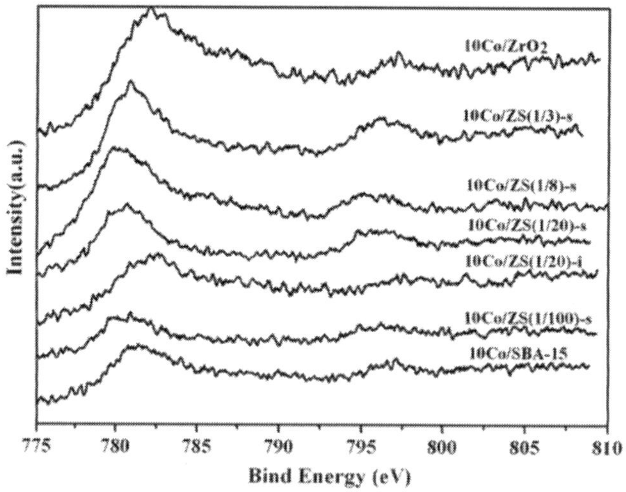

Fig.6. Co 2p XPS spectra of the catalysts

3.1.6. H_2-TPR

TPR profiles of the catalysts with different Zr/Si atomic ratios are shown in Fig. 7. For 10Co/SBA-15, three reduction peaks are observed: the peak α corresponds to the first reduction step of Co_3O_4 ($Co_3O_4 \rightarrow CoO$), the peak β is attributed to the reduction of intermediate CoO phase ($CoO \rightarrow Co^0$) and the high temperature reduction peak γ is attributed to reduction of the barely reducible cobalt silicates (Co_2SiO_4) formed by the strong interaction between cobalt and the siliceous support [12]. For 10Co/ZS(x)-s, 10Co/ZS(1/20)-i and 10Co/ZrO_2, only two peaks are identified for the reduction of Co_3O_4 to CoO (peak α) and CoO to Co (peak β), and the third peak (peak γ) could barely be identified.

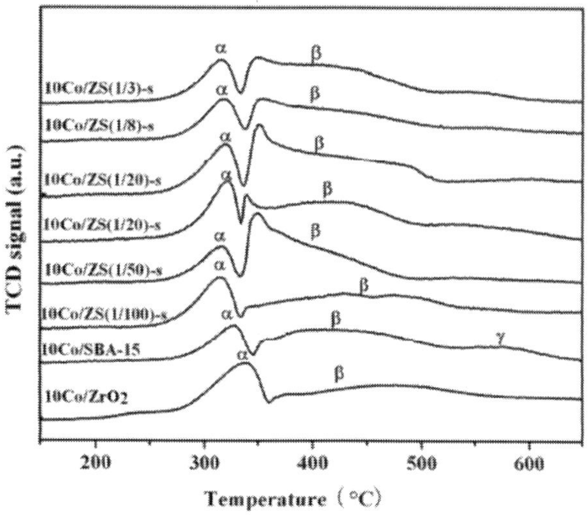

Fig.7. TPR profiles of the catalysts

It is observed that with addition of zirconium, the peak position of the first reduction of Co_3O_4 is significantly shifted to lower temperature, meaning that Co_3O_4 particles supported on ZS(x)-s or ZS(x)-i are more easily reduced to CoO species, most likely due to weakening of the cobalt - silica support interaction by the addition of zirconium, as revealed by XPS, similar findings reported by Moradi et al. [18]. The peak position of the second reduction is not shifted significantly. However, the peak area increases initially but decreases later. In supported catalysts, the reducibility of the cobalt species often depends on the extent of the metal–support interaction. On addition of zirconium at lower content, the weakening of the cobalt-silica support interaction improves the reduction of CoO to Co, while at higher zirconium content, perhaps due to the further increased concentration of cobalt species and interaction with the surface decreasing the reduction of CoO to Co, as shown in XPS results. In addition, the reduction peak at high temperature disappears, most likely due to the suppression of interaction between cobalt and silica support with addition of zirconium, as revealed by XPS. For 10Co/ZS(1/20)-s, the peak area of the second reduction is larger than that of 10Co/ZS(1/20)-i, indicating that incorporation of zirconium with isomorphic substitution was superior in improving the degree of reduction of the catalysts.

3.1.7. Hydrogen temperature programmed desorption (H_2-TPD) and O_2 titration

The results of H_2-TPD and O_2 titration are shown in Table 4. As can be seen, H_2 chemisorption decreases significantly with increasing zirconium content. It indicates that the cobalt active sites decreased when zirconium were added to the catalysts. The percentage reduction increases for catalysts with low zirconium content and decreases for those with low zirconium content. It is suggested that small amounts of zirconium suppressed the interaction between cobalt and support, and improved the reducibility of the catalysts. At higher zirconium content, the reducibility of the catalysts decreases due to the strong interaction of the cobalt species with the surface of the supports. This result was also revealed by XPS and H_2-TPR experiments. The dispersion of cobalt decreases due to the decrease of surface area and pore volume with the addition of zirconium. Furthermore, the metal cluster size increases with increasing zirconium.

Table 4
H_2-TPD and O_2 titration data

Sample	H_2 desorbed (umol/g)	D_{uncor} [a] (%)	d_{uncor} [b] (nm)	O_2 uptaked (umol/g)	Red (%)	D_{corr} [c] (%)	d_{corr} [d] (nm)
10Co/SBA-15	66.2	7.19	19.1	306.2	22.05	32.61	4.2
10Co/ZS(1/100)-s	61.0	6.74	22.3	637.4	46.32	14.55	10.3
10Co/ZS(1/50)-s	57.2	6.01	21.7	783.0	52.89	11.36	11.5
10Co/ZS(1/20)-s	54.9	5.82	22.7	888.7	58.17	10.01	13.2
10Co/ZS(1/20)-i	48.6	4.52	36.2	678.3	48.02	9.41	17.4
10Co/ZS(1/8)-s	41.3	3.46	43.6	551.8	43.71	7.92	19.1
10Co/ZS(1/3)-s	38.6	2.61	58.7	491.2	37.48	6.96	22.0
10Co/ZrO_2	32.2	1.91	64.1	364.7	28.33	6.74	18.2

[a] uncorrected catalyst dispersion
[b] uncorrected metal cluster diameter
[c] corrected catalyst dispersion
[d] corrected metal cluster diameter

3.2. Fischer-Tropsch synthesis

3.2.1. Catalytic activity

The results of FTS activity and product selectivity of the catalysts are summarized in Table 5. The following trends are apparent: (1) With increasing zirconium content, the CO conversion increases initially with zirconium increasing from 1/100 to 1/20, but the reverse is obtained when the Zr/Si ratio increases from 1/20 to 1/3; (2) With the same content of zirconium, the CO conversion of the 10Co/ZS(1/20)-s prepared by isomorphic substitution are higher as compared to 10Co/ZS(1/20)-i prepared by co-impregnation; (3) The CO conversion of 10Co/ZrO_2 is lower than that of other catalysts under the same conditions.

Table 5
Performances of the catalysts in a fixed bed reactor

Catalysts	*Xco* (%)	Products selectivity (mol %)					
		CH_4	C_2	C_3	C_4	C_{5+}	CO_2
10Co/SBA-15	15.47	18.07	0.98	4.12	5.81	71.28	1.43
10Co/ZS(1/100)s	17.62	12.01	0.86	3.01	4.21	79.01	0.86
10Co/ZS(1/50)-s	25.44	8.51	0.72	2.23	4.01	83.86	0.61
10Co/ZS(1/20)-s	26.21	7.45	0.53	1.93	2.74	86.71	0.58
10Co/ZS(1/20)-i	20.97	14.37	0.82	2.98	4.21	76.86	0.76
10Co/ZS(1/8)-s	19.48	8.69	0.74	2.86	4.88	79.91	0.86
10Co/ZS(1/3)-s	18.48	13.03	0.81	2.79	4.37	82.62	0.75
10Co/ZrO_2	8.92	7.07	0.7	4.80	9.06	78.31	0.63

Reaction conditions: 1.0MPa, 230 °C, CO/H_2=1:2, space velocity of syngas 8SLg^{-1} h^{-1}

It has been demonstrated that the catalytic activity of Co catalysts for FTS depended on the catalyst reducibility. In principle, the activity of reduced cobalt catalysts should be relevant to the concentration of surface metal cobalt sites [4-6, 33]. For 10Co/ZS(x)-s, when Zr^{4+} ions were introduced into the SBA-15 framework with the hexagonal array of uniform tubular channel configuration intact, retaining high surface area and large pore volume, the probability of Zr^{4+} ions contacting with Co_3O_4 particles was high. Further, with increasing zirconium content, the density of surface cobalt and the degree of reduction of Co increase, more accessible cobalt active surface sites were generated, thus, the CO conversion increased. However, on increasing Zr/Si atomic ratio from 1/8 to 1/3, the hexagonal array of uniform tubular channel configuration was destroyed gradually, leading to the decrease of surface area and pore volume, the strong interaction of the cobalt species with the surface of the supports, and the degree of reduction of cobalt decreases, resulting in a decrease of CO conversion.

The low activity of 10Co/SBA-15 and 10Co/ZrO_2 can be related to the low reducibility of cobalt species in the catalysts. Cobalt silicates are present in the former, as discussed in H_2-TPR results, and in the latter catalyst, the low activity may be related to the low surface area and low pore volume. At the same content of zirconium, 10Co/ZS(1/20)-s shows higher activity compared to 10Co/ZS(1/20)-i, perhaps due to higher reducibility of cobalt species in the catalysts.

3.2.2. Product selectivity

As observed in Table 5, 10Co/SBA-15 gives the highest methane selectivity and the lowest C_{5+} hydrocarbons selectivity. For 10Co/ZS(x)-s, with the addition of zirconium, C_{5+} hydrocarbons selectivity increases from 79.01 % to 86.71 %, and then declines to 79.91 %. Comparing 10Co/ZS(1/20)-s and 10Co/ZS(1/20)-i, the former shows higher C_{5+} hydrocarbons selectivity and lower methane selectivity.

The high methane selectivity was usually reported for catalysts having high metal dispersion and low cobalt reducibility [4, 12]. This resulted in the formation of unreduced cobalt oxides which catalyzed WGS reaction. As evidenced by O_2 titration, the high cobalt dispersion and the low catalyst reducibility of 10Co/SBA-15 led to high methane selectivity. It has been demonstrated that the presence of larger cobalt particles led to higher selectivity to heavier hydrocarbons [12, 34-35]. As shown in Table 4 and Table 5, it is obvious that the change in the C_{5+} selectivity corresponds to the size of cobalt clusters.

It has been known that SiO_2–ZrO_2 mixed oxides had high acidity, whereas the single oxides showed only weak acidity [36]. The number of acidic sites was perhaps higher on the ZS(x) than on SBA-15. Zhang et al. [37] reported that supports with lower acidity led to the higher activities, higher C_{5+} hydrocarbons selectivity and lower methane selectivity. The increase in acidic sites with increasing zirconium content may have also induced the poor catalytic performance.

4. Conclusions

Zirconium isomorphic substitution of SBA-15 did not affect the tubular structure of the parent SBA-15 at lower zirconium content until Zr/Si ratio of 1/20 but resulted in partial collapse of the meso-structure at higher extent of substitution.

With addition of zirconium, the density of surface cobalt increased, the barely reducible of cobalt silicates was inhibited and the reduction of Co_3O_4 to CoO was facilitated. At higher zirconium content (Zr/Si atomic ratio >1/20), the hexagonal array of uniform tubular channel configuration was partially destroyed, the surface area and pore volume decreased, and the interaction between cobalt species and supports increased leading to the reduction of CoO to Co being inhibited, resulting in the decrease in FTS activity.

Compared to incipient wetness impregnation, the isomorphic substitution was superior to retain the integrity of SBA-15 structure, improved the cobalt reducibility and showed better catalytic performances.

(69b) Flare Minimization at LNG Receiving Terminal

Chaowei Liu, Dan F. Smith Department of Chemical Engineering, Lamar University, Beaumont, TX

Jian Zhang, Dan F. Smith Department of Chemical Engineering, Lamar University, Beaumont, TX

Qiang Xu, Dan F. Smith Department of Chemical Engineering, Lamar University, Beaumont, TX

John L. Gossage, Dan F. Smith Department of Chemical Engineering, Lamar University, Beaumont, TX

The LNG (liquefied natural gas) receiving terminal is an important sector of the entire LNG value chain. The handling of boil-off gas (BOG) during LNG re-gasification at LNG-receiving terminals significantly influences the BOG flare emission and the energy consumption. In this paper, thermodynamic analysis-based design and operation methods are integrated to recover BOG with the minimum total energy consumption, a goal of which is to provide a cost-effective flare minimization strategy at LNG receiving terminals. Case studies demonstrate both design and operational improvements will provide energy-saving flare minimization opportunities. The presented rigorous models and thermodynamic analysis also provide a fundamental understanding of LNG re-gasification operations, which is valuable for industrial applications.
See more of this Session: Professor Cedomir M. Sliepcevich Memorial Session Two - LNG Regasification
See more of this Group/Topical: Topical 6: 10th Topical Conference on Gas Utilization

Timeline of Sabine Pass LNG Terminal

Darron Granger, PE
Sr. Vice President, Engineering & Construction
Cheniere LNG, Inc.
March 2010

CHENIERE

Presentation Contents

1. Timeline of SPLNG
2. SPLNG Cost
3. Project Execution Strength/Weakness
4. Design Points for Onshore Terminals
5. Facility Photographs
6. Q & A

CHENIERE

SPLNG 2003 Timeline

- January - Lease Property (853 acres) and Develop design basis **(3 month duration)**
- January - Develop relations with local community leaders
- All Year - Hold frequent community meetings to explain our plans, address concerns, educate public about safety and design of LNG facilities
- May - Contract Black & Veatch to develop a Pre-FEED **(8 month duration)**
- May - Contract TWI to test each site **(8 month duration)**
- June - Contract PBSJ to create FERC's RR-1 through RR-12 **(6 month duration)**
- June - Use portions of the Pre-FEED to create RR-13 **(6 month duration)**
- September - BV Pre-FEED documents complete
- December - Submit FERC Applications (both SPLNG and CCLNG)

CHENIERE

SPLNG 2004 Timeline

- All of 2004 - Relentlessly answer FERC questions
- January - Enhanced FEED Awarded to Bechtel **(12 months duration)**
- March - FERC Public Scoping Meeting
- August - Draft EIS Issued
- September - FERC Public Scoping Meeting
- December - FERC Order Granting Authority to exist
- December - Award LNTP Contract to Bechtel for Phase 1

CHENIERE

SPLNG 2005 Timeline

- January - Develop Design basis for SPLNG Expansion (Phase 2)
- January - LDEQ Air Permit received
- February - Soil Improvement Contractor Mobilize
- March - FERC authorization to Commence Construction of Phase 1 Received
- March - COE Permit Received
- April - Award Full NTP to Bechtel for Phase 1 Engineering
- July - Start LNG Tank Test Piling
- July - Submit EA + 13 Resource Reports for Phase 2 to FERC
- December - Tank Pile Production Begins
- December - EPC Contract award to Bechtel for Phase 1

CHENIERE

SPLNG 2006 Timeline

- May - FERC issued Environmental Assessment for Phase 2
- June - FERC issued Authorization for Construction for Phase 2
- July - EPCM Agreement with Bechtel executed
- July - Full NTP to Bechtel for Phase 2 Engineering
- July - Phase 1 Engineering 90% Complete

CHENIERE

SPLNG 2007 Timeline

- February - FERC issued Authorization to Construct Phase 2
- February - Full NTP to Bechtel for Phase 2 Construction
- May - Tank 1 and Tank 2 roof raise
- October - First of Four GTG Startup
- October - Deliver 4 tractor tugs to site
- November - Hydrotest Tank 1
- December - SP Pipeline Mechanical Completion

CHENIERE

SPLNG 2008 Timeline

- April - Initial cooldown Cargo Arrives – Tank 1 Cooldown
- April - Grand Opening Celebration
- May - Start gas to pipeline
- June - Tank 2 Cooldown
- July - Tank 3 Cooldown
- June - Performance Testing to 2.0 Bcf/d
- June - Performance Testing to 2.7 Bcf/d
- June - CMI TUA Commences (Commercial Operations)

CHENIERE

2009 Timeline

- March - Tank 4 and Vaporization RFCD
- April - Total + Chevron TUA Commences
- May - Tank 5 RFCD
- May - Performance Testing to 4.0 + Bcf/d
- July - Final Acceptance – Commercial Operation

Bottom Line: 6.5 years concept to finish for full permitted scope

CHENIERE

SPLNG Cost

- Cost for Phase 1: $900 million
- 2 berths, 3 tanks @ 160,000 cm each, 2.6 Bcf/d throughput

- Cost for Phase 2: $500 million
- 2 tanks @ 160,000 cm each, 1.4 Bcf/d throughput

- Total Cost: $1.4 Billion
- 2 berths, 5 tanks @ 160,000 cm each, 4.0 Bcf/d throughput

CHENIERE

Project Execution – Strength/Weakness

Weaknesses

- Use of unproven contractors
- Lack of O&M personnel on the Project Team in early design phase
- Complete FEED to give to the EPC contractor
- Early design changes impacted engineering efforts
- New company = no operating procedures
- New company = no design specifications
- Low resistance to design changes

CHENIERE

Project Execution – Strength/Weakness

Strengths

- Hire persons that have appropriate experience, and a can-do attitude.
- Give employees ample freedom and open forums to discuss options, approaches, strategies, and to create.
- Make decisions promptly; don't study the situation to death. Even if the decision is not optimum, the greater gain is that the project keeps moving forward.
- Revisit past decisions only if compelling rationale.
- Strong and effective relationships with regulatory Authorities.

CHENIERE

Design Points for Onshore LNG Terminals

CHENIERE

Use Safe, Proven Equipment - 1 of 6

- Use of pump/SCV train for high safety & reliability
- Use of single containment tanks
- Use of Submerged Combustion Vaporizers
- Use of 20” unloading arms with PERCs
- Use of N+1 philosophy for high reliability
- Provide 3 levels of ESD + plant emergency isolation
- Enough power to operate all installed equipment
- High and open control room with full view
- Eliminate flanges & gaskets where practicable
- Since Space = Safety; minimize confined areas

CHENIERE

Safety Considerations - 2 of 6

- Clear line of sight from control room to operations areas
- Direct communication between control room, plant areas
- Remote control cameras to eliminate “hidden” areas
- Minimal flanges for LNG service
- Stainless steel & extra heavy wall piping (2” minimum)
- The ball valve is the valve of choice
- Arrange layout to minimize impact of potential gas leaks
- Slope potential spill areas into an impoundment basin
- PSV & maintenance drains piped to a remote safe vent
- Early alarms for combustible vapor detection, smoke, heat
- Proven Emergency Shutdown (ESD) philosophy and isolation valves
- Fire fighting equipment designed for LNG emergencies
- Emergency response plan with ongoing safety training

CHENIERE

Minimize Environmental Impact - 3 of 6

- Incorporate a no-venting design for all operating modes
  - Use a vent, not a flare, for unavoidable releases
- Use closed loop vaporization to avoid water impact
- Wetlands philosophy
  - Avoid wetland impacts if possible
  - Minimize wetland impacts if unavoidable
  - Mitigate all wetlands impacted

CHENIERE

Code Approach - 4 of 6

- Take no exceptions to any code
- Use proven equipment acceptable to FERC
- Use containment under LNG pipes and equipment acceptable to FERC/DOT
- Create impoundment basins to catch a 10 minute guillotine failure of the largest line as required by FERC/DOT

CHENIERE

Uncontested Permits - 5 of 6

- Find safe-harbor sites on the beach
- People and LNG don't mix
- Use sound well proven design preferred by FERC/DOT/USCG
- Use large spacing between equipment
- Make the FERC filing as complete as possible
- Insure good community support

CHENIERE

Community Support - 6 of 6

Lessons learned in previous jobs:

DON'T Talk only to elected officials – speak to everyone

DON'T Ignore grassroots philosophy – vigorously promote it

DON'T Use corporate elbows to muscle in – be part of the community

DON'T Ignore local people – become part of the culture

DON'T Ignore local concerns – address each and every one

DON'T Talk down to local people – be considerate

DON'T Keep the technical people away from local discussions

DON'T Use all imported labor and services – Maximize local content

DON'T Keep silent about the project – speak often, speak positively

CHENIERE

SPLNG as permitted

SPLNG Photographs

Added only if time allows

CHENIERE

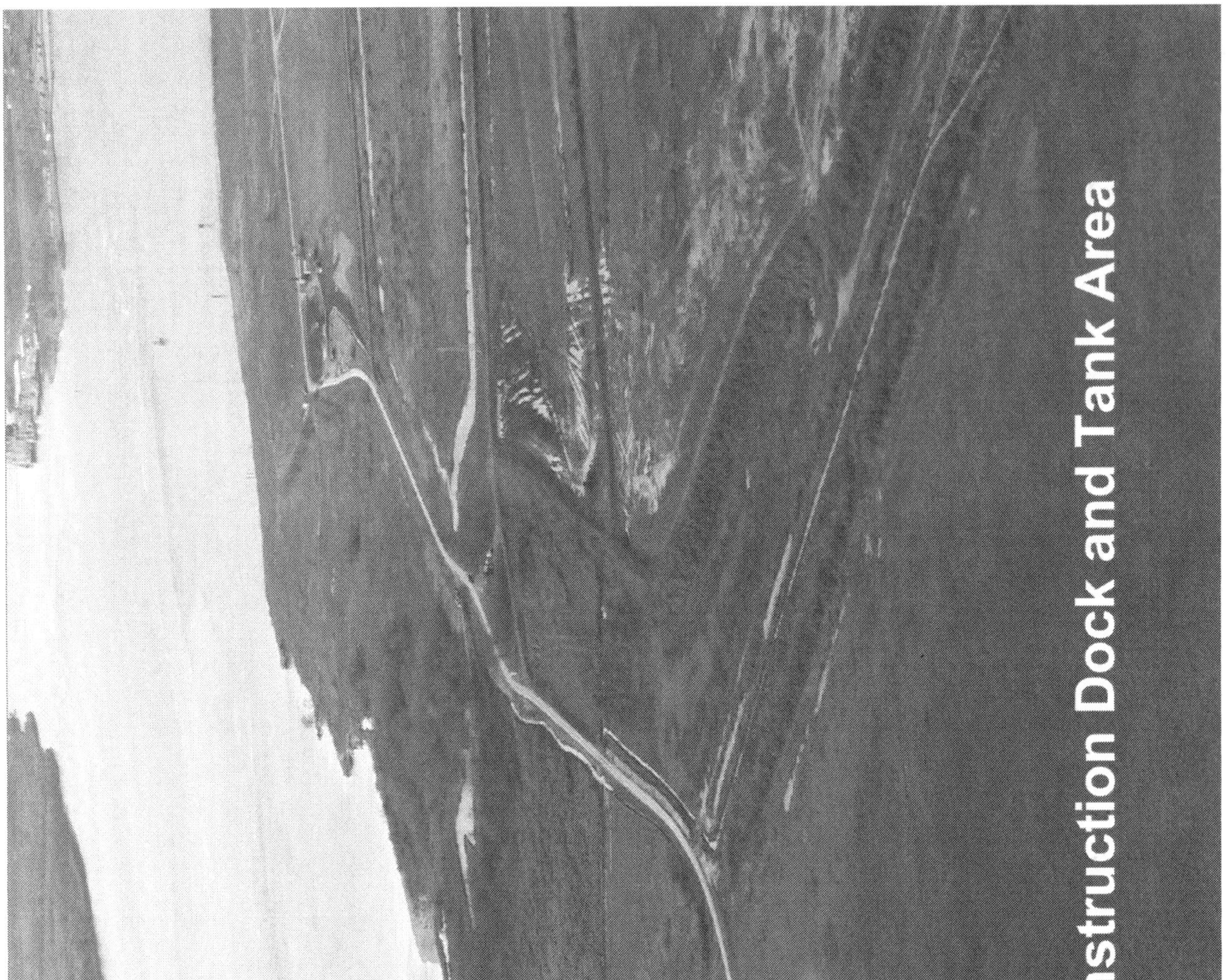
February 17, 2005 Construction Dock and Tank Area

Soil Improvement / Wick Drains

Sabine Pass LNG Construction Photos
September 2006

Aerial View of Tank 2

Aerial View of Tank 3

Aerial View of Marine Berth

Aerial View of Sabine Site

CHENIERE

Sabine Pass LNG Construction Photos April 2007

Tank 1

Tank 2

Tank 3

Aerial View of Tanks 1-3

View of Pipe Racks

Marine Berth, Unloading Docks and Pipe Racks

Aerial View of Site and Channel

CHENIERE

LNG Tank Piling and Foundation Slab

CHENIERE

LNG Tank Walls and Roof

CHENIERE

Berth Area
May 2007

Artificial Islands created by dredge spoil from SPLNG berth

CHENIERE

Sabine Pass LNG - March 2008 - Aerial view of SPLNG Terminal (Channel View).

CHENIERE

May 2007
Looking East

SPLNG NOW

SPLNG NOW
Umn Slal, world's largest LNG ship

SPLNG NOW
Looking NW

SPLNG NOW
Looking SE

SPLNG NOW
Umn Slal, world's largest LNG Carrier
166,000 CM QMAX

Ambient Air Vaporizers

CHENIERE

Ambient Air Vaporizers

On line
July 2009

Cheniere Energy is pioneering Ambient Air Vaporizers (AAVs) at its Sabine Pass LNG Terminal

Natural draft Ambient Air Vaporizers turn liquid LNG at -260°F to natural gas at +40°F using free heat from the air

CHENIERE

Ambient Air Vaporizers

Natural gas exits the finned tubes from the top of the AAV

Warm air enters the top of the AAV

- Zero CO, CO_2, NOx, VOC emissions
- No moving parts
- No fuel used
- No sea water impact
- No intermediate fluids
- Makes cold air, fresh water and fog
- Proven technology, innovative controls

Cool air and water vapor exit the bottom of the AAV

LNG enters finned tubes at the bottom of the AAV

CHENIERE

AAV's in Action

CHENIERE

Thank You
SPLNG as permitted

Ship-Shore Interaction and LNG Tank Piping Design Considerations

Chen-Hwa Chiu
Ugochukwu Diribe
Chevron Energy Technology Company

John S. Hartono
Chevron Shipping Company

AIChE Spring Meeting, March 20010
10th Topical Conference on Gas Utilization
San Antonio, 21-25 April, 2010

INTRODUCTION

The LNG storage and loading system of the LNG import or export terminal consists of the following major subsystems:

- LNG rundown system
- LNG storage tanks
- LNG loading system
- Boil-off Gas (BOG) recovery system
- Vapor handling system
- Start-up and circulation systems
- Ship/shore interface

The purpose of this paper is to offer guidance on various considerations in designing LNG storage tank piping and loading lines, including operational aspects and other issues encountered by the pipelines and equipment.

TOPOLOGY OF SHORE/LNG TANK

LNG tank topological position relative to the shore area is only of concern relative to the provision and sizing LNG booster pumps. These pumps are used to ensure that LNG loading can be accomplished from LNG plants.

Loading arms are typically drained into the jetty surge drum upon completing ship loading.

LNG RUNDOWN SYSTEM OPERATING MODES

The LNG storage and loading system operate in two main circulation modes: holding mode and transfer mode.

The holding mode of operation is maintained in the periods between LNG ship loadings. There are therefore three circulation options. These three options are:

- Normal holding mode: LNG rundown is in operation and loading arms are closed, flow goes first from one loading line to the LNG header to the jetty, and then returns to the other loading line to the LNG tanks.
- Long circulation mode: This circulation maintains the loading system and the rundown system cold during a period when all LNG trains are out of operation and there are no LNG rundown to LNG tank.
- Short circulation mode: This particular procedure is to alleviate stratification inside LNG tanks. Short circulation established around one tank via the pump minimum flow line to minimize the occurrence of stratification, thus avoid the rollover.

The BOG compressors are sized to recover the vapors produced during any of the three circulation options. Vapor produced during holding mode is compressed and exported to the HP fuel gas system. In the receiving terminal, LNG circulation is maintained in holding mode to keep lines cold.

ROLLOVER CONSIDERATIONS

Roll-over is usually a concern at peak shaving plants, where the LNG is stored for extended periods. It is not commonly a concern in storage tanks in liquefaction plants. Experience has shown that when two LNG of different composition are successively fed into a LNG tank, the two layers of LNG may not mix and stratification may occur.

After a long period of liquefaction unit outage, stored LNG becomes heavy due to heat ingress from sidewall and bottom, which causes preferential boil-off of light components and then stored LNG increases in density. After restart of liquefaction unit, the new light LNG entering the top of LNG storage creates a light layer above the old heavy layer stored in bottom. Heat leak into LNG storage tank results a rise in temperature in heavy bottom layer due to hydrostatic load of light layer above, which prevents a partial boil-off. Simultaneously, heat ingress in the light layer in upper part gives boil-off of light components, such as nitrogen and methane, resulting in an increase in the density.

Consequently, the differential density is gradually reduced until the density of LNG layer at the bottom becomes slightly lower than LNG at the top. A rollover of the two layers then occurs. The heated LNG from tank bottom reaches the surface and there is a strong increase in the boil-off rate, which create an over pressure which may open pressure relief valves and, potentially, reserve capacity relief valves.

If stratification occurs, short circulation will be established around one tank via the pump minimum flow line to mix the stratified layers. In the case of the LNG receiving terminal, during the holding, in order to alleviate any potential for the phenomena called rollover in LNG storage tanks it is usually useful to pump around LNG from one tank to another in order to avoid any stagnation that may promote the chance of roll over in the LNG storage tanks.

BOIL-OFF GAS MANAGEMENT AND VAPOR RETURN

BOG System

The Boil-off Gas (BOG) recovery system consists of a BOG header to which the vapor spaces of the LNG tanks and the vapor return line of the LNG jetty are connected and which leads to BOG compressors. These compressors route the generated BOG vapors of LNG storage and loading system to the liquefaction plant HP fuel gas system. The Boil-off Gas (BOG) System is connected to, but can be isolated from, the ship's vapor return system. The mechanical design pressure of the system is a nominal 6.3 barg. The BOG header is insulated to minimize heat in-leak.

The pressure in the vapor spaces of the LNG tanks is held constant at some 25-45 mbarg by a pressure controller on the common BOG header, which throttles the discharges of the BOG compressors. In case the BOG generation is in excess of what the BOG compressors can handle, e.g. upon a compressor trip, the excess boil-off vapors are routed to the storage and loading flare by a second pressure controller on the BOG header with a set pressure of 50 mbarg.

The system and equipment are designed to cater for an increased loading rate The BOG compressor is designed to be able to handle nominal boil-off, without quench injection and a suction temperature of –123.5°C, and any increases due to loading operations.

Vapor Return

During LNG ship unloading, the vapor returned to LNG ship is maintained at cold temperatures depending on the type of LNG ship. At the time of completing LNG discharge, the vapor line is closer to the temperature of the cold blower discharge. However, after the LNG ship departs, the vapor line warms up to near ambient temperature due to heat leakage into this line. Therefore, during the next LNG discharge, the initial vapor returning to the ship is warm since the vapor line heats it. This effect continues until the line is cooled down by the cold vapors.

LNG pumped ashore is replaced by vapor coming back from the onshore storage tanks through the return vapor line. Pump energy and friction in the transfer line, as well as any heat in-leak, temperature and pressure differences in the onshore storage tank will cause vapor flashing, which is partly led back to the process and returned to the ship's cargo tanks. Initially the vapor envelope in the ship's tanks is rather small and the return gas temperature difference is equalized by the presence of the much colder LNG. As the liquid level in the ship's tanks go down, and the vapor envelope increases, it is sometimes necessary to start the onshore return gas blower. This is a blower between the storage tanks and the ship which takes suction from the tank and discharges in the vapor return line. This should not be confused with the BOG compressor, which compresses BOG and moves it to the process.

LIQUID DRAINAGE OPERATION

In order to disconnect LNG loading arms after cargo transfer operations, the arms must be drained, purged and inerted. After the onshore block valves and the LNG carrier's manifold valves are closed, the arms are drained into a collection tank or vessel. The liquid in these containers are either allowed to vaporize or routed back to the onshore storage tanks, or is pumped back through the liquid recirculating line back to the onshore storage tanks, this approach is not normally applied, because this requires the installation of a small cryogenic pump.

The run down header shall be designed as self-draining to a drain pump, located at a suitable low point, which shall be provided to facilitate this draining and to transfer the drained LNG to storage. Draining of the rundown header shall be possible within 12 hours. Where not possible drain pumps may be provided at a suitable low point location along the loading line system.

Maintenance of the shore loading arms or regular operation of LNG loading arms will require liquid drainage from the arms during disconnections. The drainage of LNG from the LNG loading arms need not depend on slope to drain back to LNG storage tanks. Air Products designs drain LNG liquid into the jetty surge drum, from where accumulated LNG is pumped into the LNG run down line or other liquid gathering place.

LOADING SYSTEM SURGE MANAGEMENT

A hydraulic transient, or surge analysis, should be performed for the LNG loading arm feed piping. The surge analysis will analyze the effects of quick closing of a valve on either the shipside piping system or the shore side piping system. Actual valve closing time furnished by the supplier of loading arm shall be used in this analysis. Upon completion of surge analysis, the LNG project team will provide the loading arm supplier with the anticipated flange loads on the arms and ship's manifold.

The LNG loading arm supplier shall review the impact of these flange loads and shall ensure that the loads transmitted to the manifold flange, under all circumstances are limited to the maximum given in the latest edition of the OCIMF (Oil Companies International Marine Forum) "Recommendations for Manifolds for Refrigerated Liquefied Gas Carriers".

The upper design pressure of the loading system including the liquid loading lines and the liquid loading arms shall at least exceed the maximum surge pressure generated at the fastest valve closure time of ESD. A surge relief system may be provided as an additional protection of the loading arms and manifold system. The surge relief system consists of a fast opening dump valve to a surge drum. The dump valve is opened by the second stage ESD or a high-pressure detection system at the base of the loading arm.

EMERGENCY SHUTDOWN PHILOSOPHY

LNG marine terminals, loading terminals as well as discharge terminals are equipped with an ESD system (ESDS). These systems are designed and built in accordance to SIGTTO (Society of International Gas Tanker & Terminal Operator) and OCIMF (Oil Companies International Marine Forum) as well as any applicable regulatory requirements.

The basic intent of the ESD system is to ***"shut down safely liquid and vapor flows between ship and shore in the event of an emergency during cargo transfer operations".***

The ESD system includes fast closing valves, loading arms equipped with emergency release couplers and, where required, surge protection equipment. These systems are installed in order to protect personnel and prevent material damage to equipment, such as the onshore storage tanks, the loading pipeline and valves, the loading arms and the LNGC equipment. In an emergency, prompt controlled shutdown of shipside and shore valves must take place to avoid damage to the loading arms and to minimize the risk of spillage quantities. It does this by securing the whole system or individual units before an unrecoverable incident is allowed to occur.

These systems are so designed that they operate automatically when pre-defined operational parameters are exceeded or they can be operator initiated. This can be done remotely from either the LNGC Cargo Control Room (CCR) or the onshore central control room. The ESD is also arranged in such a way that it can be operated manually from a number of control points on board the LNGC or ashore. The Emergency Shutdown System consists of the onshore ESD system and the shipboard ESD system. These two individual systems are linked by the Ship/Shore Interface link, when an LNGC is moored at the berth.

The ship/shore interface link is a system that connects the ship and onshore control center to provide for ship-shore and shore-ship mutual shutdown, communications and data transfer between the ship and shore. The ship/shore interface link on modern LNG marine berths consists of an electrical circuit and a fiber-optic circuit, with a pneumatic back-up system. In the event of excessive ship movement off the berth, the umbilical can be broken and cause both shore and ship trips almost simultaneously.

The ESD system minimizes potential risks during the transfer of liquefied gases between ship and shore loading and unloading installations. It provides a quick and safe means of stopping the transfer of cargo and isolating ship and shore cargo systems in a controlled manner either manually or automatically in the event of fault conditions that affect the ability of the operator to control the safe transfer of cargo.

The biggest challenge in the design of an LNG ESD system is the necessity to control pressure surges in the long loading line. Controlling the resultant pressure surges when hydraulic conditions are changed in the pipeline is a very complex task. Besides considering the proper sizing of the equipment, a hierarchy of actions and timing of valve closures must be developed.

Some design approaches to alleviate surge pressures in pipelines are:

- Complete computer modeling of pipeline profile during initial stages of pipeline design work.
- Stage pump shutdown sequence
- Linked ship/shore ESD (Loading and discharging LNGC's).
- Stage emergency shutdown or motor operated valve closure times.
- Select proper surge pressure relief system.

The main function of the Emergency Shutdown (ESD) system is to stop cargo liquid and vapor flow in the event of an emergency and to bring the cargo handling system to a safe, static condition. This is accomplished by either the onshore ESDS or the shipboard ESDS, which are linked during cargo transfer operations, in such way that activation from one side will also affect an activation of the other side. The Emergency Shutdown System has two stages – ESD1 and ESD2 with an alarm stage preceding. The initial action of ESD is to stop the transfer operation in a quick, safe and controlled manner. In situations involving potential damage to loading arms or LNG ship, or require rapid disconnection of the loading arms from LNG ship, then ESD2 is activated,.

Pre-ESD

The process is still located within the normal operating envelope; this predetermined point alerts the operators of an impending condition and can be triggered by a variety of reasons, such as an increase vessel movement. This can be detected by the mooring monitoring equipment, sounds audio and visual alarms on the terminal as well as on the gas carrier.

ESD-1

This is the first stage of the ESD action. ESD1 is activated in situations not requiring disconnection of loading arms from ship. An emergency shutdown system (ESD-1) is provided for safe and quick shutdown under emergency conditions while ship-loading, to avoid or minimize possible spills. The first stage ESD1 shuts down the LNG product transfer operation in a controlled manner by closing shore & ship valves and stopping pumps in the shore LNG tanks.

Ship and shore ESD-1 systems are linked so that initiation of either system simultaneously activates the other. The ship is thus able to stop the shore loading pumps and to shut the shore ESD-1 valves. This functionality is achieved by presenting the ship with a pendant box containing a shutdown button connected to the shore ESD-1 system or, depending on the ship, by means of a ship/shore link which interconnects the shipboard and shore ESD signal systems.

The ESD-1 action can thus be initiated manually from both ship, via the pendant box or ship/shore link, and shore, automatically from the various shore signals and, if the ship/shore link is provided, automatically from various shipboard signals. Where a ship/shore link is provided, initiation of an ESD-1 signal either from shore or ship will also cause the ship's manifold valves to close.

ESD-2

The second stage emergency shutdown system (ESD-2) activates the loading arm Emergency Release System (ERS) to disconnect the arm from the ship. Upon initiation of ESD2, an ESD-1 action is initiated simultaneously. Ship/shore isolation valves - one upstream and one downstream of the Emergency Release Couplers (ERC) must close fully before the uncoupling action.

The loading arms shall be provided with an ERS, which shall isolate the flow of LNG and break away from their connection to the LNGC manifold if the relative motion exceeds pre-determined safety parameters. This prevents damage to the arms, and averts the spill of cargo which would result from a broken arm. The ERS system shall be composed of two isolating valves installed on either side of a Powered Emergency Release Coupling (PERC). The valves must also be able to absorb very large pressure surges which can result from abruptly closing the valves during full rate loading operations.

The operation of the ERS is linked to the multi-stage ESD system on the LNGC as well as Shore Control, to ensure that the loading pumps are stopped and that the two isolating valves are shut prior to the opening of the PERC and the arm automatically maneuvered away from the LNGC manifold flange. The PERC shall permit a clean and safe separation of the arm from the LNG ship's loading manifold upon actuation of the second stage ESD.

The shore personnel will be able to initiate ESD-2 action via protected push buttons located in the same positions as the ESD-1 manual stations. ESD-2 cannot be initiated from the ship. Furthermore, to prevent direct rundown from LNG trains to ships (as it cannot be assumed the ship's relief valves are sized sufficient to cope with warm LNG rundown), a rundown/loading line crossover is provided

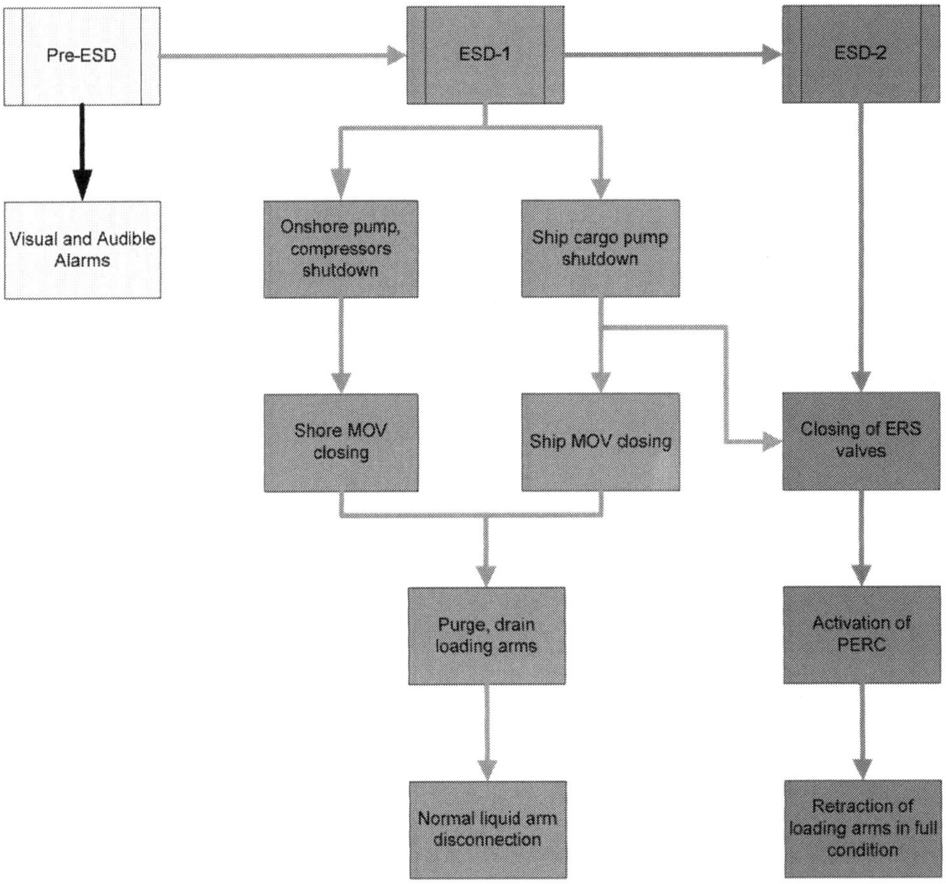

Figure 1 Schematics of Emergency Shut Down (ESD) System for LNG Marine Terminals

REFERENCE

1. D. B. Crawford, and C. A. Durr "LNG Receiving Terminal Design is Different", Pipeline & Gas Journal, pp. 37-46, December 1983.

BIOGRAPHY

Dr. Chen-Hwa Chiu is a Senior Technology Advisor of the Chevron Energy Technology Company (Chevron ETC). Dr. Chiu has served as a company representative to the Joint Industry Projects of Azure Floating LNG Chain, LNG FPSO Safety, and Offshore Cryogenic Loading for LNG transfer. He is involved in major LNG project supports, center of excellence efforts and technical training on LNG for Chevron. Dr. Chiu is an AIChE Coordinator for Chevron ETC.

Dr, Chiu is a Fellow of American Institute of Chemical Engineers and was a Chair of AIChE Fuels and Petrochemicals Division. He is the organizer and LNG session chair of LNG Symposium for AIChE Topical Conference on Gas Utilization. Dr. Chiu graduated from National Taiwan University with a B.S. and earned his Master of Engineering and Ph.D. from the University of Oklahoma, all in Chemical Engineering. He is a Registered Professional Engineer in Texas and Pennsylvania.

Dr. Chiu can be contacted at 1400 Smith Street, Houston, TX 77002.
Phone: 713-372-7039, Cell: 713-397-4097, and E-mail: Chen-Hwa.Chiu@Chevron.com.

Stability Analysis of Safety Valve

Hisao Izuchi
PLE Technology Research Center, Chiyoda Advanced Solutions Corporation, Japan

AIChE Spring Meeting, April 2009
10th Topical Conference on Gas Utilization
San Antonio, Texas, March 21-25, 2010

Abstract

The stability was investigated for a safety valve mounted to a vessel through an inlet pipe by experiments and analysis. It was found that the dynamic instability was caused by the interaction effect between the pressure wave propagation through the inlet pipe and the valve disc motion. This dynamic instability is mitigated by the extension of the inlet pipe length because the oscillating motion of the valve disc is attenuated before the pressure wave generated by the safety valve opening returns back to the safety valve. This dynamic instability is completely different from the static stability caused by the pressure drop of the inlet piping system because longer inlet piping means larger pressure drop. The effect of the outlet area ratio to the orifice area for the safety valve was also investigated. The study results showed that the possibility of instability tends to increase if the outlet area ratio is lower than 6.0. Stability analysis based on the neutral stability searching method was carried out, and it can explain the instability for the system with relatively short pipe length of the safety valve inlet. However, it cannot explain the stable condition for the system with relatively long pipe length. Numerical simulation is effective to predict all types of the instabilities, that is, related to the excessive pressure drop of the piping system, the effect of pressure wave propagation in the inlet pipe, and the effect of the outlet area ratio of the safety valve.

Key Words
Safety valve, chatter, flutter, dynamic instability, neutral stability searching method, pressure wave

Nomenclature

A : Flow area of pipe [m^2]
A_D : Outlet area of safety valve
A_H : Area of disc holder = Pressure acting area [m^2]
A_O : Orifice area [m^2]
A_S : Inlet area of safety valve [m^2]
c : Sound speed of gas [m/s]
C_d : Discharge coefficient of safety valve nozzle
C_{dD} : Discharge coefficient of safety valve outlet
C_s : Damping coefficient [Ns/m]
C_{sv} : Additional damping coefficient to satisfy neutral stability condition [Ns/m]
C_v : Specific heat at constant volume [J/kg/K]
D : Pipe internal diameter [m]
d_H : Diameter of disc holder [m]
e : Specific internal energy [J/kg]
E : Total energy in one segment of pipe [J]
f : Lift force function
$F1$: Pressure term [m/s^2]
$F2$: Convection term [m/s^2]
$F3$: Friction loss term [m/s^2]
K : Spring constant [N/m]
$K_1 = (Z + Z_s)/Z$: Parameter which shows safety valve characteristics
L : Length of inlet pipe [m]
M : Mass of gas in one segment of pipe [kg]
M_B : Mass of gas in safety valve body [kg]
M_s : Mass of moving part of safety valve [kg]

M_W : Molecular weight of gas [kg/kmol]
P : Pressure in pipe or at safety valve inlet [Pa]
$\overline{P}$: Average pressure at safety valve inlet [Pa]
P_{atm} : Atmospheric pressure [Pa]
P_B : Pressure in safety valve body [Pa]
P_c : Critical pressure [Pa]
$P_D = P_{atm}$: Pressure at safety valve outlet [Pa]
P_d : Pressure at downstream segment of pipe [Pa]
P_u : Pressure at upstream segment of pipe [Pa]
R : Gas constant [= 8314 J kg/kmol/K]
s : Parameter for Laplace transformation
t : Time [s]
T : Temperature of gas in pipe [K]
T_B : Temperature of gas in safety valve body [K]
T_0 : Total temperature of gas in pipe [K]
U : Gas velocity in pipe [m/s]
U_d : Gas velocity at downstream segment of pipe [m/s]
U_u : Gas velocity at upstream segment of pipe [m/s]
V_B : Volume of safety valve body [m^3]
W : Flow rate of pipe or safety valve inlet [kg/s]
$\overline{W}$: Average flow rate of safety valve inlet[kg/s]
W_D : Flow rate of safety valve outlet [kg/s]
W_{in} : Inflow rate of pipe segment [kg/s]
W_{out} : Outflow rate of pipe segment [kg/s]
x : Coordinate along to inlet pipe [m]
Δx : Divided length of inlet pipe [m]
z : Compressibility factor of gas
Z : Displacement of valve disc = Lift of valve disc [m]
$\overline{Z}$: Average displacement of valve disc [m]
Z_s : Initial displacement of spring [m]
$\beta = \overline{P}A_H / K$: Parameter which shows safety valve characteristics
$\phi = \omega_n L/(\pi c)$: Non-dimensional inlet pipe length
η : Non-dimensional time
κ : Specific heat ratio of gas
λ : Friction coefficient of pipe
$\mu = \overline{P}A/(\overline{W}c)$: Parameter which shows safety valve characteristics
ν : Non-dimensional frequency of oscillation
ω_n : Natural angular frequency of safety valve disc [rad/s]
$\xi = x/L$: Non-dimensional coordinate along inlet pipe
ρ : Density of gas in pipe [kg/m^3]
ψ : Flow coefficient of safety valve nozzle
ψ_c : Critical flow coefficient
ψ_D : Flow coefficient of safety valve outlet
$\zeta =$: Damping ratio
ζ_s : Damping ratio (original one for valve disc)
ζ_{sv} : Additional damping ratio to satisfy neutral stability condition

Introduction

The unstable motion of the safety valve such as chatter or flutter would result in not only the damage of the safety valve itself but also the damage of the related piping system due to the vibration excited by the pressure pulsation which is generated by the valve disc oscillating motion. In the baseload LNG plant, hundreds of safety valves are installed and their role in the plant is essential from view point of safe plant operation. Therefore, the stability of the safety valves is one of the most important issue from view point of the plant operability and reliability. The risk relating to the safety valve stability tends to increase for the recent baseload LNG plant corresponding to the larger plant capacity requirement resulting in the safety valve size increase. API RP 520 [1] describes that the pressure drop of the inlet piping system should not exceed 3 percent of the set pressure to prevent the unstable motion of the safety valve because excessive pressure drop of the piping system around it results in lack of pressure difference between its inlet and outlet to keep its stable opening. It is also known that interaction between the pressure wave propagation in the piping system and the valve disc motion would cause the safety valve instability. There were many researches on the safety valve instability on this interaction effect [2]-[5], however, these researches could not derive adequate unstable conditions which are useful to establish design procedure to prevent the safety valve instability. In addition to the instability caused by the excessive pressure drops of the piping system and interaction between pressure wave propagation and the valve disc motion, there is another possibility of the safety valve unstable condition derived from small outlet area ratio to the orifice area of the safety valve. High build up back pressure of the safety valve would result in unstable motion of the safety valve and high back pressure acting to the valve disc would be caused by not only larger pressure drop of the outlet piping system but also smaller outlet area ratio to the orifice of the safety valve. The outlet area ratio of the safety valve tends to become smaller as it nominal size becomes larger as specified in API526 [6]. Recently, Dossnea [7] explains that pressure accumulation in the safety valve body for larger size of the safety valve would result in insufficient valve lift and capacity due to lack of the lift force in case of relatively high set pressure based on the static force balance obtained by CFD analysis. However, it seems that there is no publication in which the effect of outlet area ratio is precisely investigated on the safety valve instability.

In this study, experiments were carried out for the safety valve mounted to a vessel through a inlet pipe to investigate the stability of the safety valve. Stability analysis was carried out using the neutral stability searching method which is recently introduced by Hayama [8] for a pressure-open-type valve whose structure is similar to the conventional type safety valve. Also numerical simulation was executed to investigate the stability of the system.

Experimental Facility

Figure 1 shows the experimental facility. A safety valve is mounted to a vessel of 0.5m^3 through a inlet pipe. Air is supplied to the vessel by a compressor and the safety valve opens if the pressure in the vessel rises to its set pressure of 2 MPaG. Two sizes of the valves, 1E2 and 1.5F2 were used in the experiment. Blowdowns of these safety valves were adjusted to 10%. The inlet pipe lengths were varied from 0m to 20m. A reducer was installed to the outlet of the safety valve to adjust the outlet area ratio to the orifice area to simulate small area ratios of larger safety valves. Pressure sensors with strain gage type were installed in the vessel, the inlet of the safety valve, and safety valve body. A non-contact displacement meter with a laser sensor was used to measure the lift of the safety valve.

Figure 1 Experimental Facility

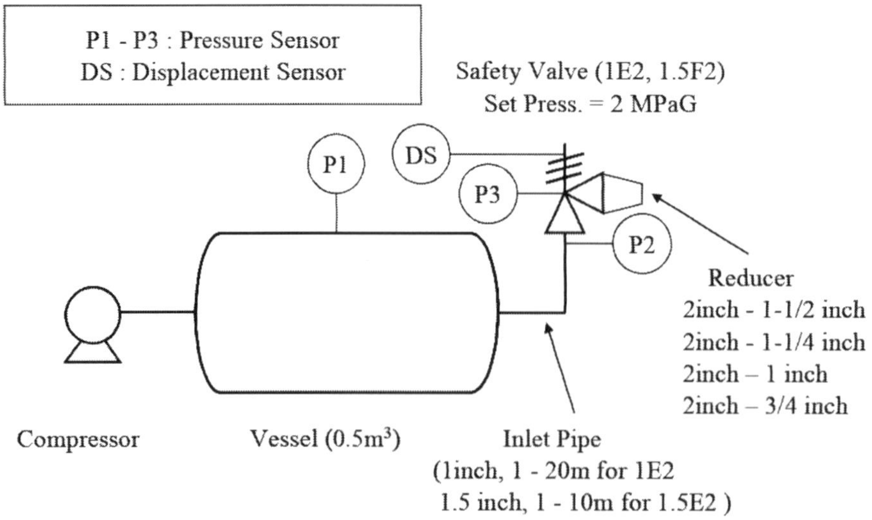

Method of Analysis

1. Stability Analysis

Hayama had recently introduced a stability theory for a pressure-open-type valve using the neutral stability searching method with additional virtual damping which gives the neutral stability condition [8]. Since the structure of the pressure-open-type valve is similar to the safety valve, the stability analysis was carried out using the neutral stability searching method introduced by Hayama with modification to the safety valve in this study.

The equation of the motion of the valve disc can be expressed by equation (1).

$$M_S\ddot{Z} + C_S\dot{Z} + KZ = (P - P_B)A_H\left\{1 + f(Z)\left(\psi/\psi_c\right)^2\right\} - KZ_s \tag{1}$$

Here, ψ and ψ_c are expressed as follows:

$$\psi = \sqrt{\frac{2\kappa}{\kappa-1}\left[\left(\frac{P_B}{P}\right)^{\frac{2}{\kappa}} - \left(\frac{P_B}{P}\right)^{\frac{\kappa+1}{\kappa}}\right]} \quad \text{for } P_B > P_c$$

$$\psi = \psi_c = \sqrt{\kappa\left(\frac{2}{\kappa+1}\right)^{\frac{\kappa+1}{\kappa-1}}} \quad \text{for } P_B \le P_c$$

$$P_c = P\left(\frac{2}{\kappa+1}\right)^{\frac{\kappa}{\kappa-1}} \tag{2}$$

The first term of the right hand side of equation (1) expresses the lift force acting to the valve disc and this term becomes zero for the pressure-open-type valve in the Hayama's stability theory. The lift force function $f(Z)$ expresses the effect of reaction force of the flow through the nozzle. Usually the velocity through the nozzle reaches the sound speed and $(\psi/\psi_c)^2$ in the lift force becomes 1.0. If the pressure in the valve body, P_B becomes higher than the critical pressure, P_c, the reaction force reduces from that at the critical flow condition, and this effect of the force reduction is expressed by the term of $(\psi/\psi_c)^2$ proportional to the momentum energy of the flow through the nozzle. Lift force function $f(Z)$ was obtained from the experiment in which the spring is removed from the safety valve and the lift force was measured by a load cell. Figure 2 shows the lift force function obtained by the measured data for 1E2 safety valve.

Figure 2 Lift Force Function of 1E2 Safety Valve

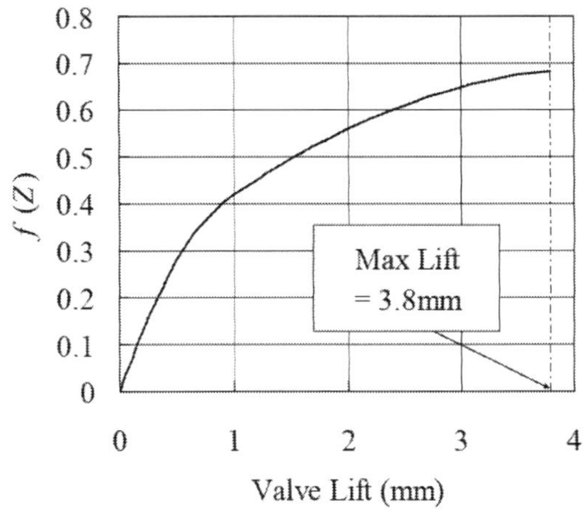

Figure 3 shows the comparison between measured lift force and calculated lift force from the lift force function with no reducer and a reducer of 1" at the safety valve outlet. As shown in figure 3 the lift force tends to increase as the lift increase in case of no reducer at the safety valve outlet. However, the lift force tends to decrease as the lift increase for larger lift region in case of 1" reducer at the safety valve outlet to simulate smaller outlet area ratio. Including these lift force characteristics the calculated lift forces have good agreement with the measure data. Therefore, the lift force expression in equation (1) is considered to be suitable to present the actual lift force characteristic of the safety valve.

Figure 3 Lift Force of 1E2 Safety Valve

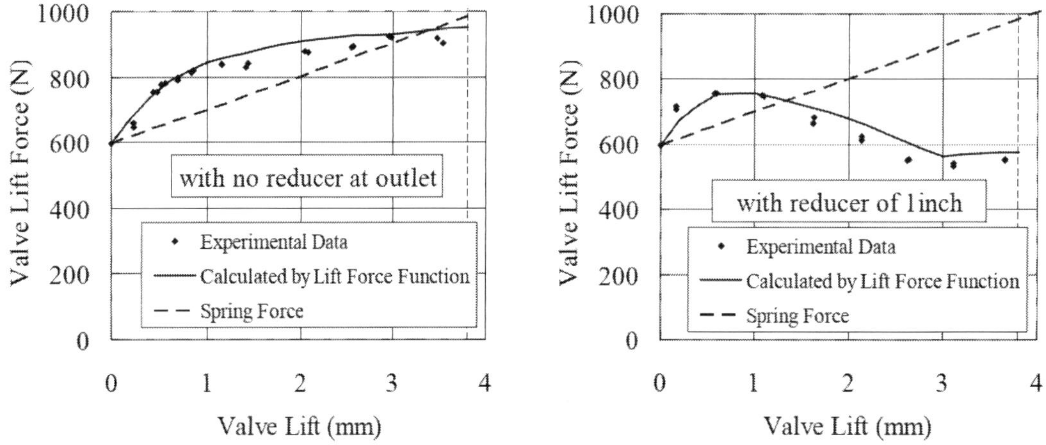

To introduce the stability condition it is assumed that the safety valve could be neutral stability condition in which the valve disc would have oscillating motion and the amplitude of the oscillation would be constant. In this neutral stability condition, each variable can be expressed as the sum of the average value and variable part as follows:

$$Z=\bar{Z}+\tilde{Z},\quad P=\bar{P}+\tilde{P},\quad W=\bar{W}+\tilde{W},\quad f=\bar{f}+\tilde{f}=\bar{f}+\frac{df}{dZ}\tilde{Z} \qquad (3)$$

Here, "-" and "~" show the average value and variable part, respectively. In order to satisfy the neutral stability condition, a virtual damping coefficient, C_{SV}, is introduced and the equation of the motion of the valve disc (1) can be transformed to the following equation at the neutral stability condition:

$$M_S\ddot{\tilde{Z}}+(C_S+C_{SV})\dot{\tilde{Z}}+K\tilde{Z}=\tilde{P}A_H\left(1+\bar{f}\right)+\bar{P}A_H\frac{df}{dZ}\tilde{Z} \qquad (4)$$

P_B is assumed to be constant in introducing equation (4). Using the relationships expressed by the equations (5), equation (6) can be obtained.

$$\omega_n^{\ 2}=K/M_s,\quad \omega_n t=\eta,\quad \beta=\frac{\bar{P}A_H}{K},\quad K_1=\frac{\bar{Z}+Z_f}{\bar{Z}},\quad \zeta=\zeta_S+\zeta_{SV}=(C_S+C_{SV})/2M_s\omega_n \qquad (5)$$

$$\frac{\ddot{\tilde{Z}}}{\bar{Z}}+2\zeta\frac{\dot{\tilde{Z}}}{\bar{Z}}+\left(1-\beta\frac{df}{dZ}\right)\frac{\tilde{Z}}{\bar{Z}}=K_1\frac{\tilde{P}}{\bar{P}} \qquad (6)$$

Equation (6) shows the natural frequency of the valve disc motion decreases due to the effect of df/dZ. As the weight flow rate through the safety valve is proportional to the product of the upstream pressure and the valve lift, the following equation is derived:

$$\frac{\widetilde{W}}{\overline{W}} = \frac{\widetilde{P}}{\overline{P}} + \frac{\widetilde{Z}}{\overline{Z}} \tag{7}$$

Using equations (6) and (7) and introducing the Laplace transformation the following equation is derived:

$$\left[s^2 + 2\zeta s + \left(1 - \beta\frac{df}{dZ}\right)\right]\frac{W(s)}{\overline{W}} = \left[s^2 + 2\zeta s + \left(1 - \beta\frac{df}{dZ}\right) + K_1\right]\frac{P(s)}{\overline{P}} \tag{8}$$

The general solution of the wave equation for the inlet pipe with the length L can be expressed by the following equation with the constant pressure at $x = 0$:

$$P(\xi,\eta) = F(\eta - \pi\phi\xi) - F(\eta + \pi\phi\xi), \quad W(\xi,\eta) = (A/c)[F(\eta - \pi\phi\xi) + F(\eta + \pi\phi\xi)] \tag{9}$$

Here, ϕ is the ratio of the duration in which the pressure wave propagates from the safety valve to the upstream vessel and returns back to the safety valve to the natural period of the disc of the safety valve and it expresses a non-dimensional pipe length. The Laplace transformation of the equation (9) at $\xi = 1$ where the pipe connects to the safety valve derives the following equation:

$$\frac{W(s)}{P(s)} = -\frac{A}{c}\coth(\pi\phi s) \tag{10}$$

From the equation (7) and (10), the following equation is derived:

$$\left[s^2 + 2\zeta s + \left(1 - \beta\frac{df}{dZ}\right) + K_1\right] + \mu\left[s^2 + 2\zeta s + \left(1 - \beta\frac{df}{dZ}\right)\right]\coth(\pi\phi s) = 0 \tag{11}$$

Here $\mu = \overline{P}A/(\overline{W}c)$ is a parameter which shows one of the safety valve characteristics. Since the root of s in the equation (11) shall be a purely imaginary number at the neutral stability condition, it can be expressed as $s = i\nu$. In this condition, both of real and imaginary parts of the equation (11) shall be equal to zero, simultaneously. Therefore, ν and ζ can be obtained as follows:

$$\nu = \sqrt{\left(1 - \beta\frac{df}{dZ}\right) + \frac{K_1}{1 + \mu^2\cot^2(\pi\phi\nu)}}, \quad \zeta = -\frac{K_1\mu\cot(\pi\phi\nu)}{2\nu(1 + \mu^2\cot^2(\pi\phi\nu))} \tag{12}$$

Here, ν and ζ are the non-dimensional frequency and the damping ratio to satisfy the neutral stability condition, respectively.

2. Numerical Simulation

The weight flow rates through the orifice and the outlet of the safety valve are expressed as follows:

$$W = \frac{A_S C_d \psi P}{\sqrt{zRT_0/M_w}}, \quad A_S = \min(\pi d_H Z, A_O), \tag{13}$$

$$W_D = \frac{A_D C_{dD} \psi_D P_B}{\sqrt{zRT_0/M_w}} \tag{14}$$

Hear, ψ_D is the flow coefficient of the safety valve outlet and obtained by equation (2) with substitution of $P_B = P$ and $P_D = P_B$.

The mass conservation equation in the safety valve body is expressed as follows:

$$\frac{dM_B}{dt} = W - W_D, \tag{15}$$

Using the equation of state for gas the pressure in the safety valve body can be expressed as follows:

$$P_B = \frac{M_B zRT_B}{V_B M_w} \tag{16}$$

The inlet pipe is divided into the segments as shown in figure 4. The following continuity equation, the energy equation, and the equation of motion are derived for each segment:

$$\frac{dM}{dt} = W_{in} - W_{out} \tag{17}$$

$$\frac{dE}{dt} = W_{in}\left(e + \frac{P}{\rho} + \frac{U^2}{2}\right)_{in} - W_{out}\left(e + \frac{P}{\rho} + \frac{U^2}{2}\right)_{out}, \quad E = M\left(e + \frac{U^2}{2}\right), \quad T = \frac{e}{C_v} \tag{18}$$

$$\frac{dU}{dt} = F1 + F2 + F3, \quad F1 = -\frac{1}{\rho}\frac{\partial P}{\partial x} = -\frac{P_d - P_u}{\rho \Delta x}, \quad F3 = -\lambda \frac{1}{D} \cdot \frac{U|U|}{2},$$

$$F2 = -U\frac{\partial U}{\partial x} = -U_u \frac{U - U_u}{\Delta x} \quad \text{or} \quad F2 = -U\frac{U_d - U}{\Delta x} \tag{19}$$

Figure 4 Divided Segment of Pipe

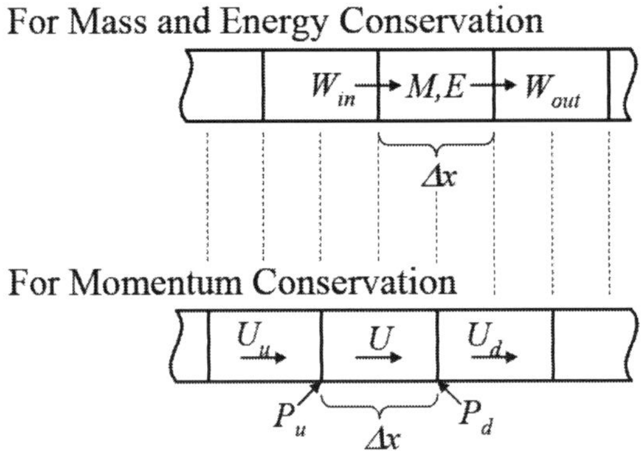

The upwind method is applied for the convection term, *F*2, in the equation (19). Using the equation of state for gas the pressure in the pipe can be expressed as follows:

$$P = \frac{MzRT}{M_w A \Delta x} \tag{20}$$

The motion of the valve disc can be obtained by the numerical integration of the differential equations (1), (15), (17), (18) and (19) using other related equations. The Runge-Kutta method with forth order was used in the numerical integration in this study.

Results and Discussion

1. Experiment

Figure 5 shows the experimental results for 1E2 safety valve with no reducer at its outlet and inlet pipe length of 1m. It is observed that the valve lift first rises rapidly with popping action then oscillating motion occurs. The phase difference between the valve lift and the pressure at the safety valve inlet becomes around 180 degrees. If the static force balance would be assumed between the spring force of the safety valve and the pressure force at the safety valve inlet this phase difference should be around 0 degree. Therefore, it is considered that the dynamic instability would be generated due to the interaction effect between the pressure at the safety valve inlet and the valve disc motion. Figure 6 shows the experimental results for 1E2 safety valve with no reducer at its outlet and inlet pipe length of 10m. It is observed that the valve lift first rises around 1mm then oscillating motion is attenuated. The second popping action occurs at around 70 milliseconds after the first popping action then the disc oscillates with attenuation. After that disturbance is observed every 70 milliseconds, however, the valve lift finally becomes constant. Since 70 milliseconds is equal to the duration in which the pressure wave propagates from the safety valve to the upstream vessel and returns back to the safety valve, it is considered that the oscillating motion of the valve disc can be attenuated sufficiently until the negative pressure wave generated by the popping action returns back to the safety valve, therefore, the safety valve could be stable. On the other hand, the safety valve disc could become unstable in case of the inlet pipe length of 1m because enough attenuation of the valve disc motion could not be achieved when the reflection wave returns back to the safety valve as shown in figure 5. This interaction effect between the valve disc motion and the pressure wave propagation through the inlet pipe would result in the oscillating motion of the valve disc in case of relatively short inlet pipe length. As shown in figure 6, the valve lift does not smoothly reduce after its maximum position. It is suggested that the mechanical frictional force of the moving part of the safety valve would act to suppress the movement of the valve disc and result in the nonlinear characteristics of the valve disc motion.

Figure 5 Experimental Result (1E2, Inlet Pipe Size/Length = 1"/1m, No Reducer at Outlet)

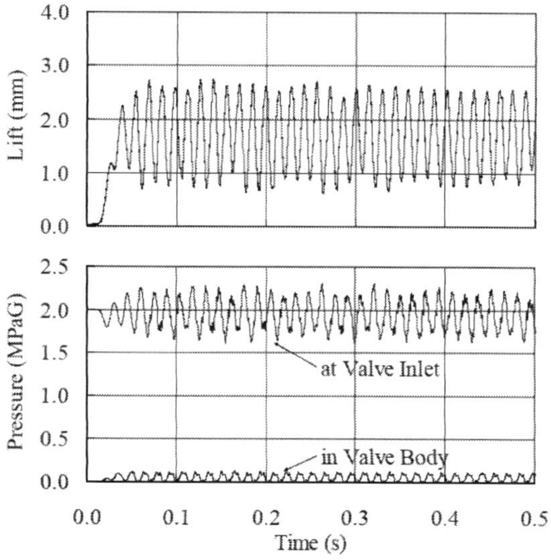

Figure 6 Experimental Result (1E2, Inlet Pipe Size/Length = 1"/10m, No Reducer at Outlet)

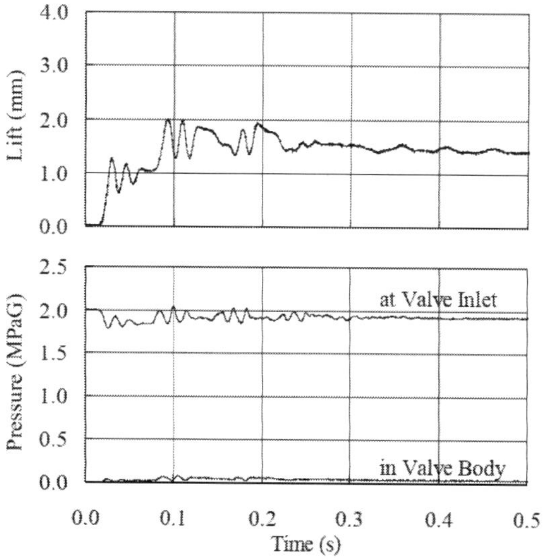

Figure 7 shows the valve lift for the inlet pipe length from 0m to 20m. As shown in figure 7, the safety valve becomes unstable in case of the inlet pipe length from 1m to 5m, and becomes stable for the inlet pipe length of 0m and more than 10m. The duration between the first and second popping action increases as the inlet pipe length increases because the time for pressure wave propagation through the inlet pipe increases as its length increases. Table 1 shows the summarized results of the experiment concerning to the effect of the inlet pipe length. As shown in table 1, similar experimental results were obtained for both of 1E2 and 1.5F2 safety valves on the effect of the inlet pipe length.

Figure 7 Experimental Result (1E2, Inlet Pipe Size = 1", Effect of Inlet Pipe Length)

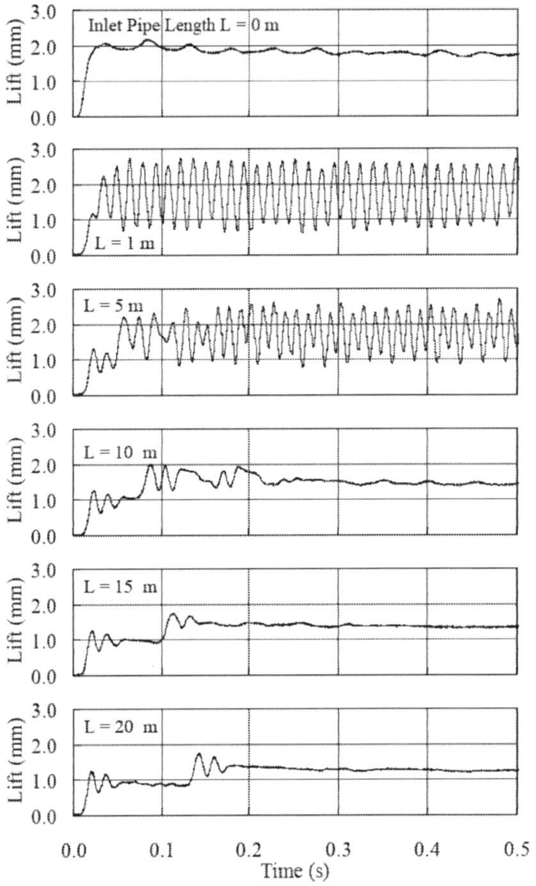

Table 1 Summarized Results of Experiment (Effect of Inlet Pipe Length)

Valve Size	Inlet Pipe Length (Inlet Pipe Size = 1")				
	1m	5m	10m	15m	20m
1E2	Unstable (55 - 68 Hz)**	Unstable (79-104Hz)**	Stable	Stable	Stable
1.5F2	Unstable (42 - 59 Hz)**	Stable / Unstable * (66-70 Hz)**	Stable	-	-

* : Stable for some experiments and unstable for other experiments
** : Frequency for dominant component of oscillation
(Natural frequency of valve disc with spring is 75Hz for both of 1E2 and 1.5F2)

Figure 8 shows the experimental results for 1E2 safety valve with 1 inch reducer at its outlet and no inlet pipe. As shown in this figure, it is observed that the pressure in the valve body rises as the lift increases then the lift decease and the valve closes. This cyclic motion repeats thus the chatter of the safety valve occurs. It is considered that the insufficient outlet area of the safety valve outlet would result in the rise of the pressure in the valve body and this pressure rise would be the cause of the unstable motion of the safety valve. Table 2 summarizes the experimental results on the effect of the outlet area ratio to the orifice area. From this table the unstable motion of the safety valve could occur if the outlet area ratio is less than 6.0. Since the actual outlet area ratio of the safety valve specified in API526 [6] is lower than 6.0 for larger sizes of the safety valve, the further investigation of this effect shall be desired to be executed for large size of the valve in the future.

Figure 8 Experimental Result (1E2, Inlet Pipe Length = 0m, 1inch Reducer at Outlet)

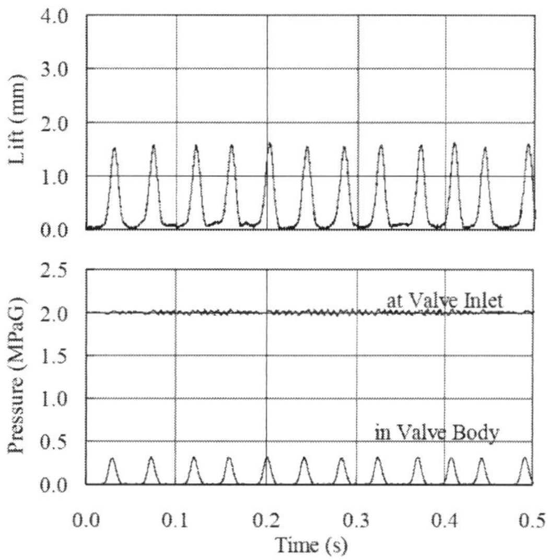

Table 2 Summarized Results of Experiment (Effect of Outlet Area Ratio)

Valve Size	Reducer Size at Outlet				
	No	1-1/2"	1-1/4"	1"	3/4"
1E2	Stable (11.2)***	Stable (7.5)***	Stable (5.5)***	Unstable * (3.3)***	-
1.5F2	Stable (8.3)***	Unstable** (5.6)***	Unstable** (5.1)***	Unstable (2.5)***	Unstable (1.5)***

* : Repeated popping action is observed
** : Unstable just before closure of safety valve
*** : Figure in parenthesis is outlet area ratio to orifice area

2. Stability Analysis

Figure 9 shows the non-dimensional frequency ν and damping ratio ζ satisfying the neutral stability condition obtained by equation (12) for the lift of 1mm with 1E2 safety valve. The first to forth modes are depicted in this figure and each mode corresponds to the acoustic mode in the inlet pipe with the closed and open ends for the safety valve and the vessel, respectively. From this figure, the number of the exciting modes increases as non-dimensional pipe length, ϕ, increases. The damping ratio, ζ, becomes the maximum when ν is closed to 1.0, and this means that the acoustic mode whose frequency is closed to that for the natural frequency of the safety valve could be excited. ζ is around 0.8 to 1.0 except ϕ is closed to zero, and this means that the safety valve becomes stable if it has damping ratio larger than 0.8 to 1.0. As mentioned later, the damping ratio of the safety valve using at the experiment is expected to closed to 0.2, therefore, the safety valve becomes unstable with long inlet pipe from this stability analysis even though the experimental results shows that it becomes stable in case of the inlet pipe length of 10m or longer.

As shown in figure 10 constant amplitude is assumed in the stability analysis. On the other hand, the actual amplitude reduces before the reflected wave reaches to the safety valve. From this difference in the attenuating characteristic during pressure wave propagation through the inlet pipe, the stability analysis cannot explain the stable condition of the safety valve with relatively long inlet pipe length.

Figure 11 shows the comparison between the frequencies obtained by the experiment and the stability analysis. The frequencies of the experiment are obtained by the spectrum analysis of the measured pressure data. Here, the inlet pipe length is modified taking the length of the safety valve mount into account. Since the both of the frequencies have good agreements as shown in figure 11 the stability analysis can be considered to express accurately the actual valve oscillating motion.

As shown in figure 9, the safety valve becomes stable when ϕ is small enough. This stable condition is derived from equation (12) and expressed as follows:

$$L \leq \frac{\pi c}{2\omega_n \sqrt{K_1}} \tag{21}$$

Figure 9 Non-dimensional Frequency and Damping Ratio at Neutral Stability Condition (1E2, Inlet Pipe Size = 1", Lift = 1.0mm)

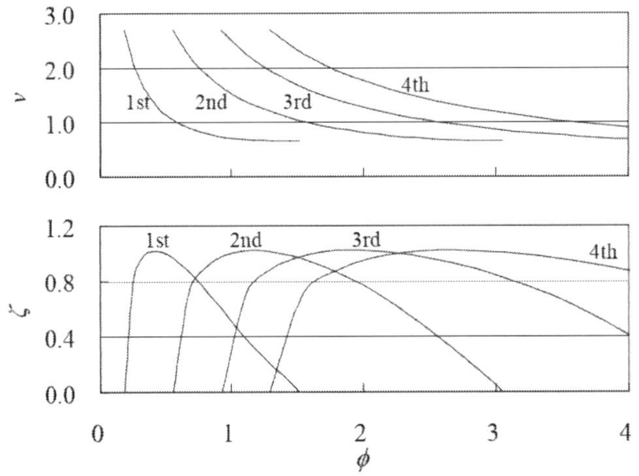

Figure 10 Difference between Oscillating Motions for Stability Analysis and Actual Condition Obtained by Experiment

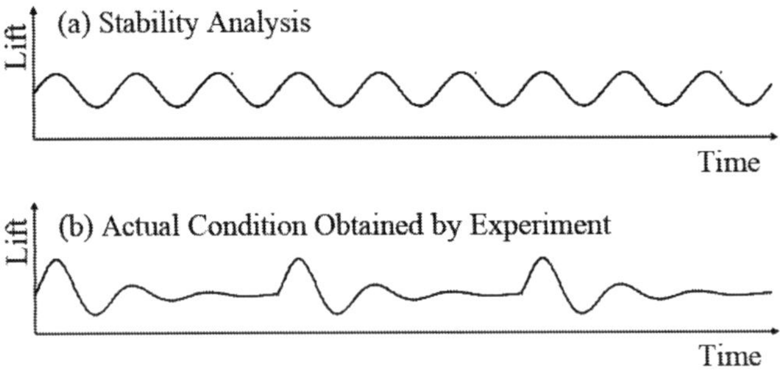

3. Numerical Simulation

Figures 12 and 13 show the comparisons between the simulations and the experiments for 1E2 safety valve with no reducer at its outlet and inlet pipe length of 1m and 10m, respectively. The damping ratio of the safety valve in the simulation is taken to be 0.2 from the comparison with the experimental data. From figure 12 for the inlet pipe length of 1m, the unstable motion of the safety valve can be expressed by the simulation even though there is no good agreement in the amplitudes. From figure 13 for the inlet pipe length of the 10m, the simulation can express the actual safety valve motion including the pressure wave propagation through the inlet pipe. The oscillating motion of the valve disc remains at 0.5 seconds in the simulation for the inlet pipe length of 10m though it is completely attenuated in the experiment as shown in figure 13. This difference is considered to be caused by the reason that the mechanical frictional force acts in the experiment when the velocity of the disc is small and this characteristic cannot be expressed by the linear damping model in the simulation.

Figure 11 Comparison between Stability Analysis and Experiment (1E2, Inlet Pipe Size = 1")

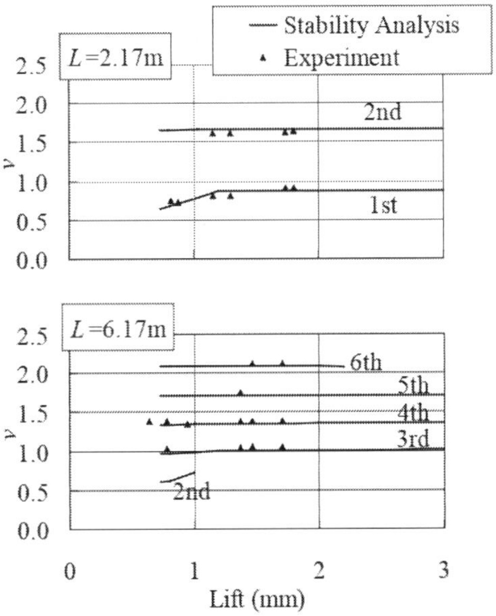

Figure 12 Comparison between Simulation and Experiment (1E2, Inlet Pipe Size/ Length = 1"/1m, No Reducer at Outlet)

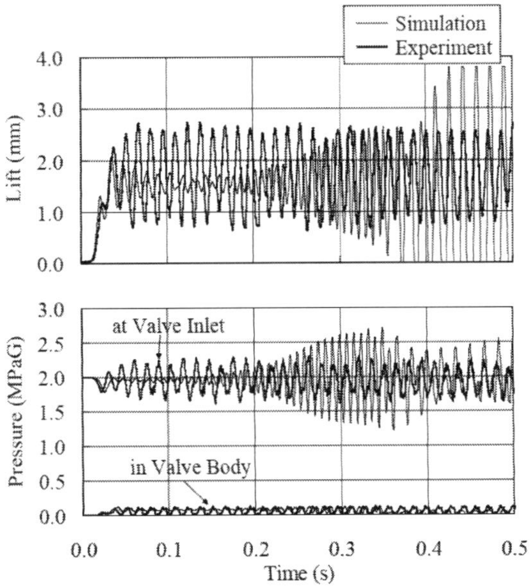

Figure 13 Comparison between Simulation and Experiment (1E2, Inlet Pipe Size/ Length = 1"/10m, No Reducer at Outlet)

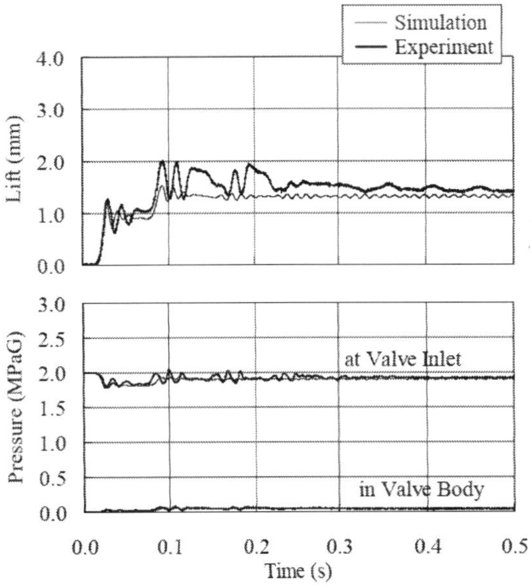

Figures 14 and 15 show the comparisons between the simulations and the experiments for 1E2 safety valve with 1-1/4 and 1inch reducers, respectively, in case that the safety valve mounted to the vessel with no inlet pipe. As shown in these figures the simulation can express the stable and unstable conditions of the safety valve related to the small outlet area ratio to the orifice area.

4. Effect of Pressure Drop of Inlet Pipe

Figure 16 shows the simulation results to investigate of the effect of inlet line pressure drop for 1E2 safety valve with inlet size of 1". As shown in this figure, in case of the inlet pipe length of 100m, larger pressure drop of the inlet pipe results in the insufficient inlet pressure of the safety valve, and cyclic open and close operation of the valve occurs because the inlet pressure reaches to the reseating pressure of the safety valve during opening condition. On the other hand, in case of the inlet pipe length of 10m, stable operation of the safety valve is observed since the inlet pressure is kept above the reseating pressure. As shown here, the numerical simulation is quite effective to investigate the safety valve stability relating to the excessive pressure drop of the inlet pipe in addition to the pressure wave propagation through the inlet pipe and relatively small outlet area ratio to the orifice area of the safety valve.

Figure 14 Comparison between Simulation and Experiment (1E2, Inlet Pipe Length = 0m, 1-1/4inch Reducer at Outlet)

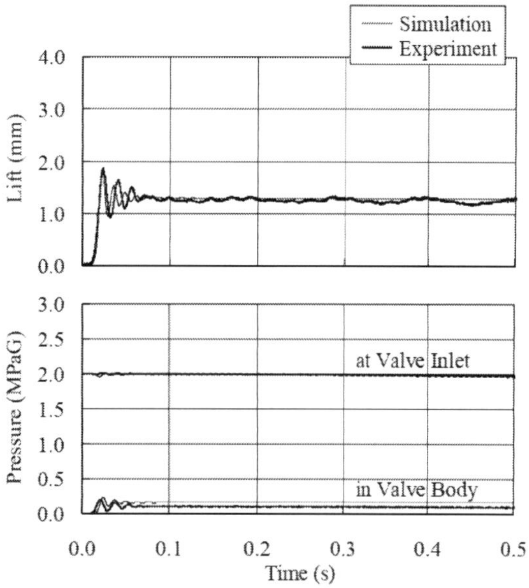

Figure 15 Comparison between Simulation and Experiment (1E2, Inlet Pipe Length = 0m, 1inch Reducer at Outlet)

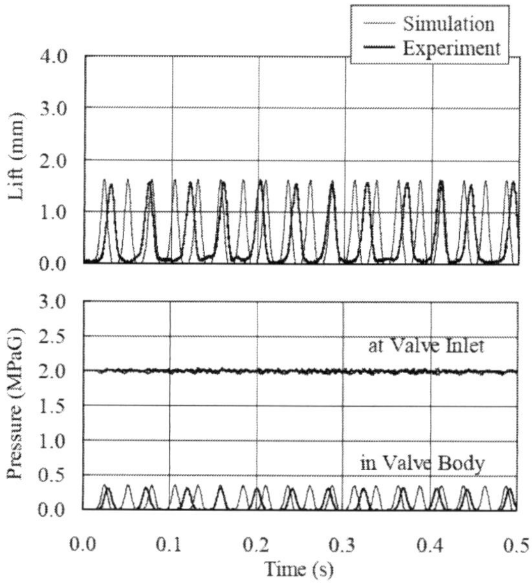

Figure 16 Effect of Pressure Drop of Inlet Line (Simulation Results, 1E2, Inlet Pipe Size = 1", No Reducer at Outlet)

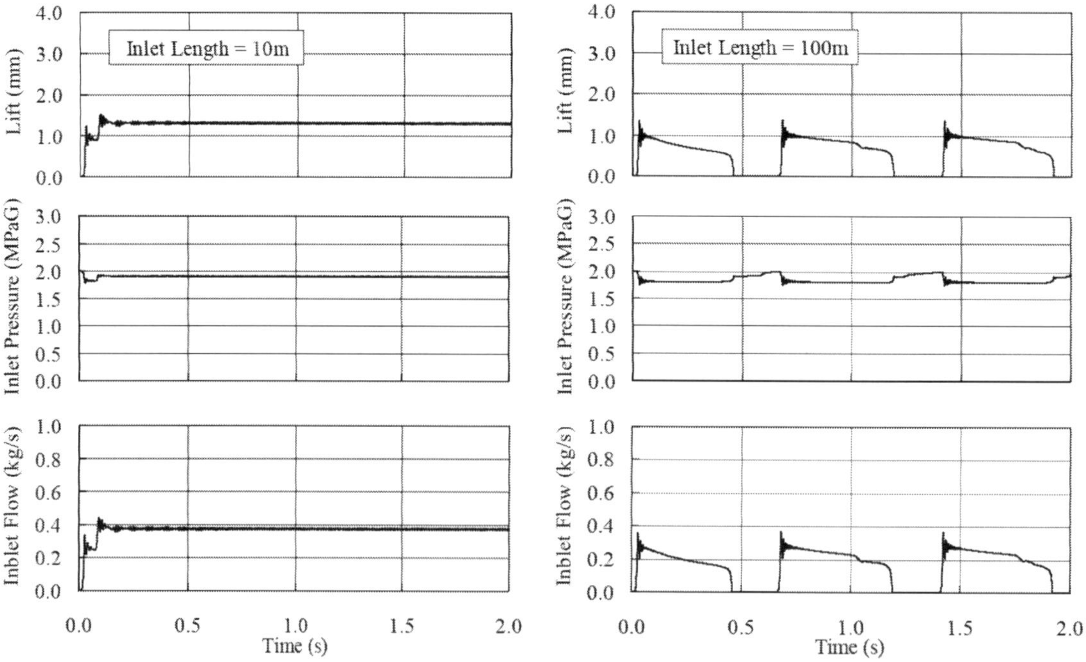

Conclusion

The stability of the safety valve mounted to a vessel through a inlet pipe was investigated by the experiments, the stability analysis, and the numerical simulation. As the results the following conclusions were obtained.

(1) The safety valve becomes unstable due to the dynamic instability caused by the interaction effect between the valve disc motion and pressure wave propagation.
(2) The dynamic instability does not occur in case that the inlet pipe length is enough short or relatively long.
(3) If the oscillating motion of the valve disc can be sufficiently attenuated before the pressure wave returns back to the safety valve in case of relatively long inlet pipe length, the safety valve becomes stable.
(4) The attenuation of the valve disc motion is caused by the mechanical friction force of the moving part of the safety valve and has nonlinear characteristic.
(5) The safety valve could be unstable if the outlet area ratio to the orifice area is less than 6.0.
(6) The stability analysis is effective to explain the instability of the safety valve and the stable condition for the inlet pipe length can be expressed by $L \le \pi c / \left(2\omega_n \sqrt{K_1}\right)$.
(7) The stability analysis can not explain the stable condition in case of sufficiently long inlet pipe length.
(8) The numerical simulation is effective to show the instability of the safety valve for the following three types of safety valve instabilities:
    (a) Excessive pressure drop of the inlet pipe
    (b) Instability caused by the effect of the pressure wave propagation in the inlet pipe
    (c) Relatively small outlet area ratio of the safety valve

Further investigation of the safety valve stability will be desired to execute for the following view points:

(1) The dynamic instability of the safety valve caused by pressure wave propagation in the inlet pipe for wide range of the safety valves including the effect of set pressure, safety valve size, etc.
(2) The instability caused by the effect of small outlet area ratio to the orifice area for large size of the safety valve.

Acknowledgment

The author gratefully acknowledges to Professor Hayama for valuable advises to proceed the study and Mr. C. Hasegawa and Mr. T. Imanishi, Fukui Corporation for executing the valuable experiments. The author also acknowledges to Mr. T. J. Bevilacqua, ExxonMobil Development Company for advising the publication of the study.

Literature

[1] API RP520, Sizing, Selection, and Installation of Pressure-Relieving Devices in Refineries, Part II – Installation, 2003

[2] A. Singh, A. Singh, A. M. Hecht and M. E. Teske, "A Model for Predicting the Performance of Spring-Loaded Safety Valves, A. M. Hecht and M. E. Teske, , ASME Paper, 82-WA/NE-12, 1982

[3] J. M. Henry and P. Coppolani, Use of Damping Device to Control the Stability of Spring Loaded Safety Valves Operated in Liquid Service, Proc. Developments in Valves and Actuator for Fluid Control, 1988, 103-114

[4] I. Harris etc., Dynamic Simulation of Safety Valve Installations, AIChE Symp. Ser., 80-236, 1984, 171-178

[5] K. A. Kastor, Relief Valve Chatter Testing, ASME PVP, 237-2, 1992, 133-139

[6] API Std526, , Flanged Steel Pressure Relief Valves, 2002

[7] V. Dossena, F. Marinoni, B. Paradiso, Valve Size Influence on the Discharge Capacity of Spring Loaded Safety Valves, Paper 722, Valve World Conference 2007

[8] S. Hayama, Stability Analysis of a Piping System Equipped with a Pressure-Open-Type Valve at the Exit, Proc. D&D, JSME, 309, 2006 (Japanese)

Methods for Evaluating and Reducing the Risks of Pneumatic Pressure Testing of Vessels and Piping

Victor H. Edwards[1,2], Don Sanford[2], Brooke Bonstead[2], and Larry Skoda[3]

ABSTRACT

Hydrostatic testing of piping and vessels designed to handle LNG or other cryogenic liquids is problematic because of the need to remove residual water.

Pneumatic pressure testing of piping and vessels at moderate-to-high test pressures or at low test pressures with high volume is more hazardous than hydrostatic pressure testing because the stored energy is much greater with compressed gases. Hazards from loss of containment during pneumatic pressure testing include both blast overpressure and missiles.

In applications where pressure testing with liquids is undesirable, such as in cryogenic piping systems and vessels, pneumatic pressure testing can only be justified when care in fabrication and in non-destructive examination of vessels and piping reduces the probability of loss of containment to such a small value that risk is acceptable.

This paper outlines methods for evaluating the risks of pneumatic pressure testing of vessels and piping, as well as methods of risk reduction.

1 – To whom correspondence should be addressed; vic.edwards@akersolutions.com

2 - Aker Solutions US Inc., 3600 Briarpark Drive, Houston, TX 77042-5206

3 – Current address: Chevron, Houston, TX

Paper accepted for presentation at the 10th Natural Gas Utilization Symposium and the Spring National AIChE Meeting, March 21-25, 2010, San Antonio, TX.

INTRODUCTION

The pressure testing of the process piping and vessels on an LNG or other cryogenic facility is essential prior to the introduction of LNG or other cryogens. Reported here are methods used to minimize the risks associated with pneumatic pressure testing.

The atmospheric boiling point of LNG is approximately minus 160°C. Residual water left by hydraulic pressure testing could result in operational problems if not completely removed. Complete removal of residual water can be very difficult and time consuming. Pneumatic pressure testing avoids these problems, and is frequently used for LNG and other piping and vessels in which moisture is undesirable.

Hydraulic pressure testing with water is much more common than pneumatic pressure testing with a gas because the stored energy of compressed gas can be roughly 200 times the stored hydraulic energy for the test pressures in the range of 100 barg. Therefore, rupture of a piping test system during a pneumatic pressure test can release much more energy. Damage due to rupture can result from shock waves, flying projectile fragments from the ruptured piping, and unrestrained movement of piping and equipment propelled by escaping gas.

The cryogenic gas industries have experienced pneumatic pressure testing incidents in the past, sometimes resulting in serious injuries and major equipment damage.

To address the risks associated with pneumatic pressure testing, many companies attempt to limit the amount of stored energy in the test system to a prescribed maximum value by limiting the size of each test system. This approach is often not practical for high pressure piping of typical diameters because of the severe limitation that it places on the size of each test system. Consequently, this approach may result in an unreasonably large number of test systems. Attempting to isolate and test a large number of test systems may prove to be impractical. When the approach of limiting the amount of stored energy becomes impractical, an alternative approach such as that described herein may offer the best option. Regardless of which approach is taken, many of the considerations outlined herein must be considered to safely execute high pressure pneumatic pressure testing.

A variety of measures can increase the safety of pneumatic testing. Of first importance are measures to ensure the mechanical integrity of the vessels and piping systems undergoing test. These measures include design, fabrication, and inspection methods. It is also necessary to exclude personnel from exclusion zones surrounding the vessel or piping system under test, and to conduct tests at night or on weekends when few people are in the vicinity of the test site.

HAZARDS FROM OVERPRESSURE FROM A RUPTURED VESSEL OR PIPING SYSTEM

Rupture of a Piping System under Pressure Produces a Blast Wave

The pneumatic pressure testing planned for the piping systems for one LNG terminal were as high as 121 Barg, depending on the class and size of piping under test. Pressures this high can produce a damaging overpressure in the atmosphere within the exclusion zone due to the blast wave or shock wave that results upon rupture of a piping system under test. Lower test pressures can also present significant hazards. For example, rupture of a specific eight inch diameter pipe segment at a test pressure of 18 barg could result in a blast overpressure of 0.5 psig (0.0345 barg) at a distance of 28 meters.

Overpressure Can Injure Personnel and Damage Facilities

Overpressure is the localized increase in the atmospheric air pressure associated with the passage of a shock wave.

The overpressure that accompanies failure of a piping system causes harm that is a function of the magnitude and the duration of the shock wave. Typical damage from overpressures is listed below (Mannan, 2005):

- 0.4 psig - Limited minor structural damage to buildings
- 0.5 to 1 psig - Glass shattering with body penetrating velocities
- 0.7 psig - Minor damage to house structures
- 1 psig - Partial damage of house structures; made uninhabitable
- 1 psig - 95% eardrum protection with ear plugs
- 1 psig - People knocked down with potential of significant resulting injuries

Overpressure has the potential to affect most of the nearby area surrounding the piping under test. *Therefore, the minimum exclusion zone in this work is defined here as a zone within a radius beyond which the overpressure from rupture of the piping system under test will not exceed 0.5 psig (0.0345 barg).*

Stored Pneumatic Energy Powers the Shock Wave

Part of the energy that went into compressing a gas up to the test pressure of a piping system is released suddenly if the piping system ruptures. Several different ways of estimating the theoretical amount of stored energy that is converted into blast overpressure and acceleration of fragments of the ruptured vessel have been proposed. The most common methods of estimating stored energy release assume isentropic expansion (United Kingdom Health and Safety Executive [HSE], 1998; American Institute of Chemical Engineers Center for Chemical Process Safety [CCPS], 1999; Mannan, 2005), isothermal expansion (CCPS, 1999; Mannan, 2005), and/or thermodynamic availability (Crowl, 1992; CCPS, 1999; Mannan, 2005). The isentropic expansion model gives the lowest estimate; the isothermal model gives the highest estimate, and the availability model gives an intermediate value (Mannan, 2005). The isothermal model can give energy release estimates twice as large as the isentropic model.

Various equations have been developed to estimate the energy release, but most of these equations are based on the ideal gas law. Perhaps the simplest equation was initially proposed by Brode (1959), and is based on the energy required to raise the pressure of the gas at constant volume from atmospheric pressure to the piping burst pressure (CCPS, 1999):

$$\mathbf{E = [(P_1 - P_o)V/(\gamma - 1)]} \qquad (1)$$

Where **E** = Energy Released Upon Piping System Rupture

P1 = Piping Burst Pressure

Po = Atmospheric Pressure

V = Volume of Piping System Under Test

γ = Ratio of heat capacity of gas at constant pressure to heat capacity of gas at constant volume

The ideal gas law is not accurate and is non-conservative for the upper range of planned test pressures, so isentropic fluid expansion energies based on measured thermodynamic properties of nitrogen gas (United Kingdom Health and Safety Executive [HSE] 1998) are tabulated in the summary of pneumatically tested piping spreadsheet in Table 1. Refer to HSE (1998) for an explanation of the calculation method.

Computation of Overpressure from Rupture of a Piping System

Published methods used to estimate the overpressure of a blast wave produced by rupture of a pressure vessel are usually based on the stored energy of the total volume of compressed gas contained within the vessel. The total gas volume has been used traditionally because the length-to-diameter ratio of the vessel is usually small enough that all of the compressed gas contributes to the blast wave. In contrast, the LNG piping systems may have a large length-to-diameter ratio, and only a portion of the piping system near the initial point of failure is expected to contribute to the initial blast wave.

Blast calculations for LNG piping systems in one project were based on the stored energy available in a volume equal to a 40 ft length of pipe plus an additional 24D (D=Pipe Diameter). This assumes a longitudinal seam failure of one 20 ft pipe section (i. e., one single random length). An additional 20 ft was added to account for flow from branches in or adjacent to the failed section, and 24D has been added to account for flow originating from the header piping adjacent to the failed section. The maximum 20 ft weld length failure is based on the premise that a propagating crack will stop when it meets the resistance of a circumferential weld. For simplicity, the same volume basis was used to evaluate piping systems fabricated from seamless pipe.

The assumptions stated above with regard to the amount of stored energy in adjacent sections of piping that contribute to overpressure in the exclusion zone and projectile propulsion must be reviewed on each individual project with consideration given to the physical dimensions and layout of the piping.

An American Institute of Chemical Engineers (AIChE) spreadsheet (Center for Chemical Process Safety [CCPS] 1999) that is based on the method of Baker, et al. (1983) was used to estimate overpressure from rupture of a piping system by gas pressure. Baker's method bases the stored energy of the gas on the Brode equation (Equation 1), but then applies correlations based on experimental data. The resulting exclusion distances to keep overpressures below 0.5 psi (0.0345 barg) are summarized in Table 1 in the column labelled R_{blast}. Overpressures were calculated for maximum test pressures that specific piping systems could experience, as well as lower increments of piping system test pressures.

A representative output from the CCPS/Baker spreadsheet (CCPS, 1999) is presented in Table 2. The spreadsheet example in Table 2 was for a 36" carbon steel, schedule 80 pipe that has broken at a test pressure of 101 Barg. The resulting $\mathbf{R_{blast}}$ for an overpressure of 0.5 psig (0.0345 barg) was 156 meters. For comparison, the range predicted by Equation (4) for fragment range for brittle fracture, $\mathbf{R_{brittle}}$ was 908 meters

Exclusion zones were chosen to insure that overpressures from rupture of the piping system could not exceed 0.5 psig (0.0345 barg) at the boundary of the exclusion zone radius, regardless of the mode of fracture of the piping system.

HAZARDS FROM PROJECTILES FROM A RUPTURED VESSEL OR PIPING SYSTEM

Rupture of a piping system may occur due to either ductile fracture or brittle fracture. Under the planned test conditions for one LNG project, it was determined that the carbon steel systems might exhibit ductile or brittle fracture and that the stainless steel piping systems might exhibit ductile fracture. The number of fragments and the methods of estimating the velocities of the resulting projectiles differ for brittle and ductile fracture.

Rupture of a Piping System under Pressure May Produce Projectiles

Here are Mannan's (2005) introductory remarks on fragmentation resulting from either brittle or ductile fracture:

"The number of missiles formed in an explosion involving rupture of containment varies widely. At one extreme is the bursting of a weapon such as a high explosive shell or grenade which normally gives a large number of fragments. Large numbers of missiles are also produced by fragmentation of a gas filled pressure vessel. At the other extreme is the ejection of a single item such as a valve component due to failure in a high pressure system."

"Of particular interest here is rupture of a pressure vessel. This may involve either brittle or ductile fracture. ...In general, failure is more likely to be ductile. Ductile failure does not usually produce missiles, but if it does they are likely to be small in number but may have the potential to do severe damage. It is brittle fracture which is most likely to produce failures in which quite a large number of fragments are generated."

Theoretical Maximum Velocity and Range of Projectiles

One method of estimating the theoretical maximum velocity is to assume that the pressure energy stored in a piping system is completely converted into kinetic energy of the piping system under test (HSE, 1998; CCPS, 1999). Then

$$\mathbf{E = (m * v^2)/2} \qquad (2)$$

Where **E** = Energy released upon piping system rupture

m = total mass of projected piping system fragments

v = velocity of each fragment of the piping system

TABLE 1 – TYPICAL BLAST AND FRAGMENT RANGES FOR PIPING SECTIONS

OD (Inches)	Wall Schedule	Design Pressure (Barg)	Max. Test Pressure (Barg)	Selected Length of Test Pipe (ft)	Stored Energy (MJ)	V_{max} (m/s)	R_{max} (m)	$R_{brittle}$ (m)	$R_{ductile}$ (m)	R_{blast} (m)	Exclusion Zone (m)
Carbon Steel Pipe											
8.625	STD	19.6	22	57	1.746	68.6	480	192	394	30	30
8.625	XS	91.0	101	57	8.534	123.0	1543	617	891	47	47
24	STD	19.6	22	88	22.777	109.7	1228	491	716	69	69
24	80	91.0	101	88	104.900	133.0	1805	722	1448	108	108
Stainless Steel Pipe											
8.625	20	19.0	21	57	1.751	76.8	602	241	383	29	29
8.625	60	109.8	121	57	10.688	150.3	2304	922	985	51	51
24	10S	19.0	21	88	22.519	131.9	1775	710	691	68	68
24	60	99.3	110	88	119.084	156.4	2497	999	1614	113	113

Exclusion Zone = The zone within the radius R_{blast}

MJ = Mega Joules

R_{blast} = Radius within which overpressure exceeds 0.5 psig (0.0345 barg).
R_{max} = Theoretical maximum projectile range based on 100% conversion of released stored energy into kinetic energy [ref. Eq. (3)].

$R_{brittle}$ = Estimated maximum projectile range resulting from brittle fracture based on HSE (1998) report.
$R_{ductile}$ = Estimated maximum projectile range of a blind flange resulting from ductile fracture based on HSE (1998) report.
V_{max} = Theoretical maximum velocity of projectiles based on 100% conversion of released stored energy into kinetic energy [ref. Eq. (2)].

TABLE 2 – TYPICAL RESULTS FROM THE APPLICATION OF BAKER'S METHOD FOR ESTIMATION OF BLAST OVERPRESSURE (CCPS, 1999)

Input Data:		
Vessel burst pressure:	102	bar abs
Distance from vessel center:	156	m
Vessel volume:	18.84	m**3
Final pressure:	1.01325	bar abs
Heat capacity ratio:	1.4	
Molecular weight of gas:	28	
Gas temperature:	298	K
Speed of sound in ambient gas:	297.5	m/s

Calculated Results:

Energy of explosion using Brode's equation for constant volume expansion:

Energy of explosion:	475.65	MJ
TNT equivalent:	101.42	kg TNT
Effective energy of explosion (x 2):	951.30	MJ
Scaled distance:	7.39	
Interpolated scaled overpressure:	0.024384	
Interpolated scaled impulse:	0.006893	

Vessel shape:	Spherical		Cylindrical	
Overpressure multiplier for vessel shape:	1.1		1.4	
Corrected scaled overpressure:	0.0268		0.0341	
Actual overpressure:	0.0272	bar	0.0346	bar
	0.39	psi	0.50	psi
Impulse multiplier for vessel shape:	1		1	
Corrected scaled impulse:	0.0069		0.0069	
Actual impulse:	39.31	kPa - ms	39.31	kPa - ms

The theoretical maximum range (neglecting air friction) is then calculated assuming that the projectiles are launched at a 45 degree angle from the horizontal. An angle of 45 degrees is the optimum angle for the maximum projectile travel distance. Equation (3) is based on Newton's Second Law of Motion and Newton's Law of Gravitation; its derivation is found in many college level physics textbooks.

$$R_{max} = v^2/g \qquad (3)$$

Where R_{max} = theoretical maximum range

g = acceleration due to gravity

Distances calculated with this conservative method are very large, as can be seen in Table 1 (R_{max} in Table 1). Tabulated values range from 35 meters to 2497 meters.

The HSE (1998) report indicates that only about 40% of the pressure energy would be translated into kinetic energy for brittle fracture of a cylindrical container during test and less energy would be translated into kinetic energy for ductile fracture of the container. Thus for brittle fracture, Equation (4) gives a better estimate of the theoretical range of missiles.

$$R_{brittle} = 0.4 * R_{max} \qquad (4)$$

Brittle fracture ranges estimated with Equation (4) are tabulated in Table 1 in the column labeled $R_{brittle}$.

Predicted Velocities and Ranges of Projectiles from Brittle Fracture

Naturally, most fragments would not launch at the optimum angle and air friction would reduce the range as well. Finally, when two or more fragments result from rupture of a piping system, the resulting projectiles will usually have a distribution of sizes and initial velocities.

Baker (1983) developed methods to estimate fragment range taking into account shape and air friction. Spread sheets developed by the CCPS (1999) to apply these methods were used to develop the following results for a 20" carbon steel, schedule 80 pipe that has broken into two fragments at a test pressure of 101 Barg. Each of the two fragments is a section of pipe flying end wise with a lift-to-drag ratio of zero.

TABLE 3 – TYPICAL FRAGMENT RANGES FOR BRITTLE FRACTURE BY METHOD OF BAKER (1983)

Fragment Mass Fraction	Fragment Mass (kg)	Exposed Area (m2)	Velocity (m/sec)	Range (m)
0.1	186	0.2027	109	1093
0.2	372	0.2027	80.5	597
0.5 (2 of equal size)	930	0.2027	64.4	429
0.6	1117	0.2027	63	419
0.7	1303	0.2027	62.2	401
0.8	1489	0.2027	61.8	399
0.9	1675	0.2027	61.5	397

For comparison, the range predicted by Equation (4) is 709 meters.

Representative values calculated with Equation (4) in Table 1 are used as estimates of the maximum range of missiles from brittle fracture.

Field Data on Projectiles from Ruptured Vessels

Mannan (2005) presents observed fragment ranges from BLEVE's (Boiling Liquid Expanding Vapor Explosions) of LPG vessels (see Figure 1 below). Although most fragments traveled 700 meters or less, one fragment traveled about 1000 meters (ref. Curve "d" for spherical vessels). A BLEVE might result in substantially larger distances than a compressed gas explosion of the same container at the same burst pressure. This is because the liquid-filled vessel contains a much larger mass of liquid and a significant fraction of that liquid will vaporize upon vessel failure (boiling liquid). The vapors will then burn and expand even further. Although the LPG vessels probably failed at pressures well below the test pressure of 100 Barg, the BLEVE mechanism may account for the large projectile ranges relative to pneumatic pressure testing.

Mannan (2005) does not indicate the failure mechanism in these incidents, but it was probably ductile failure because the LPG vessels failed during external fire exposure.

Figure 2 is a map of the fragment location from a test at Sandia National Laboratories (Baker, 1985). An 8:1 scale model of a nuclear reactor containment vessel with a design pressure of 40 psig (2.76 barg) was pneumatically pressurized to failure with nitrogen gas. The vessel failed at a pressure of 195 psig (13.4 barg). Figure 2 shows the locations of twelve resulting fragments, as indicated by X's and an identification number. Fragment number 8 traveled 408 m (1338 ft) from the original location of the test vessel. Baker (1985) reports that the failure was ductile.

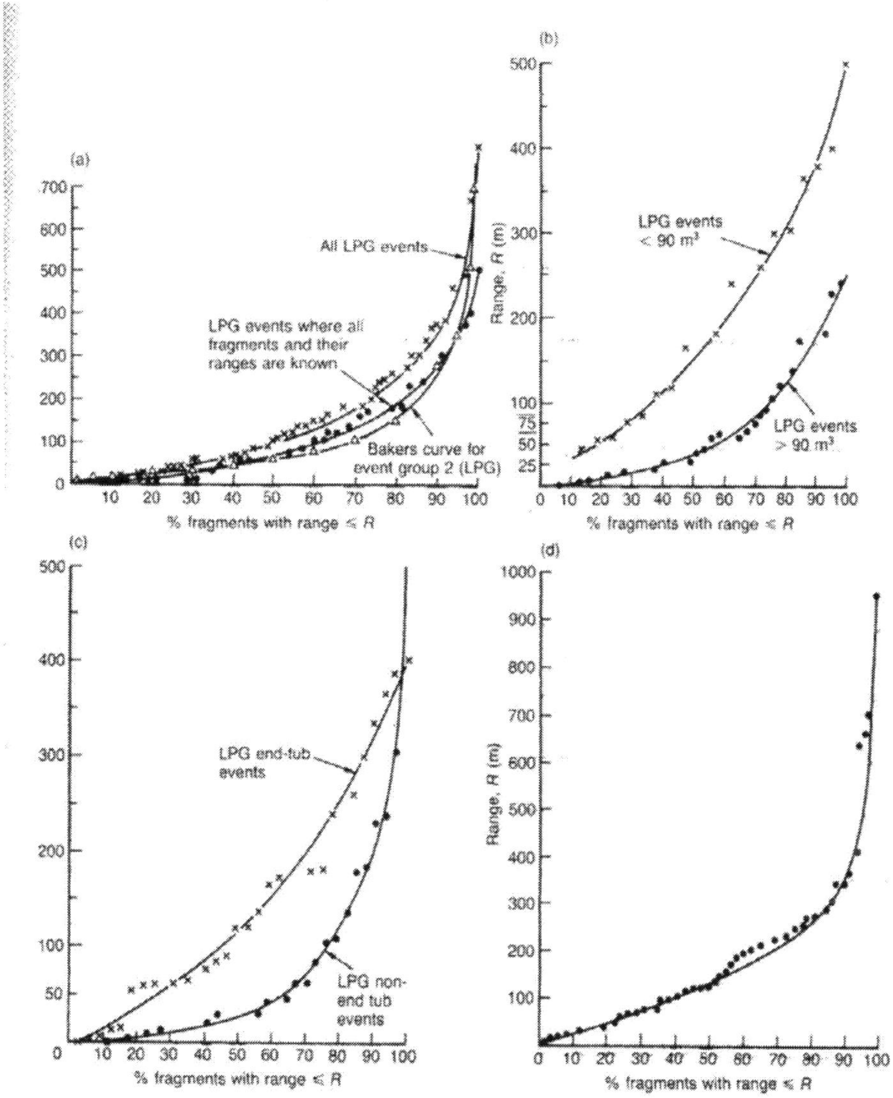

Figure 17.101 Missile effects of explosions: distance travelled by missiles from bursting of LPG vessels (Holden and Reeves, 1985): (a) all events; (b) effect of vessel size; (c) end tub and other fragments and (d) spherical vessels (BLEVEs) (Courtesy of the Institution of Chemical Engineers)

FIGURE 1 – RANGE OF MISSILES FROM EXPLOSIONS OF LPG VESSELS (From Mannan, 2005)

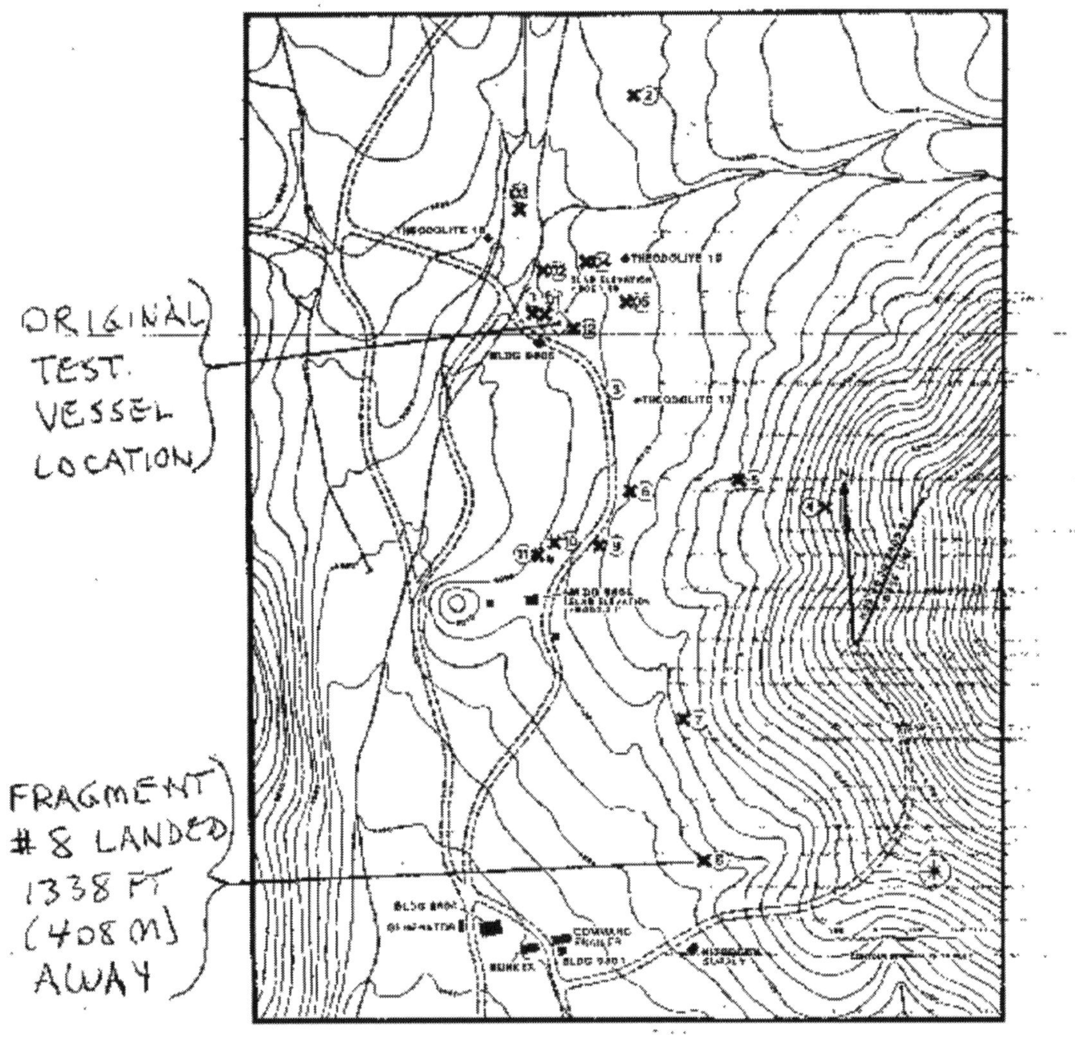

MISSILE MAP SHOWING POST TEST LOCATION OF SHELL FRAGMENTS.

FIGURE 2 – MISSLE MAP FROM TEST BURST AT 13.4 BARG (195 PSIG) OF STEEL VESSEL (Baker, 1985)

Range of Projectiles from Ductile Fracture

HSE Guidance Note GS4 (1998) gives the following equation for calculation of maximum fragment velocity for ductile fracture of a container.

$$\mathbf{v^2 = [(2 * d * P_1 * A)/m_f]} \qquad (5)$$

where **d** = diameter of hole left behind (Note: For small fragments such as plugs or small closures, it should be assumed that d = twice the diameter of the hole left behind.)

A = Area of ejected fragment

$\mathbf{m_f}$ = mass of fragment

Assuming an optimum projectile launch angle of 45 degrees, the maximum range of the fragment is then calculated with Equation (6).

$$\mathbf{R_{ductile} = v^2/g} \qquad (6)$$

These equations lead to projectile ranges similar in magnitude to brittle fracture, but the projectile ranges are highly dependent on the size and shape of the ejected fragment. For example, for a 100 Barg test pressure leading to the failure of a 1 ½ inch 600 lb nozzle, an exclusion zone of 22.4 meters was calculated in a different study. That work was based on assuming that the nozzle fragment included [(Boss) + (0.3m pipe) + (gate valve) + (flange) + (blind flange)].

Lighter fragments would fly farther (for example, just the blind flange). Table 1 presents values of $\mathbf{R_{ducile}}$ calculated with Equation (6) for blind flanges. (Blind flanges were arbitrarily chosen to calculate $\mathbf{R_{ductile}}$ because their size is well-defined for a given pipe class and size, and because they project a large area this is exposed to the escaping gaseous energy.) These distances are much larger than those in the different study cited above, and similar in magnitude to those calculated for brittle fracture using Equation (4).

Note that fragment ranges for small fragments can be significantly larger than that of blind flanges.

Probability of Personnel Injury by Projectiles

Because of the large potential ranges of projectiles from rupture of a piping system, the choice of exclusion zones for pneumatic testing could be quite large if they were based on potential fragment ranges. However, most fragments from rupture of a piping system will fall near their original location (See Figures 1 and 2 above). One rough approximation proposed here to the distribution of fragments is to assume an exponential decay with distance.

$$\text{[Probability of a fragment at radius r]} = \mathbf{e^{-2*r/R}} \qquad (7)$$

Where **R** = characteristic range

Then, the probability of a fragment falling outside an exclusion zone **Z** becomes:

[Probability of a fragment falling outside **Z**] = $\int e^{-2*r/R}\, d(2*r/R)$

(Integral evaluated between **2*Z/R** and infinity) so

[Probability of a fragment falling outside **Z**] = $e^{-2*Z/R}$ (8)

If the exclusion zone **Z** is chosen to be equal to the characteristic range **R** (as calculated in Table 1 for the appropriate failure mode), then the probability of a fragment falling outside the exclusion zone is e^{-2} or 0.135 or 13.5%. The total number of fragments falling outside the exclusion zone would thus be 13.5% of the total number of fragments formed.

The problems in applying this concept to a quantitative estimation of the probability of a person being injured are (1) accurately estimating the number of fragments formed when a piping system ruptures, (2) accurately estimating the characteristic range **R**, and (3) accurately estimating the number of people in the area.

The probability of one injury during a pneumatic test at the facility depends on the probability of failure during the test and the probability of injury during any single failure:

[Probability of injury during a pneumatic test] =

= [Probability of piping failure] x

x [Probability of Injury outside exclusion zone **Z** when pipe fails] x

x [Probability that a fragment falls outside **Z**] x

x [Number of fragments formed by failure] (9)

The Gas Research Institute sponsored a study of human and equipment failure rates in the LNG industry by Atallah, et al. (1990). Their data includes failure rates of piping systems. Applying data from plant operation to pneumatic testing prior to initiation of plant operation is questionable, but may be roughly indicative of the probability of a failure during test. Atallah, et al. (1990) presented a mean time between failure (MTBF) for LNG piping systems of 5.8×10^8 ft-hr.

[Probability of One LNG Piping System Failure During Pneumatic Testing] =

= [(Number of Pneumatic Test Systems) x (Length of Pipe Per System) x

X (Duration of Each Pneumatic Test)]/ (MTBF) (10)

The test duration used in one project was approximately two hours (1 hour at lower pressures, 15 minutes at $\mathbf{P}_{design}$, 30 minutes at $\mathbf{P}_{test}$, and 15 minutes at $\mathbf{P}_{design}$.

Assuming 200 LNG pneumatic test systems and an average of 300 ft of pipe per test system, the probability of failure of one LNG piping system among all those tested during LNG pneumatic testing is approximately one in 5000. Because of the "bath tub" shape of failure rate data, the probability of failure of an LNG piping system during test may be somewhat higher. (Graphs of failure rates versus time typically resemble a bath tub, with a high initial failure rate, then a constant and low failure rate during the expected life of the product, and finally an increasing failure rate at end-of-life.)

In the unlikely event of a failure of an LNG piping system during pneumatic pressure testing, all personnel would be protected from blast overpressure by the conservative exclusion zones in use. Although one or more fragments might fall outside the exclusion zone, most of the fragments would fall within the exclusion zone, and the probability of a person being struck by a fragment is very low.

A confidential and detailed industry quantitative risk assessment of a proposed pneumatic test of a 6 km LNG pipeline also reached the conclusion that the risk of pneumatic testing would be acceptably low.

When rigorous precautions have been taken to establish safe testing procedures and to ensure the integrity of the piping systems being tested, the risk of pneumatic pressure testing of selected LNG piping systems can be made acceptable. As a minimum, rigorous attention needs to be given the essential considerations listed in Appendix A herein.

SAFE EXCLUSION ZONES FOR PNEUMATIC TESTING OF PIPING AND VESSELS

The shock wave from rupture of a piping system under test could affect much of the immediate area surrounding the test, so the exclusion zone for the test should be at least large enough to keep the overpressure from exceeding the criterion selected here of 0.5 psig (0.0345 barg). The intent is that only those people conducting the pneumatic pressure test would be allowed to enter the exclusion zone and then only under carefully prescribed conditions.

Conversely, projectiles resulting from fragmentation of the piping system will be relatively few in number, and most would fall relatively close to their origin. The remaining few would be distributed over a comparatively large area and would have only a very low probability of hitting a person. These considerations form the basis of the logic of using overpressure only to establish exclusion zones.

APPENDIX A – PRECAUTIONS IN PNEUMATIC PRESSURE TESTING

The following are considered essential to minimize the risk of failure and injury during high pressure pneumatic pressure testing:

- Comprehensive testing and safety procedures must be formalized and implemented, and documentation must be maintained certifying that all personnel involved in the testing activities have attended training sessions in which the project procedures have been reviewed. Safety procedures should include requirements that test personnel shelter to the maximum extent possible behind either heavy equipment or barricades during the tests.
- Test procedures must clearly define the points in time during the test when test personnel are permitted to leave sheltered areas and enter the exclusion zone. Prior to proving the system at the full pneumatic test pressure, it is suggested that personnel enter the exclusion zone only when the circumferential stress in the piping is no greater than 50% of the specified minimum yield strength (SMYS) of the piping materials in the test system, and only after prescribed holding times at such pressures. When the system test pressure results in circumferential stresses that exceed 50% SMYS, test personnel should enter the exclusion zone only after the test system has first been proven for the prescribed

time period at full test pressure. Test personnel should never enter the exclusion zone when the test pressure exceeds the piping system design pressure.

- Formalized check lists must be utilized to document that pre-test preparations have been completed.
- Schedule tests at optimum times to ensure safety. The risk of injury resulting from a test system failure can be dramatically reduced by testing at night or on weekends when fewer personnel are on (and possibly off) site.
- Materials must be procured only from manufacturers that have been determined to adhere to suitable, documented quality control and production processes.
- Materials received must be carefully checked to ensure compliance with material specifications (including review of all MTR's received).
- The project must have a suitable positive material identification (PMI) procedure in place that effectively ensures proper materials in the fabricated piping.
- Pressure design calculations for both operating and test pressures must be documented and checked for all piping.
- The use of non-destructive examination (NDE) must be maximized to ensure the quality of all welded joints in the system. Butt welded joints should be 100% ultrasonically or radiographically tested.
- The intended use of "golden welds" (i.e., welds that are not proven by pressure testing) to join sections of pre-tested piping must be carefully reviewed to ensure guaranteed quality of the weld joint. It should be noted that in some jurisdictions may require approval for their use (e.g. jurisdictions in which compliance with the European Pressure Equipment Directive is mandated).
- The limits of each test system must be clearly defined in test package documentation, and the methods of isolation (test blinds, etc.) must be well defined. Test blind designs must be checked to ensure suitability for the system test pressures. The proposed design of test systems must be checked by qualified personnel.

REFERENCES

Sami Atallah, Jatin N. Shah, and Mariassunta Betti, "Reduction of LNG Operator Error and Equipment Failure Rates", Gas Research Institute Report No. GRI-90/0008, Chicago, IL (1990)

W. E. Baker, P. A. Cox, P. S. Westine, J. J. Kulesz, and R. A. Strehlow, *"Explosion Hazards and Evaluation"*, Elsevier, New York, NY (1983).

W. E. Baker, "Post-Test Assessment of Blast and Fragment Effects of Explosive Failure of a Large Steel Containment Shell Model During Pneumatic Testing", Conference Paper, pages 289-296 (1985).

H. L. Brode, "Blast Wave from a Spherical Charge", *Physics of Fluids,* Volume 2, page 217 (1959).

CCPS (Center for Chemical Process Safety), *"Guidelines for Process Equipment Reliability Data with Data Tables", American Institute of Chemical Engineers"*, New York, NY (1989).

CCPS (Center for Chemical Process Safety), *"Guidelines for Consequence Analysis of Chemical Releases", American Institute of Chemical Engineers"*, New York, NY (1999).

D. A. Crowl, "Calculating the Energy of Explosion Using Thermodynamic Availability", *Journal of Loss Prevention in the Process Industries,* Volume 5, pp. 109-118 (1992).

Health and Safety Executive, *"Safety in Pressure Testing"*, Guidance Note GS4, UK Health and Safety Executive, Sudbury, Suffolk, United Kingdom (1998).

M. Sam Mannan (Editor), *"Lee's Loss Prevention in the Process Industries"*, 3rd Edition, 3 volumes, Elsevier Butterworth-Heinemann, Oxford, United Kingdom (2005)

G. Saville, S. M. Richardson, and B. J. Skillerne de Bristowe, *"Pressure Test Safety"*, Contract Research Report 168, Health and Safety Executive, United Kingdom (1998).

SPEAKER – VIC EDWARDS

Victor H. Edwards, Ph. D., P. E. is Director of Process Safety for Aker Solutions US, Inc., 3600 Briarpark Drive, Houston, TX 77042 (713/270-2817; Fax: 713/270-3195; e-mail: vic.edwards@akersolutions.com). Vic has worked for Aker Solutions for 27 years. His experience includes process engineering, process safety management, and process, biochemical and environmental technologies. His project assignments have included biochemical and chemical processes, biomass energy resources, liquefied natural gas, chemical weapons demilitarization, hazardous waste disposal, and offshore hydrocarbons facilities.

Dr. Edwards has received three DuPont awards for safety and environmental engineering excellence and two DuPont awards for health, safety, and environmental excellence during 27 years as an alliance engineering contractor. In 1998, Kvaerner Engineers and Constructors named him employee of the year. In 2007, Vic's Houston HSE team was recognized by Aker Kvaerner with the Process and Construction Excellence Award for HSE in Design.

Vic's earlier experience included Assistant Professor of Chemical Engineering at Cornell University, a one-year assignment at the National Science Foundation (on leave from Cornell), pharmaceutical research at Merck, alternate energy research at United Energy Resources, Visiting Professor of Environmental Engineering at Rice University, and process engineering at Fluor.

Vic has contributed more than sixty publications to the engineering literature, including several book chapters. He chaired the 1992 Process Plant Safety Symposium and the 1997 International Plant Operations and Design Conference. Vic received the 2003 Service Award from the Mary K O'Connor Process Safety Center at Texas A & M University, where he chairs the Technical Advisory Committee. Vic was recently elected to the Centennial Council of the Department of Chemical and Biomolecular Engineering at Rice University. He also serves on the Editorial Advisory Board of CHEMICAL PROCESSING magazine. Edwards earned his B. A. from Rice University and his Ph. D. from the University of California at Berkeley; both degrees in chemical engineering. A registered professional engineer in Texas, he is an AIChE Fellow, and a member of ACS, AAAS, NFPA, NSPE, and the New York Academy of Sciences.

(69f) Power Recovery in LNG Regasification Plants

Chen-Hwa Chiu, Process Engineering, Chevron Energy Technology Company, Houston, TX

Yoshitsugi Kikkawa, Process Technology Division, Chiyoda Corporation, Yokohama, Japan

Masaaki Ohishi, Division Technology & Engineering, Chiyoda Coorporation, Yokohama, Japan

Steve Rush, Sales & Marketing, Ebara International Corporation, Sparks, NV

Hans E. Kimmel, Ebara International Corporation, Sparks, NV

Regasification plants vaporize liquefied natural gas using mainly high-pressure LNG pumps and heat exchangers. The vaporization pressure corresponds to the natural gas pipe line pressure.

Due to the large temperature difference between the LNG feed and the pipe line gas, a substantial power recovery is suggested. This paper proposes and describes in detail a Modified Brayton Cycle to recover power from the regasification plant.

See more of this Session: Professor Cedomir M. Sliepcevich Memorial Session Two - LNG Regasification
See more of this Group/Topical: Topical 6: 10th Topical Conference on Gas Utilization

(91a) Analysis of Greenhouse Gas (GHG) Regulation Impacts to the U.S. Refining Industry

Ian M. Glasgow, Mustang Engineering, L.P., Houston, TX

Stan Polcar, Mustang Engineering, L.P., Houston, TX

R.E. (Ed) Palmer, Mustang Engineering, L.P., Houston, TX

President Obama has stated that reduction of greenhouse gases (GHG) and reduced dependence to foreign oil is part of his immediate legislative agenda. At the time this abstract is written, the Waxman-Markey bill that will reduce greenhouse gases (GHG) emissions was passed by the House in June 2009; the Senate has passed a similar bill (Kerry-Boxer bill) out of Energy & Environment committee. The EPA has determined that endangerment to human health and welfare exists due to climate change from GHG and is actively pursuing options to utilize the Clean Air Act for GHG emission reductions. Updated CAFÉ standards have been established that will take effect in 2016 to increase light duty vehicle efficiency to 35.5 mpg and include GHG emissions as a part of vehicle standards. Previously passed legislation from the 2007 Energy Independence Act (EIA) will also impact refiners with a mandated 36 billion gallons per year of biofuel produced by 2022 under the Renewable Fuel Standard (RFS-2).

In this paper various legislation on GHG and energy are reviewed, with the focus being on the proposed Waxman-Markey American Clean Energy and Security Act and the EPA currently proposed RFS-2 that is a part of the 2007 EIA.

A brief overview of Life Cycle Assessment (LCA) and cap-and-trade systems are presented for information purposes. Case studies for a notional 150 MBPSD refinery are analyzed to evaluate the trends and impacts to a refinery resulting from a cap-and-trade system (Waxman-Markey) where product emissions are included. Comparison of the Waxman-Markey bill to the increased CAFÉ and RFS-2 regulations is also provided. The major GHG manufacturing emissions sources of a notional refinery are identified and a preliminary discussion of potential solutions to reduce emissions including the use of petroleum coke gasification for CCS and offsetting hydrogen plant CO2 emissions is highlighted.

See more of this Session: Greenhouse Gases Sequestration
See more of this Group/Topical: Topical 6: 10th Topical Conference on Gas Utilization

(91b) High Capacity Regenerable Sorbent for Pre-Combustion CO_2 Capture

Gökhan Alptekin, TDA Research Inc., Wheat Ridge, CO

Ambalavanan Jayaraman, TDA Research Inc., Wheat Ridge, CO

Steve Dietz, TDA Research Inc., Wheat Ridge, CO

Margarita Dubivik, TDA Research Inc., Wheat Ridge, CO

Lauren Brickner, TDA Research Inc., Wheat Ridge, CO

TDA is developing a novel sorbent that removes CO_2 via physical adsorption from both synthesis gas, while the relatively strong affinity of the sorbent for CO_2 enables effective operation at temperatures up to 300°C (well above the dew point of synthesis gas stream generated by most commercial gasifiers). However, because the sorbent and the CO_2 do not form a true covalent bond, the energy needed to regenerate our sorbent (5.4 kcal per mol of CO_2) is much lower than that observed for either chemical absorbents (e.g., 29.9 kcal/mol CO_2 for sodium carbonate) or amine-based solvents (e.g., 14.2 kcal/mol CO_2 for monoethanolamine). Our sorbent can be regenerated isothermally and CO_2 can be recovered at pressure (~150 psia). Thus, the energy needed to regenerate the sorbent and compress the CO_2 for sequestration is significantly lower than that for any technology reported to date. TDA's CO_2 sorbent could be used to capture CO_2 in a Integrated Gasification Combined Cycle (IGCC) power plant after the water-gas-shift (WGS) reaction when most of the carbon is in the form of CO_2. The high surface area and favorable porosity of the sorbent also provide a unique platform to introduce additional functionality, such as active groups to catalyze the or to remove trace metals (e.g., Hg, As).

We carried out bench-scale evaluations of TDA's CO_2 sorbent in both fixed-bed adsorber and a thermo gravimetric analyzer (TGA). The sorbent achieved high CO_2 capacity, high removal efficiency and excellent durability. The sorbent showed stable performance in both these systems under simulated synthesis gas (44 cycles in the bench-scale flow system and over 1,644 cycles in the TGA). We also studied the effect of operating parameters like temperature, pressure, and cycle time on the sorbent performance. The results from our on-going sorbent evaluations will be presented at the meeting.

See more of this Session: Greenhouse Gases Sequestration
See more of this Group/Topical: Topical 6: 10th Topical Conference on Gas Utilization

(91c) The Third C in CCS - Compression

Dipanjan Bhattacharya, Bechtel Oil and Gas, Houston, TX

Ramachandra P. Tekumalla, Advanced Simulation & Analysis, Bechtel Corporation, Houston, TX

David Messersmith, Bechtel, Houston, TX

Lee Schmoe, Bechtel Corporation, Houston, TX

Tomas Maramba, Bechtel, Houston, TX

The capture and sequestration of carbon dioxide emitted from process plants is one of the more pressing issues for the energy industry. While much has been reported on the capture of CO2, little has been published on the compression options and issues. This paper summarizes the results of a Bechtel investigation for three different plant applications: a pulverized coal-fired power plant, an LNG plant processing high CO2 natural gas, and an IGCC complex. Working closely with equipment suppliers, Bechtel examined capital and operating costs for several configurations including a series of compressors, a combination of compressor and pumps (supercritical and refrigerated liquids), and different dehydration designs.

The paper provides a framework for the readers to more-effectively initiate and complete their own project-specific designs.

See more of this Session: Greenhouse Gases Sequestration
See more of this Group/Topical: Topical 6: 10th Topical Conference on Gas Utilization

Study of the effect of water on Fischer-Tropsch synthesis over iron based catalysts

Venkat Ramana Rao Pendyala, Gary Jacobs, Janet C. Mohandas, Wenping Ma, Muthukumaran Gnanamani, and Burtron H. Davis*

Center for Applied Energy Research, University of Kentucky, 2540 Research Park Dr. Lexington, KY-40511, USA

Introduction:

Conversion of syngas ($CO + H_2$) to liquid fuels and chemicals has great potential for utilization of the large world reserves of natural gas and coal [1]. Recently, there has been a renewed interest in Fischer-Tropsch synthesis (FTS) as it presents an attractive way to produce chemicals and transportation fuels with low sulfur and low aromatics from carbon sources alternative to crude oil. Although several metals are active for FTS, only iron and cobalt catalysts appear to be economically viable on an industrial scale. The Fe-based FTS catalysts also catalyze the water-gas shift reaction (WGS) under typical reaction conditions and therefore these materials are of prime interest for the conversion of hydrogen lean synthesis gas types, like those derived from coal and biomass [2].

In the absence of the water gas shift reaction, water is the main product of Fischer-Tropsch CO-hydrogenation. Its presence can affect a catalyst's performance (activity and selectivity) in several ways. Iron based catalyst studies dealing with the kinetics of syngas consumption generally show that the consumption rate decreases as water concentrations increase [3,4]. Satterfield et al. [5] reported the effect of co-feeding water and syngas with a potassium promoted iron catalyst in a continuously operated slurry reactor at 248°C. They found a decrease in reaction rate by water. The effect was completely reversible. Furthermore water increased the rate of the water gas shift reaction, oxygenates, olefin and α-olefin selectivities. In further water co-feeding studies on FT synthesis with iron catalysts no sufficient selectivity data had been determined. Karn et al. [6] reported that 10-30% H_2O added to syngas with a 1:1 H_2/CO had relatively little effect on CO conversion using a fixed bed reactor and a nitride fused iron catalyst. They observed that in the presence of water at low conversion of syngas that the methane selectivity was lower than prior to water addition, while at higher conversion levels it was much higher, than in the absence of water. On the other hand Dry [7] found no significant change in the methane selectivity in studies of water addition up to 17 mol% using a promoted fused iron catalyst in a fluidized bed at 320°C.

Variations among these studies could certainly be attributed to reactor type or catalyst used. In this contribution, the goal is to examine the influence of alkali promoters on the relative rates of FT synthesis and WGS under realistic FT conditions and attempt to assess not only the impact of alkali promoter basicity (e.g., Li, K, and Rb were assessed). That is, a fundamental question that remains to be answered is whether the FT and WGS active sites are the same, such that changes in the selectivity of one impact the other, or whether the sites are different, so that the rate of WGS can be tuned without significant suppression of the FT rate. The impact of reaction temperature was also explored during the addition of water. The CSTR is well-suited to this type of experiment because all catalyst particles experience the same gas composition during a set of conditions.

Experimental:

Precipitated iron catalysts were prepared using a ferric nitrate solution obtained by dissolving $Fe(NO_3)_3.9H_2O$ in deionized water, and then tetraethylorthosilicate (TEOS) was added to provide the desired Fe:Si ratio. A flow of the TEOS and iron nitrate mixture was added to a CSTR precipitation vessel together with a stream of 30% NH_4OH that was added at a rate to maintain a pH of 9.0. The final filter cake was dried for 24 h in an oven at 110 °C with flowing air. For this study, the Fe:Si catalyst base powder was then impregnated with the proper amount of aqueous alkali nitrate (e.g., $LiNO_3$, KNO_3, or $RbNO_3$) and $Cu(NO_3)_2$ solution to produce the desired composition of Fe:Si:alkali:Cu = 100:4.6:1.25:2.0 (atomic ratios). A high alpha catalyst (100Fe:4.6Si:5.0K:2.0Cu) was also prepared. Catalysts were dried at 110 °C overnight followed by impregnation, drying, and calcination at 350 °C in an air flow for 4 h.

The FTS experiments were conducted in a 1 L CSTR equipped with a magnetically driven stirrer with turbine impeller, a gas-inlet line, and a vapor outlet line with a stainless steel fritted filter placed external to the reactor. Carbon monoxide was passed through a vessel containing lead oxide on alumina to remove any traces of iron carbonyl. The mixed gases entered the CSTR below the stirrer, which was operated at 750 rpm. The reactor temperature was maintained constant (± 1 °C) by a temperature controller. A series of measurements were conducted so that the pressure of carbon monoxide and hydrogen in the feed remained constant and was 70% of the total pressure, with the remainder being inert gas. A high precision, high pressure ISCO syringe pump was used to add water to the reactor. When water was added, a fraction of the inert gas was replaced by water.

Results and Discussion:

We studied the effect of water during FT synthesis over alkali promoted precipitated Fe catalysts by co-feeding different amounts of water (5.0 to 25 mole %) added into the feed gas. The effect of water on CO conversion for potassium promoted low alpha iron catalyst is shown in Figure 1. After establishing the steady state conversion level, water was introduced to the feed gas. The externally added water induced a positive effect on CO conversion for all three catalysts, which differed by alkali metal type (i.e., Li, K, or Rb), when the reaction was carried out at 270°C. CO conversion was found to increase with increasing volume % of co-fed water up to 10 - 15 mole % and then there were slight decreases with further increases of water added to the feed gas. This effect was completely reversible upon terminating the water addition. Similar findings were observed for all the promoted iron catalysts. The added water to the reaction feed may react with CO (i.e., resulting in a decrease in the partial pressure of CO in the reactor) to produce CO_2 and H_2 via the WGS reaction.

The effect of water on CO conversion for a high alpha iron catalyst carried out at 230°C is shown in Figure 2. Steady state CO conversion (~ 45%) was reached in 210 h of time on-stream. The external addition of water caused a significant decrease in CO conversion; however the CO conversion was not recovered when water addition was terminated, indicating that the catalyst had deactivated permanently. The apparent negative influence of water on CO conversion may be ascribed to the competitive adsorption between water and CO on the catalyst surface. It is likely that the iron carbide was oxidized during exposure to the higher water partial pressure, which is likely accompanied by the complete transformation of iron carbide to the magnetite (Fe_3O_4) phase. That is, the addition of water vapor increases the Fe_3O_4 concentration at the expense of iron carbide. In FT synthesis H_2O/H_2 or CO_2/CO ratios are high enough that iron carbides can be reoxidized to Fe_3O_4 [7,8].

Figure 3 shows the XANES spectra at Fe K-edge of iron catalyst for reactions carried out at 230 and 270°C. The proposal of deactivation by oxidation was tested by retrieving catalyst samples from the reactor after sealing the catalyst particles in their in-situ state by cooling them in the wax product. As shown in Figure 3, the trend in oxidation, as characterized by the intensity of the white line, agrees with the trend in deactivation rate observed with water addition. The sample corresponding to higher temperature FT, 270°C, and having the lowest deactivation rate, had the lowest white line intensity, indicating the highest extent of reduction. Recall that the catalyst corresponding to low temperature FT, 230°C, exhibited the highest deactivation rate with water addition. It also displayed the highest white line intensity in XANES, corresponding to a higher extent of oxidation.

The absolute rates of FT and WGS reactions as a function of water amount for potassium promoted low alpha iron catalyst are shown in Figure 4A. The impact of the added water is to increase the rate of the WGS reaction up to 10 to 15 mole % of added water, leveling off at further increases in the water amount. The relative rates (selectivity) of FT and WGS reactions as a function of water amount for the same catalyst are shown in Figure 4B. FT selectivity decreases and WGS selectivity increases with increasing the water amount. But, when we are adding the water amounts to the feed gas, the CO conversion was found to increase. In FT synthesis, selectivity depends on the conversion level and it is therefore important to compare the catalysts at constant CO conversion levels. We can not definitively compare the results at different conversion levels, and it is difficult to maintain the constant conversion levels during addition of external water to syngas. It appears that if we were able to maintain constant conversion levels before and after water addition that the relative rate of FT is suppressed by increasing the water amounts and that WGS increases. However, due to the increasing CO conversion by co-feeding the external water, it is difficult to make a definitive conclusion regarding the active site for FT and WGS reactions.

Figure 4C shows the selectivity to C_{5+} at different amounts of water added to the feed gas for alkali promoted iron catalysts. The results reveal that the selectivity to C_{5+} decreases with increasing water amount up to 15 mole %, but after that point, increases somewhat with further increases in the water concentration. H_2/CO ratio increases inside the reactor with increasing the water amounts up to 15 mole%, then decreases with further increases in the water concentration. The higher selectivity to C_{5+} in the case of low H_2/CO ratios might be ascribed to a high surface concentration of CO, which would favor the chain propagation reaction [9]. Lower partial pressures of CO should tend to lower the surface coverage of adsorbed CH_2 species hence lowering the probability for desorption of the $n(CH_2)$ species. On the other hand, higher H_2 partial pressures favor the termination of surface species to paraffins. Thus, one could expect that as the H_2/CO ratio increases inside the reactor the selectivity would shift to lighter and more saturated hydrocarbons. The catalyst promoted with Rb displays the highest C_{5+} selectivity among the catalysts, whereas the Li-promoted catalyst produces the highest C_{5+} selectivity at higher amounts of water due to decrease in the WGS rate after passing through a maximum.

At the baseline FT synthesis conditions (i.e., in the absence of co-fed water), the Rb-promoted catalyst (i.e., most basic) displays the lowest methane selectivity (Figure 4D) and the Li- promoted catalyst (i.e., most acidic) exhibits the highest selectivity to methane. A similar trend results with water addition. Methane selectivity increases with increasing the water amount up to 15 mole % and decreases with further increases of added water for all the promoted catalysts, and it is directly correlated to CO conversion. A major reason for the increase in methane production with increasing CO conversion is that the concentration of water has increased so that the WGS reaction rate increases, producing more CO_2 and H_2; thus, there is an increase in the H_2/CO ratio inside the reactor. Higher hydrogen partial pressures in the reactor favor the formation of methane [10]. An increase in the effective H_2/CO ratio usually leads to higher methane selectivity and lower chain growth probability. Likewise, the decrease in methane selectivity after passing through the maximum in CO conversion with higher amounts of water (i.e., above 15 mole % H_2O) can be attributed to a decrease in the H_2/CO ratio.

Conclusions:

The externally added water had a positive effect on CO conversion for all alkali promoted precipitated iron catalysts at 270°C, where as at 230°C was found to decrease CO conversion and deactivates the catalyst. Due to the increasing conversion level by co-feeding the water amount, it is difficult to make a definitive conclusion regarding active sites for FT and WGS reactions. Increasing the water amount in the feed gas caused the fraction of CO converted into hydrocarbons decrease; subsequently WGS activity and the oxygenate selectivity increases. Common for all the alkali promoted iron catalysts was a decrease in C_{5+} selectivity and an increase in the methane selectivity occurs with increasing that amount of water up to 10 to 15 mole %; with further increases of water, the C_{5+} selectivity increases and methane selectivity decreases. With increasing the reaction temperature, an intensive change in the relative concentrations of iron carbide and iron oxide resulting from the establishing of more stable iron phase equilibrium. The oxidation state trends from XANES spectroscopy of used catalyst samples sealed in wax were in agreement with the deactivation trends. That is, the catalyst sample corresponding to the higher temperature 270°C condition displayed the highest extent of reduction. On the other hand, the catalyst sample corresponding to the low temperature 230°C, displayed the highest extent of oxidation.

Acknowledgements:

This work was supported by the Commonwealth of Kentucky and a grant from the Kentucky Governor's Office of Energy Policy (KY - GOEP) Energy Research and Development Program entitled "Towards understanding how alkali promoters alleviate CO_2 production and enhance chain growth probability over iron Fischer-Tropsch synthesis catalysts."

References:

1. I. Wender, Fuel Proce. Technol. 48 (1996) 189.
2. E. van Steen, M. Claeys, Chem. Eng. Technol. 31 (2008) 655.
3. M.A. Vannice, Catal. Rev. 14 (2) (1976) 153.
4. G.P. van der Laan, A.A.C.M. Beenackers, Appl. Catal. A 193 (2000) 39.
5. C.N. Satterfleid, R.T. Hanlon, S.E. Tung, Z. Zou, G.C. Papaefthymiou, Ind. Eng. Chem. Prod. Res. Dev. 25 (1986) 407
6. F.S. Karn, J.F. Shultz, R.B. Anderson, Actes. Congr. Int. Catal. 2 (1961) 2439.
7. M.E. Dry, In Catalysis-Science and Technology, J.R. Anderson, M. Boudart, Eds.; Springer-Verlag: (1981) New York, vol, pp 196-198.
8. R.J. O'Brien, L. Xu, R.L. Spicer, B.H. Davis, Energy Fuels 10 (1996) 921
9. E. Iglesia, S.C. Reyes, R.J. Madon, S.L. Soled, Adv. Catal. 39 (1993) 221.
10. X. Huang, C.B. Roberts, Fuel. Proc. Technol. 83 (2003) 81.

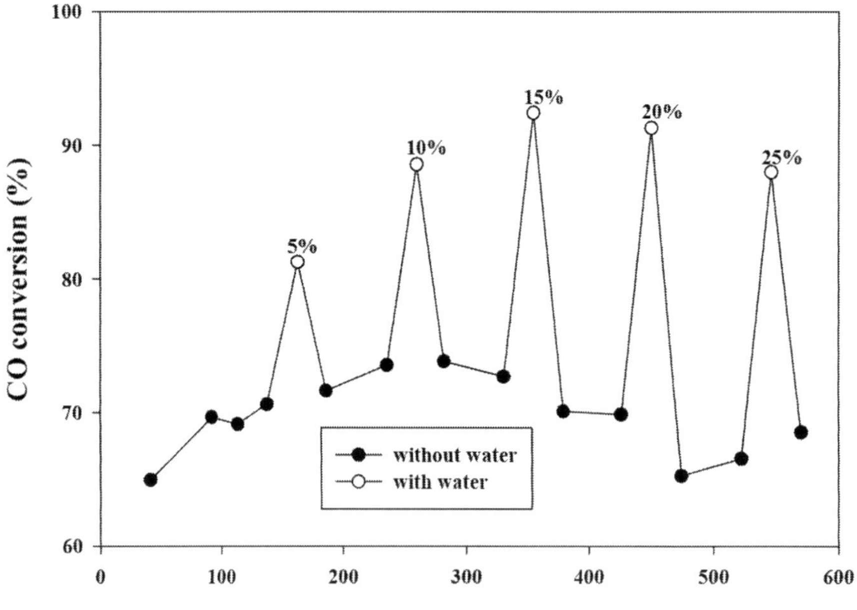

Figure 1: Effect of water on CO conversion for potassium promoted low alpha iron catalyst at 270°C.

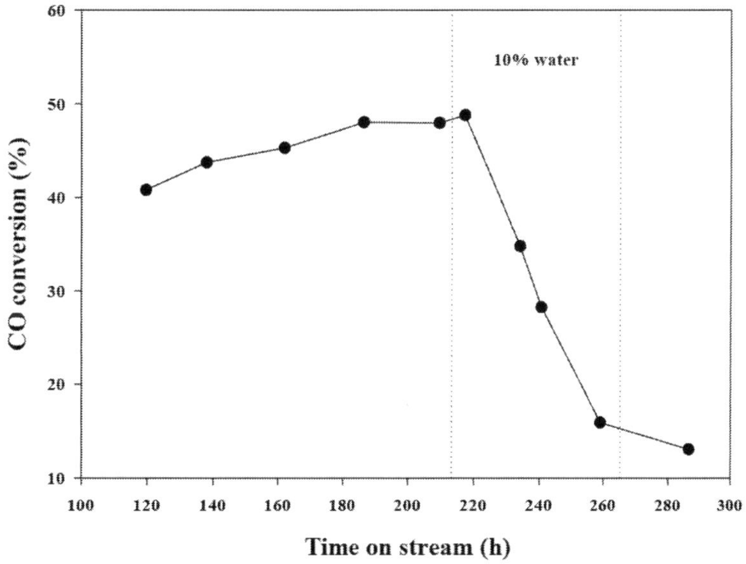

Figure 2: Effect of water on CO conversion for potassium promoted high alpha iron catalyst at 230°C.

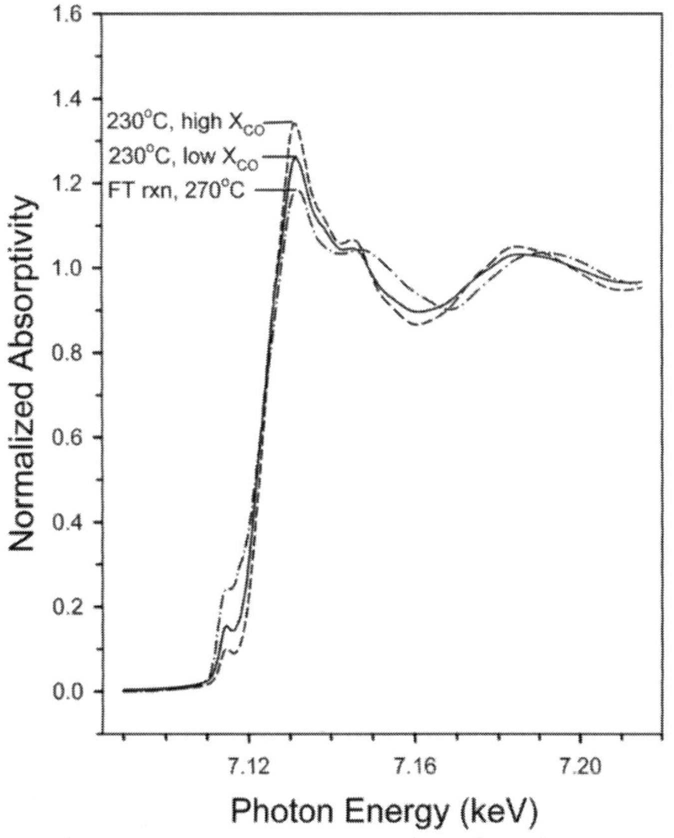

Figure 3: Normalized XANES spectra at Fe K-edge of used catalyst for the reactions carried out at 230 and 270°C.

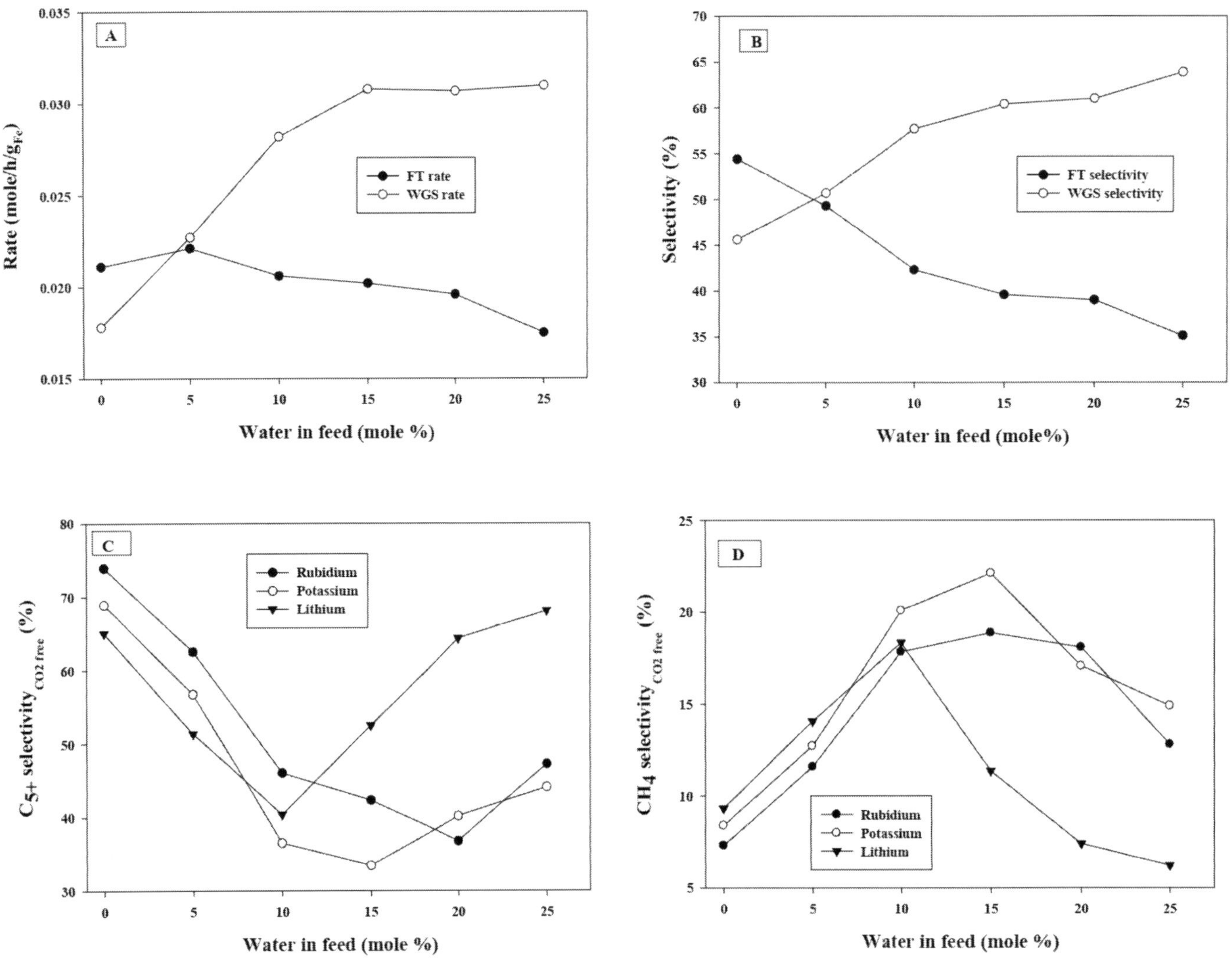

Figure 4: Effect of water on reaction rates and selectivity for alkali promoted iron catalysts
(A) Absolute rates of FT and WGS (B) Relative rates (Selectivity) of FT and WGS
(C) C_{5+} selectivity (D) CH_4 selectivity

(92c) Fischer Tropsch Synthesis On CNT Based Catalysts

Matthew Jeffers, Department of Mechanical Engineering and Energy Processes, Southern Illinois University, Carbondale, IL

Kanchan Mondal, Department of Mechanical Engineering and Energy Processes, Southern Illinois University, Carbondale, IL

Saikat Talapatra, Physics, Southern Illinois University, Carbondale, IL

Jitendra Gautam, Department of Mechanical Engineering and Energy Processes, Southern Illinois University, Carbondale, IL

Carbon nanotubes (CNT) have been found to have extraordinary physical properties. As a result of these properties, they have been used as supports in several reactions. Recent publication have shown the utility of CNTs as supports for syngas conversion to ethanol. However, the role of CNTs in the overall reactions have not been evaluated. In this paper we present the activity of as-produced and purified CNTs towards Fischer Trospch synthesis. The results from prametric studies on Fischer Tropsch synthesis in the presence of CNT catalysts are presented. The use of as-produced CNT catalysts significantly improved the CO conversion rate per unit mass of catalyst by over 100 fold while the same value obtained when employing purified CNTs was nearly 1500 as high as an Fe-Zn-K/alumina catalyst. While it can be argued that the iron (which is used as a catalyst in the production of CNTs) may act as a catalyst, it should be noted that the amount of Fe on these catalysts (<1%) as compared to Fe (15 %) loaded on the alumina supports, shows that Fe alone cannot be responsible. In addition it was found that a;though SWNTs provided high CO conversion, the primary product was found to be methane. On the other hand MWNTs produced liquid hydrocarbons and oxygenates. This clearly demonstrates that the structure of the CNT is an important factor to the degree of CO conversion and reaction products. The product distribution from the parametric studies will be presented. It was observed that the temperature plays a key role in the liquid product distribution. It was observed that below a certain temperature, the primary products are hydrocarbons with no oxygenates. However, above this temperature, the amount of oxygenates increase reaching 100 % alcohols within a 100 oC rise in temperature. The results from studies conducted on other process parameters such as syngas composition, pressure and syngas flow rate will also be presented. This research was made possible with support, in part, by the Illinois Department of Commerce and Economic Opportunity through the Office of Coal Development and the Illinois Clean Coal Institute.

See more of this Session: Kinetics, Design and Modeling of Fischer-Tropsch Catalysts and Reactions II

See more of this Group/Topical: Topical 6: 10th Topical Conference on Gas Utilization

Effect of cobalt particle size on the catalyst intrinsic activity for Fischer Tropsch Synthesis

Khalid Azzam, Gary Jacobs, Burtron Davis

Center for Applied Energy Research, University of Kentucky, 2540 Research Park Dr. Lexington, KY 40511, USA

Introduction:

Fischer Tropsch synthesis (FTS) is a major process in Gas-to-Liquids (GTL), Coal-to-Liquids (CTL) and Biomass-to-Liquids (BTL) technologies. In the FTS, syngas produced from natural gas, coal, or biomass can be converted into valuable hydrocarbons by hydrogenation of carbon monoxide. Co supported catalysts are known as effective catalysts for the FTS reaction. It has been reported by Iglesia [1] that the FTS turnover rate is independent of Co dispersion for the Co particle size varied from 10 to 210 nm using -Al_2O_3, -TiO_2, and -SiO_2 Co based catalysts. Later, Lok [2] showed using Co/Al_2O_3 catalyst that the FTS turnover rate is independent of Co dispersion even at Co particle sizes of 3 nm. In contrast with that, Bezemer et al. [3] and Barbier et al. [4] recently reported that the FTS turnover rate of Co is a function of its particle size when Co particle sizes are lower than 6 nm [3,4].

The aim of this work is to study the effect of Co particle size on the FTS turnover rate. The catalyst with ultra-small particles was prepared using the KL-zeolite as a non-acidic support in order to produce very tiny Co particles residing inside the L-Zeolite lobes using a chemical vapor deposition (CVD) method. The catalyst was compared with a conventional Co/alumina catalyst.

Experimental:

Catalyst preparation: A sample of potassium exchanged zeolite LTL (Süd-Chemie, Inc.) was calcined in flowing air (100 cm^3/min) at 400°C for 4 h. After cooling, the zeolite LTL was transferred to a glove bag filled with inert gas, carefully mixed with a mixture of platinum acetylacetonate (Alfa Aesar) and cobalt acetylacetonate (Alfa Aesar), and loaded into a sublimation tube, sealed at one end. The other end was attached to a high vacuum system, equipped with an oil diffusion pump and capable of achieving a vacuum of $< 10^{-6}$ Torr. The sublimation tube was then held at room temperature for 4 h, and slowly ramped (1°C/min) to 70, 80, 90, and then 100°C, holding at each temperature for 1 h. Finally, the temperature was ramped to 130°C, held for 15 min, and cooled to room temperature. To decompose the platinum acetylacetonate (Alfa Aesar) and cobalt acetylacetonate compounds, the catalyst was then calcined at 350°C in flowing air for 4 h. A 1°C/min ramp was used.

Extended X-ray Absorption Fine Structure (EXAFS) and X-ray Absorption Near Edge (XANES) spectroscopies: The 0.5%Pt-5%Co/KL catalyst was activated in H_2 at 500°C and cooled to liquid nitrogen temperatures for analysis. The catalyst was evaluated by XANES and EXAFS spectroscopies at Brookhaven National Laboratory (beamline X-18b), Upton, New York. EXAFS/XANES spectra were taken in transmission mode at liquid nitrogen temperatures.

The beamline is equipped with a Si(1 1 1) channel cut monochromator. A crystal detuning procedure was used to help minimize glitches arising from harmonics. The X-ray ring at the NSLS has a flux of 1×10^{10} photons per second at 100 mA and 2.8 GeV, and the usable energy range at X-18b is from 5.8 to 40 keV. The spectra were recorded near the boiling temperature of liquid nitrogen to minimize contributions to the dynamic Debye-Waller factor. Sample thickness was determined by calculating the amount in grams per square centimeter of sample, w_D, by utilizing the following thickness equation: $w_D = \ln(I_o/I_t)/ \text{sum} \{(m/r)_j w_j\}$, where m/r is the total cross section (absorption coefficient / density) of element "j" in the sample at the absorption edge of the EXAFS element under study in cm^2/g, w_j is the weight fraction of element j in the sample, and $\ln(I_o/I_t)$ was taken over a typical range of 1 to 2.5. An average value of w_D from inputting both values was employed. Based on the calculation for w_D, and the cross-sectional area of the pellet, the grams were calculated. A smooth self-supporting pellet of catalyst and boron nitride, free of pinholes, was pressed and loaded into the in-situ XAS flow cell, and the hydrogen/helium (50 ccm/ 200 ccm) treatment gas was directed right to the sample area.

EXAFS data reduction and fitting were carried out using the WinXAS [5], Atoms [6], FEFF [7], and FEFFIT [8] programs. The k and r-ranges were approximately 3 - 15 $Å^{-1}$ and 1.3 - 3.0 Å. Details of the analysis procedures are reported elsewhere [9].

Activity test: The FTS reactions were conducted in 1 L continuously stirred tank reactor (CSTR) at 220°C, 250 psi, H_2/CO ratio of 2 (v/v) and 3L/g/h space velocity. Prior to loading into the CSTR, the calcined catalyst (17g of powder) was reduced ex situ in a fixed bed reactor at 500 °C for 10 h in pure hydrogen at a flow rate of 1 L/min. Details of the experimental setup and operation procedures can be found elsewhere [10].

Results and discussion:

Figure 1 compares the reduced 0.5%Pt-5%Co/KL catalyst with the cobalt metal foil and the CoO reference. There is a very small white line, suggesting that the catalyst was significantly reduced. A linear combination XANES fitting suggested that the cobalt in the catalyst was ~75% reduced.

Results of the EXAFS fitting are provided in Figure 2 and Table 1. The first Co-Co metal shell was fitted for the cobalt metal foil and the 0.5%Pt-5%Co/KL catalyst. The r-factors were very low and well below 0.02, indicative of excellent fits in each case. The coordination number of the first shell in the catalyst was ~5.5. Assuming complete reduction and a spherical morphology, the Co cluster size would be about 1 nm. Thus, the Co clusters are well dispersed with dispersion close to 85%.

Table 2 shows the dispersion, FTS turnover frequency (TOF), selectivity to CH_4 and selectivity to C_5^+ for the 0.5%Pt-5%Co/KL and 0.5%Pt-25%Co/Al_2O_3 catalysts. The results showed that the FTS turnover rate (TOS) is indeed a strong function of the Co particle size. The 0.5%Pt-5%Co/KL catalyst showed very low activity despite the fact that the Co particles are extremely dispersed (~ 85% dispersion). Moreover, the selectivity toward high carbon chain products (C_5^+) was less than that for 0.5%Pt-25%Co/Al_2O_3 catalyst and the selectivity to CH_4 was higher, on the other hand. In case of 0.5%Pt-25%Co/Al_2O_3, the catalyst has relatively low Co dispersion, high TOF, and high selectivity to the desired products (C_5^+). Our results indicate that the FTS activity does depend on the Co particle size when the Co particles are highly dispersed in the catalyst. This finding is in agreement with the finding of Bezemer et al. [3] and Barbier et al. [4] and in disagreement with the results of Lok [2].

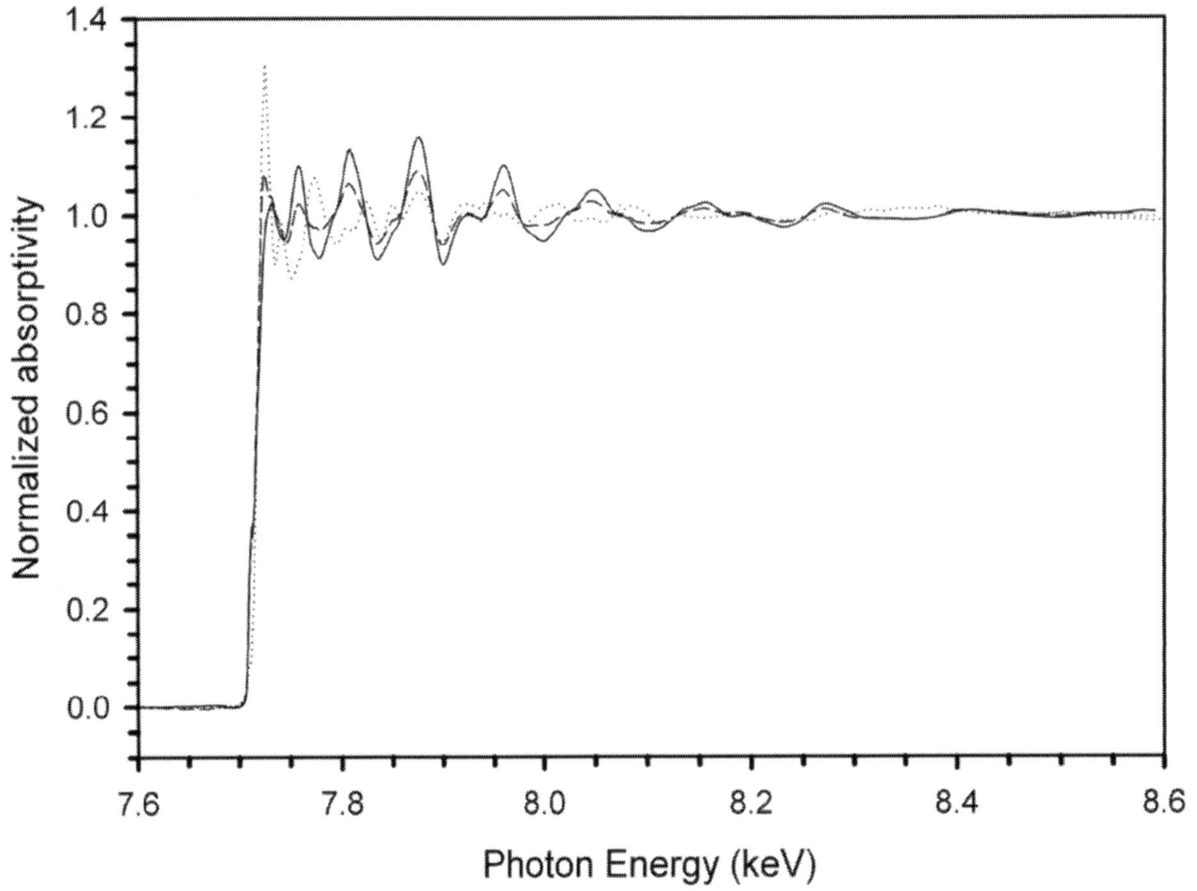

Figure 1: Normalized spectra of (solid) Co metal foil; (dashed) 0.5%Pt-5%Co/KL catalyst; and (dotted) CoO oxide. The catalyst is significantly reduced.

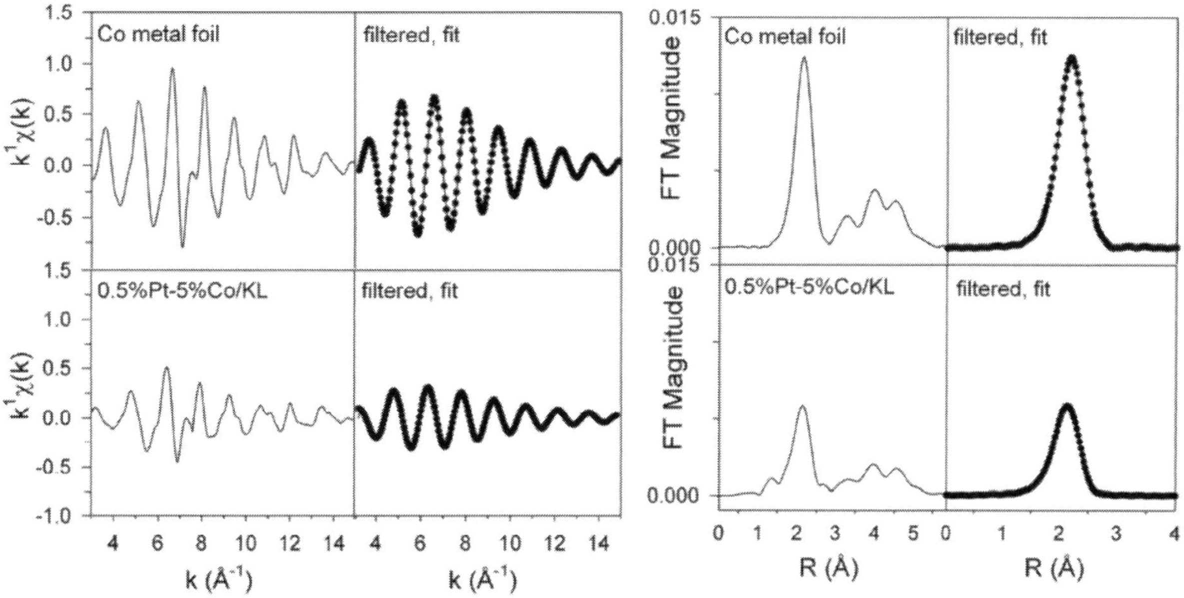

Figure 2: EXAFS results at the Co K-edge for (top) Co metal foil and (bottom) 0.5%Pt-5%Co/KL catalyst after reduction at 500°C, with spectra for (left) the raw k^1 weighted $\chi(k)$ versus k, (middle) the filtered $\chi(k)$ versus k and (dotted) the fitting; and (right) the raw Fourier transform magnitude spectrum and the filtered spectrum and (dotted) result of the fitting. Both reference and catalyst were scanned at liquid nitrogen temperatures.

Table 1: Results of EXAFS fitting parameters for references acquired near the Co K edge. The fitting ranges were approximately $\Delta k = 3 - 15$ Å^{-1} and $\Delta R = 1.3 - 3.0$ Å. $S_0^2 = 0.9$.

Sample Description	N* Co-Co metal	R Co-Co (Å) metal	e_0 (eV)	σ^2 (Å^2)	r-factor
Cobalt metal Co^0	**12**	2.4900 (0.000756)	7.6912 (0.4325)	0.00333 (0.000157)	0.00397
0.5%Pt-5%Co/KL	5.49 (0.323)	2.4927 (0.00414)	-3.8726 (0.7043)	0.00267 (0.000458)	0.00458

* Fixed parameter – first coordination shell.

Table 2: Comparison of Co dispersion, FTS activity and selectivity to CH4 and C^{5+} for 0.5%Pt-5%Co/KL and 0.5%Pt-25%Co/Al_2O_3 catalysts.

Catalyst	Co dispersion (%)	TOF (h^{-1})	Selectivity-C^{5+} (mol%)	Selectivity-CH_4 (mol%)	Selectivity-CO_2 (mol%)
0.5%Pt-5%Co/KL	85	2	69	24	2
0.5%Pt-25%Co/Al_2O_3	10	274	87	7	1.5

References:

[1] E. Iglesia, Appl. Catal. A. 161 (1997) 59

[2] C. M. Lok, Stud. Surf. Sci. Catal. 147 (2004) 283

[3] G. L. Bezemer, J. H. Bitter, H. P. Kuipers, H. Oosterbeek, J. E. Holewijn, X. D. Xu, F. Kapteijn, A. J. van Dillen, K. P. de Jong, J. Am. Chem. Soc. 128 (2006) 3956.

[4] A. Barbier, A. Tuel, I. Arcon, A. Kodre, G. A. Martin, J. Catal, 200 (2001) 106

[5] Ressler, WinXAS 97, Version 1.0 (1997).

[6] B. Ravel, J. Synch. Rad. 8 (2001) 314.

[7] J.J. Rehr, S.I. Zabinsky, R.C. Albers, Phys. Rev. Lett. 69 (1992) 3397.

[8] Newville, B. Ravel, D. Haskel, E.A. Stern, Y. Yacoby, Physica B 208/209 (2005) 154.

[9] G. Jacobs, A. Sarkar, Y. Ji, M. Luo, A. Dozier, B.H. Davis, Ind. Eng. Chem. Res. 47 (2008) 672.

[10] G. Jacobs, M. C. Ribeiro, W. Ma, Y. Ji, S. Khalid, P. T.A. Sumodjo, B. H. Davis, Appl. Catal. A 361 (2009) 137

SIMULATION OF RATE AND PRODUCT IN THE FISCHER-TROPSCH SYNTHESIS BASED ON THE DYNAMIC KINETICS

K. Fujimoto[1], *M. Murayama*[1], *Y. Suehiro*[2]
[1]*Department of Chemical Processes and Environments, Faculty of Environmetal Engineering, The University of Kitakyushu, 1-1 Hibikino, Wakamatsu-ku, Kitakyushu, Fukuoka, 808-0135, Japan, fujimoto@env.kitakyu-u.ac.jp*
[2]*Oil & Gas Technology Research & Development, Japan Oil, Gas and Metals National Corp., 1-2-2 Hamada, Mihama-ku, Chiba, 261-0025, Japan*

Abstract

A new dynamic kinetic model which simulate the carbon number (c-number) distribution and extract a kinetic parameters was developed. The main unit reaction are the chain propagation, and the hydrogenation of 1-olefins or alkyl species whose rates increased with c-number. The simulated result revealed that both chain growth and hydrogenation increase with carbon number. The ratio chain growth / hydrogenation increased also with c-number.

The effect of added 1-olefin, which promoted the formation of higher hydrocarbon could be well simulated by the same rate parameters to those of non-added system.

Keywords: Fischer-Tropsch synthesis; Co/SiO_2 catalyst; Co-fed olefins; Reaction model; Carbon number distribution; Simulation

1. Introduction

For a long period of time the carbon-number distribution of the Fisher-Tropsch Synthesis (FTS) product has been simulated by the Anderson-Schulz-Flory (ASF) distribution which is based on the concepts that (1) the chain growth probability is independent on the c-number of the intermediate, and (2) products are never subjected to the secondary reactions which change the c-number. For this ideal ASF the product distribution is characterized by a single parameter, α. This type of ideal distribution can only be obtained if the kinetic environment is identical and constant at each catalytic site of the synthesis. It is widely considered that the occurrence of the deviation of the real data from the ideal data can be attributed to the secondary reactions (reinsertion into the chain growth process, hydrogenolysis and isomerization) gives the most reasonable explanation for these deviations.

Studies on the produced and co-fed olefins have previously shown that the reaction of 1-olefins in the chain initiation and growth cause an increase in the heavier hydrocarbon yields, the hydrogenolysis shorten the chain of hydrocarbons by successive demethylation, as well as their extensive conversion to the corresponding paraffin by secondary hydrogenation. To this point, addition of 1-olefin to FTS can be utilized for controlling the carbon number (c-number) distribution towards desired products.

In the previous study, we proposed a new basic reaction model to express the carbon number distribution and analyzed it by the simulation method as 1-olefin was added in the n-paraffin solvent on FTS. In this study, we show a reaction model and analyzed it by a simulation method to elucidate the rate parameters which control the carbon number distributions, and also the effect of added olefins using the parameters.

2. Experimental

2.1 Catalyst preparation

The 20 wt.% Co/SiO_2 catalyst was prepared by an incipient wetness method. The cobalt nitrate was used as the precursor of cobalt and a commercially available silica gel (Fujisilica Q-15, surface area: 200 m^2/g, average pore diameter: 15 nm, pore volume: 1.0 ml/g) was used as the support. The catalyst was pretreated at 673 K for 3 h in situ in a flow of hydrogen. The detail and the catalyst has been written elsewhere.

2.2 Reaction procedure and product analysis

The configuration of reactor for the conventional or olefin-added FT reaction is shown in Fig.1. Syngas and solvent (or co-fed olefin in the solvent) were fed to the vaporizer by a mass flow controller and a high pressure liquid pump, and then injected into the reactor

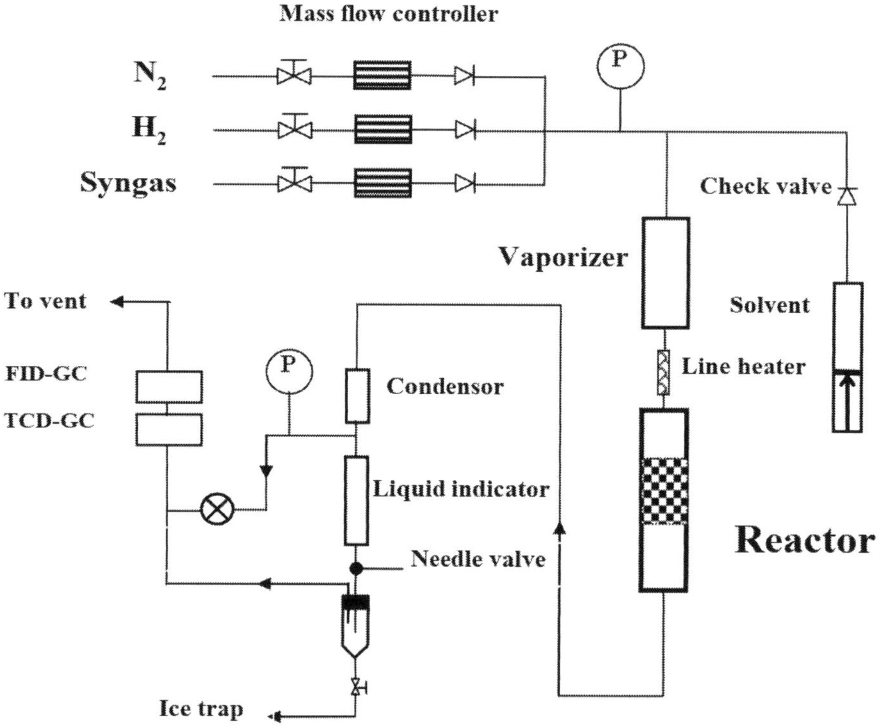

Fig.1 Schematic flow diagram of FT synthesis apparatus

concurrently. The effluent gas and liquid from the reactor passed through an ice-cooled condenser, where the solvent and FT products were condensed continuously, and uncondensed products with unreacted syngas were depressurized and led to an online gas chromatograph. N-dodecane (T_b, 216.3 ℃) was utilized as the solvent fluids. 1-decene (T_b 172 ℃) was selected as the co-fed olefin. Typical reaction conditions were: P_{total} = 2.1 MPa, W/F_{syngas} = 5 g/(h/mol) , F_{syngas} = 81 ml/min, T = 230 ℃, Mole ratio of co-fed olefin to CO = 0.16, $F_{co\text{-}fed\ olefin}$ = 1.1-mmol/h. Also, an agitated slurry bed reactor was used with the same catalyst and solvent. The effluent gaseous products were analyzed with two on-line gas chromatographs (GC). The chain growth probability (α) of the products was defined followed by the classical ASF model, where the carbon number extended from C_4 to C_{20}.

2.3 Reaction model and simulation software

Basic concept of simulation model

We have shown reaction network as Fig.2 where CH_2 species play important roles in chain growth and chain degradation. However, since it is impossible to measure CH_2 species on the catalyst directly, we propose a modified reaction model which utilize stable molecules, instead. Basic model is shown in Fig 3.

(1) CH_2 species on the catalyst is substituted by H_2 and CO such as $2H_2+CO \rightarrow {}^*CH_2+H_2O$ (① in Fig3).

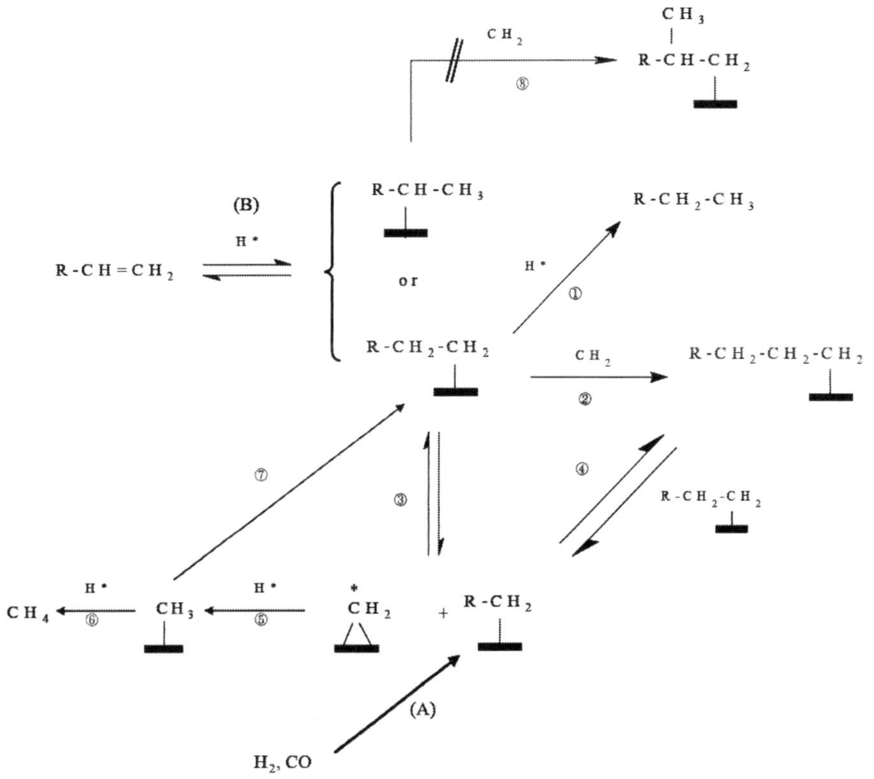

Fig.2 Scheme of the reaction network model

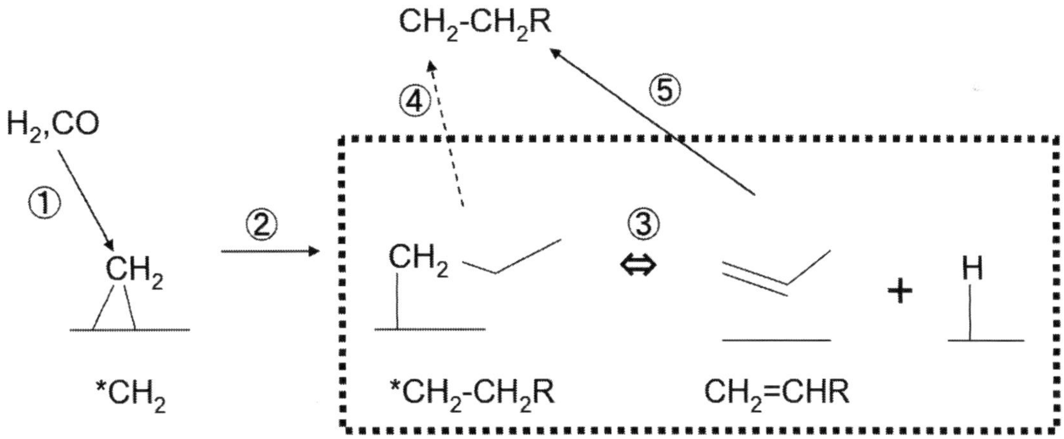

Fig.3 simplified reaction network

(2) Chain growth and chain degradation on the catalyst is proceeding such as $*CH_2+*CH_2R \Leftrightarrow *(CH_2)_2R$ (② in Fig3).

(3) Olefin species is pseudo-equilibrium state such as $*(CH_2)_2\text{-}R \Leftrightarrow CH_2{=}CHR+*H$(③ in Fig 3).

(4) Paraffin species is produced by $*(CH_2)_2\text{-}R+*H \rightarrow CH_3\text{-}CH_2R$(④in Fig 3) or $CH_2{=}CHR + H_2 \rightarrow CH_3\text{-}CH_2R$(⑤ in Fig 3).

(5) We focused olefin species in ③ (either produced in the FTS reaction or added). $*(CH_2)_2R$

species are in the pseudo-equilibrium state with CH_2=CHR:
$2H_2$+CO+CH_2=CHR⇔CH_2=CH-CH_2R+H_2O, and H_2+CH_2=CHR→CH_3-CH_2R.

Reaction model for calculation is shown schematically in Fig3 and the rate equation is Langmuir-Hinselwood-Hougen-Watson (LHHW).

2.4 Simulation software

The used simulation software REX developed by Optience Corporation. It facilitates reasoning of the experimental data through kinetic modeling.

3. Results and discussion

Fig. 4 shows the product distribution in the tricle bed reaction on a 20wt% Co/Q-15 catalyst.

The carbon number distribution with different contact time (W/F) shows that with increasing contact time, the carbon number (c-number) distribution shift to higher side. The phenomenon means that c-number distribution (or average molecular weight) grows higher from the inlet of the catalyst bed to the exit, showing the intermediate product grows to higher hydrocarbon. The rate constant of (1) chain growth (k_p), (2) chain degradation (k_d) and (3) chain termination (hydrogenation) were calculated from the simulated data, shown in Fig.4 (b) (k_d << k, k_H). It is clear from the data that the rate constant of either chain growth or hydrogenation increased with the increasing c-number of the product. The ratio of k / (k + k_h), which correspond to the α value of ASF plot, increased from 0.8 (C=5) to 0.92 (C=20). This means the deviation of the c-number distribution of ASF plot from the real data come from the different chain growth probability of each olefin.

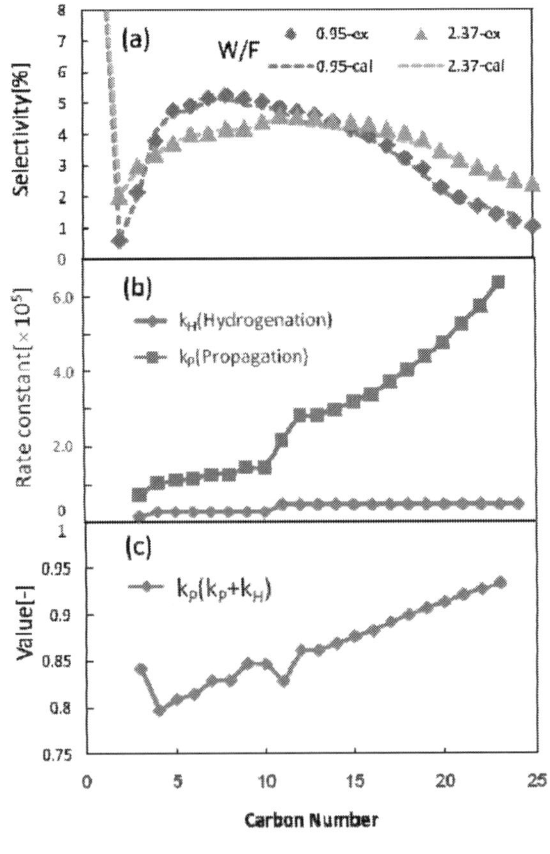

Fig. 4 Product distribution and kinetic parameters on 20wt% Co/SiO_2

3.1 Effect of co-fed 1- olefin on the carbon number distribution

In this study, we used 1-decene as 1-olefin. As displayed in Fig.5, we can see that the CO conversion and 1-olefin selectivity was promoted and the lighter hydrocarbon products in C_1-C_9 range were suppressed by the addition of 1-decene. The simulation which used the same rate parameters without 1-decene addition showed excellent agreement of the experimental results and calculated data including the phenomenon that (a) the formation of C_{11}+ hydrocarbon was effectively promoted and (b) the formation of C_9- product was suppressed. The phenomena suggest;

(1) Added 1-decene is adsorbed on the catalyst and consume CH_2 for chain growth or emitted it for chain degradation.

(2) Added olefin is also subjected to the hydrogenation to paraffin to stop all successive reaction.

(3) As the result, if the rate of chain growth is higher than that of the de-methylation CH_2 species (chain degradation) on the catalyst is consumed to decrease its concentration to result in the suppressed C_9- product.

Results of similar experiments in the slurry system to those of tricle bed system are shown in Fig.6. It is clear that similar analysis to that of tricle bed can be applied to slurry system.

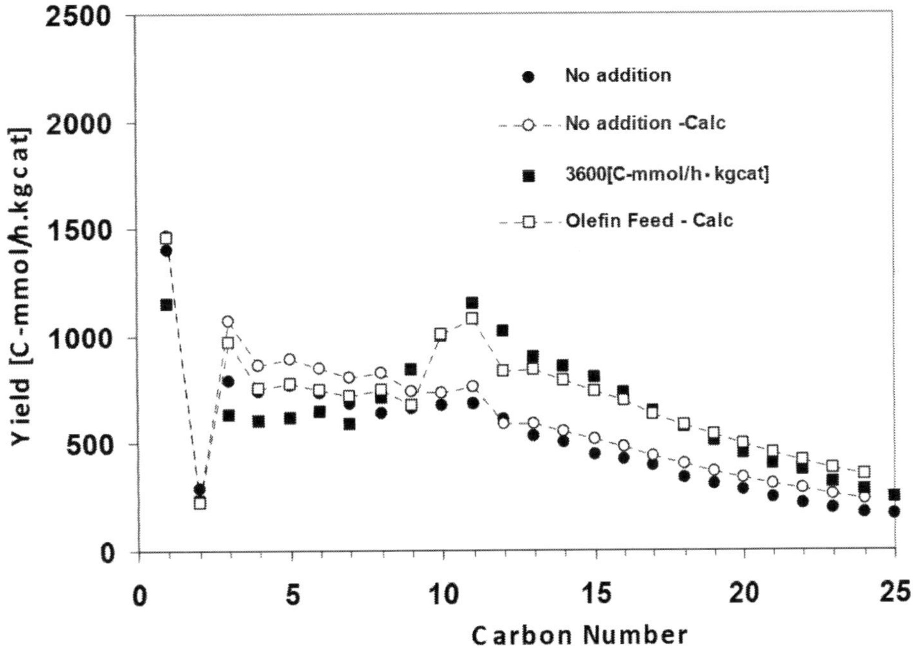

Fig.5 Experimental data and simulation result

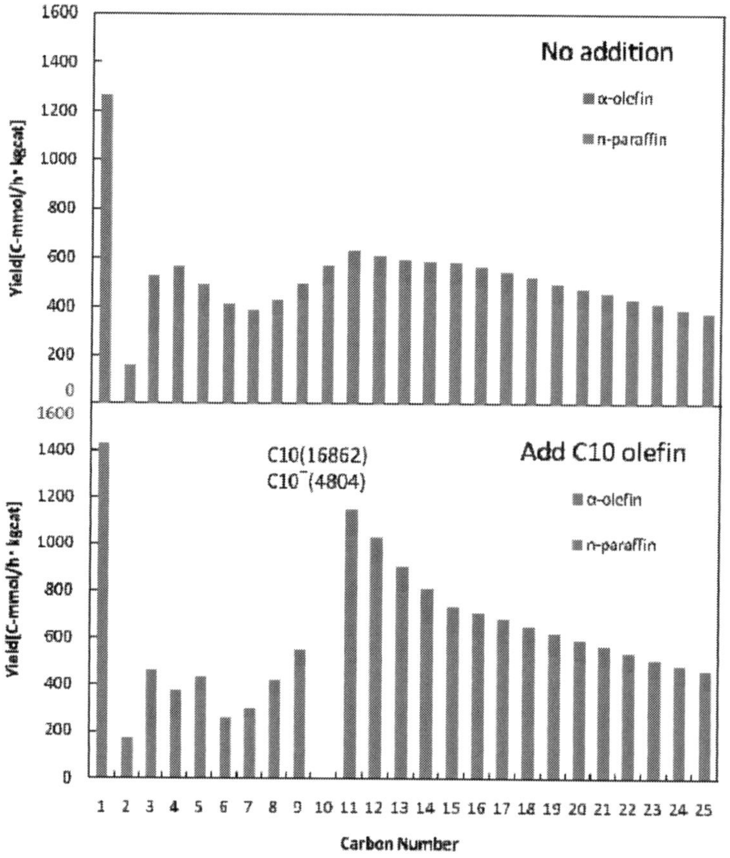

Fig.6 Product distribution in the slurry reactor on the Co/SiO_2 catalyst

4. Conclusions

1. We proposed new reaction model and simulated both olefin-added and conventional FTS by using the same rate parameters.
2. The essence of new model was (1) using stable molecules, (2) $*CH_2$ species included indirectly and (3) olefin as key reaction medium.
   The concept of new reaction model is as follows:
   $*(CH_2)_2R$ species are in the pseudo-equilibrium state with $CH_2{=}CHR$:
   $2H_2{+}CO{+}CH_2{=}CHR \Leftrightarrow CH_2{=}CH{-}CH_2R{+}H_2O$, and $H_2{+}CH_2{=}CHR \rightarrow CH_3{-}CH_2R$.
3. Both hydrogenation and chain growth rate decreases with increasing c-number of the products, but the rate constant of them increased with c-number.
4. The simulation results revealed that the rate of chain growth is faster than that of hydrogenation especially at high c-number products.
5. The rate of chain degradation is quite slow with $C_{10}+$ species.
6. Product distribution is mostly controlled by the rate ratio of the hydrogenation and chain growth, whose value was close to the α -value of ASF distribution.

Acknowledgment

The authors wish to acknowledge the financial support from Japan Oil, Gas and Metals National Corporation(JOGMEC).

We also would like to acknowledge their technical discussion of Dr. Subash Balakrishna, Dr. Xiaohao Liu, Dr. Wensheng Linghu, Professor Xiaohong Li, and Professor Kenji Asami.

Studies on the La_2O_3-doped Al_2O_3 Supported Cobalt Catalyst for Fischer-Tropsch Synthesis

*Zhe Cai, Kongyong Liew, Jinlin Li**

1. Introduction

Fischer-Tropsch synthesis (FTS) attracts much attention as an option for productions of clean transportation fuels and chemicals via conversion of natrual gas or coal[1,2]. Cobalt based catalysts are one of the most important catalysts for the Fischer-Tropsch synthesis for high molecular weight paraffinic waxes which could be hydrocracked to produce lubricants and diesel fuels[3]. To increase the number of cobalt active sites, the cobalt is dispersed as clusters on a high surface area support such as SiO_2, TiO_2 and Al_2O_3. Alumina is often used due to its favorable mechanical properties[4]. Previous studies indicated that cobalt supported on alumina was not completely reduced to the metallic state because of significant metal-support interactions[3,5]. Chen et al.[6] reported that when the metal loading was lower than the dispersion capacity in the γ-Al_2O_3 supported transition metal (Ni, Co or Cu) oxide catalysts, the supported M^{2+} ions were preferentially incorporated into the tetrahedral vacancies of γ-Al_2O_3, and that the amount of M^{2+} ions incorporated into the octahedral vacancies increasesd with increased M^{2+} loading. Since octahedral M^{2+} ions are easier to be reduced to the metallic state than the tetrahedral M^{2+} ions, reseachers[7] tried to increase the octahedral ratio of the M^{2+} (Ni^{2+} in the study) to increase the degree of reduction. Similarly, many reserchers[8-10] reported that the octahedrally coordinated cobalt FTS catalyst could also be more easily reduced than the cobalt in the tetrahedral environment, hence, showing higher catalytic properties. Rare earth metal oxides are often used as promoters in cobalt[11-13] and iron[14] based CO hydrogenation catalysts. It has been reported that rare earth metal oxide promotion of cobalt catalyst is beneficial for Fischer-Tropsch synthesis[15]. Vada et al.[12] studied a La^{3+}/Co/Al_2O_3 system and found that La^{3+} increased the overall activity and chain growth probability for a low La^{3+} loading. Ren et al.[16] studied NiO/Al_2O_3 catalysts with La^{3+} promoter, and found that the dispersed La^{3+} species on γ-Al_2O_3 increased the ratio of octahedral Ni^{2+} ions to the tetrahedral ones, and thus increased the degree of reduction of the catalysts. Ledford et al.[13] studied two methods of preparing lanthanum promoted Co/Al_2O_3 catalysts. They observed promotional effects only when La^{3+} was impregnated first followed by cobalt, and for higher La^{3+} loadings, La-Co mixed oxide were suggested to be formed and metallic cobalt dispersion was shown to be enhanced. In the present study, γ-Al_2O_3 doped with La_2O_3 by co-precipitation has been prepared followed by cobalt loading. It was expected that when the lanthana was doped on γ-Al_2O_3 by co-precipitation, more octahedral sites formed in the γ-Al_2O_3 crystallites were left vacant for the Co ions in the subsequent impregnation than when the lanthana doping was carried out by impregnation, La species occupied the tetrahedral vacancies of γ-Al_2O_3, hence preventing these sites from the cobalt ions later. A series of alumina supports were doped by lanthana using impregnation or co-precipitation. Cobalt catalysts were supported onto the La_2O_3-doped or undoped alumina by incipient wetness impregnation. Nitrogen adsorption-desorption, powder X-ray diffraction (XRD), hydrogen temperature programmed reduction (H_2-TPR), hydrogen temperature programmed desorption (H_2-TPD) and O_2-titration and diffuse reflectance infrared Fourier transform spectroscopy (DRIFTS) were used to characterize these catalysts. The Fischer-Tropsch synthesis reaction was carried out in a fixed bed reactor.

2. Experimental

2.1 Catalyst preparation

2.1.1 Preparation of supports

The La_2O_3-doped alumina support was prepared by co-precipitation as follows: known amounts of aluminum nitrate and lanthamum nitrate solutions were mixed in apropriate porportions. Ammonium hydroxide was added to the mixed solution to precipitate the hydroxide. The pH value of the solution was kept constant at 9.0±0.1. The obtained support is denoted as LAP (2.9 wt % La_2O_3). Alumina support without La_2O_3 prepared with the precipitation method was labeled as AP.

The support with the same loading of La_2O_3 was also prepared by incipient wetness impregnation with the desired amount of aqueous lanthanum nitrate on AP. The excess water was evaporated in a rotary evaporator at around 363 K. This support was named LAI. A commercial La_2O_3-doped supports (SASOL Germany GmbH), labeled as LAS, was used as the reference.

All the materials were dried in an oven overnight at 393 K, crushed and sieved to obtain 120 mesh particles, followed by calcination at 1023 K for 6 h.

2.1.2 Preparation of cobalt catalysts

The 12 wt.% cobalt catalysts were prepared by impregnating the different synthesized supports with cobalt nitrate solutions. The samples were dried at 393 K and calcined in air at 623 K for 6 h, hereafter denoted as CAP, CLAI, CLAS and CLAP, respectively.

2.2 Catalysts characterization

2.2.1 Powder X-ray diffraction (XRD)

The XRD measurements for the samples were carried out on a Brukers D8 powder X-ray diffractometer with monochromatic CuKa radiation and a VANTEC-1 detector within a 2-theta ranges from 20° to 80° with 0.0167° steps and from 35° to 40° with 0.008° steps for the crystallite size of cobalt oxide. Crystallite phase was detected by comparing the patterns with those in the standard powder XRD file compiled by the Joint Committee on Powder Diffraction Standards (JCPDS). The average Co_3O_4 crystallite size was calculated using the Scherrer equation[17].

2.2.2 Nitrogen adsorption-desorption

The pore texture properties were measured by nitrogen adsorption-desoprtion at 77 K with a Quantachrome Autosorb-1-C-MS. Before the measurement, the samples were outgassed at 473 K for > 6 h. The surface area was obtained using the Brunauer-Emmett-Teller (BET) model for adsorption in a relative pressure ranging from 0.05 to 0.30. The total pore volumes were calculated from the amount of N_2 vapor adsorbed at a relative pressure of 0.99. The pore size distribution were evaluated from the desorption branches of the isotherms using the Barrett-Joyner-Halenda (BJH) model.

2.2.3 Hydrogen Temperature programmed reduction (H_2-TPR)

Reduction properties of catalyst were measured by temperature programmed reduction experiments carried out on a Zeton Altamira AMI-200 unit. Calcined catalysts (0.10g) were placed in a U-shape quartz reactor, with a thermocouple for continuous temperature measurement. Before measurements, the samples were decontaminated by flowing high purity argon at 423 K, then cooled down to 323 K. Then 10 % H_2/Ar (constant flow rate of 30 ml/min) was switched on and the temperature was raised from 323 to 1073 K at a rate of 10 K/min, and held for 30 min at 1073 K. The H_2 consumption (TCD signal) was recorded automatically by a computer.

2.2.4 Hydrogen temperature programmed desorption (H_2-TPD)

The dispersion and crystalite size of cobalt were measured by hydrogen temperature programmed desorption in a U-shaped quartz reactor with the Zeton Altamira AMI-200 unit. The catalysts (0.15 g) were reduced at 723 K for 12 h and were then cooled down to 373 K in the environment of high purity hydrogen. Prior to increasing the temperature from 373 K to 723 K at the rate of 10 K/min, the samples were held at 373 K for 1 h under argon stream to drive away weakly bound physisorbed species. Then the catalysts were held at 723 K under flowing argon to desorb the remaining chemisorbed hydrogen, meanwhile, the TCD detector began to record the signal untill it returned to the baseline. The amount of desorbed hydrogen was calculated by comparing the integrated TPD spectrum with the mean areas of calibrated hydrogen pulses. The formula for the calculation has been shown in previous works[18].

2.2.5 O_2-titration

The degree of reduction was measured by O_2-titration done in the Zeton Altamira AMI-200 unit. The samples which had been reduced in pure hydrogen at 723 K for 12 h were reoxidized at 723 K by introducing pulses of high purity oxygen, until there was no further consumption of O_2, detected by the thermal conductivity detector located downstream. The reference gas for TCD was high purity helium. All flow rates were set to 30 ml/min. The extent of catalyst reduction was calculated assuming stoichiometric reoxidation of Co to Co_3O_4. The formula for the calculation has been shown in previous works[18].

2.3 Catalytic evaluation

Fischer-Tropsch synthesis was performed in a fixed bed reactor (id=12 mm) at 503 K and 1.0 MPa. The catalysts (ca. 0.5 g) were mixed with 5 g carborundun to minimize the temperature gradient and reduced in situ with H_2 at atmosphere pressure. The reactor temperature was raised in a H_2 stream of 6 NL h^{-1} g^{-1} (298 K, 0.1 MPa) from room temperature to 373 K at 2 K/min, then increased to 723 K at a rate of 1 K/min and held for 10 h. Subsequently, the reactor was cooled down to 373 K. Then, the syngas,with space velocity of 4 SL g^{-1} h^{-1} (273 K, 0.1 MPa), was switched on and the pressure was increased to 1.0 MPa. The reactor temperature was increased to 503 K in 16 h and the reaction was performed at 503 K. Typically, the reaction products were collected after more than 80 h of operation to achieve a good mass balance at the steady state. The carbon monoxide conversion and hydrocarbon selectivities were measured at steady state average over the period of operation. The discharged gases were analyzed online by an Agilent MicroGC 3000A. The analysis of the liquid product, which was collected in the cold trap (269 K), was performed with an Agilent 6890N GC equipped with a FID detector. The solid wax which was collected in a hot trap (373 K) was analyzed with an Agilent 7890A GC.

3.Results and Discussion

3.1 sample porosity

The surface area, average pore diameter and pore volume data of the supports and the cobalt catalysts are listed in Table 1. The La_2O_3-dopping has a small influence on the pore volume. However, the porosity of the supports was significantly affected by the preparation method. LAP has the highest BET surface area, meanwhile the LAS has the lowest surface area but the largest pore size. As can be seen in the Table 1, La_2O_3-doped by the method of impregnation increased the average pore diameter from 7.81 for support AP to 9.58 nm for LAI, indicating that the impregnation of lanthanum blocked up smaller pores of AP. The impregnation with cobalt lowered both the support surface area and the pore volume, and did not change the average pore size, which suggests that some cobalt species have entered the pore of the supports.

Table 1 Surface area, average pore diameter and pore volume.

Sample	Surface area(m^2/g)	Pore volume(cm^3/g)	Average pore diameter(nm)	Crystallite diameter(nm)[a]
AP	168.8	0.438	7.81	—
LAI	160.0	0.417	9.58	—
LAS	143.9	0.548	12.37	—
LAP	199.7	0.497	7.81	—
CAP	143.6	0.335	7.86	14.02
CLAI	127.7	0.306	9.55	14.14
CLAS	121.9	0.430	12.38	14.17
CLAP	173.2	0.377	7.81	14.49

[a] Average Co_3O_4 particle diameter obtained from XRD data

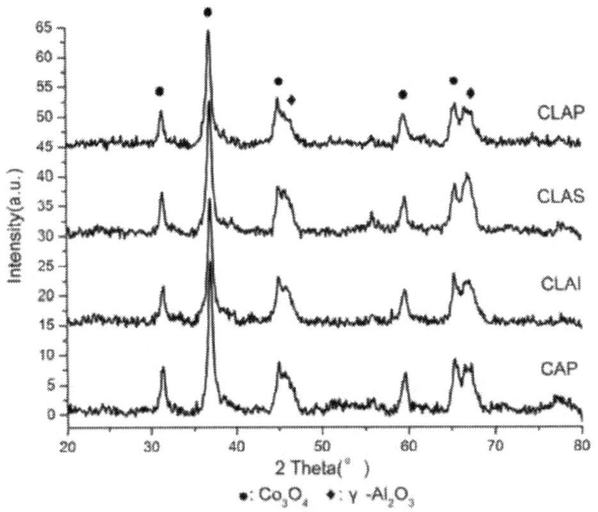

Figure 1. X-ray diffraction patterns for calcined catalysts

3.2 X-ray diffraction results

XRD patterns of the catalysts are presented in Figure 1. The diffraction peaks at 45.7° and 66.6° are attributed to the γ-Al_2O_3 support and the peaks at 31.4°, 36.9°, 45.0°, 59.5° and 65.5° are attributed to the Co_3O_4 phase, which exists on all the catalysts[15]. The diffraction peak of La_2O_3 phase could not be detected from the XRD patterns of catalysts, indicating that the La_2O_3 was highly dispersed on the support[19]. The crystallite size data of the Co_3O_4 phase are shown in Table 1. It can be seen that the Co_3O_4 crystallite size increased slightly on doping with La_2O_3, suggesting that La_2O_3 may have influenced the Co_3O_4 crystals growth.

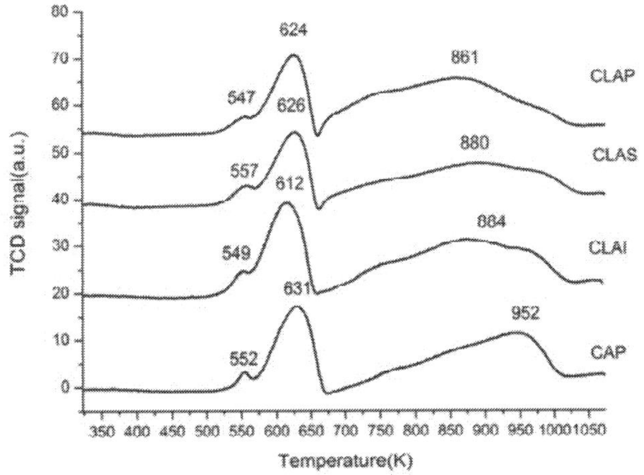

Figure 2 TPR spectra of calcined catalysts

Table 2 H_2-temperature programmed desorption and O_2 pulse re-oxidation results for catalysts.

Sample	$H_{2\ desorbed}$ (μmol/g)	$d_{uncorrected}$ (%)[a]	$D_{uncorrected}$ (nm)[b]	O_2 uptake (μmol/g)	$d_{corrected}$ (%)[c]	Reducibility (%)	$D_{corrected}$ (nm)[d]
CAP	183.2	17.99	5.7	579.3	42.37	42.46	2.4
CLAI	189.0	18.56	5.6	734.3	34.49	53.82	3.0
CLAS	193.5	19.00	5.4	736.8	35.19	54.00	2.9
CLAP	197.5	19.39	5.3	813.6	32.51	59.64	3.2

[a] The uncorrected catalyst dispersion.

[b] The uncorrected cobalt metal cluster diameter.

[c] The corrected catalyst dispersion.

[d] The corrected cobalt metal cluster diameter.

3.3 H_2-temperature programmed reduction

The TPR spectra of catalysts are shown in Figure 2. Three reduction peaks for Co/Al_2O_3 are observed at temperature ranges of < 573K, 573-673K and > 673K, respectively. The peak at 573-673K is attributed to the reduction of Co_3O_4 to CoO[17]. The broad peak at high temperature (> 673K) is ascribed to the subsequent reduction of CoO to Co^0 [20]. Besides two main reduction peaks, a small peak can be observed below 573 K. This peak could be attributed to the reductive decomposition of redidual $Co(NO_3)_2$[21].

All reduction peaks of CAP appeared at higher reduction temperature than the catalysts doped with La_2O_3, indicating that La_2O_3-doping increased the catalysts reducibility. The methods of La_2O_3-doping markedly affected the reduction properties of Co/Al_2O_3 catalysts. It should be noticed that the second reduction step (> 673K) of several catalysts was comprised of broad peaks, which is supposd to be the effect of different degrees of interaction between cobalt and support[22]. Also, the existence of a cobalt aluminate could not be excluded. The reduction peak above 673 K shifted to a lower temperature for catalyst CLAP compared with the catalysts CLAI and CLAS, and the intensity of this broad peak increased. All the above results indicated that CLAP exhibits the best reduction properties among the cataltysts studied.

3.4 H_2-temperature programmed desorption and O_2-oxidation

Table 2 presents the H_2-temperature programmed desorption and O_2-oxidation data. The La_2O_3-doping increased the degree of reduction of the catalysts significantly, and the method of La_2O_3-doping affected the metal cobalt cluster size, dispersion and reducibility. Catalyst reducibility increased in the order of CLAI, CLAS to CLAP. Furthermore, the cobalt cluster size increased by the doping of La_2O_3 and was slightly affected by the doping method. Khassin et al.[9] reported that the Co^{2+} cations which entered the support structure, and occupied the octahedral positions could be reduced at lower temperature. For Mg^{2+} and Zn^{2+} promoted catalysts, promoter cations occupy the tetrahedral sites of the supports and caused the Co^{2+} cations occuping mostly the octahedral sites, and thus increase the reduction of the catalyst. The resulting performance of these promoted catalysts are far better than non-promoted catalysts. In the present work, it could be considered that in CLAP, the La_2O_3 doping process carried out at the same time with the formation of γ-Al_2O_3 crystallite phase and the dispersed lanthanum species on γ-Al_2O_3 might inhibit the incorporation of cobalt species from entering the tetrahedral vacancies of γ-Al_2O_3. When the La_2O_3 doping process was carried out after the formation of the support crystallite phase in the CLAI catalyst, the impregnated La_2O_3 could not occupy the tetrahedral vacancies at the same degree as in CLAP catalyst, and thus has higher degree of reduction [6,9,10]. Ren et al.[16] also reported the same results in the studies of the La_2O_3 promoted NiO/Al_2O_3.

3.5 Fischer-Tropsch synthesis

Results of the FTS tests are displayed in Table 3. CO conversion and selectivities to heavy hydrocarbons increased with doping of La_2O_3.The activity of the La_2O_3-doped catalysts increased in the order of CLAI < CLAS < CLAP which is consistent with the order of reducibility. Selectivity for methane and light weight hydrocarbons (C_2-C_4) showed an opposite trend to heavy hydrocarbons. CLAP catalyst displayed the highest FT activity and the highest C_5+ selectivity among the catalysts studied.

Previous studies on cobalt based catalysts for Fishcer-Tropsch synthesis found that high reducibility provided more cobalt active sites and hence, increased the catalytic activity[23-28]. In the present work, it can be seen that lanthanum oxide facilitated the reduction of the cobalt oxide dispersed on the supports, and the performance was more dramatic in CLAP catalyst leading to an increase in catalytic activity. Methane selectivity is at 13.19% for unpromoted catalyst, with La_2O_3-doped, the methane selectivity decreased significantly to 9.14% for CLAS and 8.05% for CLAP. High methane selectivity has been reported for low reducibility and high dispersion[29] attributed to the presence of unreduced cobalt oxides which catalyzed the WGS reaction, resulting in a high H_2/CO ratio on the surface Co^0 sites, which favors hydrogenation of the adsorbed species. It should be mentioned that CO_2 selectivity was 0.49% for CLAP, much lower than those of CLAI (1.50%) and CLAS (1.64%), showing low WGS reactivity which is consistent with the low CH_4 selectivity.

Table 3 FTS reaction CO conversion and product distribution

Catalysts	CO conversion(%)	CO_2 selectivity(%)	Hydrocarbon selectivity(%)				
			CH_4	C_2	C_3	C_4	C_5+
CAP	28.04	1.88	13.19	1.44	2.96	5.26	75.28
CLAI	32.76	1.50	8.57	0.83	2.31	4.01	82.79
CLAS	33.06	1.64	9.14	1.15	2.32	3.99	81.75
CLAP	34.92	0.49	8.05	0.91	2.39	3.81	84.36

The catalysts were reduced in H_2 at 723 K for 10 h before FTS.
Reaction conditions: $H_2/CO=2$, space velocity of syngas was 4 SL $g^{-1}h^{-1}$ (273 K, 0.1 MPa), temperature was 503 K; pressure was 1.0 MPa.

4. Conclusion

The effect of preparation methods on La_2O_3-doped Co/Al_2O_3 FTS catalysts has been investigated. The following conclusions can be draw.

The doping of La_2O_3 strongly affected the catalytic performance. The catalyst with the La_2O_3-doped support prepared by co-precipitation is more easily reducible, higher FTS activity and product selectivity. For this catalyst, the co-precipitated La_2O_3 occupied the tetrahedral vacancies of the γ-Al_2O_3 crystallites and favoured the incorporation of cobalt species into the octahedral vacancies of γ-Al_2O_3 crystallites in the subsequent Co impregnation. When the La_2O_3 doping was carried out after the formation of γ-Al_2O_3, both the tetrahedral and the octahedral sites will be partially occupied by the lanthana so that vacant octahedral sites available will be decreased for Co in the subsequent impregnation, and thus has lower reducibilty and lower activity.

Reference

(1) Schulz, H. *Appl. Catal.* **1999**, 186, 3-12.
(2) Anderson, R. B. Fischer-Tropsch Synthesis. *Academic Press NY*, **1984**.
(3) Jacobs, G.; Das, T. K.; Zhang, Y.Q.; Li, J. L.; Racoillet, G.; Davis, B. H. *Appl. Catal., A* **2002**,

233, 263-281.
(4) Iglesia, E.; Soled, S. L.; Fiato, R. A.; Via, G. H. *J. Catal.* **1993**, 143, 345-368.
(5) Wang, W. J.; Chen, Y. W. *Appl. Catal.* **1991**, 77, 223-233.
(6) Chen, Y.; Zhang, L. F. *Catal. Lett.* **1992**, 12, 51-62.
(7) Ren, S. B.; Qiu, J. H.; Wang, C. Y.; Xu, B. L.; Fan, Y. N.; Chen, Y. *Chin. J. Catal.* **2007**, 28, 651-656.
(8) Khassin, A. A.; Yurieva, T. M.; Kaichev, V. V.;Bukhtiyarov, V. I.; Budneva, A. A.; Paukshtis, E. A.; Parmon, V. N. *J. Mol. Catal., A.* **2001**, 175, 189-204.
(9) Khassin, A. A.; Yurieva, T. M.; Kustiva, G. N.; Itenberg, I. S.; Demeshkina, M. P.; Krieger, T. A.; Plyasova, L. M.; Chermashentseva, G. K.; Parmon, V. N. *J. Mol. Catal., A.* **2001**, 168, 193-207.
(10) Khassin, A. A.; Yurieva, T. M.; Kustiva, G. N.; Plyasova, L. M.; Itenberg, I. S.; Demeshkina, M. P.; Chermashentseva, G. K.; Anufrienko, V. F.; Zaikovskii, V. I.; Larina, T. V.; Molina, I. Y.; Parmon, V. N. *J. Mol. Catal., A.* **2001**, 168, 209-224.
(11) Barrault, J.; Guilleminot, A. *Appl. Cayal.* **1989**, 21, 307-312.
(12) Vada, S.; Chen, B.; Goodwin, J. G. *J. Catal.* **1995**, 153, 224-231.
(13) Ledford, J. S.; Houalls, M.; Proctor, A.; Hercules, D. M. *J. Phys. Chem.* **1989**, 93, 6770-6777.
(14) Pour, A. N.; Shahri, S. M.. K.; Bozorgzadeh, H. R.; Zamani, Y.; Tavasoli, A.; Marvast, M. A. *Appl. Catal., A.* **2008**, 348, 201-208.
(15) Haddad, G. J.; Chen, B.; Goodwin, J. G. *J.Catal.* **1996**, 161, 274-281.
(16) Ren, S. B.; Qiu, J. H.; Wang, C. Y.; Xu, B. L.; Fan, Y. N.; Chen, Y. *Chin. J. Inorg. Chem.* **2007**, 23, 1021-1028
(17) Cullity, B. D. Elements of X-ray Diffraction, third ed. *Additon-Wesley. Reading. MA*, **1967**
(18) Zhang, Y. H.; Xiong, H. F.; Liew, K. Y.; Li, J. L. *J. Mol. Catal.* **2005**, 237, 172-181.
(19) Jongsomjit, B.; Panpranot, J.;Goodwin, J. G. *J. Catal.* **2003**, 215, 66-77.
(20) Bao, A.; Liew, K. Y.; Li, J. L. *J. Mol. Catal., A.* **2009**, I304, 47-51.
(21) Lapidus, A.; Krylova, A.; Kazanskii, V.; Borovkov, V.; Zaitsev, A.; Rathousky, J.; Zukal, A.; Jancalkova, M. *Appl. Catal.* **1991**, 73, 65-81.
(22) Borg, Ø.; Blekkan, E. A.; Eri, S.; Akporiaye, D.; Vigerust, B.; Rytter, E.; Holmen, A. *Top. Catal.* **2007**, 45, 39-43.
(23) Martinez, A.; Lopez, C.; Marquez, F.; Diaz, I. *J. Catal.* **2003**, 220, 486-499.
(24) Zhang, J. L.; Chen, J. G.; Ren, J.; Sun, Y. H. *Appl. Catal., A.* **2003**, 243, 121-133.
(25) Belamble, A. R.; Oukaci, R.; Goodwin Jr., J. G. *J. Catal.* **1997**, 166, 8-15.
(26) Li, J. L.; Coville, N. J. *Appl. Catal., A.* **1999**, 181, 201-208.
(27) Wang, Y.; Noguchi, M.; Takahashi, Y.; Ohtsuka, Y. *Catal. Today* **2001**, 68, 3-9.
(28) Khodakov, A. Y.; Bechara, R.; Griboval-Constant, A. *Stud. Surf. Sci. Catal.* **2002**, 142, 1133-1140.
(29) Reuel, R. C.; Bartholomew, C. H. *J. Catal.* **1984**, 85, 63-77
(30) Sun, S. L.; Fujimoto, K.; Tsubaki, N.; Yamada, M. *Catal.Commun.* **2003**, 4, 361-364.

(98g) Advances in LNG Expander Technology

Joel Madison, Ebara International Corporation, Sparks, NV

The first use of cryogenic liquefied gas expanders dates back to the late nineties. The proven advantages of incorporating liquid expanders in the LNG liquefaction process lead the way to equip every new liquefaction plant with LNG expanders. The basic design concept of cryogenic expanders has remained unchanged since their inception, but several improvements in the detailed design have been successfully accomplished. This paper presents an overview of the advanced design of LNG expanders.

See more of this Session: Professor Cedomir M. Sliepcevich Memorial Session Three - LNG Equipment
See more of this Group/Topical: Topical 6: 10th Topical Conference on Gas Utilization

Improved Centrifugal Pump Performance with Counter Helical Inducer Housing Grooves

Sarah D. Alison-Youel
R&D Engineer
Ebara International

Abstract

Net positive suction head required (NPSHR) is closely related to inducer performance and is an important requirement for overall pump performance. The ability of an inducer to propagate fluid towards the impeller inlet is greatly diminished as the fluid begins to rotate with the impeller blades. By machining helical grooves in the inducer housing that are in a counter rotation to the inducer blades, the fluid that has rotational energy is channeled up the grooves and towards the impeller. The helical groove concept was tested on an axial centrifugal pump and yielded positive results. NPSHR level was improved with the modified grooved inducer inlet housing without significant change in the head, power, or efficiency when compared to the same pump without the modification.

Background

As fluid is propagated up inducer blades the fluid begins to take on some rotational energy and therefore begins to rotate with the inducer blade. When this occurs the inducer is no longer able to move the fluid towards the impeller inlet. This is in direct correlation to the point at which the pump reaches the NPSHR limit. It is proposed that if there were grooves machined on the inside of the inducer housing where the grooves were in a counter rotational direction as compared to the inducer blades that these grooves would assist in transmitting the fluid towards the impeller inlet. The grooves would "catch" the rotating fluid which would be pushed up the groove by the rotation of the inducer blade resulting in more fluid translated from the inducer inlet towards the impeller inlet and threfore yielding a lower NPSHR level. An important feature to the counter helical grooved inducer concept is the easy retrofit for pumps already in service.

Design

A model 6-ECC-122 retractable in-tank pump designed and manufactured by Ebara International was used for testing of the grooved inducer housing concept. This particular pump is designed for 330 m^3/hr flow and 266 m of head at 3000 RPM and for service in liquid natural gas (LNG). The pump that was used for testing had two inlet housings, one per standard design and one per standard design that was machined with helical grooves as shown in Fig. 1. The helix of the grooves is in counter rotation to the inducer blades.

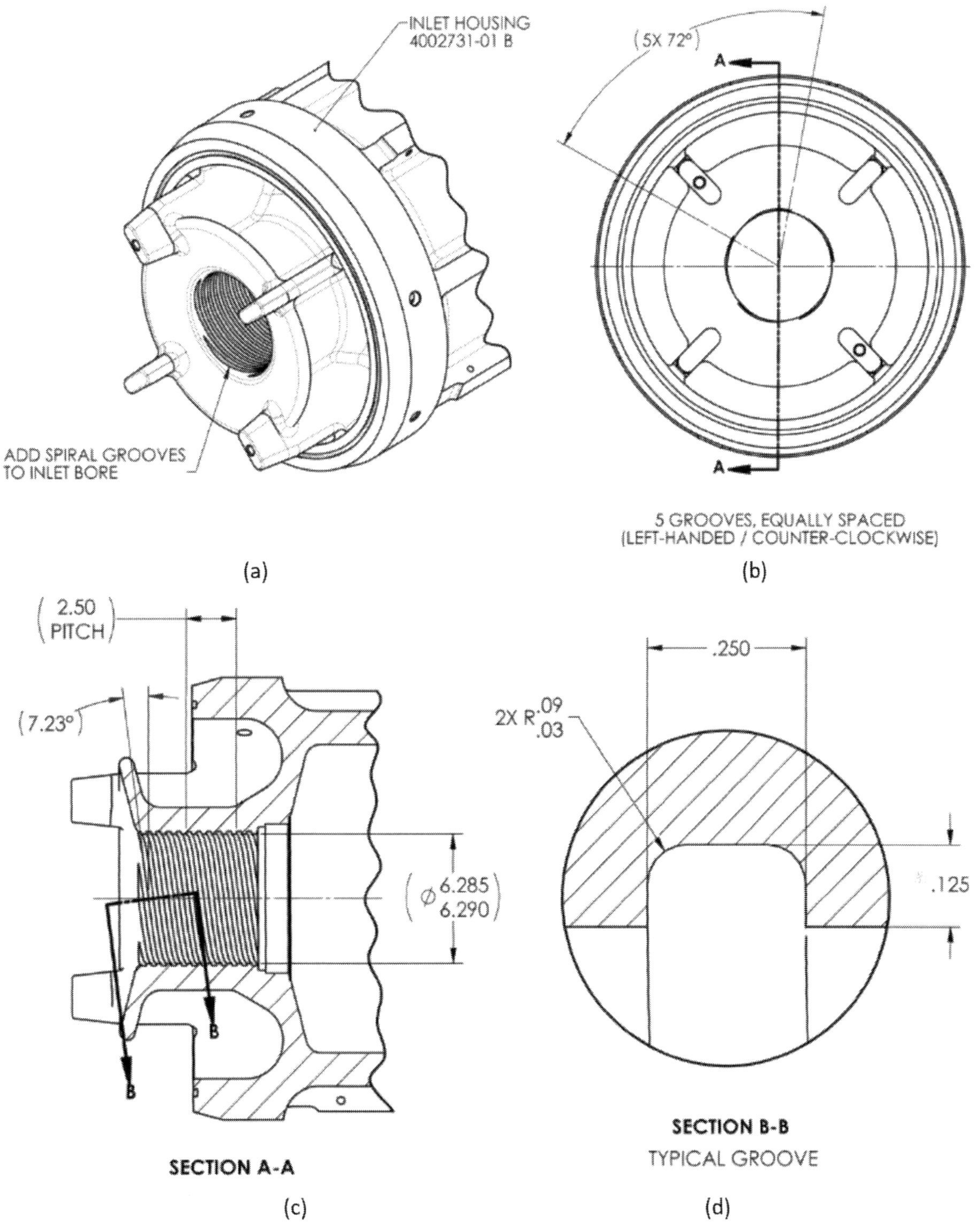

Fig. 1 (a)-(d): Machine Drawing of the Modified Inducer Housing with Counter Helical Grooves

The counter helical grooves were machined to be equally spaced with a 2.5 inch pitch as shown in Fig. 1 (c). The grooves had a depth of 0.125 in, a width of 0.250 in and the corners of the grooves were rounded as shown in Fig. 1 (d). The pump was configured and tested using the unmodified and the modified inlet housing.

Testing and Results

As discussed in the previous section one pump was tested with two different inlet housings, one per standard design (unmodified) to provide a baseline and one with the counter helical grooves (modified). Tests were performed at Ebara International's Test Stand located in Sparks, NV. Each configuration was subjected to an NPSHR test to identify the impact of the grooves on NPSHR levels and each configuration was also tested per standard testing procedures to investigate how the grooves affected overall performance including head, power and efficiency.

Figure 2 is a comparison graph showing the NPSHR curves for both the modified and the unmodified inducer housing. The pump with the unmodified inducer housing is denoted by the blue data points and curve and the pump with the modified grooved inducer housing is denoted by the green data points and curve. This graph shows that the modified inducer housing had a significant positive impact on the NPSHR performance of the pump. At flow rates close to the rated 330 m^3/hr the modified inducer housing improved the NPSHR by 1 m of head which corresponds to an 11% improvement. This is a considerable increase when related to how much more fluid may be pumped out of a tank which will increase profits.

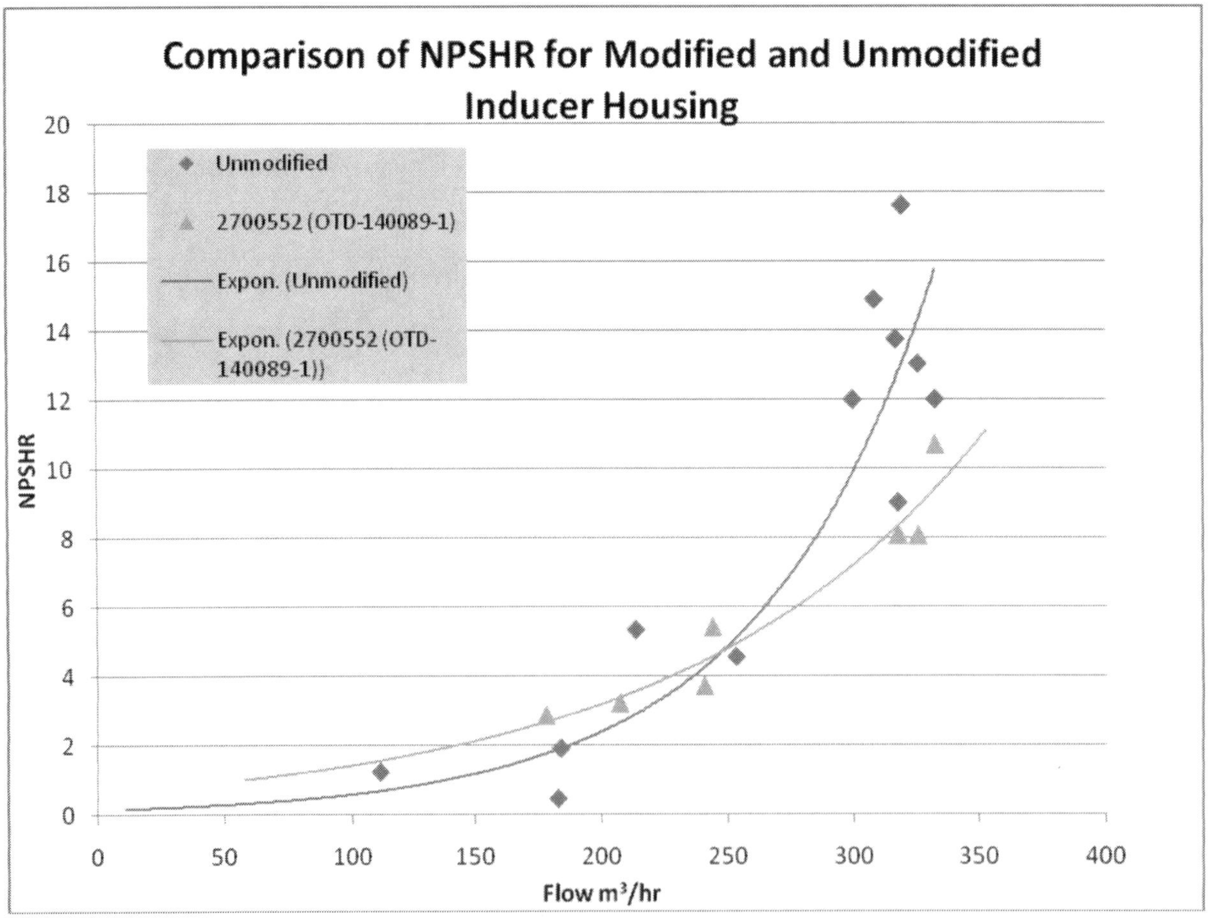

Figure 3 shows the pump characteristics curves for both the modified configuration denoted by smooth curves and the unmodified configuration denoted by the dashed lines and data points. As shown in the curves the pump power is the same between both designs. From approximately 250 m^3/hr to 375 m^3/hr the efficiency of the modified grooved inducer housing is increased. This includes the BEP and rated flow. The head curve doesn't see a significant shift in performance; however, at relatively low flow the modified inducer housing exhibited a reduced head curve that is reversed at relatively higher flows when compared to the unmodified baseline. There is an interesting "flat" portion of the modified head curve between approximately 75 m^3/hr and 225 m^3/hr. This is an interesting restult and is currently under further investigation.

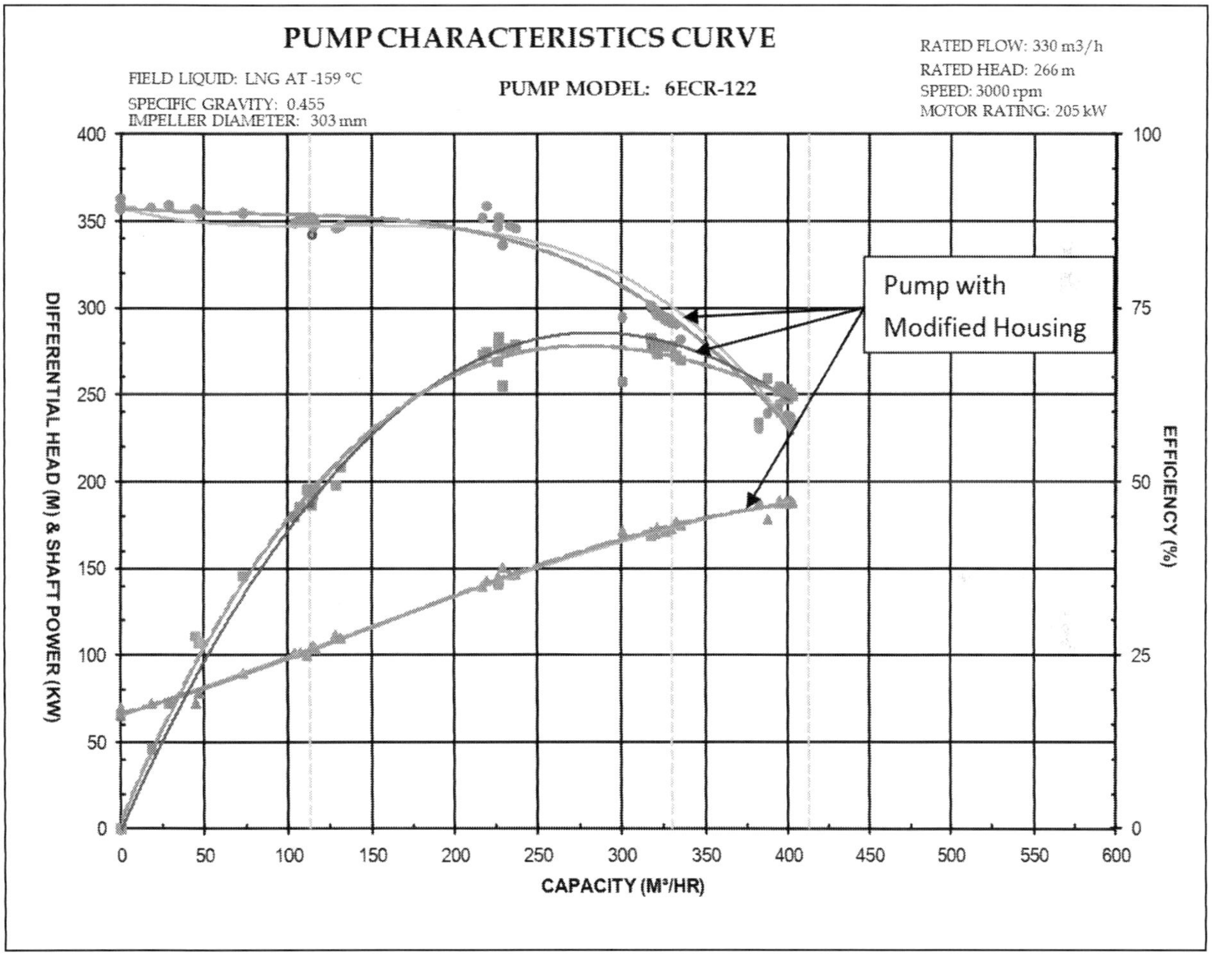

Future Work

The results of this initial test are inciting and deserving of further investigation. Future tests and computational fluid dynamics (CFD) of variations of the groove concept are under investigation. The concept is also patent pending.

(98b) Development of a Novel Full-String Test Facility for Two-Phase LNG Expanders

Christopher D. Finley, Ebara International Corporation, Sparks, NV

With the increasing demand of liquefaction process efficiency improvements, two-phase LNG expanders have been utilized to replace inefficient Joule-Thompson valves. To date, the performance of two-phase LNG expanders have only been verified after being installed on site due to the inability of current testing facilities to simulate on site conditions. This paper presents a method to modify existing LNG expander test facilities to simulate on-site conditions including the full expansion of LNG into the two-phase state.

See more of this Session: Professor Cedomir M. Sliepcevich Memorial Session Three - LNG Equipment
See more of this Group/Topical: Topical 6: 10th Topical Conference on Gas Utilization

(98d) Design (and Verification) of Controls Parameters to Minimize Emissions Due to Valve Retrofit

Hiroki Hayashi, Tokyo Gas Corporation, Tokyo, Japan

Kazuhiro Hayase, Tokyo Gas Corporation, Tokyo, Japan

Kenichi Matsuoka, Yamatake Corporation, Tokyo, Japan

David Hill, CHEMCAD Technical Support, Chemstations, inc., Houston, TX

Design (and Verification) of Controls Parameters to minimize emissions during valve retrofit

Abstract : The Tokyo Gas Corporation's Sodegaura Terminal is now one of the world's largest receiving terminals in the world. After the LNG is regasified, it is delivered to customers in the Tokyo cosmopolitan area by way of high pressure gas pipelines. From an economical and global warming point of view, the plant's operations and controls should have a minimal effect on the environment. This can be accomplished by limiting amounts of natural gas expelled into the atmosphere. The existing pressure control valve (CV) used for start-ups and emergency cases is being replaced. This change is necessary due to expanded process requirements, design conditions, etc. This means that CV characteristics and controllability will be changed as a result. Fine-tuning of PID control parameters will be required; historically this has been done through actual operation of the plant. Prior to installation, we tuned the PID control parameters to optimum values through rigorous dynamic process models. This tuning was performed without any actual live operations. This allows us to reduce the amount of LNG vented to atmosphere during installation and commissioning of the new control valve. Proper tuning also allows us to minimize the amount of LNG vented during startup operations. This paper discusses the techniques used to examine controls scenarios, in this project. Part of this work was creation of a 'virtual plant' in a software model. When remodeling an existing plant, it is useful to have a virtual plant environment. Both the validity of the plant engineering design, and the chance to obtain an optimum adjusted parameter, are increased due to the use of a virtual plant environment for parameter tuning.

See more of this Session: Professor Cedomir M. Sliepcevich Memorial Session Three - LNG Equipment
See more of this Group/Topical: Topical 6: 10th Topical Conference on Gas Utilization

Material Selection for the Propane systems CS or ITCS?

J.C. Kuo, P.E.

Chevron Energy Technology Company
1400 Smith Street
Houston, TX 77002
Phone Number: 713-372-7048
Fax Number: 713-372-1800
Email: JKuo@chevron.com

Presentation at the AIChE 2010 National Spring Meeting
LNG Technology Session
March 21 - 25
San Antonio, Texas USA

1. Executive Summary

Standards and specifications used by specific projects will dictate the material of construction. If a project is using US based standards like ASME, most of the propane system can be constructed out of carbon steel. The exception is the equipment relating to the lowest pressure stage (heat exchanger, suction vessel, piping, etc) and the de-inventory pump. This equipment must use impact tested carbon steel due to the operating pressure near atmospheric and the potential of vacuum transient conditions. Under European and Australian based standards, the material dictated is all impact tested carbon steel in the propane system.

Most of the equipment in the propane refrigeration system can use normal carbon steel, with the exception of E-1, V-1 and P-1, as they are located at the lowest stage suction of the propane compressor where the normal operating temperature is below -30°C. The pressure may temporarily drop below atmospheric depending on the operation of the compressor. ITCS must be used for this equipment. All other equipment (E-2, E-3, E-4, E-5, V-2, V-3, V-4, V-5 and P-2) can use regular carbon steel. The equipment can be operated at lower temperature (-45.6°C) during depressurization as long as the stress level is not exceeded. The same conclusion can be applied to the depropanizer system.

The difference in material cost for a nominal 5 MTPA LNG plant with the majority of equipment using carbon steel, and only the equipment at the lowest suction stage using ITCS when compared to the case with all equipment using ITCS is around 1-2 MM US$. Since the cost impact is minor compared to the total cost of the project, ITCS could be used for the entire propane refrigeration and depropanizer systems.

The impact of using impact tested carbon steel in the propane refrigerant system is minimal, and the materials of construction will be dictated by standards and specifications utilized by specific projects.

2. Introduction and Objective

2.1 Introduction

Modern day propane refrigeration systems and/or propane fractionators (depropanizers) require emergency shutdown and depressurization systems for fire prevention. The design codes (ASME) dictate the procedure(s) for material selection, carbon steel (CS) or Impact Tested Carbon Steel (ITCS) used for process equipment and piping.

Since propane's boiling point temperature is -42°C (design for -46°C) at atmospheric conditions after depressurization, the selection of the design temperature (DT) and design pressure (DP) can influence the material selection and related cost. Moreover, the issue of cold repressurization and startup can impact the plant operability and availability.

2.2 Objective

The purpose of this study is to analyze the impact of material selection for a typical propane refrigeration system and propane fractionation system corresponding to a nominal 5.0 MTPA (million tons per annual) LNG liquefaction plant.

3. Basis and Assumptions

3.1 Basis

The cost estimate was developed in accordance with the following bases and assumptions:

1. General
   - Project LNG Capacity is nominally 5.0 MTPA
2. Scope of facilities include carbon steel process piping Propane Refrigeration Facility
   - Scope of Piping includes:
     - pipe, fittings and valves materials are per "Reference Project" materials quantities
     - standard factored costs for welding rods at 1% of piping materials costs
     - standard factored costs for bolts and gaskets at 5% of piping materials costs.
   - Scope excludes:
     - fabrication labor for shop fabricated piping
     - all field fabrication and installation labor
     - Control systems valves (control valves, relief valves, etc.
     - permanent plant equipment
     - piping specialties items
     - all other bulk materials, such as insulation, concrete, civil, site work, steel, field painting, pipe supports
     - standard factored costs for: hangers & supports, materials handling labor, scaffolding labor, scaffolding materials, testing and inspection
3. Direct Costs
   - Piping material quantities are based on the "Reference Plant" quantities.
   - Piping material pricing is based on the "Reference Plant" unit pricing, which is 1st Quarter 2006.

4. Indirect Costs

- All indirect costs are excluded, such as freight, construction indirect costs, commissioning & startup, home office services, insurances, etc.

5. Escalation/Contingency/EPC Costs

   - Escalation beyond "Reference Plant" pricing (current-day 1st Quarter 2006) is excluded.
   - Contingency is excluded.
   - EPC Contractor Overhead, Risk, Fee is excluded.

3.2 Assumptions, Clarifications, Exclusions

1. Cost estimates are for piping materials costs only and do not represent a total installed cost. Note that a significant cost has not been considered by excluding spool pipe fabrication, installation labor and indirect costs
2. Materials unit pricing is based on vendor quotations received for "Reference Plant" cost estimate of 1st Quarter 2006 for the specific location, scope and specifications.
3. Pipe fabrication costs have been excluded, which vary from project to project, and could impact overall spool piping materials costs.
4. Installation labor has been excluded.
5. Materials quantities are based on specific "Reference Plant" quantities. This could change depending on specific site layouts.

4. Process Description

4.1 Typical Propane System

Please refer to the process flow diagram (Figure 4.1.1-1) for a description of a typical propane refrigeration system used in the production of LNG with a nominal production rate of 5 MTPA.

The system will generally consist of one or two compressor trains in parallel with gas turbines as the drivers. Propane refrigeration can be provided at three levels:

- High level (HP) at 17.7°C and 7 barg
- Intermediate level (IP) at -3°C and 3.3 barg
- Low level (LP) at -35.6°C and 0.3 barg.

The discharge from the third stage propane compressors is desuperheated and condensed by ambient air-cooling in the Condenser (E-4). A Propane Accumulator (V-4) is provided and is equipped with a refrigerated refluxed purge column for the removal of light components (C_1 and C_2). Propane is condensed at 13.4 barg and 41°C by ambient air at 27°C. Propane liquid from V-4 is sub-cooled first in the Subcooler (E-5) and then flows into the shell side of the HP STG Chiller (E-3). Vapor from E-3 flows to the HP Suction Drum (V-3) before being fed to the suction side of the third stage compressors. Propane liquid from E-3 flows into the shell side of IP STG Chiller (E-2). Vapor from E-2 flows to the IP STG Suction Drum (V-2) before being fed to the suction side of the second Stage Compressors. Propane liquid from E-2 flows into the shell side of LS Chiller (E-1). Vapor from E-1 flows through the LS Suction Drum (V-1) before being fed to the suction of the1st Stage Compressor.

A Propane De-inventory Pump (P-1) is provided to de-inventory from the shell side of the chillers to the Refrigeration Storage Tank (V-5). A charging pump (P-2) is provided to fill the system by pumping liquid from the Refrigeration Storage Tank to the Accumulator (V-4).

4.1.1 Typical Material

There are essentially two design pressure levels for the system. The high pressure level is set at 22.4 barg and includes equipment E-4, V-4, E-5, and P-2. The equipment E-3, E-2, V-2, V-3, and V-5 are designed at 17.24 barg. All of the above-mentioned equipment and piping are designed with carbon steel.

ITCS is used for E-1, V-1 and P-1, which are also designed at 17.24 barg. The normal operating temperatures of these equipment and associated piping are below -30°C and are located at the low stage compressor suction. The pressure may decrease into a vacuum condition and the associated temperature may drop below -45°C. Therefore, ITCS must be used for these items of equipment.

The minimum design temperature for E-2, E-3, E-4, E-5, V-4, V-3, V-2, V-5 and P-2 is -45.6°C. This temperature only occurs when the equipment is depressurized to atmospheric pressure. Since the pressure under this condition is much lower than the operating pressure, calculations indicate that the equipment still operates below the allowable stress level with regular carbon steel material. Therefore, normal carbon steel can be used for those items of equipment.

4.2 Material Required for Very Cold Plant Locations

In certain areas, such as Arctic locations, where the minimum ambient temperature is lower than -45oC (-49oF), ITCS shall be used for the entire system.

Figure 4.1.1-1: PFD of Propane Refrigerant System

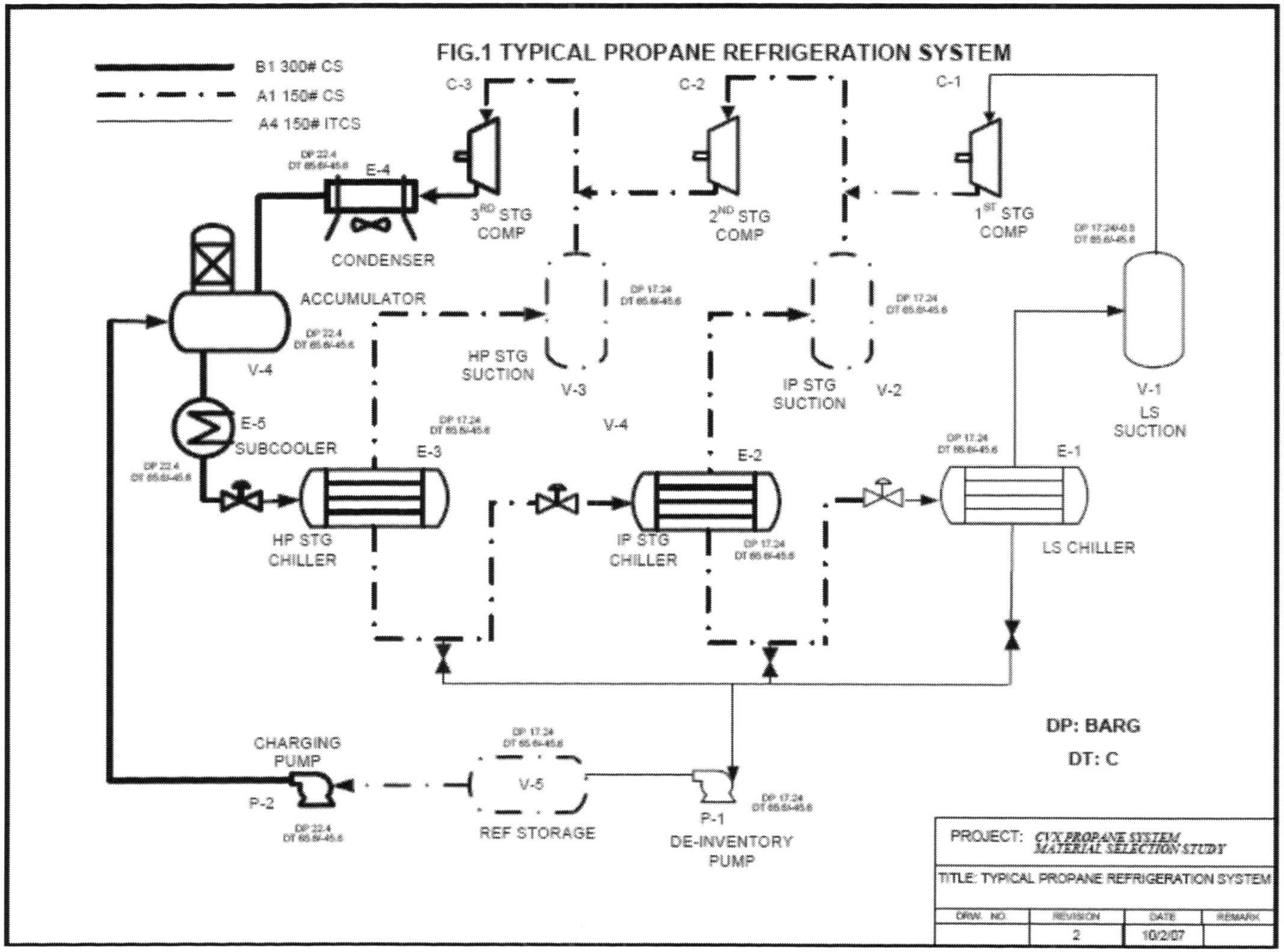

4.3 Typical Depropanizer System

Please refer to Figure 0-1 which depicts a typical depropanizer system. This system takes feed from the bottom of the previous distillation tower, usually the bottom of a deethanizer operated at 30 barg and 105°C. This feed stream is introduced to the middle of the depropanizer (T-1), where propane and lighter components are distilled to the overhead of the tower and are totally condensed by an air cooled condenser (E-6). A reflux drum (V-6) and reflux pumps (P-3A/B) are equipped to send the reflux back to the top tray of the depropanizer and the propane product to storage. The overhead product primarily consists of 95 mol% propane when the condensing temperature is 52°C, and the depropanizer is operated at 17.8 barg. A reboiler is located at the bottom of the column where heat is normally supplied by hot oil. The design pressure of the column, condenser, reflux drum and reflux pumps is 20.7 barg, and the minimum design temperature is -45°C. The reboiler is designed at 22 barg and the minimum design temperature is -29°C.

4.3.1 Typical Material

Although the minimum design temperature is -45°C for certain equipment, it only occurs when the equipment is depressurized to atmospheric pressure. Since the pressure under this condition is much lower than the operating pressure, calculations indicate that the equipment still operates below the allowable stress level with regular carbon steel material. Therefore, normal carbon steel can be used for those items of equipment.

The entire system uses regular carbon steel, unless the minimum ambient temperature is lower than -45°C (-49°F). ITCS shall be used for the entire system in certain Arctic locations.

Figure 0-1: PFD of Depropanizer System

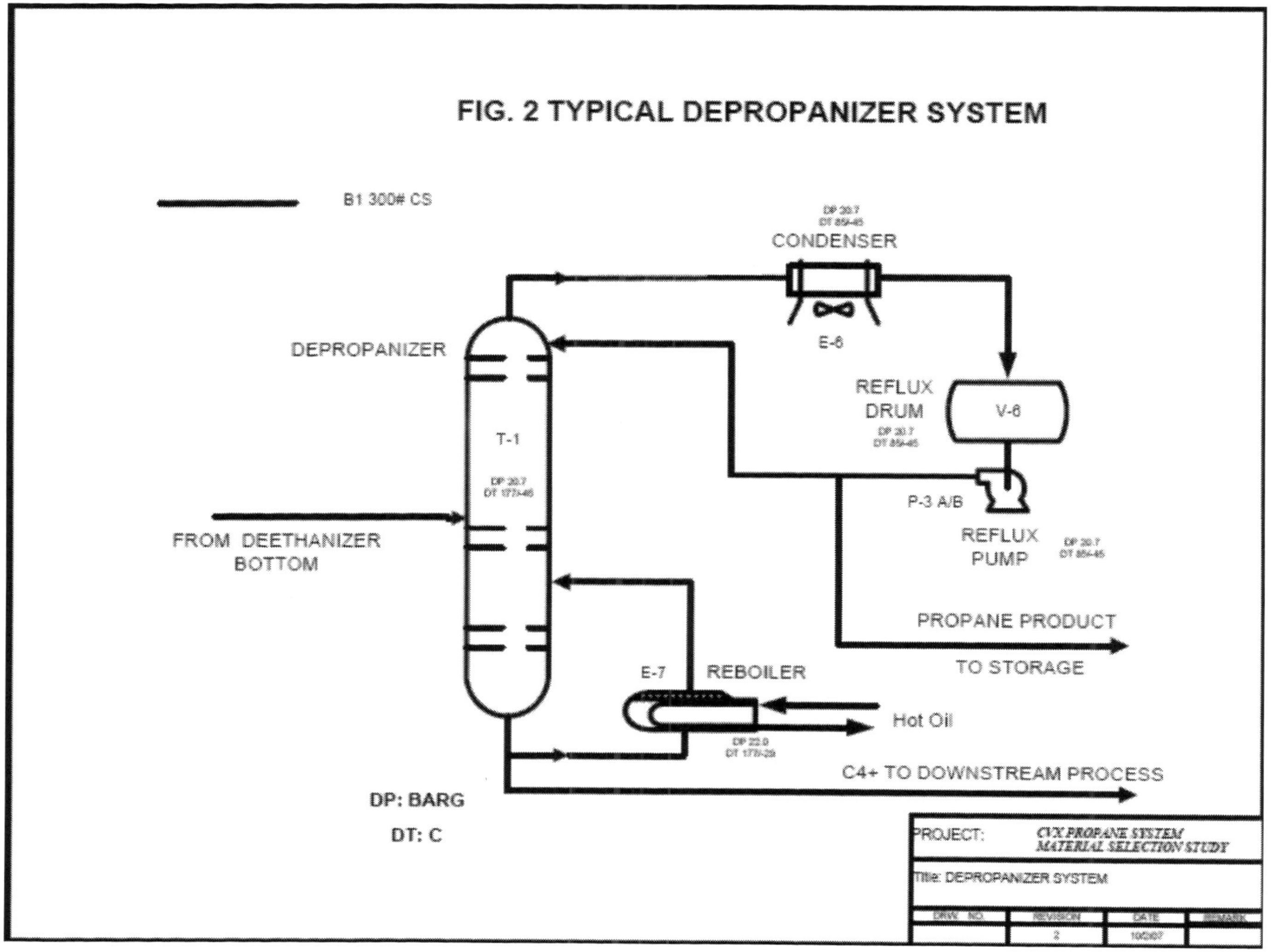

5.0 Material Selection

The Materials of Construction selected for equipment and piping are based on information shown on the Process Flow Diagram (PFD) and Heat and Material Balance.

In the case of the Propane Refrigeration System, the gas being processed has no contaminants, which could cause corrosion and has gone through dehydration, so the only concern is the temperature limitation. The PFD's generally reflect operating temperature whereas design pressure temperatures (DP, DT) are done later on. Material selection is based on operating temperature with a suitable margin. In certain cases, based on the plant location where the minimum atmospheric temperature is –45°C (-49°F) or lower, ITCS shall be specified.

ASME Codes for both Vessels and Piping have guidelines for use of non-ITCS. In the case of piping, Appendix A of B31.3 specifies minimum temperature limitations. In an LNG plant, usually A 106 Grade B seamless, or A 672 Grade 70 welded is specified. A 106 Grade B is good to –29°C (-20°F) with a maximum thickness of 0.5 inch, whereas in the case of A 672, based on thickness and heat treatment material can be used below –29°C (-20°F) without any impact testing. Materials are classified as A through D. Paragraph 323.3 covers guidelines for materials where impact testing is required.

ASME Figure 323.2.2A has curves plotted for materials classified as A through D with nominal wall thickness and minimum design temperatures. Materials in Curve D are considered the best option when temperatures are anticipated to drop below –29°C (-20°F). Classification of these materials is based on the steel making process and subsequent heat treatment. Based on industry data, killed steels[1] that are normalized exhibit the best Charpy values. Table 323.2.2A has tabular values based on types of materials as well as thickness. To satisfy B31.3 requirements, it is preferable to use A 106 Grade B down to -29°C (-20°F) with a maximum thickness of 0.5 inch and switch to A 672 for thicknesses exceeding 0.5 inches.

Another option would be to verify each pipe class for maximum thickness and minimum temperature. This could result in various pipe classes with each having restrictions on temperature/thickness. For this approach, Charpy impact testing of regular carbon steel would be required based on thickness/temperature, resulting in long term planning and extra field management procedures.

In the case of pressure vessels, ASME Section V111 has similar guidelines for selection of material as the B31.3 piping code. Industry practice is to specify A 516 material for vessels classified as fine grain steel, which is suitable down to –46°C in normalized steel up to 4 inch thickness. For vessels, it is more cost effective to get the vessel designed and stamped with full pressure and lowest design temperature (-46°C). The only associated cost is normalizing the plates, impact testing of plates and welding procedures qualifications. UG-20 of ASME Section V111, gives guidelines for the selection of carbon steel materials for low temperature applications.

J.C. Kuo (aka Chen Chuan J Kuo) is a veteran of the Gas Processing, Gas Treating and LNG industry, with 35 years of experience in these fields.

As senior advisor for Chevron's. Energy Technology company, he has served as the Process Manager/Process lead for many projects, including the Wheatstone LNG, Gorgon LNG, Delta Caribe LNG, Casotte Landing and Sabine Pass LNG terminal projects. He has also served as the Technical Process Reviewer for Angola, Olokola, Algeria and Shtokman LNG projects.

Before working with Chevron, J.C. was the Technology Manager for IPSI/Bechtel and served as the Process Manager/Process lead for the Pemex Catarell offshore project, the Egyptian LNG (Idku) train 1/2, China Shell Nan Hai, Chevron Venice gas plant de-bottleneck Tunisia NRU and Australian SANTOS Projects.

The Egyptian LNG train 1 was delivered more than 3 months ahead of an aggressive schedule of 36 months, passed the performance test, and was in full operation in two months. Train 2 was delivered 9 months ahead of schedule. Not only did he help Train 1 and 2 avoid up to 150 Million dollars in schedule penalties and another 150 Million dollars in performance penalties, both trains actually made bonuses.

J.C. is a frequent speaker and presenter to the international conference such as AIChE, Gas Processing/Treating Conferences, and LNG summit. He has contributed to the gas processing and LNG technology improvements through a patent, a book, and many papers. He has also served as co-chair of the AIChE LNG sessions for the Topical conferences on natural gas utilization. He is a member of the steering committee for the North American LNG summit.

His degrees include a BS from Chung Yuan Christian University, Taiwan, a MS from the University of Houston, both for Chemical Engineering, and a MS in Environmental Engineering from Southern Illinois University. He is a registered Professional Engineer in Texas and a member of AIChE. He is the president of the 99 Power Qi Qong Texas divisions.

(98f) An Alternative Formulation of the Standard Orifice Equation for Natural Gas and Pure Compounds

Diego E. Cristancho, Chemical Engineering, Texas A&M University, College Station, TX

Luis A. Coy, Chemical Engineering, Texas A&M University, College Station, TX

Diego O. Ortiz-Vega, Chemical Engineering, Texas A&M University, College Station, TX

Gustavo A. Iglesias-Silva, Departamento de Ingeniería Química, Instituto Tecnológico de Celaya, Celaya, Guanajuato, Mexico

Kenneth R. Hall, Chemical Engineering, Texas A&M University, College Station, TX

This paper presents a detailed analysis of the standard orifice equation reported in the American Gas Association (AGA) Report No. 3 using data for natural gas measured by the Southwest Research Institute (SRI), data for nitrogen measured by the National Institute of Standard and Technology (NIST), and air data from the Colorado Engineering Experiment Station, Inc (CEESI). It appears that the standard orifice equation does not fit satisfactorily all of the data. A bias of about 0.5% exists for the natural gas data and 4% for the air data. In this paper, an alternative scheme to determine the expansion factor coefficient leads to a single curve for all the data with an uncertainty of ±0.02% at 3ó. Finally, a simple fit for the discharge coefficient using the new expansion factor provides a correlation that fits the natural gas data within ±0.3%.

See more of this Session: Professor Cedomir M. Sliepcevich Memorial Session Three - LNG Equipment
See more of this Group/Topical: Topical 6: 10th Topical Conference on Gas Utilization

Research of preparation of liquid hydrocarbons by Supercritical Fischer-Tropsch synthesis on Bifunctional catalyst $Co/Zn/Mn/K/Ni/SO_4^{2-}$

Yongming Zheng[1], Jingchang Zhang[*], Weiliang Cao[1]

1 Institute of Modern Catalysis, Beijing University of Chemical Technology, State Key Laboratory of Chemical Resource Engineering, Beijing 100029, China.

*Corresponding author. Tel.:+86 1 64434904; fax: +86 1 64434898.

E-mail address: zhangjc1@mail.buct.edu.cn (J. Zhang).

Present/permanent address:Beijing University of Chemical Technology,15 Beisanhuan, Dong Lu, Beijing 100029, China .

Abstract:.

Nano-catalysts (Co/Zn/Mn/K/Ni)(SCFD) and Bifunctional nano-catalyst (Co/Zn/Mn/K/Ni/SO_4^{2-})(H-SCFD) impregnated with 0.4 mol/L H_2SO_4 of Fischer-Tropsch synthesis have been prepared. Catalysts were characterized by means of XRD, TEM, XPS and FTIR. And Catalysts were investigated in a slurry-bed reactor under Supercritical FTS conditions(T=240 ℃,P=4.5 MPa). Results Show: the liquid hydrocarbons products selectivity and content of isoparaffin on Co/Zn/Mn/K/Ni/SO_4^{2-} catalyst were even higher than those of the Co/Zn/Mn/K/Ni catalyst. The highest selectivity of liquid hydrocarbons products (96.66% from 91.8%) and isoparaffin (72.19% from 50.04%) could be demonstrated over Co/Zn/Mn/K/Ni/SO_4^{2-} catalyst. In addition, the lower content of CH_4 (0.09% from 3.48%)and CO_2 (0.03% from 0.42%) could also be demonstrated over Co/Zn/Mn/K/Ni/SO_4^{2-} catalyst.The CH_4 as cosolvent adding amount to achieve a supercritical fluid were investigated. The result shows that CH_4 addition can enhance activity of catalyst, decrease the content of CH_4 and temperature, and increase the isomarization degree for a supercritical phase reaction. As the CH_4 pressure reaching 1.0 MPa, for Co/Zn/Mn/K/Ni/SO_4^{2-} catalyst the conversion of CO increased from 88.45 to 98.29%,and isoparaffin products was 85.25%, content of CH_4 was 0.012%, while the temperature was 230℃, respectively.It can be seen that the supercritical Fischer–Tropsch reaction with CH_4 as cosolvent over Co/Zn/Mn/K/Ni/SO_4^{2-} nano-catalyst was the best reaction process. It was benefical to the increase of production isoparaffin, thereby increase the quility level of the final prodcution.

Keywords: nanosized catalyst; supercritical combined technology; Bifunctional catalyst ;isoparaffin

1. Introduction

Fischer-Trøpsch synthesis (FTS) has been well known as a process of converting synthesis gas ($CO + H_2$) into hydrocarbons and oxygenates[1]. Facing up with the dwindling petroleum reserves and the increasing demand for high-quality and environmentally friendly fuels, an effort for developing the novel FTS catalysts has been made by academic communities and institutes[2-9].

As a result of the satisfactory activity, high selectivity towards C_5^+ products and ideal resistant to catalyst deactivation, cobalt-catalyst was widely accepted as a promising industrial FTS catalyst. However, these FTS products contain large amounts of normal paraffins, of which have low octane number near to 0. In order to obtain the ideal gasoline components from syngas, it is of great significant to obtain the isoparaffin from FTS. Some new catalysts have been developed for one-step isoparaffin synthesis. Tsubaki *et al* [10]once prepared a three-component hybrid catalyst (ZSM-5, zeolite, Pd/SiO_2) for direct synthesis of isoparaffin.

In this paper a novel acidic cobalt-based catalyst prepared by supercritical fluid dry method was introduced and evaluated in a supercritical phase slurry-bed reactor with CO_2 and CH_4 as cosolvent. The result shows that the addition of CO_2 and CH_4 into the supercritical phase reaction system can inhibit the formation of CO_2 and CH_4 and enhance the selectivity of C_5^+ and isoparaffin products. It was demonstrated that the present work provide an effective way for isoparaffin synthesis.

2. Experimental

2.1 Preparation of catalysts

2.1.1 Catalyst preparation by the supercritical combined technology

Cobalt nitrate, zinc nitrate and manganese nitrate were dissolved into the deionized water with a weight ratio Co/Zn/Mn/Ni=100/50/10/6 using AEO-3 as a surfactant (0.2 vol.%). NaOH solution (1 mol/L) was added gradually to maintain the pH at a constant value of 8.5±0.1. The water gel was stirred continuously for approximately 1 h. After the precipitation, the water gel was aged for 24 h. The precipitate was filtered and washed repeatedly by deionized water in order to remove the Na^+ and NO_3^- ions. Subsequently, the water solvent in gel was exchanged by ethanol to make an ethanol gel. The ethanol gel was further transferred into the high-pressure autoclave, with the addition of ethanol as a supercritical fluid. Finally, the gas gel was prepared by supercritical fluid drying process. The required amount of catalyst precursor, K_2CO_3 solution and deionized water were added into the gas gel to obtain the desired weight ratio Co/Zn/Mn/Ni/K=100/50/10/6/7. After drying at room temperature (T=298 K), The catalyst precursor was then calcined at 773 K for 2 h and denoted as SCFD.

2.1.2 Preparation of acidic cobalt-based catalyst

The catalyst precursor prepared through method as described in section 2.1.2 followed by an impregnation with 0.4mol/L H_2SO_4 with a similar volume was calcined at 773 K for 1 h and denoted as H-SCFD.

2.2Characterization techniques

The crystal structure of the nanosized catalysts was analyzed on a HR-6000 X-ray diffractometer with Cu Kα radiation (40.0 KV, 30.0 mA, 2θ=20°~80°), the average particle sizes of Co_3O_4 were calculated using the Scherrer equation. The morphology, dispersibility and particle size of catalysts were characterized by using a HITACHI-800 transmission electron microscope (TEM). The surface acidity was determined by using a Shimadzu 8400S IR spectrometer with in-situ NH_3 adsorption inside a DRIFT reactor.

2.3 Catalytic experiments

2.3.1 The supercritical phase reaction

The FTS performance of catalysts was tested in a 100 ml stirred slurry batch tank reactor with n-hexane as supercritical solvent. 1 g of catalyst was loaded into the reactor and reduced in-situ by H_2 during co-feeding of the supercritical solvent (n-hexane) for 6 h at 663 K. After reduction, the temperature was declined to 293 K. Subsequently H_2 was replaced by syngas (CO : H_2 : N_2=1 : 2 : 1) under a total pressure of 1 MPa for FTS reaction. Nitrogen was used here as an internal standard gas. Gaseous compounds were analyzed online by gas chromatography (GC); the light hydrocarbons (C_1-C_5) were analyzed by an Agilent 6820N GC with an Al_2O_3 capillary column (GS-Alumina, 50 m) and flame ionization detector (FID); CO, CO_2, and CH_4 were analyzed by HP 1490 GC equipped with a thermal conductivity detector (TCD) and separated by a filled column (Supel-1-2392-U). The oil-phase compounds (C_5^+) were analyzed off line by an Agilent 6820 N GC with a capillary column (SE-54, 30 m) and FID.

2.3.2 The supercritical phase reaction with CO_2(or CH_4)

The catalyst was reduced through a process as mentioned in Section 2.3.1. After reduction, the temperature declined to 293K. Subsequently H_2 was replaced by syngas (CO : H_2 : N_2 =1 : 2 : 1) and CO_2(or CH_4) for FTS reaction. Nitrogen was used here as an internal standard gas. Gaseous compounds were analyzed online by gas chromatography (GC); the light hydrocarbons (C_1-C_5) were analyzed by an Agilent 6820N GC with an Al_2O_3 capillary column (GS-Alumina, 50 m) and FID. CO, CO_2, and CH_4 were analyzed by HP 1490 GC with a filled column (Supel-1-2392-U) and TCD. The amount of CO_2(or CH_4) adding into the reaction system and the CO_2 (or CH_4)formation during reaction were calculated by internal standard N_2. The oil phase compounds (C_5^+) were analyzed off line by an Agilent 6820 N GC with capillary column (SE-54, 30 m) and FID. A flow scheme of apparatus for supercritical Fischer–Tropsch synthesis is shown in figure1.

THE GRAPHIC ON THIS PAGE WAS NOT AVAILABLE ON THE E-MEDIA VERSION OF THIS PAPER.

3. Results and discussion

3.1 X-ray diffraction (XRD) analysis of catalysts

XRD patterns of catalysts prepared by SCFD and H-SCFD are shown in Fig. 2. It was found that the main phase of catalyst prepared by SCFD was assigned to Co_3O_4 crystallite. The absence of diffraction bands related to Mn and K suggests that they were well dispersed over surface or their amounts are below the detection limit. The particle size of SCFD calculated via the Scherrer equation was around 15 nm. The crystalline phase of H-SCFD catalyst with a particle size of around 12 nm (calculated from Scherrer equation) was mainly attributed to Co_3O_4 and CoO.

THE GRAPHIC ON THIS PAGE WAS NOT AVAILABLE ON THE E-MEDIA VERSION OF THIS PAPER.

3.2 TEM analysis of catalysts

Fig. 3 shows the TEM images of catalysts prepared by SCFD and H-SCFD. The nanosized particle prepared by SCFD method is well dispersed and the particle size is approximately 15-20 nm (Fig. 3a),. Catalyst prepared by H-SCFD shows slightly agglomeration state and the particle size is also around 15-20 nm (Fig. 3b), of which was in a good agreement with the XRD patterns (Figure 2).

THE GRAPHIC ON THIS PAGE WAS NOT AVAILABLE ON THE E-MEDIA VERSION OF THIS PAPER.

3.3 NH_3-DRIFT spectroscopy analysis of catalysts

DRIFT spectroscopy associated with NH_3 adsorption is a powerful technique for investigating the surface acidity of catalysts. Fig.4 presents a comparation of surface acidity between SCFD and H-SCFD. The FTIR bands at 1520 cm^{-1} is usually attributed to the L-acid sites, while the ones at 1690cm^{-1} was assigned to the B-acid sites. It is clearly seen from Fig. 4 that the catalyst with a H_2SO_4 impregnation (H-SCFD) has more surface acidity than untreated one (SCFD).

THE GRAPHIC ON THIS PAGE WAS NOT AVAILABLE ON THE E-MEDIA VERSION OF THIS PAPER.

3.4 F-T synthesis performance of SCFD and H-SCFD catalysts

A comparision of FTS performances between SCFD and H-SCFD catalysts was conducted as shown in Table 1. The activity of catalyst prepared by H-SCFD (88.45%) was lower than that of catalyst prepared by SCFD (94.30%). By contrast, isoparaffin yield of H-SCFD catalyst was higher with respect to the SCFD catalyst.Isoparaffin yield of C_5^+ increased from 50.04 % to 72.19 %. This would ascribed to the influence of surface acidity of catalyst after H_2SO_4 treatment.

THE GRAPHIC ON THIS PAGE WAS NOT AVAILABLE ON THE E-MEDIA VERSION OF THIS PAPER.

3.5 Effect of CO_2 addition into the reaction system for F-T synthesis over H-SCFD catalyst

The activity, selectivity and product distribution were evaluated by SCFD catalyst with adding different amounts of CO_2 (0.5 MPa, 1 MPa, 1.2 MPa) in syngas (CO : H_2 : N_2= 1 : 2 : 1, 1.5 MPa) and summarized in Table 2. With the increase of CO_2 amount, the CO conversion was slightly changed. CO conversion over H-SCFD catalyst achieved under a CO_2 partial pressure of 1.2 MPa is 92.01%, which is higher than that of the values achieved under other CO_2 partial pressures. However, the highest selectivity of C_5^+ products (97.76%) was achieved under 1.0 MPa CO_2 pressure over H-SCFD catalyst, along with the highest isoparaffin yield (84.35%), relatively low CO_2 (0.03%) and CH_4 (0.34%) formation. It thus can be deduced from Table 2 that CO_2 introduction can increase the system pressure, which leads an enhancement of the selectivity to C_5^+ products and isomerization of liquid products. On the contrary, the formation of CO_2 was inhibited by the existence of large amount of CO_2.

THE GRAPHIC ON THIS PAGE WAS NOT AVAILABLE ON THE E-MEDIA VERSION OF THIS PAPER.

3.7 Effect of CH_4 addition into the reaction system for F-T synthesis over H-SCFD catalyst

In order to investigate the impact of FTS catalytic performance over H-SCFD catalyst due to CH_4 addition into the supercritical reaction system, different amount of CH_4 (0.5 MPa, 1 MPa, 1.2 MPa) were introduced and the corresponding catalytic performances were compared as shown in Table 3. CO conversion was hardly increased simutanously with the addition of 0.5 MPa CH_4 and then gradually decreased with the further increase of CH_4 pressure. This may be explained by that the overfull CH_4 was most likely reducing the chemisorbed CO over surface through a competitive adsorption. The formation of CH_4 was also suppressed gradually accompanying with the increasing CH_4 content. Moreover, it can be seen that, the lowest selectivity towards CH_4 (0.02%) together with a high yield of liquid products (96.45%) and isoparaffin (85.25%) was achieved under 1.0 MPa CH_4..

THE GRAPHIC ON THIS PAGE
WAS NOT AVAILABLE ON
THE E-MEDIA VERSION OF
THIS PAPER.

3.8 Comparison of the FTS performance over SCFD and H-SCFD catalysts with CO_2 and CH_4 Table 4 indicates a comparison of FTS performance over H-SCFD catalysts under a CO_2 or CH_4 partial pressure of 1.0 MPa. With the same amount of CO_2 addition, CO conversion and selectivity of liquid products obtained over H-SCFD catalyst is lower than those obtained over SCFD catalyst. However, yield of isoparaffin achieved over H-SCFD catalyst was higher than those achieved over SCFD, increased from 50.04% to72.19%,which was ascribed to the higher surface acidity beneficial for the hydrogenation of 1-olefin product.

THE GRAPHIC ON THIS PAGE
WAS NOT AVAILABLE ON
THE E-MEDIA VERSION OF
THIS PAPER.

4. Conclusions

(1) Cobalt-based catalyst was prepared by supercritical fluid dried method, the main phase of SCFD catalyst was Co_3O_4 with a particle size of about 15 nm. The crystalline phase of H-SCFD catalyst was mainly attributed to Co_3O_4 and CoO crystallite, having a particle size of about 15 nm. NH_3-DRIFT result shows that catalyst treated by H_2SO_4 impregnation has more surface acidity.

(2) CO conversion over H-SCFD catalyst varies upon the CO_2 addition amount. 1.0 MPa CO_2 addition results in a maximum of 92.01%. Interestingly, maximum yield of C_5^+ products (97.76%) was achieved under 1.0 MPa CO_2, accompanied with the highest isoparaffin yield (77.41%), comparable low CO_2 (0.03%) and CH_4 (0.34%) formation.

(3) CH_4 addition also effects the activity and CH_4 selectivity obtained over H-SCFD catalyst. When CH_4 content was 1.0 MPa, the lowest selectivity to CH_4 (0.012%) was observed. And yields of liquid products (94.80%) and isoparaffin (83.79%) were relatively higher than values achieved under 1.0 MPa CO_2 pressure.

(4) The isoparaffin yield on H-SCFD catalyst was higher than that of on the SCFD catalyst . When CH_4 content was 1.0 Mpa, the isoparaffin selectivity of C_5^+ was increased from 50.04% to 85.25% on SCFD and H-SCFD, respectively.

Acknowledgments

The work was supported by the National Development Project of High Technology (Project 863) (2006AA03z412).

References

1. Khodakov AY, Chu, W, Fongarland, P. Advances in the Development of Novel Cobalt Fischer-Tropsch Catalysts for Synthesis of Long-Chain Hydrocarbons and Clean Fuels. Chem Rev 2007;107: 1694-1695.

2. DRY ME. The Fischer-Tropsch process: 1950-2000. Catal Today, 2002, 71: 227-241.

3. Li, J M ; Zhang, J C; Zhang, R D; Cao, W L. Nanosized cobalt-based catalyst prepared by supercritical phase condition for Fischer-Tropsch synthesis. Journal of Natural Gas Chemistry 18(2009) 325–330

4. Linghu W, Li XH, Asami K, Fujimoto K. Supercritical phase Fischer-Tropsch Synthesis over Cobalt catalyst. Fuel Process Tech 2004;85:1121-1138.

5. Yan SR,,Fan L, Zhang ZX, Zhou JL, Fujimoto K. Supercritical-phase process for selective synthesis of heavy hydrocarbons from syngas on cobalt catalysts. Appl Catal A 1998;171:247-254.

6. Qiu XQ, Tsubaki N, Sun SL, Fujimoto K. Promoting effect of noble metals to Co/SiO_2 catalysts for hydroformylation of 1-hexene . Catal Commun 2001;2:75-80.

7. Fan L, Yoshii K, Yan SR, Zhou JL, Fujimoto K.Supercritical-phase process for selective synthesis of wax from syngas: Catalyst and process development . Catal Today 1997;36:295-304.

8. Fan L, Fujimoto K. Fischer–Tropsch synthesis in supercritical fluid: characteristics and application . Appl Catal A 1999;186:343-354.

9. Huang, X.W; Roberts, C.B. Selective Fischer-Tropsch synthesis over an Al_2O_3

supported cobalt catalyst in supercritical hexane. *Fuel Process Tech* 2003;83:81-99.

10. Tsubaki N, Yoneyama Y, Michiki K, Fujimoto K. Accelerated methanol synthesis in catalytically active supercritical fluid . Catal commun 2003;4:108-111.

11. Bechara R, Balloy D, Dauphin JY, Grimblot J. Influence of the Characteristics of γ-Aluminas on the Dispersion and the Reducibility of Supported Cobalt Catalysts. Chem Mater 1999;11:1703-1711

HYDROCRACKING OF COBALT-BASED FISCHER–TROPSCH WAX OVER Pt/ZSM-5 AND Pt/SiO_2-Al_2O_3 CATALYSTS

Wenping Ma, Janet C. Mohandas, Keogh Robert, Gary Jacobs, and Burtron H. Davis
Center for Applied Energy Research, University of Kentucky, 2540 Research Park Drive, Lexington, Kentucky 40511, USA

Introduction

The rapidly increasing demand for high-quality ultra clean fuels have greatly stimulated the development of coal- and gas-to-liquid (CTL and GTL) technologies in the past decades. The hydrocracking process is an effective route in CTL and GTL to upgrade the heavy waxy hydrocarbons produced by Fischer-Tropsch (FT) reaction over commercial iron or cobalt catalysts into a variety of high value fuels such as gasoline (C_5–C_{11}) and middle distillate (C_{10}–C_{19}) [1]. As reviewed in the scientific literature, the advantage of diesel fuels derived from the FT and hydrocracking processes over those from petroleum refining is its ultra clean characteristics, which means that the FT synfuels are virtually free of sulfur (~5 ppm), nitrogen and aromatic compounds (~1%), and typically have a high cetane number of over 70 [2-9]. The sulfur levels are significantly below the maximum required sulfur value for conventional diesel (~10 ppm in 2009), and the cetane number is well above the minimum classification for high cetane number, which is 51 [10]. However, the cold flow properties of the synfuls are poor [5].

Conventional bifunctional catalysts, consisting of transition metals or noble metals supported on acidic oxides have been found to be effective for hydrocracking FT wax and crude oil. A bifunctional mechanism in which the metals play a hydrogenation/dehydrogenation role for forming olefinic precursors to carbenium ions was proposed. The most conventional hydrocracking catalysts used for the production of gasoline or diesel from crude oil are NiMo and NiW catalysts supported on alumina silica–alumina, or zeolite [3,6,8,9,11,12]. The acidity of the supports has been studied by NH_3-TPD or pyridine adsorption. FT-IR studies of pyridine adsorbed on different supports showed that the trend in the strength of the Brønsted acid sites follows the order: $SiO_2 < \gamma$-Al_2O_3 < silisilicated alumina < SiO_2-Al_2O_3 < H-ZSM-5 [13]. For the hydrocracking of FT wax, transition metals and noble metals such as Pd or Pt with or without modification by Co or Mo oxides loaded on those acidic supports have been investigated [3, 5,7,9,14]. Different catalysts displayed different activity and selectivity features. Ni supported catalysts have been reported to yield relatively more light hydrocarbons [9], while noble metals like platinum were relatively more active and may provide a favorable balance with the acidity of catalyst supports [5]. However, even though hydrocracking of heavy hydrocarbons has been studied extensively, most studies have based on the cracking of a specific hydrocarbon by largely varying temperatures (320-380 °C) and pressures (3-7 MPa) and H_2/wax ratios (3-100) without altering liquid space velocity, which could determine optimized yield of the fuels [8]. The Sasol group recently reported results for the hydrocracking of LTFT Fe-wax using commercial bifunctional catalysts. The recycle operation yielded a hydrocracking product spectrum of 80% diesel, 15% naphtha, and 5% gaseous products at 3.5 MPa [6].

The present study investigated the hydrocracking of FT wax to gasoline and middle distillate using a noble metal on ZSM-5 and SiO_2-Al_2O_3 catalyst. The catalysts were prepared by the traditional ion-exchange method. The hydrocracking severity (activity, selectivity and yield) on the two catalysts has been investigated by employing a wide range of reaction conditions, (T, P, H_2/wax ratio and LHSV). BET and FTIR spectroscopy of chemisorbed pyridine were conducted to investigate the surface porosity and acid property of the catalysts. The effects of reaction conditions and acidity of the catalysts on catalytic behavior are discussed.

Experimental

Pellets of HZSM-5 from Süd-Chemie, Inc. (~300 m^2/g) and Davison 135 SiO_2-Al_2O_3 (434 m^2/g) were used as supports for loading Pt. $(NH_3)_4Pt(NO_3)_2$ was used as the precursor for Pt. 0.5 % Pt was loaded to the supports by using ion exchange in an aqueous solution. The process lasted for 3 days. Following the ion-exchange, the catalysts obtained were dried in an oven at 110 °C for 12 h and calcined in flowing air at 350 °C for 4 h.

The surface area, pore volume, and average pore radius of the supports and catalysts were measured by BET using a Micromeritics Tri-Star 3000 gas adsorption analyzer system. IR spectra were collected at a resolution of 4 cm^{-1}; 1024 scans were taken for each spectrum. A reference spectrum was collected for the untreated catalyst at room temperature. Before pyridine adsorption, the catalyst (~40 mg) was reduced *in situ* using a H_2/He gas mixture at 400 °C for 1 h. Pyridine adsorption on the catalyst surface was performed by flowing He (25 cm^3/min) through a pyridine saturator prior to flowing into the DRIFT chamber, which was maintained at 150 °C; this was carried out for 15 min, during which the changes in concentration of adsorbate and gaseous species were recorded by DRIFTS. Note that for the purpose of comparison, IR spectra of the clean catalyst surface were also collected at 150 °C prior to pyridine adsorption (i.e., these served as background spectra).

Hydrocracking tests using wax obtained from cobalt-based FT were performed in a fixed-bed (FB) reactor unit made from 314 stainless steel (SS); the reactor is of 1 inch inner diameter and is 22 inches in length. The 2.5 ml of received catalyst was diluted by 10 ml of glass beads (3 mm diameter). Ultra high purity H_2 gas from a cylinder passes through a reducing valve body and its flow was adjusted by a pre-calibrated Brooks mass flow controller before entering the reactor. Cobalt FT wax was melted in a 600 ml heating vessel first and suctioned into a high temperature ISCO syringe pump. The hot liquid wax (130 °C) was pumped to the FB reactor at a desired flow rate through a tape heated line using the syringe pump. All catalysts for the hydrocracking reaction tests were activated at 400 °C, 15 Nl/h H_2 1 atm for 12 h. After activation, reactor temperature was slowly dropped to 300 °C and reactor pressure was increased to 500 psig. During the hydrocracking run period, the flow rate of H_2 was adjusted to give a range of H_2/wax molar ratios between 5 and 30, reaction pressures in the range of 400-700 psig, and reaction temperatures in the range of 200-350 °C. After exiting the reactor, the vapor passed through warm (175 °C) and cold (0 °C) traps to condense out the liquid products. Lighter gas products were sent through a ventilation line. Gas flows were measured using a 100 ml bulb meter; gases were collected in a 1-L gas sample bag for analysis using a Micro GC typically used for FT tailgas analysis. Liquid samples were collected and analyzed using the same HP 5890 and HP 5790 GCs as used to analyze FT liquid products. Generally, the sampling interval between the hydrocracking gas and liquid samples was 15-24 h. A typical hydrocarbon distribution of Co wax is shown in Figure 1. It can be seen that the products cover the hydrocarbon range from C_{12} to C_{57}.

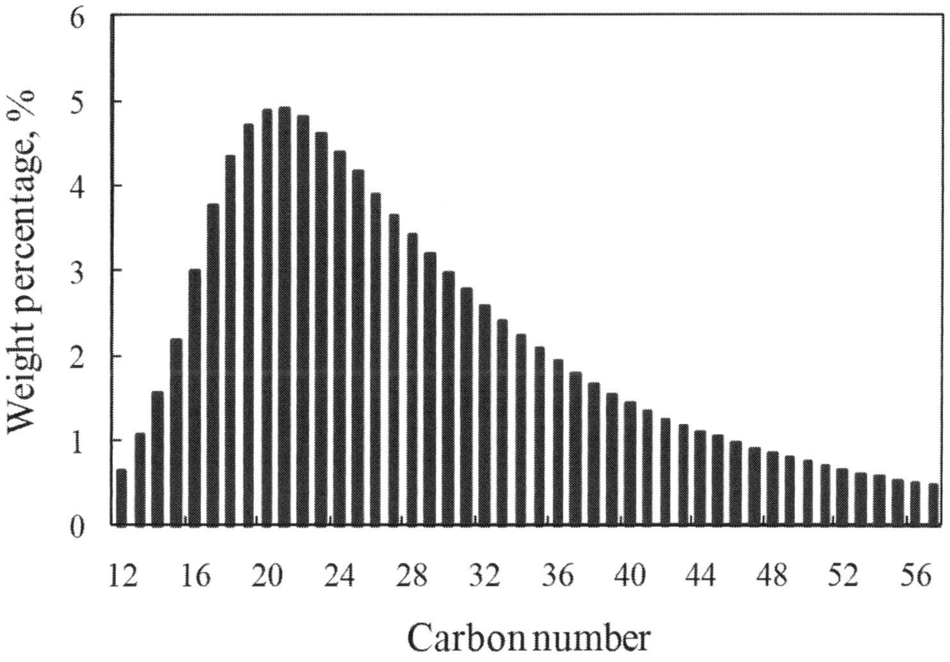

Figure 1. Typical hydrocarbon distribution of FT-wax formed on cobalt catalyst

Results and discussion

BET measurement results

The BET measurement results over the SiO_2-Al_2O_3 Davison 135 support, and the supported 0.5%Pt/ZSM-5, and 0.5%Pt/SiO_2-Al_2O_3 catalysts are shown in Table 1. The SiO_2-Al_2O_3 support and its supported Pt catalyst exhibited larger BET surface areas (403-420 m^2/g) compared to the 0.5%Pt/ZSM-5 catalyst. A similar trend was also observed from the pore volume data, corresponds to a larger average pore diameter for the 0.5%Pt/ZSM-5 catalyst.

Table 1. BET measurement results for supports and catalysts

Composition	SA (m^2/g)	SPV (cm^3/g)	APV (cm^3/g)	DPV (cm^3/g)	SPR (nm)	APR (nm)	DPR (nm)
SiO_2-Al_2O_3 Davison 135	419.9	0.69	0.71	0.71	3.33	2.82	2.54
0.5 % Pt/ZSM-5	335.65	0.38	0.34	0.35	2.29	3.1	2.71
0.5% Pt/SiO_2-Al_2O_3	403.2	0.68	0.7	0.7	3.53	2.83	2.56

SA: BET Surface area; SPV: Single point pore volume; APV: Adsorption cumulative pore volume; DPV: Desorption cumulative pore volume; SPR: Single point adsorption average pore volume; APR: Adsorption average pore radius; DPR: Desorption average pore radius

FTIR results

Pyridine adsorption results for the two catalysts are shown in Table 2. Brønsted acid sites and Lewis acid sites were detected on both catalysts. However, the ratio of Brønsted/Lewis (B/L) sites calculated from the ratio of the peak heights of bands at 1540 and 1445 cm^{-1} for the two catalysts displayed significant differences. The 0.5%Pt/SiO_2-Al_2O_3 contains relatively more Bronsted acid sites than 0.5%Pt/HZSM-5 (B/L, 0.4 vs 0.28). This may be due to the greater amount of surface hydroxyl groups present on the silica surface which forms pyridinium ions.

Table 2. DRIFT intensity summary of pyridinium ion on Bronsted sites and covalently bound pyridine on Lewis sites

Catalyst	IR Intensity of Pyr-B (1540 cm^{-1})	IR Intensity of Pyr-L (1445 cm^{-1})	B/L intensity ratio I_{1540}/I_{1445}
0.5% Pt/HZSM-5	0.048	0.171	0.281
0.5% Pt/SiO_2-Al_2O_3	0.067	0.166	0.403

Hydrocracking results over 0.5%Pt/ZSM-5 and 0.5%Pt/SiO_2-Al_2O_3 catalysts

Figure 2 shows the hydrocarbon distribution after the hydrocracking reaction at the same reaction conditions (300 °C, 500 psig, H_2/Wax ~ 17, LHSV = $1h^{-1}$) on the 0.5%Pt/ZSM-5 and 0.5%Pt/SiO_2-Al_2O_3 catalysts. The ZSM-5 supported catalyst produced predominantly a narrow range of hydrocarbons, C_1-C_{11}, while the distribution of the hydrocracking products produced on the SiO_2-Al_2O_3 supported catalyst is broad,C_1-C_{37}, among which, the diesel fraction reaches 27%. Although the hydro-cracking products on 0.5%Pt/ZSM-5 could cover a wide range carbon numbers when a high liquid hourly space velocity (LHSV) was used, the liquid composition was mainly gasoline ranged hydrocarbons (results not shown). Compared with Figure 1 (raw cobalt wax distribution, C_{12}-C_{57}), only products below C_{11} were present in the case of 0.5%Pt/ZSM-5, implying 100% of conversion. This is an apparent indication of the greater activity for the ZSM-5 catalyst for hydrocracking wax compared to the SiO_2-Al_2O_3 catalyst.

The effects of reaction conditions including temperature, pressure, LHSV and H_2/Wax ratio on hydrocarbon conversion, selectivities and yields to light hydrocarbons, C_4-C_{11} and middle distillates on the two catalysts are summarized in Table 3. On both catalysts, wax conversions increase with increasing temperature. But, at relatively low temperature, the 0.5%Pt/ZSM-5 displayed much higher wax conversion than 0.5%Pt/SiO_2-Al_2O_3, which is in line with the results shown in Figure 2. At 300 °C, about 94% gasoline hydrocarbons and 5.3% C_1-C_3 was produced on the 0.5%Pt/ZSM-5 and diesel selectivity is rather low. In the case of 0.5%Pt/SiO_2-Al_2O_3, the diesel fraction increased to 27%, which corresponds to a large drop in the selectivities to C_1-C_3 and C_4-C_{11}. It appears that increasing the H_2/wax ratio led to slightly increased wax conversion and selectivities to C_4-C_{11}, but less middle distillates.

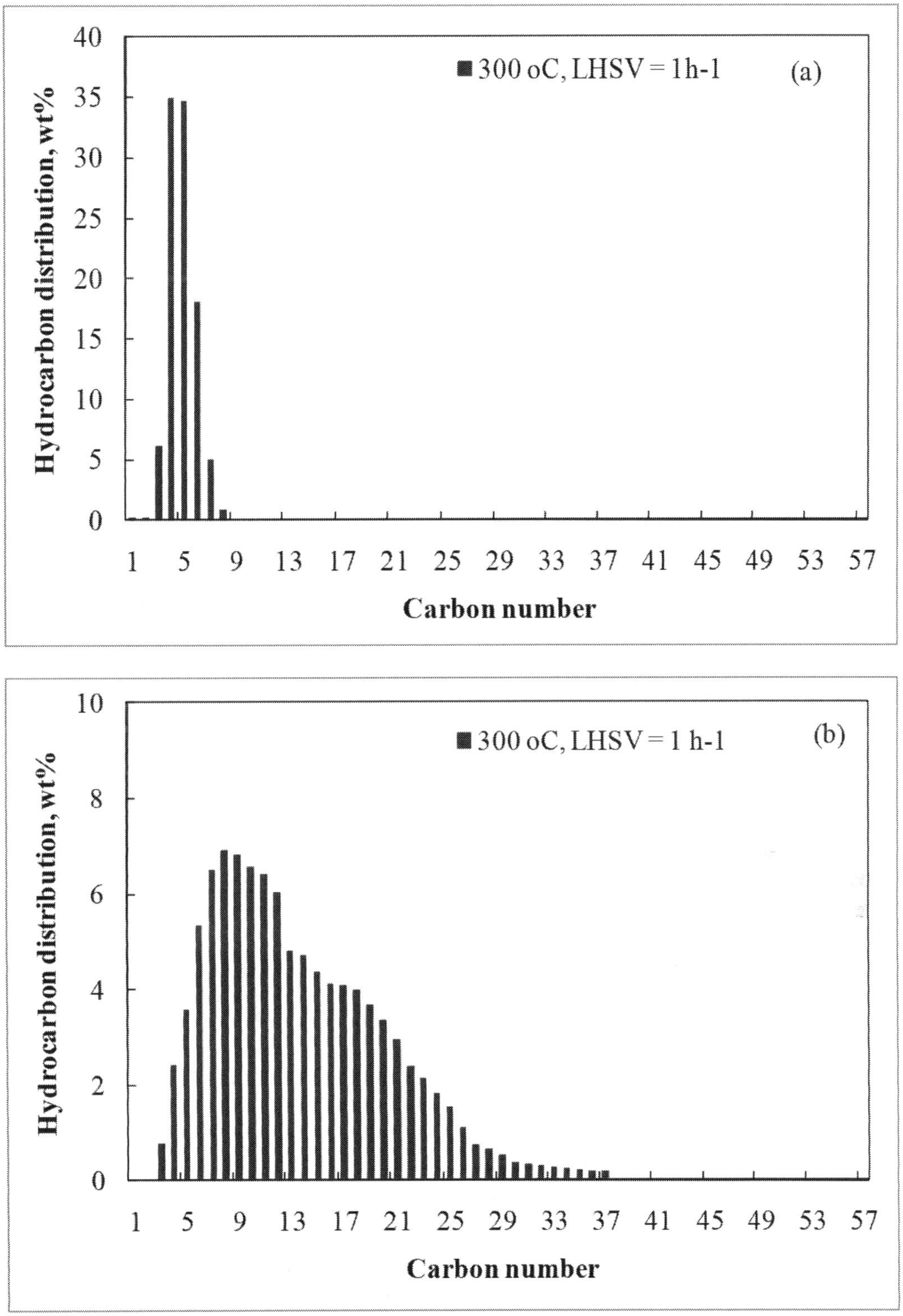

Figure 2. Hydrocracking products' distribution (a) 0.5%Pt/ZSM-5 (b) 0.5%Pt/SiO_2-Al_2O_3 (300 °C, 500 psig, H_2/Wax = 16-17, LHSV = $1h^{-1}$)

The effect of LHSV on wax conversion, selectivity and yield to gasoline and diesel hydrocarbons was examined over a large range of LHSV, 1-30 h^{-1}, for 0.5%Pt/ZSM-5. As expected, the wax conversion dropped to 41% from 100% with increasing LHSV. It is interesting that the selectivity and yield to C_{12-20} increased with LHSV first, reached maxima at 5, and then after that point, declined with increasing LHSV. Although the middle distillates were low in the entire range of LHSV, the trend in selectivity to middle distillate with LHSV demonstrated that optimizing LHSV to achieve the highest diesel yield is necessary. For the 0.5%Pt/ZSM-5, the highest yield of diesel fraction can be achieved when LHSV is set to 5.0. Furthermore, from Table 1, iso-hydrocarbons comprised 50-60% of the total products, indicating that isomerization and hydrocracking took place simultaneously over the ZSM-5 supported catalysts. Considering more L-acid sites present on the catalyst surface (Table 2), it is presumed that L-acid sites play a major role during isomerization over the catalyst.

Table 3. Summary of hydrocracking results on 0.5%Pt/ZSM-5 and 0.5%Pt/SiO_2-Al_2O_3

	0.5%Pt/ZSM-5								0.5%Pt/SiO_2.Al_2O_3			
Temp. °C	280	300	330	300	300	300	300	300	300	340	340	340
Pressure, psig	450	450	450	500	500	500	500	500	500	500	500	500
LHSV, h^{-1}	1	1	1	1	3	5	10	30	1	1	1	1
H_2/Wax ratio	16.10	16.10	16.13	17.11	17.12	17.12	17.14	17.12	15.06	15.06	15.06	29.50
HC conversion, % (X_{C13+}, % = 100*(C_{13+} in feed-C_{13+} in prod)/C_{13+} in feed)									X_{C16+}			
C_{13+} or C_{16+}	96.70	95.85	100.00	100.00	81.55	61.55	57.31	41.28	73.23	93.23	93.23	94.80
Sel, wt% = 100*(prod-feed)/(C_{13+} in feed-C_{13+} in prod)												
C_1	0.08	0.10	0.23	0.19	0.05	0.00	0.00	0.00	0.00	0.03	0.03	0.00
C_2-C_3	8.93	5.30	9.08	6.32	5.38	6.71	6.82	12.33	1.24	2.89	2.89	4.19
C_4-C_{11}	90.04	94.37	90.70	93.48	93.66	90.48	92.23	87.44	71.59	93.08	93.08	95.07
C_{12}.C20	0.95	0.23	0.00	0.00	0.91	2.82	0.95	0.24	27.16	4.00	4.00	0.74
C_{12}.C_{20}/C_4-C_{11}	0.01	0.00	0.00	0.000	0.010	0.031	0.010	0.005	0.379	0.043	0.043	0.008
Yield, g/ml-cat/h												
C_4-C_{11}	0.64	0.67	0.67	0.692	1.696	2.060	3.911	8.014	0.388	0.642	0.642	0.667
C_{12-20}	0.01	0.00	0.00	0.000	0.016	0.064	0.040	0.022	0.147	0.028	0.028	0.005
Ratio of Iso- to-normal hydrocarbons												
C_4-C_{11}				1.58		1.60						
C_{12}-C_{16}				1.17		0.99						

Conclusions

The pyridine adsorption results indicate that ZSM-5 and SiO_2-Al_2O_3 supported Pt catalysts contain both B- and L-acidic sites. A higher fraction of L-acid sites was observed on both catalysts.

The 0.5%Pt/ZSM-5 is more active than 0.5%Pt/SiO_2.Al_2O_3 for the hydrocracking of wax. However, the 0.5%Pt/SiO_2.Al_2O_3 catalysts produced a greater fraction of middle distillates and less light hydrocarbons. This is probably due to the support effect.

The optimized LHSV for achieving the highest middle distillate yield on the 0.5%Pt/ZSM-5 is 5.0. Increasing temperature between (280-340 °C) or H_2/wax ratios (17-33) increased wax conversion, but significantly decreased the yield and selectivity to diesel fractions.

Isomerization and hydrocracking were taking place simultaneously on the catalysts. A greater amount of isomerized hydrocarbons in the range of C_4-C_{16} were formed. L-acidic sites appear to play an important role in the isomerization reaction.

Acknowledgments

This work was supported by NASA contract, #NNX07AB93A and the Commonwealth of Kentucky.

References

1. Sie, S.T., Senden, M. M. G. and Wachem. and Van H. M. W. *Catalysis Today*, 8 (1991) 371-394.
2. Moller, K., Grange, P. L. and Accolla, C. *Industrial Engineering Chemical Research*, 48 (2009) 3791-3801.
3. Hwang, S. H., Lee, J. Y., Park, S.Y., Park, D. R., Jung, J. C., LeE, S. B. and Song, I. K. *Catalysis Letters*, 129 (2009) 163–169.
4. Vincenzo, Calemma., Chiara, Gambaro., Wallace O. Parker Jr., Rosa, Carbone., Roberto, Giardino. and Pietro, Scorletti. *Catalysis Today* 149 (2010) 40–46.
5. Leckel, D. *Energy & Fuels*, 19 (2005) 1795-1803.
6. de Haan, R., Joorst, G., Mokoena, E. and Nicolaides, C.P. *Energy & Fuels*, 21 (2007) 1425-1431.
7. Zhou, Z., Zhang, Y. L., Tierney, J. W. and Wender, I. *Fuel Processing Technology* 83 (2003) 67– 80.
8. Fernandes, F. A. N. and Teles, U.M. *Fuel Processing Technology* 88 (2007) 207–214.
9. de Haan, R., Joorst, G., Mokoena, E. and Nicolaides, C.P. *Applied Catalysis A: General*, 327 (2007) 247–254.
10. http://www.diesenet.com/standards/eu/ld.php.
11. Scherzer, J. and Gruia, A. *Hydrocracking Science and Technology*, Marcel, Dekker Inc., New York, 1996 (Chapter 3). (1998) 1.
12. Morawski, I. and Mosio-Mosiewski, J. *Fuel Processing Technology*, 87 (2006) 659–669.
13. Trombetta, M., Busca, G., Rossini, S., Piccoli, V., Cornaro, U., Guercio, A., Catani, R. and Willey, R. J. *Journal of Catalysis*, 179 (1998) 581-596.
14. Leckel, D. *Industrial Engineering Chemical Research*, 46 (2007) 3505-3512.

Fischer-Tropsch Synthesis: Deuterium Labeled Ethanol Tracer Studies on Iron Catalysts

Muthu K. Gnanamani[1], Robert A. Keogh[1], Wilson D. Shafer[1], Buchang Shi[2], Gary Jacobs[1] and Burtron H. Davis[1]

[1]*Center for Applied Energy Research, University of Kentucky, 2540 Research Park Dr., Lexington, KY, 40511*

[2]*Department of Chemistry, Eastern Kentucky University, Richmond, KY, 40475*

Introduction

The Fischer-Tropsch synthesis (FT) can be briefly defined as the catalytic conversion of carbon monoxide and hydrogen to hydrocarbon products. It has long been known that smaller 1-alkene and alcohol molecules may be able to act as a chain initiator under normal FT synthesis conditions [1]. The results from the addition of a C_2 monomer during FT has been reviewed by a number of workers [2-4], who advanced a number of difference surface species such as vinyl (CH_2=CH-$_{(ad)}$) and vinylidene (CH_2=C=$_{(ad)}$). The pioneering works in this field were performed by Emmett and coworkers [5,6]. ^{14}C labelled ethanol (1.5 %) was added to the syngas feed to an iron catalyst operated at 508 K and 1 atm, using a plug-flow reactor. Based on the observed constant radioactivity (^{14}C) of hydrocarbons, the authors concluded that ethanol or the adsorption complex derived from the ethanol, could primarily act as a chain initiator. Schulz et al. [7] probed the mechanism of the FT reaction for a cobalt-thoria-kieselguhr catalyst that was operated at 458 K and atmospheric pressure using ^{14}C alkenes and found that nearly 30% of the ethene and propene added were incorporated in to FT products. Co-feeding ^{14}C-labelled ethene with the syngas over cobalt catalysts demonstrated that the alkene added at concentrations lower than 2 mol% can act as an initiator and also acts as a propagator, when the ethane concentration exceeds 5 mol% [8]. Tao et al. [9] reported that 1- and 2-Propanol incorporate into the products of the FT synthesis and claimed that the surface species generated from 1- and 2-Propanol are distinct and these initiate chain growth to form different products. Hanlon and Satterfield [10] have reported that the addition of ethanol or ethene to the feed to an iron catalyst increased the olefin to paraffin ratio of the products and observed only a slight degree of chain incorporation upon additions of 1-olefin or ethanol in major amounts. Therefore, it is not only the quantity of the labeled compound that is co-feed along with a syngas during FT synthesis but also the synthesis conditions that may determine the rate of incorporation of the added compound into the FT synthesis products.

In 1960's, Krykov et al. [11] used α-deuteroethanol as a tracer to study the mechanism of the FT reaction. The authors concluded that intermediate surface compounds obtained from the alcohol taking part in the formation of hydrocarbons and that they retain at least a part of the deuterium of the added alcohol. Brundage et al. [12] obtained a two-unequal-hump response for d_1- and d_2-propionaldehyde formation from D_2 and C_2D_4 pulses during ethylene hydroformylation using Mn-Rh/SiO_2 at 513 K and 0.1 MPa. The authors demonstrated using FT-IR/mass spectrometry that the different deuterium incorporation pathways into the propionaldehyde possibly due to H/D exchange on adsorbed acyl and acyl hydrogenation. Our previous study added D_2O as the tracer to elucidate the FT mechanism for a cobalt catalyst and these show that there is no measurable H/D exchange in alkanes and the exchange in alkenes appears to be limited to the vinyl positions [13]. In the present study, partially deuterated ethanol (i.e., CD_3CD_2OH and CD_3CH_2OH) was used as a tracer to investigate the iron catalyzed Fischer-tropsch mechanism at different synthesis conditions. An attempt has been made to identify the surface intermediates formed from the added d-ethanol through an analysis of FT hydrocarbon products.

Experimental

The Fischer-tropsch runs were conducted using 1-L CSTR. The experimental apparatus and materials used here are described elsewhere [14]. The iron catalyst has a composition of 4.4 wt % Si, 56 wt % Fe, 0.5 wt % K (i.e., a low α catalyst). The ethanol-D_5 (CD_3CD_2OH) and ethanol-D_3 (CD_3 CH_2 OH) were purchased from Cambridge Isotope Laboratory, Inc., with an isotopic purity of 98%. For each run, about 20 g of crushed catalyst was mixed with 310 g of C_{30} oil and pretreated for 24 hrs using a syngas (H_2/CO:0.7) at 543 K, 0.10 MPa. The initial conditions for the FT reaction were 0.8 MPa and 543 K using a syngas feed at 60 slph with a percentage composition of 41.2 slph, H_2 and 58.8 slph, CO. About 1.50 mol% of ethanol-D_5 or dthanol-D_3 (5 g of labeled ethanol was mixed with 17 g of n-heptane) was introduced into the reaction system during a period of 300 minutes using a MiltonRoy mini-piston pump. The D-ethanol was introduced into the reaction system along with a syngas stream after reaching a steady CO conversion (~40%) at a feed rate of 0.9260 slph. In order to see the effect of pressure on D-distribution of hydrocarbon products, a separate FT reaction was performed at the same reaction conditions as mentioned above except that the pressure which was set at 2.2 MPa. The gaseous products were analyzed by an online-gas chromatograph during and following D-ethanol addition. Hot (373 K) and cold (273 K) traps were emptied before the addition of deuteroethanol. At the end of the addition, both oil and wax products were collected from cold and hot traps, respectively, and further separated from the aqueous phase. The aqueous phase was analyzed by GC using Q-porpack column at 373 K. The detailed experimental procedure was described elsewhere [14]. Quantitative analysis of ethanol in an aqueous solution was determined using a standard calibration procedure and the D-distribution in ethanol was quantified using GC-MS. It is known that fragments of deutero-ethanol could interfere with a parent peak of another D isotopic isomer of D-ethanol [15], a correction factor was introduced based on the GC-MS pattern obtained using standard D_5-ethanol and D_3-ethanol compounds. The D-distribution of various hydrocarbons present in the oil and wax samples were analyzed using GC-MS. Complications are involved in analyzing C_1-C_4 gaseous hydrocarbon products due to exchange and scrambling of deuterium among FT products during the FT reaction. The D isotopomers for different carbon numbers were calculated after correcting ^{13}C for their parent ion peak (m^+ ion) and the details were described elsewhere [16].

Results and discussion

There is a significant increase of olefin-to-paraffin ratio for C_2, C_3, and C_4 with D_5-EtOH addition as shown in Table 1 and Figure 1; this suggest that deuteroethanol can compete with hydrogen for adsorption sites on the catalyst surface. Kummer et al. [5] also noted this decrease in the hydrogenation activity of the catalyst with added ethanol, which causes an increase in the C_2, C_3, and C_4 olefin/paraffin ratios. Nearly constant CO conversion and a linear increasing trend in the selectivity to methane were observed upon D_5-ethanol addition and this suggests that D_5-ethanol is not involved in C-C bond breaking. These results are consistent with the conclusions made from other studies that small amount of added ethanol did not significantly affect the CO conversion and methane selectivity [10]. A considerable increase in selectivity to ethylene with D_5-ethanol addition (Table 2) suggests that this is due to the dehydration of D_5-ethanol.

It is well-known that among ethanol and ethylene co-feeding studies during FT synthesis using iron catalysts that the rate of incorporation of ethanol is much higher than the rate of incorporation of ethylene [10]. In the present system, about 55.4% of added D_5-ethanol was recovered from the aqueous phase; thus upto 44.6% of D_5-ethanol could be converted into various hydrocarbons by entering into the growing hydrocarbon chain. It is hard to predict the exact amount of D_5-ethanol that is incorporated into the growing hydrocarbon chains as evidenced by the lack of increase of C_3 hydrocarbon selectivity (Table 1) with added D_5-ethanol. Kokes et al. [17] applied ^{14}C tracer experiments to define the mechanism for a doubly promoted iron catalyst used at varying pressures and temperatures and found that the percentage incorporation of ethanol falls from 18% to 7%, and then from 7% to 2.2% as the pressure increased from 0.1 to 0.75 to 2.1 MPa, respectively. The authors compared their results to Kummer's work and concluded that less incorporation occurred over a doubly promoted catalyst containing potassium and magnesium than over a singly promoted catalyst without potassium. In our system, a single promoted Fe catalyst at low potassium content and at low pressure (0.8 MPa) may favor more D_5-ethanol incorporation.

If the added D_5-ethanol is only involved in initiating the hydrocarbon chain then one would expect the maximum number of deuterium atoms an alkane or 1-olefin can have is 5; however, our experimental results show that there is no evidence for the presence of D_5 isotopic isomer in any of the hydrocarbons. The D/molecule for each hydrocarbon can be calculated based on the GC-MS analysis of oil and wax samples and is shown in Figure 2. The D/molecule for alkanes was nearly constant (~1.4) up to C_{10} and may increase slightly with increasing carbon number after C_{10}. However, D/molecule for 1-olefin exhibits an increasing trend starting from 1.2 for C_5 and increases to 1.5 for C_{12}. The D_0 isotopomer for each compound comes from two sources: a normal FT synthesis during the tracer period and also from the accumulated product. The predominance of the D_1 isotopic isomer in all hydrocarbons suggests that considerable H/D exchange occurred during the interaction of D_5-ethanol on the catalyst surface and this would create a pool of deuterium which may eventually be introduced into hydrocarbon products.

If the C_2 species derived from D_5-ethanol acts as an initiator, the D/molecule must be a constant for all compounds. The data obtained in this study display a constant D/molecule for alkanes indicating that the added D_5-ethanol could initiate the carbon chain growth, but after extensive H/D exchange on the catalyst surface. The D-content per molecule for the 1-olefin increased linearly from 1.2 to 1.5 with increasing carbon number in the C_5-C_{12} range, suggesting that 1-olefin readsorption that was followed by some H/D exchange could cause this observation. It is known that iron catalysts readily effect dehydrogenation-hydrogenation of alcohols and so it is common to relate the participation of alcohol in the synthesis of hydrocarbons and H/D exchange to reverse dehydrogenation of alcohol. Schme 1 shows that there are more than one step likely exists where H/D exchange could takes place at α and β position of adsorbed d_5-ethanol. Thus, the intermediate compound (1), (2) will readily exchange deuterium with adsorbed hydrogen on the catalyst surface leading to a spectrum of D-ethanol's (D_0, D_1, D_2, D_3, D_4 and D_5). As shown in Figure 3, the recovered ethanol contains D_0, D_1, D_2, D_3, and D_5 isotopomers. After the rapid elimination-addition reactions, the ethanol that participates the chain growth is actually the partially deuterated ethanol (C_2H_5DO, $C_2H_4D_2O$, $C_2H_3D_3O$) and not the added D_5-ethanol. GC-MS analysis of ethanol after the FT run shows the presence of the following d isotopic isomers (mole, %): C_2H_6O = 22.8, C_2H_5DO = 14.3, $C_2H_4D_2O$ = 9.8, $C_2H_3D_3O$ = 7.1, $C_2H_2D_4O$ = 0, C_2D_5HO = 42.9. It reveals that hydrogenation-dehydrogenation takes place readily during adsorption of D_5-ethanol (CD_3CD_2OH) and this leads to deuterium exchanging with hydrogen of the reaction mixture on the catalyst surface.

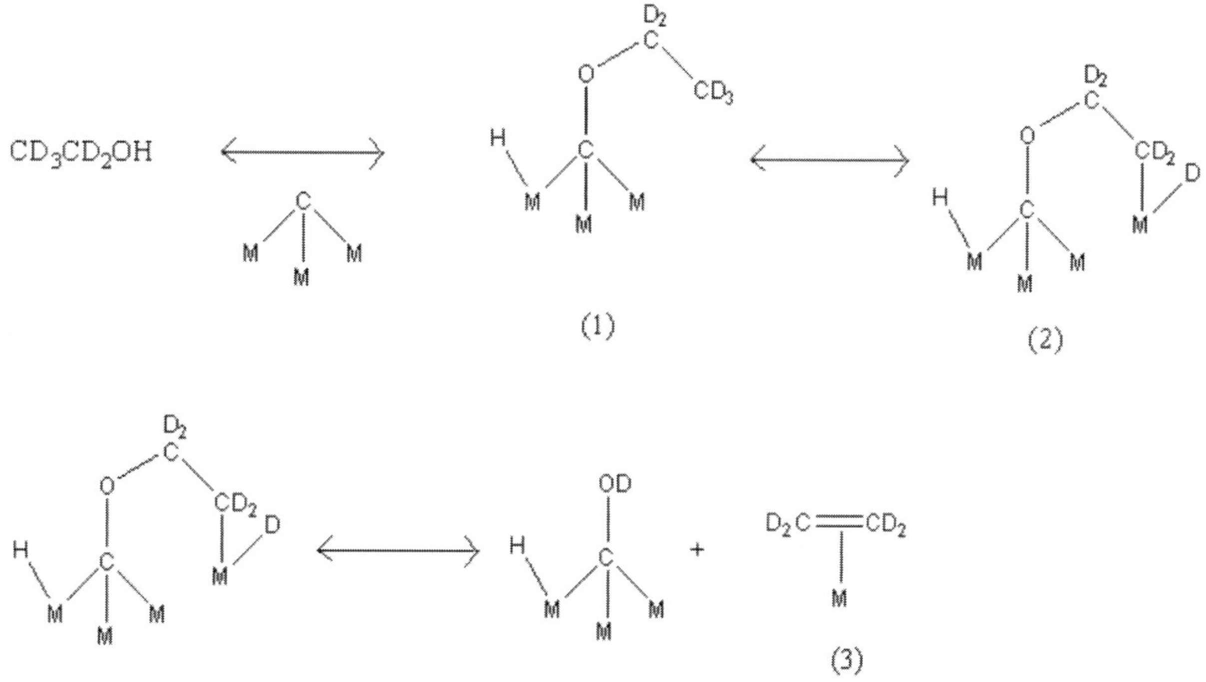

Scheme 1. Hydrogenation-dehydrogenation of d_5-ethanol on Fe catalysts

It is also possible that intermediate surface species could be an ethene π-complex derived from the dehydration of added deuteroethanol (scheme 2). After the metal bonds to one carbon in ethane to produce (4), it then undergoes a hydrogen atom shift to produce (5). Finally, the carbon-metal double bond is formed (6), which can initiate the chain growth. The intermediates (4) and (5) can also undergo H/D exchange with surface adsorbed hydrogen atoms and this leads to a total deuterium atoms are less than 4 in the C_2 unit. Furthermore, the co-feeding experiments [18] have revealed that ethanol and propanol are to a considerable extent converted to ethene and ethane or to propene and propane, respectively. It has been reported [17] that ^{14}C labeled ethene addition also initiates chain growth but that the molar radioactivity in the products is significantly smaller than observed in ethanol co-feeding runs. These results suggest that it is more difficult to form the chemisorbed surface structure (3) in Scheme 2 in the case of ethene addition than for ethene produced from added ethanol. Another possibility is that the ethanol incorporation mechanism may follow more than one pathway. At present, we do not have any direct experimental evidence to show the specific C_2 intermediate derived from the added deuteroethanol which is incorporated into growing hydrocarbon chain. However, the fact that under the same reaction conditions ethanol initiates chain growth about 50 – 100 times greater than ethane suggests that most of the H/D exchange occurs prior to loss of the oxygen – carbon bond.

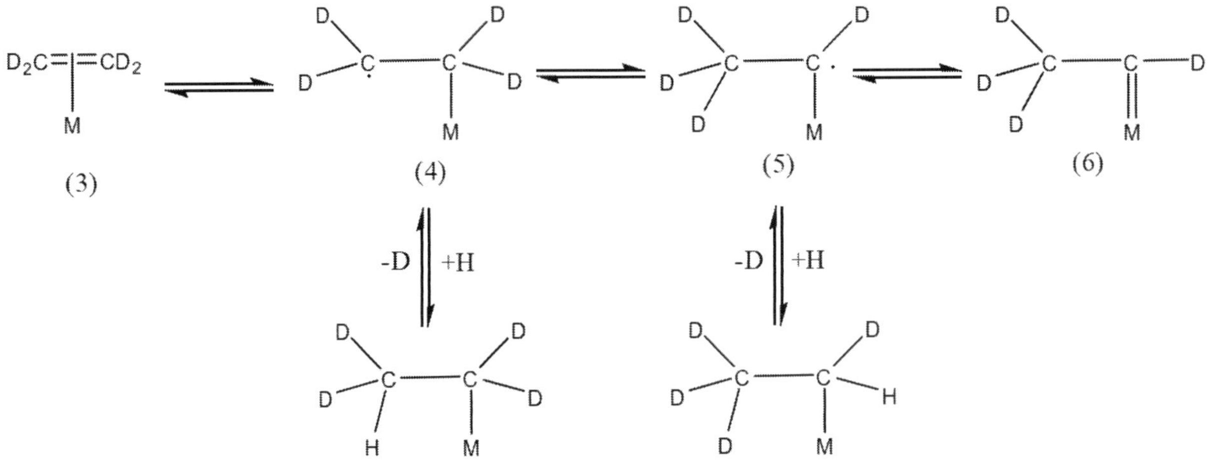

Scheme 2. The possible pathway for the formation of C_2 species

Quantitatively, the picture for D_5-ethanol co-feeds at high pressure (2.2 MPa) is not significantly different from the results obtained at lower pressure 0.8 MPa. As shown in Figure 4, there is a depression of methane formation at high pressure which could suggest that some hydrogenation sites were occupied by the added deuteroethanol. Hanlon and Satterfield also observed a similar decrease in methane selectivity with the addition of ethanol co-feed FT synthesis using a fused magnetite catalyst containing potassium at 521 K and 0.78 MPa. As described earlier, the C_2 intermediate species derived from the adsorbed D_5-ethanol are responsible for the observed D-distribution in the hydrocarbon products. A noticeable difference can be seen for D_0 and D_1 distribution between low and high CO conversion. As CO conversion increases the relative mol % of deutero compounds (D_1, D_2, and D_3) decreased irrespective of carbon number due to the fact that most of the catalyst surface will be covered with intermediates which are derived from the normal FT synthesis. This reveals that the possibility for H/D exchange is lower at high pressure and could be a dominating process occurring at low CO conversion. Indeed, the overall d/molecule remains more or less the same at 1.35, which is slightly lower than the d/molecule (1.45) obtained at low CO conversion. It is likely that there might be a considerable difference between the interaction of added D_5-ethanol and the catalyst surface that impacts the D-distribution among hydrocarbon products. Therefore, the probability that D_5-alcohol participates in FT synthesis decreases with increasing pressure, in agreement with previous studies [10]. Similar results were obtained for a CD_3CH_2OH tracer run. As shown in Table 3, the maximum number of deuterium atoms in the hydrocarbon is 4. The D/molecule exhibits a constant value of 1.4, indicating that the C_2 species derived from the ethanol acted as an initiator for the FT synthesis.

Conclusions

Partially deuterated D_5- and D_3-ethanol were used as a tracer to study the Fe-FT synthesis mechanism. A significant amount of deuterium was found in the hydrocarbon products (alkane, 1-olefin) irrespective of carbon number. Addition of D_5-ethanol during the FT synthesis produces more D_1, D_2, D_3 and to some extent D_4 isotopomers than for D_3-ethanol. The tracer runs (high and low CO conversions) yielded nearly a constant D/molecule for alkanes and 1-olefins, suggesting that the C_2 species derived from the partially deuterated ethanol acted as an initiator for the FT synthesis. None of the compounds were found to contain D_5 isotopomer, indicating extensive H/D exchange of adsorbed D_5-ethanol with hydrogen on the surface of the catalyst before initiating a growing hydrocarbon chain.

Table 1. Effect of D_5-ethanol co-feeding on olefin-paraffin to ratio of hydrocarbon products of Fe-FT synthesis at 543 K and 0.8 MPa using a syngas (H_2/CO: 0.7)

Runs	Sample no	Olefin/Paraffin ratio		
		$C_2^=/C_2$	$C_3^=/C_3$	$C_4^=/C_4$
without C_2D_5OH	3	1.59	5.80	6.5
	4	1.62	5.70	6.6
with C_2D_5OH	5	1.98	7.01	7.41
	6	1.98	6.20	7.02
without C_2D_5OH	7	1.64	5.80	6.54
	8	1.6	5.52	6.53
	9	1.62	5.50	6.56

Table 2. Effect of D_5-ethanol co-feeding on hydrocarbon product selectivity of Fe-FT synthesis at 543 K and 0.8 MPa using a syngas (H_2/CO: 0.7)

Runs	Sample no	Selectivity, %C						
		CH_4	$C_2^=$	C_2	$C_3^=$	C_3	$C_4^=$	C_4
Without C_2D_5OH	3	8.21	4.68	2.93	10.46	1.57	9.08	1.39
	4	7.87	4.45	2.74	9.80	1.72	8.45	1.28
With C_2D_5OH	5	8.38	5.78	2.92	10.70	1.77	8.40	1.08
	6	8.21	5.63	2.88	10.51	1.69	8.22	1.17
Without C_2D_5OH	7	8.63	4.88	2.97	10.61	1.83	9.14	1.40
	8	8.45	4.68	2.92	10.39	1.88	8.84	1.35
	9	8.40	4.66	2.87	10.28	1.93	8.75	1.34

Table 3. Distribution of deuterium in paraffins while co-feeding D_3-ethanol during Fe-FT synthesis at 543 K and 0.8 MPa using a syngas (H_2/CO: 0.7)

Paraffin Carbon no	mol%						D/molecule *
	D_0	D_1	D_2	D_3	D_4	D_5	
9	86.41	10.46	2.60	0.00	0.00	0.00	1.2
10	88.13	7.87	3.34	0.00	0.00	0.00	1.25
11	85.46	10.21	3.40	0.00	0.00	0.00	1.24
12	88.78	8.71	3.63	0.00	0.00	0.00	1.29
13	89.35	6.86	2.96	0.00	0.00	0.00	1.30
14	91.91	4.71	2.30	0.00	0.00	0.00	1.32
15	94.04	2.48	2.75	0.00	0.00	0.00	1.31

*The D/molecule were calculated without including d_0

Figure 1. Effect of D_5-ethanol co-feeding on C_2 olefin-to-paraffin ratio

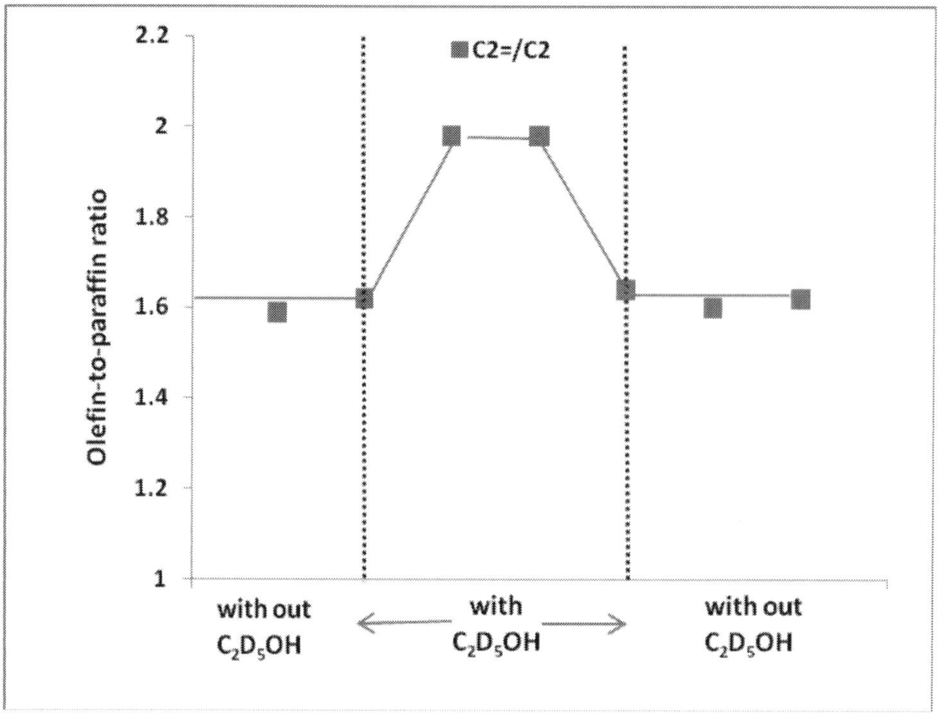

Figure 2. Variation of total deuterium content in paraffins and 1-olefins with carbon number during Fe-FT synthesis with D_5-ethanol co-feeding at 543 K and 0.8 MPa using a syngas (H_2/CO: 0.7)

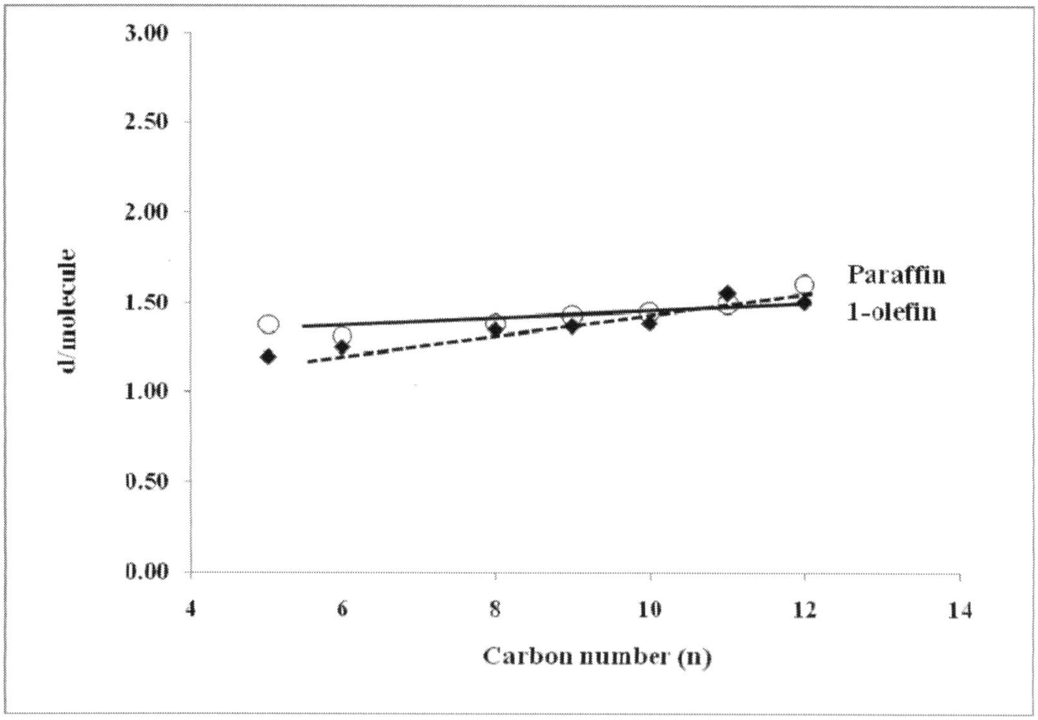

Figure 3. The D-distribution of ethanol in aqueous phase during Fe-FT synthesis D_5-ethanol co-

feeding at 543 K and 0.8 MPa using a syngas (H_2/CO: 0.7)

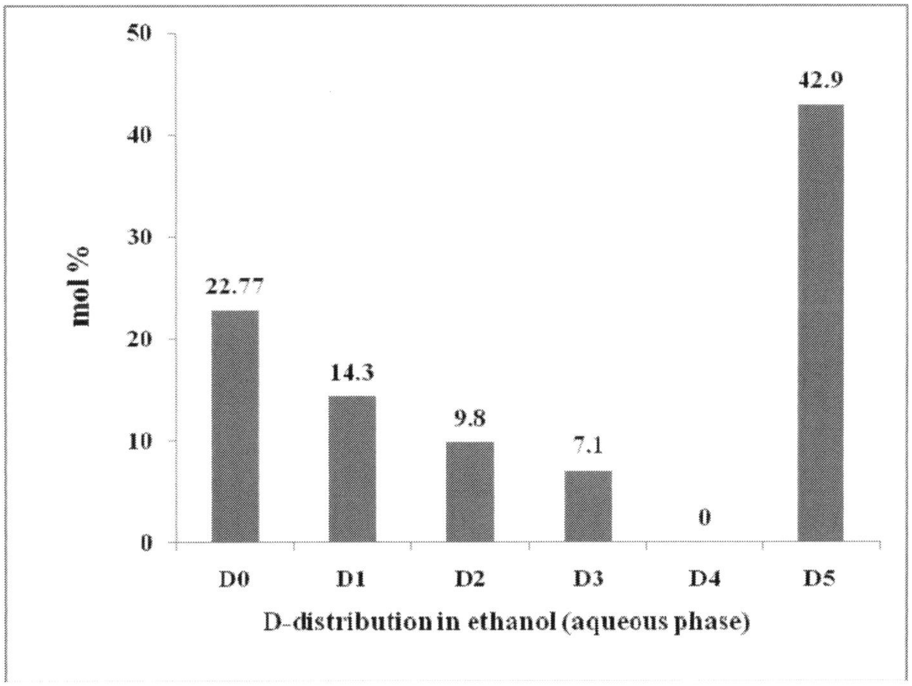

Figure 4. Effect of methane production rate during D_5-ethanol addition at 2.2 MPa and 543 K using a syngas ratio of H_2/CO = 0.7

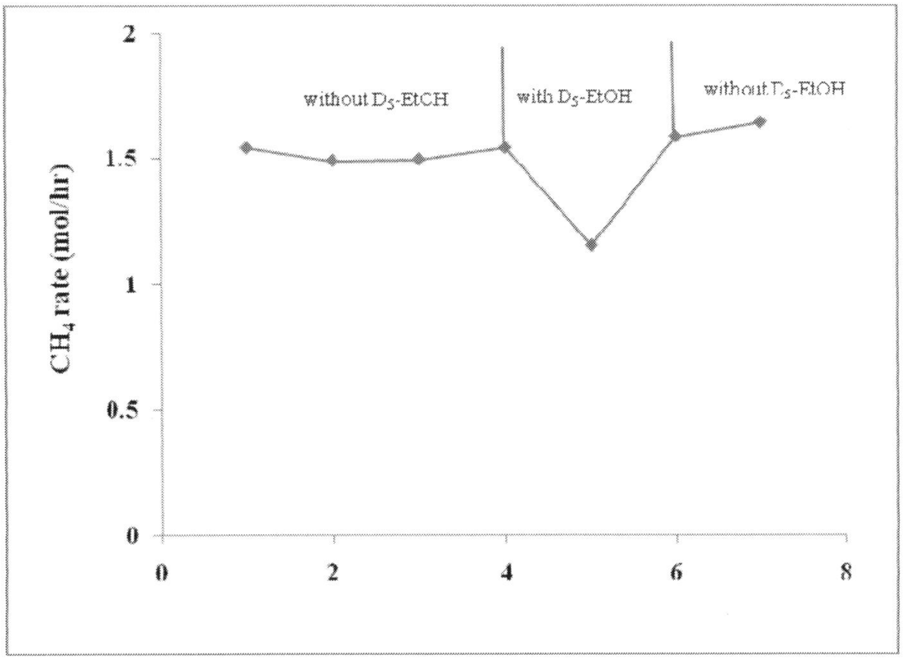

Reference

1. H. Schulz, M. Claeys, Appl. Catal. A: Gen. 186 (1999) 71.
2. M.E. Dry, J.C. Hoogendoom, Catal. Rev. Sci. Eng., 23 (1981) 265.
3. M.L. Turner, H.C. Long, A. Shenton, P.K. Byers, P.M. Maitlis, Chem.-Eur. J., 1 (1995) 549.
4. M.L. Turner, N. Marsih, B.E. Mann, R. Quyoum, H.C. Long and P.M. Maitlis, J. Am. Chem. Soc., 124 (2002) 10456.
5. J.T. Kummer, H.H. Podgurski, W.B. Spencer, P.H. Emmett, J. Am. Chem. Soc., 73 (2) (1951) 564.
6. J.T. Kummer, P.H. Emmett, J. Am. Chem. Soc., 75 (1953) 5177.
7. H. Schulz, B.R. Rao, M. Elstner, Erdol and Kohle 22 (1970) 651.
8. R. Snel, R.L. Espinoza, J. Mol. Catal. 43 (1987) 237.
9. L.-M. Tau, H.A. Dabbagh, B.H. Davis, Energy and Fuels, 4 (1990) 94.
10. R.T. Hanlon, C.N. Satterfield, Energy and Fuels, 2(2) (1988) 196
11. Y.B. Kryukov, A.N. Bashkirov, F.A. Fridman, L.G. Liberov, R.M. Smimova, A.A. Pegov, Neftekhimiya, 5 (1965) 62.
12. M.A. Brundage, S.S.C. Chuang, J. Catal. 174 (1998) 164.
13. B. Shi, J. Li, R.A. Keogh, B.H. Davis, Appl. Catal. A: Gen. 229 (2002) 283.
14. B. Shi, R.J. O'Brien, S. Bao, B.H. Davis, J. Catal. 199 (2001) 202
15. L.E. Friedman, J. Turkevich, J. Am. Chem. Soc., 74 (1952) 1666.
16. B. Shi, B.H. Davis, Appl. Catal. A: Gen. 277 (2004) 61.
17. R.J. Kokes, W.K. Hall, P.H. Emmett, J. Am. Chem. Soc. 79 (1957) 2989.
18. R. Snel, A.L. Espinoza, J. Mol. Catal. 54 (1989) 213.

(119e) Fischer-Tropsch Synthesis Using H_2/CO/CO_2 Syngas Mixtures Over Iron and Cobalt Based Catalysts: Insights Into Reaction Pathways

Yali Yao, Centre of Material and Process Synthesis, University of the Witwatersrand, Johannesburg, South Africa

Diane Hildebrandt, Centre of Material and Process Synthesis, University of the Witwatersrand, Johannesburg, South Africa

David Glasser, Centre of Material and Process Synthesis, University of the Witwatersrand, Johannesburg, South Africa

With the depleting resource of crude oil all over the world, the Fischer-Tropsch Synthesis (FTS) process, in which synthesis gas is converted into a complex multi-component mixture consisting of linear and branched hydrocarbons and oxygenated products, becomes a promising route to meet the continuously increasing demand for liquid fuels and chemical feed stocks. Among the reported FTS catalysts, iron and cobalt are used commercially at temperatures between 473K and 573K, and at 10 to 60 bar pressure. The synthesis gas, a mixture of predominantly CO and H_2 with different H_2/CO ratios, can be produced from coal, natural gas or biomass. In some cases, CO_2 may be a significant component in the synthesis gas. It is therefore interesting to gain insight into the reaction pathways for FTS using H_2/CO/CO_2 syngas mixtures over iron and cobalt based catalysts.

FTS experiments over iron and cobalt based catalysts have been carried out in a fixed-bed micro reactor. Two feed gases, one of composition H_2:CO:CO_2 = 3:0:1 and the other of H_2:CO:CO_2 = 2:1:0 were fed to the reactor and the reactions monitored. Mixtures of various proportions of the two feed gases were fed to the reactor, thus varying the ratio of CO, CO_2 and H_2 stoichiometrically. The reaction conditions for the iron based catalyst were a total synthesis pressure of 20bar, flow rate of 60ml/min/g.cat and temperature of 523K; and for the cobalt based catalyst, total synthesis pressure of 20bar, flow rate of 60ml/min/g.cat and temperature of 473K.

The results show that CO_2 rich feeds produce products that are mainly short chain hydrocarbon products while CO rich feeds shift the product composition to a Fischer"CTropsch type (mainly long chain hydrocarbons) product on both iron and cobalt based catalysts. The products distributions of hydrocarbons, described by the Anderson-Schulz-Flory (ASF) distribution, were calculated for each runs for both iron and cobalt based catalysts. The single alpha product distribution is found with CO_2 rich feeds, while a two-alpha product distribution is found when using CO rich feeds. Masuku[1] et. al. considered the effects of Vapour-Liquid Equilibrium (VLE) on the FT product distribution, and showed that at high alphas, when a liquid layer is formed on the catalyst, a 2-alpha distribution will be measured experimentally. For CO_2 rich feeds the alpha value of the product distribution is very low and all the products (short chain) are in the gas phase, and we therefore predict that the product distribution follows a single alpha distribution. However for CO rich feeds the alpha value of the product distribution is higher resulting in products in both the gas and liquid phases, so that the effect of the VLE results in a two-alpha product distribution for the total hydrocarbon products.

We find similar results to those found in the literature in that the olefin to paraffin ratio changes as a function of carbon number. Increasing the carbon number decreases the olefin to paraffin ratio (excluding O_2/P_2). However, we found that the ratio of $\frac{O_n / P_n}{O_{(n-1)} / P_{(n-1)}}$ (O-olefin, P-paraffin, n-carbon number) is equal to a constant value (β) for all experiments, irrespective of the proportions of CO, CO_2 and H_2 on the catalyst's surface for both iron and cobalt based catalysts. We therefore postulate that the product distribution may be determined by equilibrium considerations. The value of β and the FTS reaction mechanism will be discussed for both iron and cobalt based catalysts.

Keywords: Fischer-Tropsch Synthesis, CO Hydrogenation, CO_2 Hydrogenation, Product Distribution, ASF, Vapour-Liquid Equilibrium, Olefin to Paraffin Ratio, Thermodynamic Equilibrium.

Reference:

[1] Cornelius Mduduzi Masuku, Diane Hildebrandt and David Glasser, Thermodynamic Prediction of the Chain Propagation Probability (α) for the Fischer-Tropsch Reaction, 2007 AIChE Spring National Meeting.

See more of this Session: Production of Fuels and Chemicals From Fischer-Tropsch Hydrocarbons
See more of this Group/Topical: Topical 6: 10th Topical Conference on Gas Utilization

Flexibilities of LNG Storage in Lined Rock Cavern (LRC) with High Operating Pressure

Do-Youn Kim, Ph. D.
Process Engineer

Joseph H. Cho, Ph. D., P. E.
Director of Gas Tech. Center

Sang-Woo Woo
General Manager, Civil Engineering

Dae-Hyuk Lee, Ph.D.
Team leader of GSUC

SK Engineering & Construction Co., Ltd.

2010 AIChE Spring Meeting
10th Topical Conference on Gas Utilization
San Antonio, TX, March 21-25, 2010

ABSTRACT

Natural gas consumption is expected to grow significantly in the next decades. The need for building LNG import terminals with significant storage capacity is quite often a critical aspect due to restriction of land in the area of interest and environmental constraints. A new concept for the storage of LNG in underground mined rock caverns has been developed as very efficient in terms of land occupation, environmental and visual impact at ground surface, safety and cost.

The concept consists of the combination of two well-proven technologies: the storage of gas and liquid hydrocarbons in underground mined cavern and the membrane containment system used for conventional LNG tanks and ocean carriers. The advantages of underground LNG storage in rock caverns are the following: Safety – storage less vulnerable to external hazards, Security – high protection against terrorism, Footprint – very limited surface impact, Environment – no visual impact, and Size – virtually no limitation of size.

This paper describes the concept of storage system and the features of the process. The location flexibility and various unconventional LNG transferring methods to be allied to the Lined Rock Cavern (LRC) System have also been discussed. The unique features of LNG terminal process have been presented with reference to a 450,000 m^3 storage and a 750 t/h of send-out capacity.

INTRODUCTION

General trend of the public is to enjoy using energy as they need, but not allowing building of energy facilities. It is typically expressed as "Not in My Back Yard (NIMBY)". Recently, this trend has evolved into a more serious social opposition to any development for example, when project developers advocate infrastructure such as new roads, energy facility, power plants, etc. Currently, local emotional opposition against project development is well presented by the term "BANANA", an acronym for "Build Absolutely Nothing Anywhere Near Anything (or anyone)". This term is often used to criticize the ongoing opposition of certain interest groups to land development.

Every body needs energy everyday. However, if a development project of energy infrastructure is shut down by local opposition, the next question will then be "where to build energy infra". Many LNG import terminals have not materialized due to strong opposition from local people taken up to Capitol Hill.

If energy facility is far from the massive energy demanding area, energy transport costs are going to be significantly high regardless of energy forms, be it electricity, liquid or natural gas. Meeting the demand of public energy with low costs and ensuring the safety of the energy facility shall be a primary goal of those working for the energy sector.

In order to provide energy to the public safely, economically, and at the same time mitigate the public emotional reaction to oppose the project, such as NIMBY or BANANA, underground LRC LNG storage system has been developed. This storage technology adopts two well proven technologies: Underground mined rock cavern and membrane LNG storage.

Underground mined rock technology has been widely used for strategic energy storage: crude oil, gasoline, LPG, etc. The membrane technology has been used for LNG storage and LNG carriers. This new underground storage system can mitigate the public safety concerns which oppose the project development.

This paper discusses the concept of the lined rock carven system. The possible location of rock carven system is directly governed by the available rock mass and sometimes it might be far from the shore line. The longer unloading line and its associated operation issues will be also discussed. This paper also presents advantages of the flexible operating pressure of the storage system, which will be a higher operating pressure than that of the conventional storage system.

DESCRIPTION OF LINED ROCK CAVERN (LRC) LNG STORAGE SYSTEM

Underground mined rock caverns are commonly used and safely operated since many decades to store petroleum products like crude oil, propane and butane either compressed or refrigerated. The attempts to store LNG in underground rock caverns with a similar approach have not been deemed satisfactory due to large boil-off rate and the low LNG temperature acting on rock wall being liable to generate cracks in the rock mass. On the other hand, the storage of LNG using aboveground tanks and in a limited extent using in ground tanks is now a well proven technology.

The concept developed by SK E&C, Géostock, and Saipem-sa is a simple combination of both the underground mined rock cavern and the aboveground tank technologies. The underground storage is of particular interest towards reducing the land occupation, enhancing safety and security aspects. This is also economically attractive.

The concept consists of protecting the host rock against the extreme low temperature and providing a liquid and gas tight liner (see Figure 1) using insulating panels fixed on a concrete lining and a corrugated stainless steel membrane. Similar containment systems are used in LNG carriers since 30 years without any troubles. The thermal characteristics and thickness of the insulation is designed in such a way to achieve allowable minimum temperature in the rock mass for the design life of the storage. A boil-off rate around 0.05 to 0.1% per day is expected [1]. A dedicated water drainage system made of boreholes drilled from the surface and/or dedicated drainage galleries installed around the cavern allows controlling the hydrostatic pressure and the ice formation in the rock mass during the cooling down process (see Figure 2). Process and equipment to operate the storage are similar to aboveground or in-ground tanks.

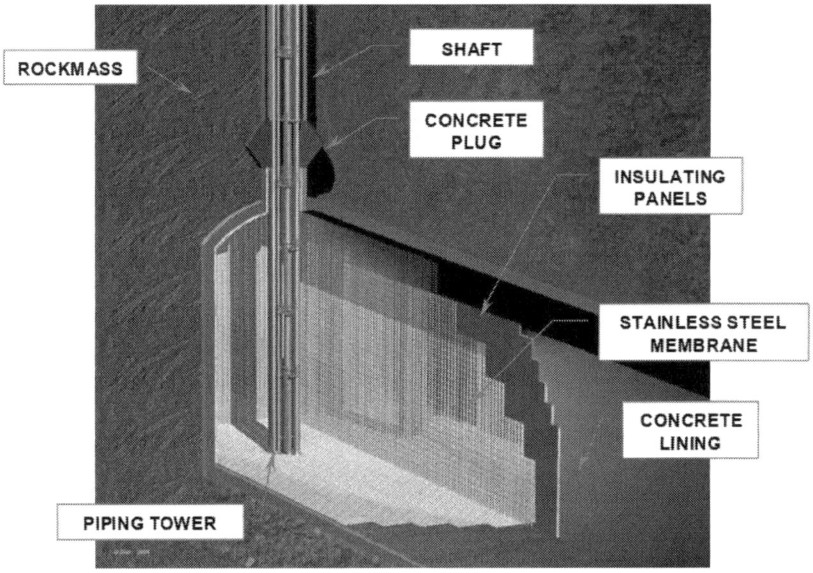

Fig. 1- Sectional View of LRC Containment System

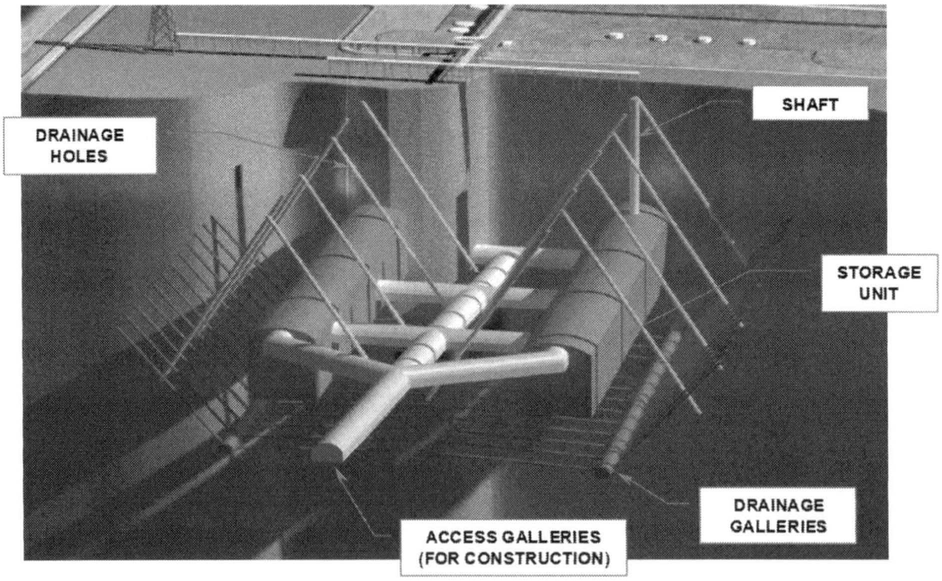

Fig. 2 - Main Components of the LRC Storage System

ADVANTAGES AND ECONOMICS OF UNDERGROUND STORAGE

There are many advantages of underground storage in terms of safety, security and environmental acceptability compared with aboveground tanks and in-ground tanks. Underground storage is much safer in consideration of fire on plant or decreased potential damages in case of industrial accident nearby. This is due to its multi-component barrier with liner and ice ring. It is less vulnerable to earthquake and typhoon. Regarding security aspects, underground storage can easily survive acts of sabotage or terrorism.

Because there is no need of large reclaimed areas and less earthworks at ground level, underground storage is environmentally friendlier, and eventually it will become a better acceptable proposition to people located nearby. The other major advantage is the minimum plot space requirement for an LNG terminal due to the fact that the LNG storage is about 50 m underground. This represents a huge cost saving especially in seashore areas where industries are already developed and the limited available real estate is very expensive. It is also the case in areas whose topography needs expensive reclaimed land.

Small galleries should be avoided wherever possible due to their poorer capacity/area ratio. Geometrical studies show that underground storage in the form of a gallery of around 20 m width by 30 m height cross section is the most favorable in terms of cost versus rock behavior [2]. Moreover, mining technologies and membrane containment system have such flexibility that unit storage capacity has no limits. As Crude oil caverns are up to 4,500,000 m^3 which are operating in Korea, it is possible that volume of LNG lined caverns can also be designed to such capacities. The comparative cost estimate between aboveground and cavern storage is only the storage itself and its equipment. It does not take into account the substantial cost saving which could be made, in the case of the cavern storage, for the safety equipment (impounding basin, peripheral retention wall, fire fighting systems, etc.) and possibly for the reduction in piping length and terminal plot area. Moreover, reduction in operation costs, including maintenance cost is also would be attractive towards cavern storage.

In 2008, a national forum for cost comparison among conventional aboveground and in-ground LNG tank, and underground storage was held in Korea with the participation of Ministry of Knowledge and Economics (MKE), Korea Gas Corp. (KOGAS), Korea National Oil Corp. (KNOC), Korea Institute of Geosciences and Mineral Resources (KIGAM), experts of engineering consultants and reputed academicians steered by Congress committee. Costs of varying stored volume from 200,000 m^3 to 1,000,000 m^3 with increment of 200,000 m^3 were evaluated and compared with reference price as of March 2006 in Korea. In all cases, underground storage is the most economic over the stored volume of 300,000 m^3, and at 400,000 m^3, cost for underground storage can be economical by 8% compared to aboveground tank [3]. The relative relationship among storage types are illustrated in Figure 3.

Intrinsically, the underground storage is cheaper than in-ground storage tank. Moreover, operation cost for underground storage units are highly competitive as compared to aboveground and in-ground tanks as systems like slab heating or fire water are not necessary or can be tremendously reduced. Based on Korean reference which has been implemented on crude oil storage by Korea National Oil Company, operation cost of underground storage is 63% less than that of aboveground one [4].

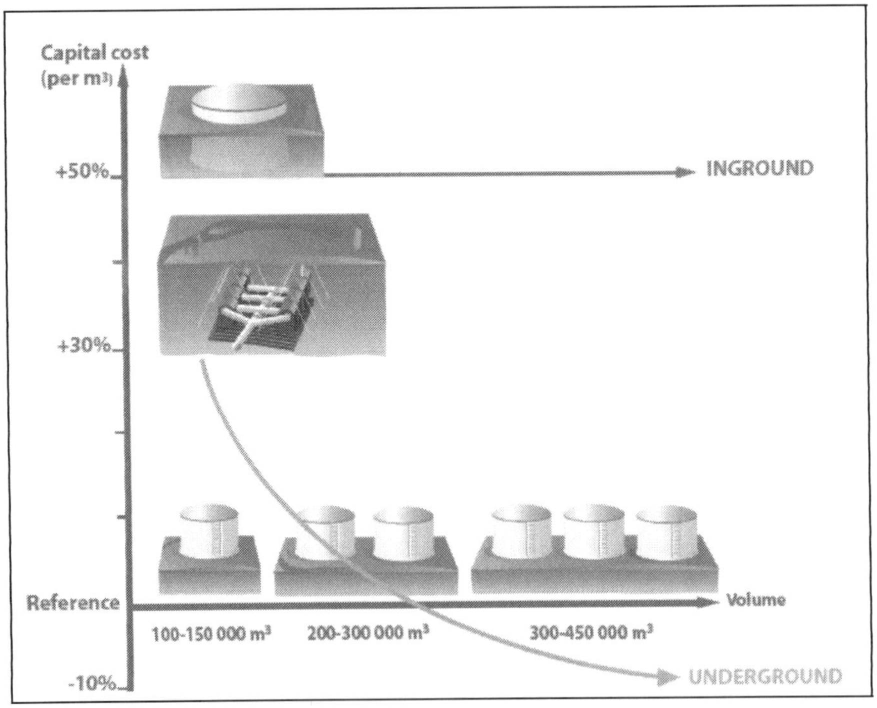

Fig. 3 - Comparison of Construction Cost

TERMINAL PROCESS AND MAIN FACILITIES

The process philosophy and the main equipment needed to operate the storage gallery and above ground facilities of the receiving terminal are discussed below. The cryogenic caverns and the above ground process shall fulfill the following basic function:

1) Cryogenic cavern

   A. Store the required Liquefied Natural Gas (LNG) volume in a safe and economical manner utilizing a dedicated containment system, designed to limit heat ingress into the storage

   B. Resist the loads generated by LNG, soil and seismic effects

   C. Control the conditions of different spaces by use of dedicated instruments

2) Above ground process

   A. Unload the LNG from carriers berthed at the jetty to fill the storage caverns

   B. Pump the LNG from the cavern to the above ground re-gasification process

   C. Vaporize the LNG and take it to the required send-out temperature

   D. Gather the Boil-off Gas (BOG) from the storage caverns and route it to the ship or recycle it into the process

   E. Provide the utilities required for the site operation

Terminal General Overview

LNG is transferred from the LNG carrier to the cryogenic caverns via unloading system with the use of the LNG carrier pumps. Unloaded LNG is stored in lined rock cryogenic caverns for an extended period of time. Each storage cavern is equipped with removable submerged LP pumps, which deliver LNG to vaporization system at the required send-out rate.

LNG from the storage caverns is routed to a recondenser vessel. It is then routed to high pressure (HP) send-out pumps that increase the LNG pressure up to the grid pressure. High pressure LNG is routed to vaporizers where LNG is heated and vaporized. Two types of vaporizers are installed in site: Fuel gas fired Submerged Combustion Vaporizer (SCV) and Open Rack Vaporizer (ORV), which derive the energy required for LNG vaporization from seawater.

BOG is naturally generated in the caverns due to heat ingress from environment and gas displacement during filling. The BOG is compressed and re-condensed into the LNG send-out stream in a recondenser vessel.

A flare or vent stack is provided to safely dispose of any emergency hydrocarbon release.

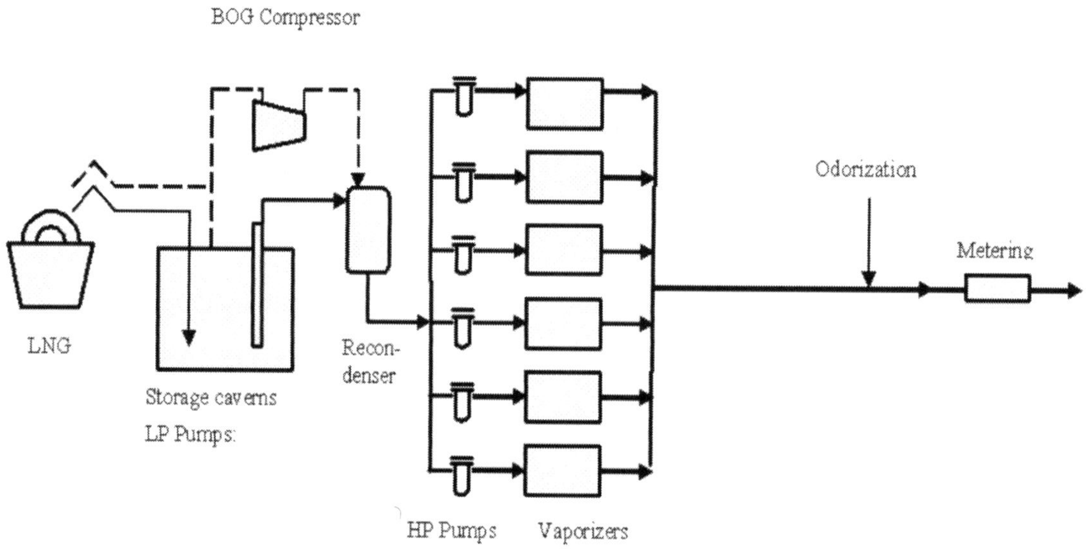

Fig. 4 - Process Flow Diagram (PFD)

HIGH OPERATING PRESSURE ALLOWS SITE FELXIBILITY

One of challenging issues on LNG storage is boil-off gas (BOG) from stored liquid. This is due to heat transfer into the storage tank from surroundings. Most of LNG tanks have a boil off rate in the range of 0.05 to 0.1%/day. The BOG rate of the LRC can also be achieved by a proper thickness of insulation panels.

However, operation of the storage system should consider not only normal BOG rate, but also the BOG generated during unloading operation. LNG will be considered saturation point when the LNG carrier is arriving at the port of the import terminal, LNG is pumped by the cargo pumps from the carrier to the onshore LNG tanks through the unloading lines.

Enthalpy of the transferring LNG increases because of heat gain from pipelines (mainly unloading line and other cargo and tankage area piping) and unloading arms. Pressurized LNG by the cargo pumps also increases its enthalpy. The increased enthalpy causes "Flash vapor" when LNG enters into the storage tanks.

Length of unloading line(s) is quite site specific. If the tide difference is considerable, the required length is long. In some cases port condition allow short jetty and trestle line, resulting in a short unloading line.

The possible location of the LRC may be near the shore or far which is depending on the available rock mass. It should be noted that with lengthy unloading lines there will be a significant amount of the heat leak into the flowing LNG.

During Unloading operation, the amount of BOG is governed by the following factors:

- Liquid displacement of unloaded LNG
- Flash Vapor
- General BOG because of heat leak through the tank roof and wall (bottom as well)
- Liquid pumping (negatively acting on the BOG rate)
- A sudden change of barometric pressure

If the flash vapor is well controlled, the amount of vapor generated during unloading operation can be significantly reduced. This can be the answer to why High operating pressure of the LRC can contribute in reducing the BOG handling system.

Maximum operating pressure and design pressure of LNG storage tanks is confined by tank type associated with its design. However, LRC can increase its design pressure up to 500 mbarg, which is the max design pressure specified in EN 14620 [5]. Table 1 summarizes the LNG tank operating pressure and design pressure.

Table 1 – LNG Tank Operating and Design Pressure

Tank Type	Operating Press	Design Pressure
Single Containment	50-75 mbarg	120-150 mbarg
Double Containment	50-75 mbarg	120-150 mbarg
Full Containment	250 mbarg	290 mbarg
Membrane Tank	170 - 250 mbarg	190 - 290 mbarg
Lined Rock Cavern	400 mbarg	500 mbarg

One of the advantages of the full containment system, which has a higher operating pressure than any other tank type (except the LRC) is to reduce BOG rate during unloading operation. As a result, the size of the BOG handling system, such as BOG compressors and the recondenser) can be significantly reduced.

As shown in Table 1, the LRC's high operating pressure can provide enough suppression pressure of flash vapor when LNG is entering into the tank. This benefit allows flexibility of the LRC site location. Our study reveals that about 10 – 12 km of LNG transfer lines does not impact BOG generation during unloading operation. The benefit can also reduce its construction cost because storage site flexibility can facilitate less expensive construction options and reduce construction infrastructure.

UNCONVENTIONAL UNLOADING LINES

When the possible LRC location is far from the shore line, the required distance of the unloading line may be considerable. Then the cost for construction of the unloading lines will be significantly high assuming that the unloading lines are built with the conventional methods: jetty and trestle, concrete support structure, steel structure lined with fire proof concrete, etc. The design of conventional unloading lines also requires consideration of LNG leaks along the lines. According to the international codes and standards generally applied to the import terminal facility design, impounding basin and associated leaked LNG path to this impounding basin are mandatory.

In order to provide high flexibility in site location of the LRC, unconventional unloading lines have been investigated. These include Vacuum Insulated Pipe (VIP) and Pipe-in-Pipe system (PIP)

Design Configuration of VIP

The configuration that is shown in Figure 5 depicts the simplest VIP configuration. The 16" LNG process pipe is jacketed with a 28" x 0.625" thick wall pipe that serves as both carrier pipe and vacuum insulation enclosure. There are two (2) major benefits in utilizing VIP™ over more traditional pipeline installations.

- The vacuum jacket / carrier pipe is combined with the process pipe. The factory assembled, insulated, and tested sections are sent to the field in 24.4 meter lengths ready for installation.
- The fact that the vacuum annulus of the 24.4 meter sections are isolated from each other gives this assembly a unique compartmentalization feature compared to previously designed subsea piping systems.

This feature, a fundamental part of the VIP™ design, shares the principle of compartmentalization which is common to marine vessels, confining potential damage to only a small section.

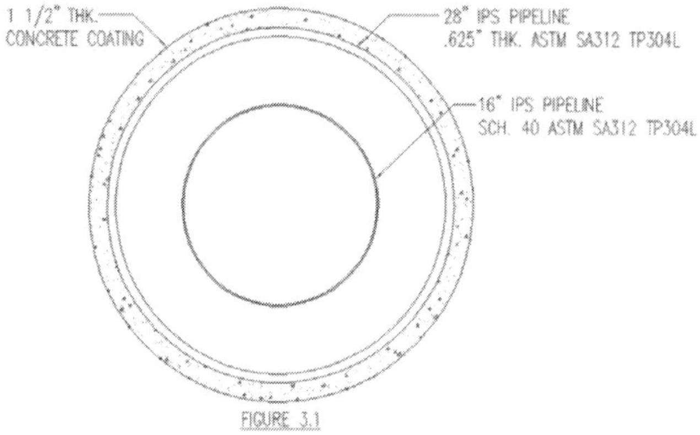

Fig. 5 - VIP™ with Combined Carrier and Vacuum Jacket

Design Configuration of PIP

The key technical features of cryogenic pipe-in-pipe systems which have been identified as being fundamental and governing the successful implementation of the concept are summarized below:

- The levels of thermal insulation required for the longer unloading lines are higher than those of the shorter above water conventional designs to limit undesirable flashing, or boil-off, to the same low levels. Heat transfer issues are therefore of fundamental importance to the design of cryogenic pipeline systems.
- The requirement for re-circulating the LNG during waiting periods between ship visits indicates provision of at least one other cryogenic line.
- All insulation materials and concepts available for cryogenic service require dry operating conditions. This dictates adoption of pipe-in-pipe configuration concepts with the provision of a steel outer pipe to provide a sealed annular environment by excluding seawater from contacting the annular insulation.
- All pipe-in-pipe designs require mechanically 'locking' or connecting the inner and outer pipes together with stiff bulkheads at least at the two ends of the pipelines, and also, sometimes at additional intermediate locations as well.

Figure 6 illustrates the simplest VIP configuration.

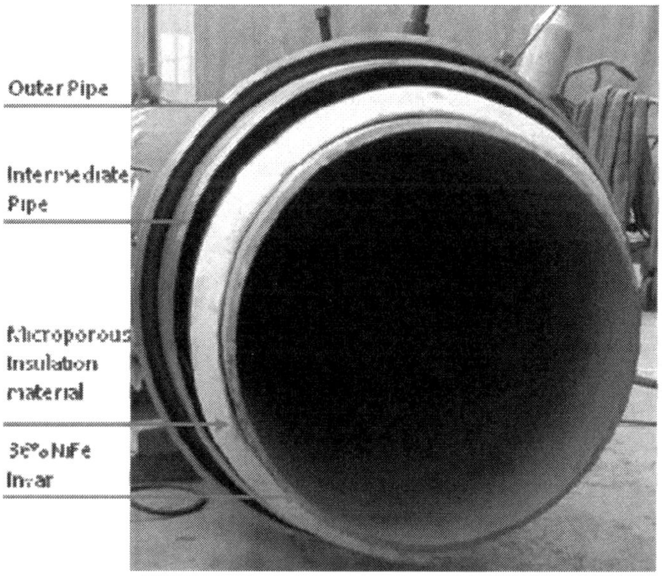

Fig. 6 – Simplest Configuration of PIP

Advantages of VIP and PIP

The advantages of long subsea cryogenic pipeline systems over that of conventional above water trestle based lines of the same length are:

- Logistics and viability
- Cost effectiveness
- Environmentally more friendly

The major advantages of using PIP or VIP to replace conventional stainless steel piping and insulation systems are:

- Reduced schedule for installing the fully insulated lines at site. Installing conventional polyurethane insulation systems require several steps (weld pipes, install special pipe supports, pressure test, paint, install insulation segments, install vapor barrier, install cladding, seal joints, etc.) which are labor intensive and involve several crafts. PIP and VIP eliminate many of these steps.
- Elimination of high density PUF pipe supports and other special supports required for conventionally insulated systems. These supports are often

difficult to install and prone to failure.

- Reduced cost due to the reduction in field labor.
- Lower heat gain and lower thermal mass compared to conventional insulation systems. This results in lower boil off which is particularly important during initial cool down of the facilities and in between ship loadings.
- PIP and VIP can be buried, and can provided an extra level of containment in sensitive areas such as road crossings, tunnels, culverts, or other areas where the public may come near.
- A much longer lifetime is expected for VIP and PIP, whereas conventional insulation systems degrade over time due to water ingress and other aging effects.

Disadvantages of VIP and PIP

The main disadvantages of replacing conventional stainless steel piping and insulation systems are:

- Limited experience in LNG applications for VIP (ALNG: 4" recirculation line, Darwin LNG: 30" and 24" for LNG and BOG line (7000-ft), Egypt LNG: 30" and 24" for LNG and BOG line, tank riser pipes and some of LNG run down pipes.
- Lack of similar design competition for either product. Both are essentially single source, although it may be possible to develop alternates without violating patents.

CONCLUSION

As compared with the conventional aboveground and in-ground storage tanks, the use of LRC LNG storage system at the LNG terminals can be more economical in terms of CAPEX and OPEX. In addition, it has also the advantage of safety, security and environmental acceptability, compared to the conventional tanks.

Lined Rock Cavern LNG storage system can be realized in due course at some countries which have suffered from the shortage of storage capacity of LNG and seasonal extreme variation of domestic demand, and where industries are already developed and remaining vacant areas are small and expensive.

High operating pressure of the LRC LNG storage system provides benefits to allow long unloading lines, which will be flexible in selecting the LRC project site. High operating pressure benefits to reduce BOG rate during unloading operation, and thus reduces Capital expenditure and Operating Cost by reducing BOG handling facility.

Lined Rock Cavern LNG storage system could be a great candidate where security of the energy facility is the national top priority because its storage system is totally located in safe underground below 30 – 50 m from the ground.

Since the LRC is an intrinsically safe LNG storage system, it can mitigate the public's concerns on facility's safety and have less local opposition to the project development.

REFERENCE

1. SKEC, Geostock, and Saipem: "Taean LNG Receiving Terminal Project Pre-Feasibility Study Report", 2007.
2. H.Y. Kim, S.W. Woo, D.H. Lee, J. Cho, "Economical and Technical Challenges in Lined Rock Cavern LNG Storage System", AIChE Spring Meeting, Tempa, USA, 2009.
3. SKEC, Geostock, and Saipem: *"Proceedings of International Symposium on LNG Storage in Line Rock Caverns"*, Seoul, Korea, 2004.
4. S.K. Chung, E.S. Park, K.C. Han, "Feasibility study of underground LNG storage system in rock cavern", presented at the*11th ACUUS Conference*, Athens, Greece, 2007.
5. European Norm 14620, "Design and manufacture of site built, vertical, cylindrical, flat-bottomed steel tanks for the storage of refrigerated, liquefied gases with operating temperatures between 0 °C and -165 °C", 2006.

Development of Technologies and Standards for the Life Time Management of LNG Storage Tanks

Seung Rim Lee, Principal Researcher
Institute of Gas Safety R&D
Korea Gas Safety Corporation, Seoul, Korea

Jang Sik Park, Chief Researcher
Institute of Gas Safety R&D
Korea Gas Safety Corporation, Seoul, Korea

Jeong Rock Kwon, Chief Researcher
Institute of Gas Safety R&D
Korea Gas Safety Corporation, Seoul, Korea

2010 AIChE Spring National Meeting
T6 Topical 6: 10th Topical Conference on Natural Gas Utilization
San Antonio, Texas, March 21-25, 2010

ABSTRACT

As the expiration date of design life (about 25 years) approaches for domestic LNG storage tanks, the issue of tank life time extension is paid more and more attention by Korea LNG industry. Recently, Korea Gas Safety Corporation is carrying out a project to develop technologies and standards for the life time management of LNG storage tanks by benchmarking those of domestic nuclear power plants. This paper summarizes the status of inspection and monitoring system for LNG storage tanks in Korea and the development course of technologies and standards for the life time management in the near future, including periodic inspection, periodic safety review and license renewal.

1. INTRODUCTION

Korea commissioned its LNG receiving terminal in 1986 and since than successfully operating the terminals over 24 years. However, among the LNG storage tanks in operation, some are aging and their life time management technology is in need of being developed to continuously operate them without compromising on their safety aspect. In addition, for the extended operation of super- sized LNG storage tanks it is required to develop and apply technologies to efficiently monitor and assess their service life time.

Regarding this issue Korea is aiming to develop new technologies for life cycle assessment and monitoring LNG storage tanks so as to consolidate relevant laws and regulations. Since systematic life time management technologies have already been developed for nuclear and petrochemical industries, similar technologies and systems applicable to new as well as aging LNG storage tanks can be developed based on these existing technologies and consultations.

2. LIFE TIME MANAGEMENT SYSTEMS BY INDUSTRIES IN KOREA

In order to develop life cycle assessment technologies for LNG storage tanks, it is necessary to investigate and analyze the life cycle assessment examples in nuclear and petrochemical industries. Especially, in the case of nuclear industry, periodic inspection, periodic safety review (PSR) and continual operation assessment systems have been systematically established and efficient life time management is carried out by means of publicly-available deterioration data and specifications for life cycle assessment of IAEA.

The systems, specifications and application examples of life cycle assessment of nuclear power plants and petrochemical plants are summarized in [Table 1]. Especially, PSR and continual operation assessment systems have been introduced and are being applied for life time management in nuclear power plants. And, in petrochemical plants, fitness-for–service conformity assessment systems have been established and are in use for management of equipment service life by means of API 579 (Fitness-for-Service) for pressure vessels, storage tanks, etc. in those plants. Meanwhile, there is no standardized national or international code available for LNG storage tanks up to now, and their life cycle assessment and life time extension are conducted by means of the in-house standards of tank operating companies. However, International Gas Union (IGU) recently organized a subcommittee in charge of the aging of LNG receiving terminals and liquefaction plants, and this study is actively in progress.

[Table 1] Status of Systems and Codes for Life Cycle Assessment of Facilities by Industries

Item	System	Related Code	Application Example
Nuclear Power Plants	Periodic Safety Review (PSR)	IAEA NS-G-2.10	All countries (except U.S.A. and Canada)
	Operation License Renewal System (continual operation assessment)	10CFR54 NUREG-1800, 1801 IAEA TEDOSs	U.S.A, Canada, Spain, etc.

Petrochemical Plants	FFS(FFP)	API 579 BS 7910, etc.	U.S.A., Japan, Korea, etc. (pressure vessels, piping, storage tanks in petrochemical industry)
LNG Plants	Life cycle assessment standards of tank operators	API579,ASMESec.Ⅷ Div.2, Nuclear power standards	France (GdF), Japan (TGC), etc.
Remarks	- IAEA NS-G-2.10(Periodic Safety Review of Nuclear Power Plants) - 10CFR54(license Renewal Application) - NUREG-1800(Standard Review Plan for Review of License Renewal Applications for Nuclear Power Plants) - U.S. NRC NUREG-1801(Generic Aging Lessons Learned (GALL) Report) - API 579(Fitness-for-Service)		

3. EXAMPLES OF OVERSEAS LIFE CYCLE ASSESSMENT OF LNG STORAGE TANKS

3.1 Tokyo Gas Company, Japan

Presently, Tokyo Gas Company is operating a total of 29 in-ground storage tanks, of which the oldest was constructed about 30 years ago. The aging problem of the major structural elements such as roofs and membranes has not known yet, but any damage due to fatigue or other reasons may have critical effect on their operation and safety. In order to inspect and diagnose these items by means of traditional technologies, an overhaul shutdown for an extended period of time is required. This will result in large amounts of expenses due to operational loss. To avoid the financial loss Tokyo Gas Company has developed a technology to inspect and diagnose LNG storage tanks without their shutdown and is practically applying it to successfully.

Tokyo Gas Corporation stipulated the aging phenomena of aboveground and in-ground storage tanks based on the inspection, diagnosis and life cycle assessment of the tanks which were put into operation in Negishi Terminal and Sodegaura Terminal in 1960s and 1970s (See [Table 2]).

[Table 2] Example of Life Cycle Assessment for LNG Storage Tanks in Japan (Tokyo Gas Corp.)

Classification	Aging Phenomenon	Measure and Treatment
Aboveground LNG Storage Tanks	Corrosion in each part	Inspection & diagnosis: Repairs
	Corrosion in piles	Inspection & diagnosis: Free of problem
	Aging of breathing tank diaphragm	Inspection & diagnosis: Replacement
In-ground LNG Storage Tanks	Fatigue in roof	Inspection & diagnosis: Free of problem
	Fatigue in membrane	Inspection & diagnosis: Free of problem
	Corrosion in each part	Inspection & diagnosis: Repairs

It was difficult to assess the aging extent of the fillet welds inside the steel roofs of the storage tanks, and Tokyo Gas Corporation established the technical means of analysis and inspection to diagnose the aging of such parts and practically applied them to the storage tanks.

The diagnosis of the roof aging was conducted in the following order: (1) Selection of parts to be diagnosed on the basis of the analysis results of fatigue crack growth, (2) Inspection of storage tanks, (3) Assessment (diagnosis of remaining service life time).

Through the above analysis, crack measurement inspection for specific parts was conducted by means of ultrasonic tests. Remaining service life time was assessed from the measured values of the cracks detected in the actual facilities and the measured values judged to be problematic on the basis of their analysis. The assessment finally concluded that there would be no possibility of roof damage due to fatigue even if the storage tanks would be operated in the present way for the next 50 years. In addition, submersible observation units were developed, and the interiors of the tanks were inspected and it was confirmed that there was no problem inside the tanks.

3.2 GdF Suez, France

GdF, which has become the fourth largest energy enterprise in the world through its recent merger with Suez, conducted life cycle assessment and life time extension through the interior-opening inspection after shutdown of a total of 4 storage tanks, including two 120,000 kℓ membrane tanks and one 9% Ni steel tank of the same capacity in Montoir Terminal and one 80,000 kℓ 9% Ni steel tank in Fos-sur-Mer terminal. The following [Table 3] shows the status of the life cycle assessment object tanks of GdF Suez.

[Table 3] Object Tanks of Life Cycle Assessment in France (GdF SUEZ)

Terminal	Aboveground Tank		Tank Type/Builder	Start of Operation (Year)
	No.	Cap. (㎥)		
Fos sur Mer	1	80,000	9%Ni / PC-CMP	1978
Montoir de Bretagne	2 1	120,000 120,000	Membrane / PC-TechniGaz/CMP 9%Ni/PC-CMP	1980/81 1983

At first, it was noticed in 2003 that there were some problems in two membrane type storage tanks in Montour Terminal as noises were generated at their primary pumps, and it was decided to open and inspect the tanks. The investigation into the integrity of concrete was started in 2006 and the inner steel tanks were inspected by in-house technical staff, while the life time of outer concrete tanks was diagnosed by a specialist consulting company (Oxand SA). The integrity of inner membranes was inspected by checking their configuration and dimensions, and insulation materials by confirming that there was no leakage in the inner insulation space (IBS) in their operation history. GdF Suez repaired some minor parts of the inner tanks such as ladders, etc. and put the four storage tanks back into operation.

4. DEVELOPMENT COURSE OF LIFE CYCLE ASSESSMENT TECHNOLOGY FOR LNG STORAGE TANKS IN KOREA

The life cycle assessment elements for LNG storage tanks include low-temperature thermal fatigue in their inner tanks, fatigue in their roof welds, loss of pre-stress and durability of concrete in outer concrete tanks, corrosion of piles and corrosion of tendons. The performance deterioration of insulation materials is also estimated to be one of major factors. The life cycle assessment elements expected by the components of LNG storage tanks are summarized in the following [Table 4].

[Table 4] Core Assessment Elements of Life Cycle Assessment for LNG Storage Tanks

Classification	Life cycle assessment Element	Description
Inner Steel Tank	Low-temperature thermal fatigue in inner tank	• 9% Ni steel / stainless steel membrane
	Fatigue in roof welds	• Dissimilar metal welds, roof fillet welds
	Fatigue in pump columns	• Pump column support structure / welds
Outer Concrete Tank	Loss of pre-stress	• Loss of pre-stress
	Durability of concrete	• Concrete strength • Covering depth • Chlorine ion concentration and salt damage diffusion factor (chloride content test, etc.) • Rebar corrosion • Creep and contraction of concrete • Carbonation
	Corrosion in piles	• Corrosion in piles
	Corrosion in tendons	• Corrosion in tendons
Insulation Materials	Deterioration of insulation performance and strength of bearing insulation materials	• PUF, cellular glass, etc.
	Deterioration of insulation performance of non- bearing insulation materials	• Perlite, glass wool, etc.

Development of life cycle assessment for LNG storage tanks will be conducted for a total of 5 years. In the first stage (April, 2009 – April, 2010), the development course of life cycle assessment technology for domestic LNG storage tanks will be established through the survey and analysis of life cycle assessment technologies.

In the second stage (April, 2010 – April, 2012), which is technology development stage, operation characteristics and deterioration mechanism will be closely examined for each tank member, and life cycle assessment and monitoring technologies will be developed for storage tanks. In the last third stage (April, 2012 – June, 2016), which is the practical utilization stage of life cycle assessment technology for LNG storage tanks, the program of and standard for life cycle assessment will be developed and they will be reflected in their final system by conducting supplementation and verification through their field application (See Figure 1).

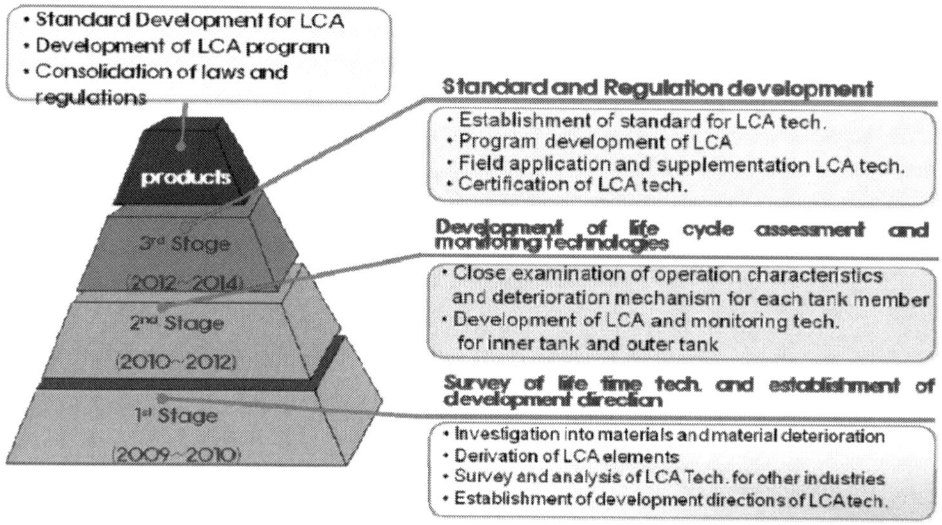

[Figure 1] Development Outline of LCA for LNG Storage Tanks

5. CONCLUSION

LNG storage tanks are parts of expensive core facilities in liquefaction plants and LNG receiving terminals. And, they are following the trend for being super-sized in step with recent enlargement of LNG carriers. While reliability standard of such important facilities is being maintained, the target to lower the operating cost through extension of their service life time should be set up, and development and standardization of life cycle assessment and monitoring technologies should be started.

The life cycle assessment technologies for LNG storage tanks can be largely divided into concrete structure technology and steel structure technology. Through the combination of the life cycle assessment technology for concrete structures in nuclear power plants and that for inner steel tanks based on API 579(Fitness for Service) and ASME codes which have been already developed and being applied in petrochemical and refinery industries, development of life cycle assessment and monitoring technologies for LNG storage tanks are to be promoted. In addition, in order to minimize economic loss due to the shutdown, the life cycle assessment and monitoring technologies should be developed based on in-service diagnosis.

On the basis of such technology development, the competitive power of super-sized LNG storage tanks to advance into overseas markets may be upgraded, and their operation efficiency can be enhanced through the life time extension based on the safety of LNG storage tanks in domestic operation.

REFERENCES

1. "Cases of Diagnoses and Life Extension Measures of LNG Facilities at Tokyo Gas Negishi and Sodegaura Receiving Terminals", Tokyo Gas Co., Ltd, LNG 14, 2004
2. "Fitness for Service and Residual Life Assessment of LNG Storage Tanks", AEA Technology, 2004
3. "Aging of installations at LNG Terminals", Takehiko Hasegawa, 22nd WGC, 2003
4. "API RP 579(Fitness for service)", API, 2000
5. "Fatigue crack growth characteristics of 9% Ni Steel for LNG tank at low temperature", Shim Kyue Taek, Kim Jae Hoon, Kim Young Kyun, Korean Society of Mechanical Engineers, Spring Papers, 2009
6. "Korean Regulatory Approach to Aging Management of NPPs", KINS, 2003
7. "Development of Deterioration Management System for Concrete Structures in Nuclear Power Plants", Jo Myeong Seok, Bang Gi Seong, Song Yeong Cheol, Korean Concrete Institute, Vol.8 No.2, pp. 546~550, 1996

AUTHOR BIOGRAPHY

Name: Seung Rim Lee
Position: Principal Researcher
Company: Institute of Gas Safety R&D, Korea Gas Safety Corporation
332-1, Deaya-dong, Siheung-si, Gyeonggi-do, 429-712, Korea
Phone: +82.31.310.1462, Fax: +82.31.315.4363
E-mail: srlee@kgs.or.kr
Cellphone : +82.10.5290.6668

Seung-Rim Lee is a Principal Researcher in the Institute of Gas Safety R&D at Korea Gas Safety Corporation. He is a registered professional engineer and has been in LNG industry for fifteen years. He has been involved in research and development projects related to domestic standardization and risk assessment in LNG industry. Especially, he has been participating in development of Korean Standards and international standards (ISO/TC67/WG10) related with LNG storage tanks and LNG terminals from 1999. He graduated from Seoul National University of Technology with a Ph.D. degree in gas engineering in 2009.

DESIGN FEATURE OF LARGE FULL CONTAINMENT LNG STORAGE TANK

Se-Jin Jeon, Ph.D., Principal Researcher, DAEWOO Institute of Construction Technology, Suwon-si, Gyeonggi-do, Korea
Young-Jin Kim, Ph.D., Research Fellow, DAEWOO Institute of Construction Technology, Suwon-si, Gyeonggi-do, Korea
Byeong-Moo Jin, Ph.D., Principal Researcher, DAEWOO Institute of Construction Technology, Suwon-si, Gyeonggi-do, Korea
Eui-Seung Park, Senior Executive Vice President, DAEWOO E&C, Seoul, Korea
Hong-Sung Kim, General Manager, DAEWOO E&C, Seoul, Korea

Abstract

A number of notable features are addressed that may be encountered in designing a large aboveground full containment LNG storage tank. The size and thickness of the inner tank, which is made of 9% Ni, is closely related to the specifications of design codes. The recent revisions of the relevant specifications and their implications toward a larger-sized tank are investigated in this study. In addition, some improvements are addressed that can be made in designing a large outer tank made of reinforced or prestressed concrete. A sample design of the world's largest LNG tank, with 250,000 m^3 capacity, is proposed through this study.

Introduction

The construction cost of the LNG tank occupies a major portion (approximately 30~40%) of the LNG terminal's total construction cost. Therefore, various efforts have been made to increase the cost efficiency of the LNG tanks. This can be largely divided into two parts:

(1) There has been a clear trend toward LNG tanks with a higher capacity, during the past decades. Through the capacity expansion, the cost per unit capacity can be reduced and the land be used more efficiently.[1,2] The trend also matches well with the increasing capacity of LNG ships.

(2) The cost efficiency as well as the shortening of the construction period has been pursued through the technological development in material and structural aspects.[3]

Table 1 shows the current record of the capacity in each type of LNG tank. 200,000 m^3 capacity has been already attempted for all types. As the capacity increases, special attention should be paid to the existing code provisions. Also, an efficient procedure needs to be established for designing the large LNG tanks that provide greater structural and cost efficiency. This paper addresses a possibility and critical points in designing a larger LNG tank, e.g., with 250,000 m^3 capacity, than the current record. The study focuses on the aboveground full containment LNG tank with a concrete roof, which is the most popular type nowadays.

Table 1. LNG storage tanks with the largest capacity

Type		Capacity (m^3)	Location
Aboveground	Concrete roof type	200,000	Tongyeong (Korea) Pyeongtaek (Korea)
	Steel roof type	200,000	Darwin (Australia)
Inground		200,000	Negishi (Japan) Incheon (Korea)
Underground		200,000	Ohgisima (Japan)

Design of 9% Ni Inner Tank

The need for revision of some code provisions has been raised, especially for the inner tank, in order to design an aboveground full containment LNG tank with a higher capacity in an efficient and economical way.[1,4]

The most commonly referenced codes for the design of inner tanks were API 620[5] and BS 7777.[6] Also, many of the domestic constructions in Japan have been based on Japanese own codes.[7] Some revisions were made in PD 7777[8] (an amendment to BS 7777) against several strict specifications of BS 7777. Recently, prEN 265002[9] provided a background for BS EN 14620[10] that is the latest Eurocode. BS EN 14620 reflected several important points required to design more reasonable dimensions for an inner tank with a higher capacity. It has now become the most general and dominant code worldwide. Table 2 compares allowable stresses, thicknesses and heights of hydrostatic test taking into account three major codes mentioned above.

Table 2. Comparison of code specifications for 9% Ni inner tank

Item		API 620	BS 7777	BS EN 14620
Allowable stress (MPa)	In service	The lesser of: $0.33f_u$, $0.67f_y$	The lesser of: $0.43f_u$, $0.67f_y$, 260	The lesser of: $0.43f_u$, $0.67f_y$, 260
	During hydrostatic test	The lesser of: $0.55f_u$, $0.85f_y$	The lesser of: $0.85f_y$, 340	The lesser of: $0.60f_u$, $0.85f_y$, 340
Maximum thickness (mm)		50	30	50
Minimum thickness (mm)		10 (For D > 67 m)	12 (For D $\geq$ 50 m)	10 (For D > 60 m)
Height of hydrostatic test (% of maximum design liquid level)	Water	60	100	60
	Conversion to LNG	125	208	125

Note 1: f_u is minimum ultimate tensile strength, f_y minimum yield strength, and D diameter of inner tank

Note 2: Specific gravity of LNG is 0.48

As shown in Table 2, included in BS EN 14620 are two noteworthy features, i.e., increased maximum thickness of the inner tank plate up to 50 mm and partial height hydrostatic testing (60% of maximum design liquid level). These features substantially relax some of the very conservative provisions specified in BS 7777, i.e., maximum thickness of 30 mm and full height hydrostatic testing (100% of maximum design liquid level). Through this, it is expected that the way toward higher capacities will be wide-open, keeping a reasonable proportion of the inner tank. BS EN 14620 reflects state-of-the-art manufacturing and welding techniques for the thicker 9% Ni plates. Given that the inner tank is designed to BS 7777, the shape gets flatter (height to diameter ratio gets smaller) as the capacity increases, resulting in an inefficient land use and an increase of the construction cost.

The height to diameter ratio of the inner tank has a primary significance since the overall proportion of the LNG tank is determined by the inner tank rather than outer tank in many cases. The following aspects can be noted in relation to the sizing matter. As for the inner tank, provided that the strength of the inner tank plate is given, the only load-resisting factor is the thickness of the plate. On the other hand, there are more load-resisting factors in the outer tank including concrete strength, thickness, and the amounts of reinforcing bars and prestressing tendons. When we take notice that both the inner and outer tanks should maintain the strength to independently resist design loadings according to a design concept of the full containment type tank, it is plausible that overall proportion of the tank is more dependent on the size of the inner tank which has less redundancies in the load-resisting mechanism.

Thus, it is recommended that in a preliminary design stage several alternatives of the inner tank dimensions (height, diameter, thickness, etc.) be compared together to select a best candidate that minimizes material cost of the outer tank as well as the inner tank.

Design of Outer Tank

Roof Dome

Many of the existing concrete roof domes of aboveground LNG tanks have a radius of curvature equal to d (diameter of wall). It corresponds to a rise to diameter ratio of 1/8, which is often recommended as an efficient structural shape for the domes when the self-weight or externally distributed load is dominant. On the other hand, a higher rise is advantageous when the dome is subjected to internal pressure, which is one of the main design loadings in LNG tank, since the higher curvature can resist the internal pressure with less tensile stresses.

Therefore, it can be mentioned that the conventional rise of the aboveground LNG tank domes is not optimal, at least from the structural aspects. When designing large LNG tanks with an increased dome span, it is recommendable that several alternatives of the dome dimensions (radius of curvature, thickness, etc.) be compared together, in order to determine the optimum design with respect to the structural safety and efficiency with a minimized amount of concrete.

No special code-related restriction is imposed on the shape of the dome, but the codes for the carbon steel liner that is attached inside the concrete dome should be followed. For example, API 650[11] specifies that the radius of curvature of the liner should range from 0.8d to 1.2d. Some of the large in-ground LNG tanks where the roof dome is exposed above the ground level have the radius of curvature close to 0.8d. Buckling safety of the carbon steel liner is also important, which varies sensitively depending on the shape and placing method of the concrete dome.[12]

Ring Beam

Ring beam is located at the joint of the roof and wall. The primary role of the ring beam is to cope with the major portion of the thrust transmitted from the roof, reducing excessive deformation of the upper part of the wall as well as the roof. Therefore, dimensions of the ring beam and the amount of prestressing tendons arranged have a close relationship with the shape of the roof. From a geometrical consideration, a higher rise dome induces less thrust to the ring beam and thus is advantageous both for the ring beam and the dome itself.

The construction sequence of roof placing and ring beam prestressing has a primary importance because the stress state of the ring beam is affected by the sequence. The procedure should be planned to avoid excessive compression or tension in the ring beam throughout all the phases.

Wall

The safety and serviceability of the wall should be checked in various stages such as construction, normal operation, and LNG leakage from the inner tank. It may be an important design issue to maintain a slender wall thickness even when the tank size gets bigger with the increase in wall height. To achieve this purpose, it is expected that the conventional design practice should be revisited and improvements be made as necessary. Some of the possible improvements are introduced below.

High strength concrete contributes to increasing the section strength, often resulting in a thinner section. For example, high strength of 60 MPa over the conventional 40 MPa was adopted for the wall of the LNG tank in Senboku, Japan. It is reported that the wall thickness was actually reduced by 30 cm (110 cm for normal strength and 80 cm for high strength) and the reduction of the number of steel piles was 4.4% through the strength increase.[3] Recently, there have been some LNG tank projects that required a more increased concrete strength than conventional one.

Prestressing tendons are usually provided in both the hoop and vertical directions to generate beneficial forces or stresses that can counteract the design loadings, such as a liquid pressure in case of the leakage of LNG from the inner tank. As the wall goes higher with the capacity increase, the conventional design practice of the hoop tendons may have a potential problem, such as congestion in arrangement at the lower part of the wall and excessive stresses at the joint of wall and bottom slab in the construction stage. A countermeasure against this matter was proposed in previous studies[13,14] by adjusting the spacing of the hoop tendons at the lower part of the wall through trial and error or optimization techniques. By adopting this strategy, a moderate magnitude of the stress can be maintained both under construction and in service, and the number of tendons required can be saved as well.

Design of an LNG Tank with 250,000 m^3 capacity

A sample design of the LNG tank with the world's largest 250,000 m^3 net capacity is proposed by reflecting the aforementioned issues, which results in the schematic sections of Fig. 1. Several notable features are addressed that were reflected in designing the largest LNG tank in a more efficient and economical way.

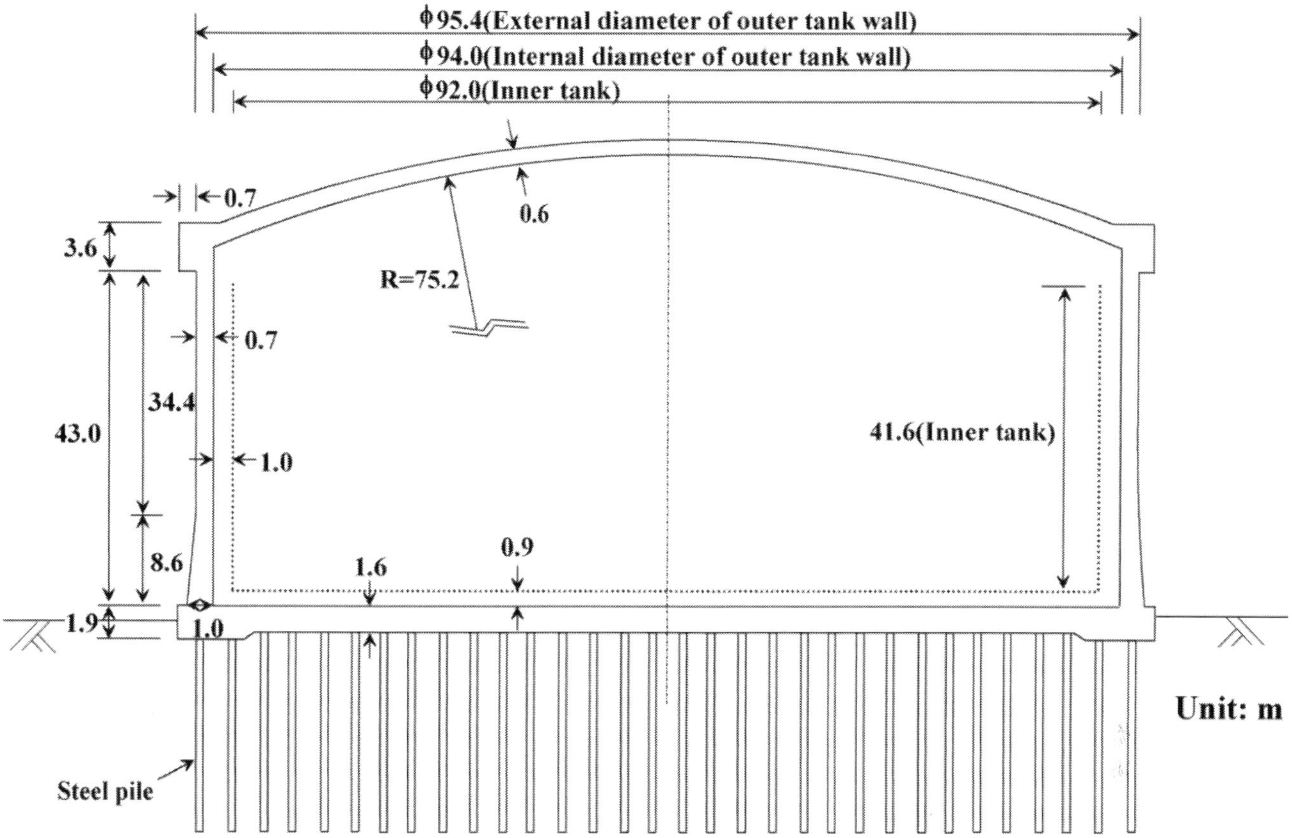

Fig. 1. Schematic dimensions of the LNG tank with 250,000 m^3 net capacity

9% Ni Inner Tank

The height to diameter ratio of the existing LNG tanks ranges from 0.35 to 0.60. As far as the quantity of 9% Ni is concerned, it is generally more economical for the ratio to be increased. Considering this aspect, along with the results of preliminary safety analyses of the outer tank, the ratio was set to 0.45. The additional conditions are as follows.

- Net capacity between maximum operating liquid level and minimum liquid level: 250,000 m^3
- Minimum liquid level: 2.0 m
- Distance between maximum design liquid level and maximum operating liquid level: 0.3 m
- Sloshing height by the earthquake: 1.0 m
- Free board: 0.5 m
- Ultimate tensile strength (f_u) of 9% Ni: 95,000 psi (654.6 MPa)
- Yield strength (f_y) of 9% Ni: 65,000 psi (447.9 MPa)
- Design code: BS EN 14620

The calculated height and diameter of the inner tank are 41.6 m and 92.0 m, respectively, as Fig. 1 shows. Thickness of each wall plate made of 9% Ni along the height of the Inner tank wall can be determined by considering the hoop stress of the cylindrical shell induced by the liquid pressure having a triangular distribution. The resulting thicknesses of the wall plate vary from 33.4 mm (the lowest wall plate) to 10 mm (upper wall plates). Table 3 compares the required maximum thicknesses of the wall plate calculated by each code when the proportion of the inner tank is the same as Fig. 1. In BS 7777, the maximum thickness is determined, not by the service stage but by the hydrostatic test, due to the severe specification for the test mentioned in Table 2. Table 3 shows that a reasonable proportion of the inner tank cannot be achieved when applying BS 7777. On the other hand, the latest BS EN 14620, adopted in this design, resulted in a minimized thickness.

Table 3. Maximum thickness of inner tank wall plate

Item	API 620	BS 7777	BS EN 14620
Dominant stage	In service	During hydrostatic test	In service
Required maximum thickness (mm)	39.8	52.8	33.4
Allowable maximum thickness (mm)	50	30	50
Safety check	O.K	N.G	O.K

Outer Tank

The outer tank is designed to ensure the structural safety and serviceability. Limit state design is widely used, while ultimate strength design is still dominant in the U.S. Main loadings considered in a design of the outer tank are self-weight of the inner and outer tanks, insulation, suspended deck and structures on the roof, liquid pressure of LNG, gas pressure, seismic load, thermal load by cryogenic LNG and by fire, prestressing load, and wind load.

The most apparent feature of this outer tank design is the application of high strength concrete. Table 4 shows the compressive strength of concrete for each part of the LNG tank. The strength can be measured by a cylindrical or a cubic specimen depending on the code or practice of the country. The strength of the cylinder is about 20% less than that of the cube, and this difference of the strength should be considered if necessary.

Table 4. Comparison of concrete strength

Classification	Relevant part	Conventional strength (MPa)	Strength of this design (MPa)
RC (Reinforced Concrete)	Roof, bottom slab (central part)	30	50
PSC (PreStressed Concrete)	Wall, ring beam, bottom slab (annular part)	40	60

By applying the higher strength concrete than the conventional practice, the sections could be designed with relatively slender thicknesses, even though the capacity considerably increased. Comparing the concrete quantity of this design using the high strength with that using the conventional strength, the quantity is reduced by 11.5%. Moreover, the required number of steel piles is reduced by 17.3% due to the reduction of self-weight. The main cost efficiency comes from the reduction of material quantities. Another factor that is conceivable to reduce the cost is to accelerate the construction by applying an advanced construction method such as slip forming system improving the conventional climbing form system.

Conclusions

There has been a clear trend toward LNG tanks with a higher capacity during the past decades, pursuing cost efficiency. The trend has been supported by the revision of codes and by the technological development in design and in material. A number of issues were addressed that may be encountered in efficiently designing the inner and outer tanks of large aboveground full containment LNG tank. A sample design of the LNG tank with the world's largest 250,000 m^3 capacity is proposed by reflecting the issues considered in this study.

References

1. Long, B. (1998), "Bigger & Cheaper LNG Tanks? Overcoming the Obstacles Confronting Freestanding 9% Nickel Steel Tanks up to and beyond 200,000 m^3," *Proceedings of 12th International Conference & Exhibition on Liquefied Natural Gas*, pp. 5.6-1-5.6-28.
2. Goto, S., Kamiya, A. and Tajima, M. (2001), "LNG Technological Progress in Japan-Three Decades of Evolution," *Proceedings of 13th International Conference & Exhibition on Liquefied Natural Gas*, pp. PO-23.1-PO-23.9.
3. Nishizaki, T. et al. (2001), "Largest Aboveground PC LNG Storage Tank in the World, Incorporating the Latest Technology-Construction Cost Reduction and Shortening of Work Period by Employing New Construction Methods," *Proceedings of 13th International Conference & Exhibition on Liquefied Natural Gas*, pp. PS6-4.1-PS6-4.11.
4. Heiersted, R. S. et al. (2001), "Capacity and Technology for the Snøhvit LNG Plant," *Proceedings of 13th International Conference & Exhibition on Liquefied Natural Gas*, pp. PS7-Sp.1-PS7-Sp.7.
5. "*API 620: Design and Construction of Large, Welded, Low-Pressure Storage Tanks*," (1996), American Petroleum Institute.
6. "*BS 7777: Flat-Bottomed, Vertical, Cylindrical Storage Tanks for Low Temperature Service*," (1993), British Standards Institution.
7. "*Recommended Practice for LNG Aboveground Storage(RPAS)*," (1981), Japan Gas Association.
8. "*PD 7777: Alternative Steel Selection and Its Effect on Design and Testing of Tanks to BS 7777*," (2000), British Standards Institution.
9. "*prEN 265002: Specification for the Design, Construction and Installation of Site Built, Vertical, Cylindrical, Flat-Bottomed Steel Tanks for the Storage of Refrigerated, Liquefied Gases with Operating Temperatures between -5 ℃ and -165 ℃*," (2001), CEN/TC 265.
10. "*BS EN 14620: Design and Manufacture of Site Built, Vertical, Cylindrical, Flat-bottomed Steel Tanks for the Storage of Refrigerated, Liquefied Gases with Operating Temperatures between -5 ℃ and -165 ℃*," (2006), CEN/TC 265.
11. "*API 650: Welded Steel Tanks for Oil Storage*," (1998), American Petroleum Institute.

12. Suzuki, H. et al. (2001), "Design and Erection of Temporary Steel Roof for Under-Ground LNG Storage Tank," *Proceedings of 13th International Conference & Exhibition on Liquefied Natural Gas*, pp. PO-27.1-PO-27.10.

13. Jeon, S. J. and Park, E. S. (2004), "Toward a Design of Larger Above-ground LNG Tank," *LNG Journal*, March/April, pp. 24-26.

14. Jeon, S. J. et al. (2002), "Basic Design for Large Above-Ground Tank," *Proceedings of GASEX 2002*, pp. A.3.4.1-A.3.4.11.

Cooldown of Large Pipes During the Commissioning Phase of Cryogenic Facilities

Nicolas F. Ponchaut, Ph.D.
Harri K. Kytömaa, Ph.D., P.E.
Exponent, Inc.
Natick, MA

AIChE Spring National Meeting, March 2010
10th Topical Conference on Natural Gas Utilization
San Antonio, Texas, March 21-25, 2010

Abstract

During day to day operations of LNG receiving terminals, process pipes are at uniform cryogenic temperatures (around -260°F) and experience relatively low thermal stresses. However, since such facilities are built at ambient conditions, fast temperature changes and high local temperature differences will occur during the initial cooldown. Pipe design and its cooldown schedule therefore need particular attention to ensure pipe integrity and to minimize leaks.

During cooldown, pipes experience large tensile stresses when they first encounter LNG. As a bigger fraction of their circumference is cooled down, pipes will tend to deform and move. Pipe motion can result in large forces that must be minimized by designing support strategies that can accommodate such pipe motion.

This paper addresses the magnitude of the stresses that can be expected during a cooldown for different pipe sizes. Cooldown strategies such as shock cooling with cryogenic liquid and precooling with vapor are also discussed.

Introduction

Most materials shrink when they are cooled down. When a steel pipe of finite length is free to move, it becomes shorter as its temperature is uniformly decreased, and as long as the temperature in the pipe is uniform, this thermal contraction does not produce any thermal stresses.

However, if the pipe ends are constrained, tensile stresses will be generated when the pipe is cooled down. These thermally induced stresses increase as the pipe temperature drops. Similarly, if a pipe cools down unevenly around the pipe circumference, warm portions of the pipe wall prevent cold portions from shrinking freely. As a result, large thermal stresses can be generated in both warm and cold regions.

During facility cooldown, the pipes are brought from ambient temperatures down to operating temperatures (typically -260°F for LNG facilities). In practice, the first introduction of LNG in the pipes creates significant temperature differentials (see Figure 1). As the LNG liquid front travels along the pipe, the bottom "cold" portion of the pipe that comes in contact with LNG cools down much faster than the top "warm" portion of the pipe in contact with only vapor. Large temperature differentials within the pipe cross-section form and can lead to substantial thermal stresses, both tensile and compressive. A carefully planned cooldown procedure is therefore critical to ensure a safe facility commissioning, including pre-cooling strategies with either LNG or LN_2 vapors.

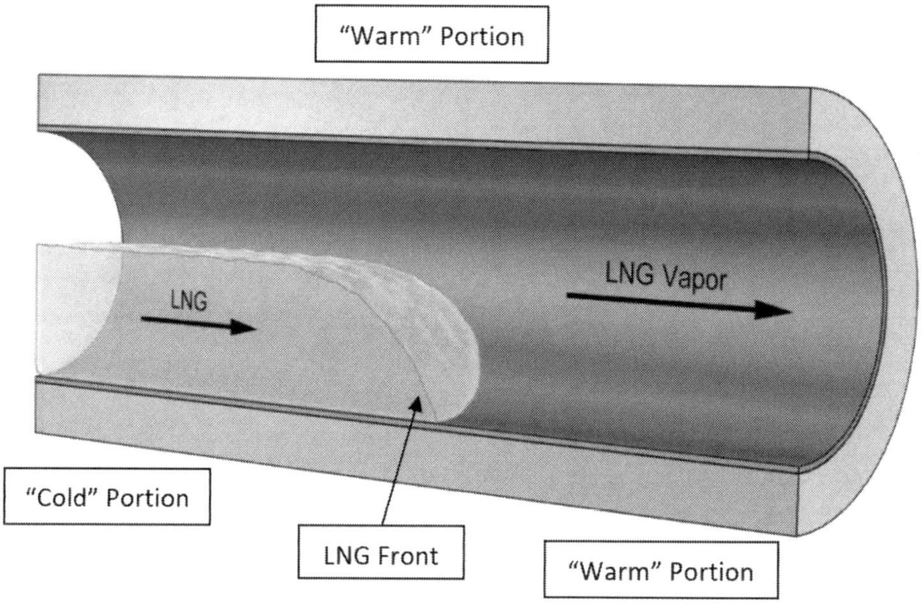

Figure 1: LNG front arrival in a pipe.

Thermal Stresses in Pipes

When LNG is first introduced in the pipes, the wet portion of the pipe cools down first. When the LNG front moves in the pipe, it first wets a small fraction of the bottom of the pipe, and rapidly cools down this wetted surface. When the wetted portion of the pipe is at the six o'clock position and only makes up a small fraction of the circumference, the contraction of the cold wall will be resisted by the warmer relatively uniform temperature portion. A consequence of the small fraction that is cold is that the pipe will not bend right away and the local temperature drop at the bottom of the pipe leads to a maximum local thermal tensile stress that is given by:

$$\sigma_{max} = E\,\alpha\,(T_0 - T_{LNG}) \tag{1}$$

where E is Young's modulus, α is the thermal expansion coefficient, and T_0 and T_{LNG} are the initial pipe temperature and the LNG temperature, respectively. This therefore leads to tensile stresses around 28,400 psi for a Stainless Steel 304L pipe at a temperature, T_0, of -160°F prior to LNG front arrival (100°F above the LNG temperature). At low temperatures, the yield strength of Stainless Steel 304L is around

32,500 psi (Shibata, et al. 2008). With a temperature differential of 100°F, thermal stresses are therefore close to reaching the yield strength of the material.

Table 1: Maximum predicted stress when LNG is introduced in a pipe at an initial temperatures of 60°F, 32°F, -10°F, -60°F, and -160°F. Only the initial temperature of -160°F will lead to thermal stresses below yield.

Pipe initial temperature, T_0	60°F	32°F	-10°F	-60°F	-160°F
Top to bottom temp. differential	320°F	292°F	250°F	200°F	100°F
Maximum tensile stress, σ_{max}	90,700 psi	82,800 psi	70,900 psi	56,700 psi	28,400 psi
% of yield stress	279%	255%	218%	174%	87%

As the LNG level rises in the pipe, a growing portion of the circumference cools down and the tensile stress at the bottom of the pipe decreases accordingly.

Equation 1 is therefore the maximum tensile thermal stress that can be encountered by a pipe during cooldown. This stress is independent of pipe size and length, of the supports, and of the facility layout but does not take into consideration over-constrained conditions on more complex pipe configurations.

The larger the initial fraction of the inner pipe circumference that is wetted, the smaller will the highest tensile stress be. Also, unwetted parts of the pipe wall will cool by conduction, which can result in smaller tensile stresses, particularly with small diameter pipes.

Temperature Profiles

Pipe stress analysis software products such as CAEPIPE,[1] CAESAR II[2] and others consider pipes as beam elements and are designed to quantify wall stresses based on the pipe weight and how it is supported. These software tools also quantify thermal stresses by assigning a uniform temperature to entire pipe segments. These pipe structural analysis software packages are not designed to accommodate complex temperature profiles across the pipe cross-section such as step wise temperature distributions or time varying conditions experienced during the cooldown of LNG terminals.

In the section below, we quantify the thermal stresses associated with a stepwise and with a linear pipe temperature distribution.

<u>Stepwise distribution</u>

In the stepwise temperature profile, the wetted portion of the pipe perimeter is at LNG temperature while the rest of the pipe wall is at the temperature that preceded the arrival of liquid. Figure 2 illustrates this temperature distribution. For a pipe that is allowed to expand or contract freely in its

[1] http://www.sstusa.com

[2] http://www.coade.com

axial direction, but that is not allowed to bow by virtue of regularly spaced supports, the thermal stresses are given by:

$$\sigma_{bottom} = E\ \alpha\ (T_0 - T_{LNG})\left(1 - \frac{\theta_0}{\pi}\right) \qquad \text{(tension)} \qquad (2)$$

$$\sigma_{top} = -E\ \alpha\ (T_0 - T_{LNG})\frac{\theta_0}{\pi} \qquad \text{(compression)} \qquad (3)$$

where θ_0 is defined by the LNG level as shown in Figure 2. When the liquid front first arrives and cools down only a narrow portion of the floor of the pipe (*i.e.* small θ_0), Equations 2 reduces to Equation 1.

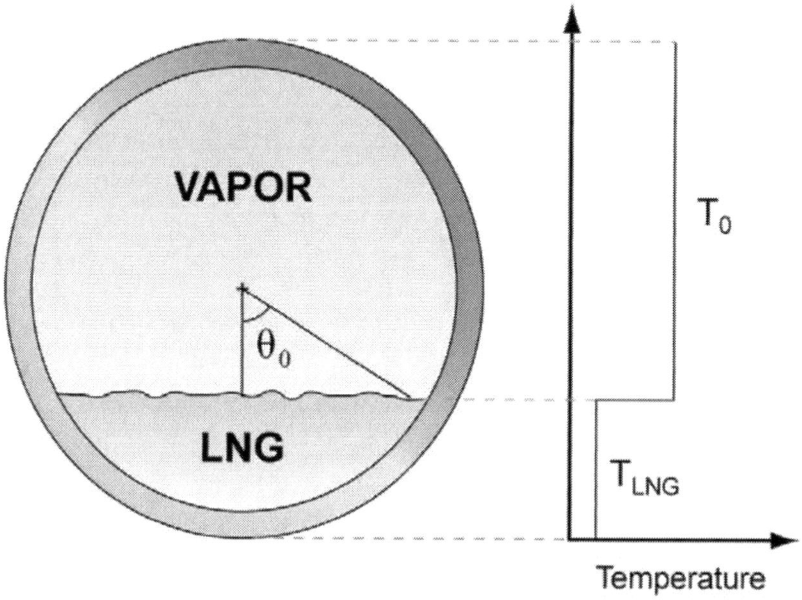

Figure 2: Stepwise temperature distribution in a pipe cross-section.

Linear distribution

Under certain circumstances where thermal conduction in the pipe smoothes out the temperature distribution, the temperature profile assumes more of a linear character, and temperature gradually changes from LNG temperature at the bottom of the pipe to the pipe temperature prior to the LNG front arrival (T_0), at the top as illustrated in Figure 3. For a pipe that is allowed to expand or contract freely in its axial direction, but that is not allowed to bow by virtue of regularly spaced supports, the thermal stresses are given by

$$\sigma(y) = -E\ \alpha\ (T_0 - T_{LNG})\ \frac{y}{D} \qquad (4)$$

where y is the height above the pipe equator and D is the pipe diameter. In this case, the thermal stresses vary linearly with the elevation, y, and are only indirectly dependent on the LNG level inside the pipe. The maximum tensile stress (at $y = -D/2$) is exactly half of the value in Equation 1.

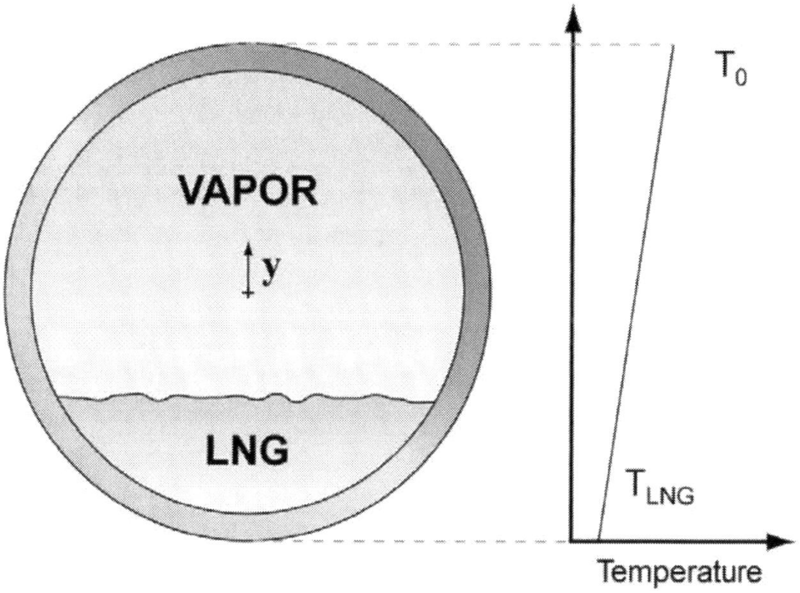

Figure 3: Linear temperature distribution in a pipe cross-section.

It is useful to recognize that these results are independent of the pipe support spacing and of the pipe thickness as long as the pipe is prevented from bowing or lifting off at supports. However, these two temperature distributions lead to significantly different results as seen in Figure 4 and it is important to choose the appropriate temperature distribution within the pipe when performing the stress analysis.

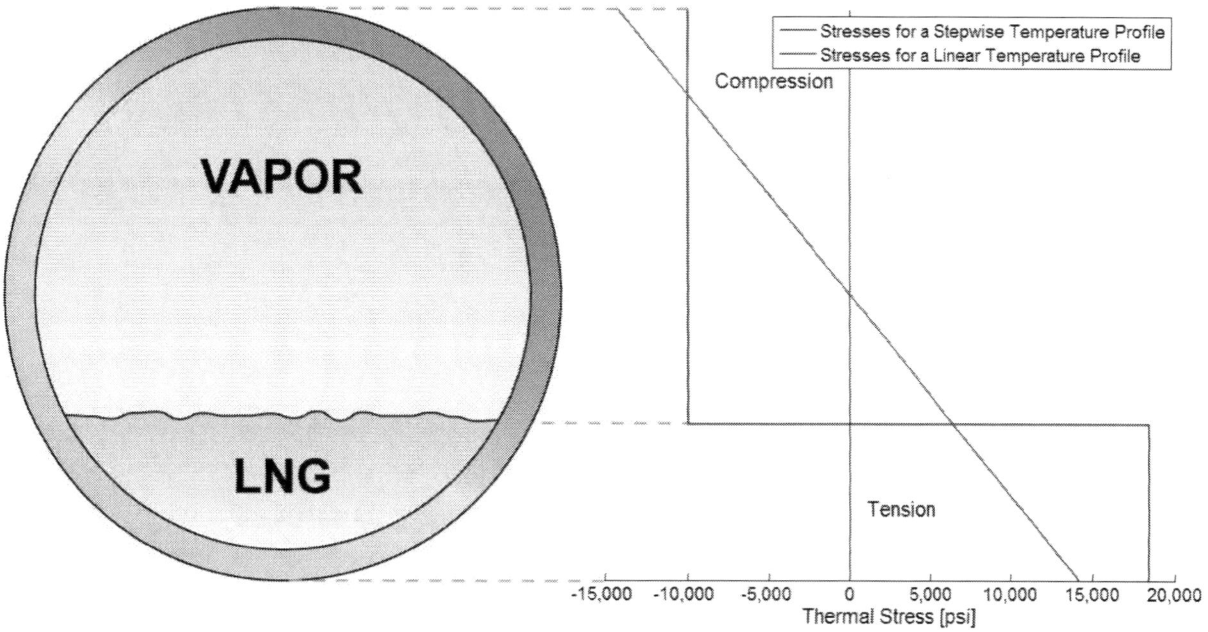

Figure 4: Stress distribution in a pipe with a top temperature of -160°F (100°F higher than the LNG temperature), and a bottom temperature of -260°F.

Large versus Small Pipes

When large pipes (typically over 24-inch diameter) are carefully cooled down with LNG, the liquid behind the wetting front typically flow along the bottom of the pipe cross section. The temperature of the bottom wetted portion of the pipe drops first. Conduction around the circumference will slowly cool the unwetted portion of the pipe wall as well. However, this process is slow particularly for large pipes. A stepwise temperature profile therefore best represents the period after the first arrival of the LNG front in large pipes.

In small pipes, the conduction within the pipe circumference will be faster in cooling the unwetted pipe wall. A stepwise temperature profile would tend to over-predict the thermally induced stresses, and a linear temperature profile is often more appropriate.

What differentiates the large pipes from the small pipes is the characteristic time associated with heat transfer along the circumference of the pipe. Table 2 summarizes the circumferential heat transfer characteristic times for various SS 304L pipe sizes. The characteristic time is the measure of how long it takes for the cold pipe bottom (-260°F at the 6 o'clock position) to have an influence on the unwetted initially warm (60°F) pipe wall at the 12 o'clock position.

Table 2: Characteristic times associated with circumferential conduction in 304L Stainless Steel[3]

Pipe Diameter	Characteristic Time
¼ in	2 sec
1 in	30 sec
4 in	8 min
8 in	35 min
12 in	1 hr 20 min
24 in	5 hr 15 min
36 in	11 hr 45 min

For 36-inch pipe, the characteristic time associated with circumferential conduction is on the order of 12 hours. This means that circumferential conduction will not tend to smooth out the stepwise temperature profile. However, if the pipe diameter is reduced, the characteristic time is reduced as well, and the circumferential conduction plays a more important role. For a ¼-inch diameter pipe, the characteristic time associated with circumferential conduction is on the order of 2 seconds, which

[3] The characteristic time is defined as $\left(\frac{\pi D}{4}\right)^2 \rho\, c/(4\, k)$, where ρ, c, and k are the density, heat capacity and conductivity of 304L Stainless Steel.

means that temperature differentials between the top and the bottom of the pipes are reduced very quickly and a stepwise temperature profile is not appropriate. For example, in the calculations presented by Liao *et al.* for a pipe diameter of 13.5 mm, the characteristic time is 9 seconds (Liao, et al. 2005). The temperature profile within the pipe cross-section is therefore reported to be nearly linear.

Unconstrained and Cantilevered Pipe

When the pipe is allowed to bow freely in response to the temperature profile within the pipe, as is the case with the cantilevered end of a pipe that is only supported at one end, the bowing deformation will relieve some of its thermal stresses, resulting in lesser stresses than the ones experienced by constrained pipes. For an unconstrained pipe, a stepwise temperature distribution leads to the following thermal stress at the bottom and at the top of the pipe:

$$\sigma(y)_{bottom} = E\,\alpha\,(T_0 - T_{LNG})\left(1 - \frac{\theta_0}{\pi} + \frac{4\,y}{\pi\,D}\sin\theta_0\right) \tag{5}$$

$$\sigma(y)_{top} = E\,\alpha\,(T_0 - T_{LNG})\left(-\frac{\theta_0}{\pi} + \frac{4\,y}{\pi\,D}\sin\theta_0\right) \tag{6}$$

Note that when the liquid front first arrives and cools down only a small portion of the pipe perimeter (very small value of θ_0), Equations 1 and 5 are equivalent.

However, in the case of a linear temperature profile with the same top to bottom temperature differential, the pipe will bend in such a way that all the thermal stresses are relieved and there are no resulting thermal stresses (*i.e.* $\sigma = 0$). These results are illustrated in Figure 5.

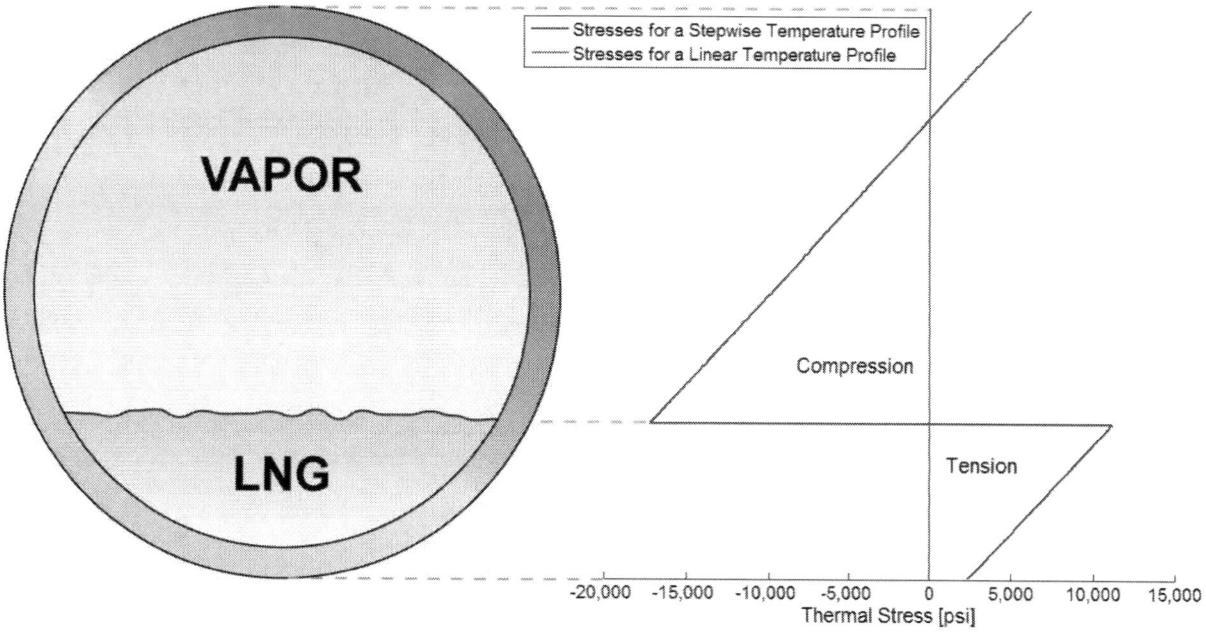

Figure 5: Stress distribution in an unconstrained or cantilevered pipe with an initial temperature of -160°F (100°F higher than the LNG temperature).

There is a fundamental difference in the stresses associated with a stepwise temperature distribution as compared to a linear temperature distribution. It is therefore critical to consider the temperature distribution so that thermal stresses are not underestimated.

Pipe and LNG Weight Influence on the Pipe Stresses

The equations introduced above only present the thermal induced stresses. In particular, they do not include the stresses resulting from the weight of the LNG and of the pipe itself. Furthermore, they do not include the thermal stresses associated with axial pipe contraction when it is constrained, since pipe supports are normally designed to allow axial motion. LNG receiving terminals, like most facilities dealing with high temperature differences make use of expansion loops to reduce and minimize thermal stresses.

During the cooldown, the weight and the reduction of the pipe length are expected to contribute very little to the stress state in the pipes. For a 304L Stainless Steel, 36-inch pipe (⅜-inch thick, 9.5 inch of foam glass insulation), supported every 25 feet and with an initial temperature of -160°F, when LNG fills half the pipe cross-section, the maximum thermal stresses are on the order of 14,200 psi while the weight contribution is only about 800 psi. In other words, the weight contributes to only 5.6% of the total pipe stresses during cooldown.

Pipe Thickness Influence

All the results presented above are independent of the pipe thickness. In practice, as long as the pipe walls are relatively thin, their thickness has very limited influence on the thermally induced stresses.

The results presented here are for uniform temperature through the pipe wall. For this to be true, the thickness must be such that the temperature equilibrates quickly across the wall. Table 3 summarizes the characteristic time associated with heat conduction across the wall of a 304L Stainless Steel pipe of various thicknesses.

These characteristic times are relatively small. And for a 36-inch pipe with a ⅜-inch wall thickness, the temperature becomes uniform across the wall in much less than a minute. During the commissioning phase of an LNG receiving terminal, pipes can remain partially filled for hours and the wall temperature is essentially constant across its thickness.

Table 3: Characteristic times associated with conduction across the walls of 304L Stainless Steel pipes[4]

Pipe Thickness	Characteristic Time
⅛ in	0.8 sec
¼ in	3.5 sec
⅜ in	7.5 sec
½ in	13 sec
¾ in	30 sec
1 in	53 sec

Vapor Precooling followed by LNG

The initial pipe thermal stresses during cooldown are influenced by the pipe temperature before the first arrival of LNG, T_0, by the Young's modulus, E, and by the thermal expansion coefficient, α, of the pipe material.

One strategy to reduce thermally induced tensile stresses involves precooling the pipes with either Liquid Nitrogen or with LNG vapors prior to the introduction of LNG. The advantage of using vapor in the initial stages of the cooldown is that both the top and bottom of the pipe will cool down uniformly. The pipes can therefore be cooled to temperatures below -160°F without introducing any local thermal stresses related to top to bottom temperature differentials. Vapors have lower heat capacity than liquids, which can make the cooling process slow. However, the heat transfer from vapors can be increased using higher vapor flow rates.

Once the vapor precooling phase is completed, it is important that LNG be introduced with as little delay as possible to avoid pipe reheating. Pipes in LNG receiving terminals are typically well insulated, but even with good insulation, pipes cooled down to -160°F will typically reheat at a rate around 3 to 10°F per hour. According to Equation (1), a 10°F pipe temperature increase results in an additional 2,850 psi of tensile stress when the LNG front arrives. It is therefore critical to minimize the time spent between cooling phases.

Once LNG is introduced in the pipes, it will create a liquid front. As the LNG travels through the pipes, a portion of the liquid vaporizes and flows ahead of the liquid front. This vaporized natural gas will assist in the cooling process if the vapor flow rate is sufficiently large. A high LNG flow rate has therefore several benefits. First, it reduces the time it takes for the liquid front to reach every part of the facility,

[4] In this case, the characteristic time is defined as $(w^2\rho\, c)/(4\, k)$, where w is the pipe thickness, and where ρ, c, and k are the density, heat capacity and conductivity of 304L Stainless Steel.

thereby minimizing the reheating duration. In addition, the vapor flow rate ahead of the liquid front assists in cooling down the pipes even more prior to the LNG front arrival.

References

Liao, Jun, Kun Yuan, Renwei Mei, James F. Klausner, and Jacob Chung. "Cryogenic Chilldown Model for Stratified Flow inside a Pipe." *Proceedings of HT2005. ASME Summer Heat Transfer Conference.* San Francisco, California, USA, 2005.

Shibata, K., T. Ogata, A. Nyilas, T. Yuri, H. Fujii, and S. Ohmiya. "Comparison of Strength and Serration at Cryogenic Temperatures among 304L, 316L and 310S Steels." *Advances in Cryogenic Engineering Materials: Transactions of the International Cryogenic Materials Conference, ICMC* 54 (2008).

Biography of the Speaker

Dr. Harri Kytömaa is Corporate Vice President and the Director of the Thermal Science Practice at Exponent. He specializes in mechanical engineering and the analysis of thermo-fluid phenomena, and performs research in cryogenic fluid mechanics, heat transfer and phase change phenomena. He has worked on numerous LNG projects with the industry in the US and internationally and has published extensively in this area. He is currently a member of the ISO TC67, WG10 working group that is developing a guidance document on the major hazards associated with the planning and design of onshore LNG facilities and associated marine activities. Dr. Kytömaa has held several positions, including that of Associate Professor of Mechanical Engineering at the Massachusetts Institute of Technology.

(39g) Marine Boil off Gas Recovery Improvement by Performing Slow Loading at Bontang LNG Terminal

Johan Anindito Indriawan, Technical Department, Production Planning Section, PT Badak NGL Plant Site, Timur, Indonesia

Introduction

During LNG loading operation at Bontang LNG Terminal, the LNG vapor called Marine Boil Off Gas (MBOG) is always generated. Ideally, this vapor should be recovered to shore as an additional fuel gas supply or to be re-Liquefied. Nevertheless, due to some constraints, only around 23-37% of the total generated MBOG during LNG loading can be recovered while the remaining is flared.

Conclusions

By performing loading at lower rate, it is expected that the MBOG generation rate will be decreased proportionally, hence reduce the flaring rate. A series of slow loading tests confirmed that the recovery rate increases to 48-69% of total generated MBOG. It will potentially save around 0.15 std. cargoes of LNG per year or equivalent with USD 2.8 million per year

References Cited 1. Rustandi, Hendi Y, 2007, "Feasibility Study of Marine Boil off Gas Recovery Improvement", PT Badak NGL. 2. Crane Engineering Department, 1988, "Flow of Fluid through Valves, Fittings, and Pipes", Crane.

See more of this Session: Professor Cedomir M. Sliepcevich Memorial Session Four - LNG Storage
See more of this Group/Topical: Topical 6: 10th Topical Conference on Gas Utilization

PROCESSES TO PRODUCE VALUE ADDED PRODUCTS FROM CO_2

D. Ferdous
B. Demirel

Rentech Inc., Commerce City, CO 80022

Biography-Deena Ferdous

Deena Ferdous obtained her bachelor degree in Chemical Engineering in 1993 from Bangladesh University of Engineering and Technology. She worked as a Research Engineer for three years at Bangladesh Council of Scientific and Industrial Research, Dhaka. She moved to Canada in 1998 to pursue higher studies and completed her masters and PhD degrees from the University of Saskatchewan in 2000 and 2003, respectively. She is currently working as a Research and Development Engineer for Rentech, Inc. Rentech is involved in the development of process and catalyst for the production of liquid fuel from synthesis gas. Her main responsibility at Rentech is the development of highly active and attrition resistant catalyst for the commercial Fischer-Tropsch plant.

Abstract

The possible ways to convert CO_2 to valuable products are (1) catalytic conversion; (2) electrocatalytic/electrochemical process; (3) plasma processes; (4) photocatalytic /photochemical process; and (5) enzymatic/biochemical process. Hydrocarbon, hydrogen and oxygenates are mainly produced by the catalytic conversion of CO_2 at the temperature and pressure of 250-1000 °C and 140-3500 psi, respectively. On the other hand, carbon monoxide, syngas, hydrocarbons and alcohols are the main products produced at low pressure and high temperature from plasma process. CO_2 is converted to hydrogen and hydrocarbons by photocatalytic/ photochemical process at the temperature and pressure of 5-400 °C and 485-1,470 psi, respectively. Electrocatalytic/electrochemical process produces alcohol, aldehydes, ketones, other oxygenated, alkenes and alkanes at very low temperature (room – 150 °C). Furthermore, bi-carbonate and alcohols are also produced from CO_2 at low temperatures and pressures using enzymatic/biochemical process. Successful implementation of the above mentioned processes could be a possible solution to reduce the amount of CO_2 emission into the atmosphere.

1. Introduction

The rapid growth in world population and industrial development has lead to massively increased usage of fossil fuels such as coal, petroleum and natural gas which resulted in the formation of carbon dioxide. Furthermore, the dynamic equilibrium of carbon dioxide formation and conversion has been seriously destroyed due to the deforestation and the reduction of rain forest. DEO report indicated that the annual worldwide production of carbon dioxide is ~6 billion tons among these US produces ~ 2 billion tons that represent the energy content of about 11 million barrels of oil per day or roughly the US daily import.[1] Suggestion has been made to sequester carbon dioxide[2]. Moreover, the ability to store several billion tons of carbon emitted in the form of carbon dioxide is questionable as there are environmental consequences. Recycling of carbon dioxide via conversion into high energy content fuel is an alternative option. However this process is energy intense.[3]

The CO_2 reduction technology can be divided into two methods (1) physics-based; and (2) chemistry based. In physics-based methods, carbon dioxide is captured from the atmosphere and then stored underground or under the sea bed using high pressure compression. From the viewpoint of equilibrium and cyclic of carbon dioxide on earth, the amount of carbon dioxide has not been reduced.[4] Therefore, a significant effort is given to the use of chemistry–based methods to convert carbon dioxide to valuable products. This paper will discuss all possible routes of chemical conversion of CO_2 to valuable products.

2. Different Processes to Convert CO_2 to Value added products

The possible ways to convert CO_2 to valuable products are (1) catalytic conversion; (2) electrocatalytic/electrochemical process;(3) plasma processes (4) photocatalytic/photochemical process; (5) enzymatic/biochemical process.

2.1. Catalytic conversion

The carbon dioxide reforming of methane has received considerable attention from both academic and industrial sectors. This reaction converts the cheapest carbon containing gases (CH_4 and CO_2) to synthesis gas with a H_2/CO ratio of 1. This process is particularly interesting environmentally when the gas fields contain a significant amount of methane and carbon dioxide. During CO_2 methane reforming the following reaction occurs.[5]

$CH_4 + CO_2 \longleftrightarrow 2CO + 2H_2$ (carbon dioxide reforming) (1)

$H_2 + CO_2 \longleftrightarrow CO + H_2O$ (reverse water gas shift) (2)

$CO_2 + 4H_2 \longleftrightarrow CH_4 + 2H_2O$ (CO_2 methanation) (3)

$CO + 3H_2 \longleftrightarrow CH_4 + H_2O$ (CO methanation) (4)

$CH_4 + 2H_2O \longleftrightarrow CO + 3H_2$ (steam reforming) (5)

$2CO \longleftrightarrow C + CO_2$ (Boudouard reaction) (6)

$CH_4 \longleftrightarrow C + 2H_2$ (methane decomposition) (7)

$C + H_2O \longleftrightarrow CO + H_2$ (coke gasification) (8)

The main drawback of this process are[5] (1) it requires temperature as high as 800 °C for high conversion; and (2) catalytic deactivation due to carbon deposition. The two main reactions of coke formation are methane decomposition (Reaction 7) and the Boudouard reaction (Reaction 6). The first reaction is favored at high temperatures and low pressures, whereas the second one is favored at low temperatures and high pressures. Both low and high temperature steam reforming have been reported in the literature. Tomishige et al.[6] studied the performance of $Ni_{0.03}Mg_{0.97}O$ catalyst for dry reforming reaction (Reaction 1) at atmospheric pressure. This catalyst was very resistant to carbon deposition. On the hand, Song et al.[7] reported a decrease in CO_2 and methane conversion and H_2/CO ratio and increase in carbon deposition when dry reforming was performed at high pressure on Ni/Na-Y and Ni/Al_2O_3 catalysts. Similar trends were observed by Shamsi and Johnson[8] when the reaction was performed at high pressure over Pt/ZrO_2, $Pt/Ce\text{-}ZrO_2$ and Rh/Al_2O_3 catalysts.

Itkulova et al.[9] studied CO_2 reforming of methane to syngas over Co-Pd supported on aluminum catalyst and then syngas was converted to oxygenates. A two step process was used which was not very attractive from the economic point of view. The disadvantages of this process presently at the industrial scale is the utilization of extra energy in the form of hydrogen, which at industrial level is obtained by decomposition of natural gas or electrolysis of water and the use of high temperature and pressure. This makes the whole technology economically unfeasible for the industries.

It is a thought that the rapid CO_2-methanation could be very useful to convert inactive CO_2 to high energy products.[10] Rapid CO_2 methanation mainly occurs in the temperature range of 250-450 °C. This reaction is very exothermic. The heat generated from this reaction can be utilized further. On the other hand, the methane formed can be used as a fuel to obtain high temperature (~1000 °C). But the disadvantage of this process is that it requires large consumption of hydrogen per unit CO_2. $Ni\text{-}La_2O_3\text{-}Ru$ or $Ni\text{-}Ce_2O_3\text{-}Pt$ supported on alumina catalyst is mainly used for this type of reaction[11]. From the point of view of energy balance, methanol synthesis by CO_2 hydrogenation is regarded as the equivalent of the hydrocarbon synthesis from syngas. Methanol synthesis from CO_2 over heterogeneous catalyst mainly occurs according to the following equations.[12]

$$CO_2 + 3H_2 \longrightarrow CH_3OH + H_2O \qquad (9)$$

Conversion to methanol is favored by increasing the pressure and decreasing the temperature.[12] Inui and co-workers[11] performed significant study on the catalyst development for CO_2 hydrogenation. They found that the yield of CH_3OH is very low when conventional Cu-Zn catalyst was used. The reaction temperatures and pressures were as identical as the methanol synthesis from the syngas. However, a higher activity of the catalyst for methanol synthesis was obtained when $Cu\text{-}ZnO\text{-}Cr_2O_3\text{-}Ga_2O_3$ catalyst was used. This catalyst also retarded the formation of CO.

2.2. Electrochemical conversion

In the electrochemical process, chemical reactions take place in a solution at the interface of an electron conductor (the electrode, composed of a metal or a semiconductor) and an ionic conductor (the electrolyte), which involve electron transfer between the electrode and the electrolyte or species in solution. Electrochemical reduction of CO_2 appears to be one of the effective methods for conversion and fixation of CO_2[13]. This reaction mainly occurs at low temperature and high pressure. Some low temperature and low pressure electrocatalysis reaction was also reported.[14] In the past, the electrochemical reduction of CO_2 in water[15] has been reported. However, the product distribution of electro catalytic reaction is influenced by the nature of noble metal.

Methanol is considered as a much better solvent for CO_2 than water because the solubility of CO_2 in methanol is about five times or more than in water at ambient temperature and pressure[16]. This type of electrolyte is mainly used to produce formic acid. Mizuno and co workers[17] have investigated the electrochemical reduction of CO_2 on various metal electrodes using ethanol at ordinary temperature and pressure. In the electro reduction system using benzalkonium chloride/methanol electrolyte, no formic acid was produced on Ag, Au, Co, Cu, Fe, Ni, Pt, Sn, Ti and Zn. In potassium hydroxide/methanol electrolyte, the formation of formic acid was observed at Au, Cu, Pd, hydrogen storing Pd, Ti, and hydrogen storing Ti electrodes. This group observed that even under optimum conditions, the faradaic efficiency for formic acid formation was relatively low (21% on Cu electrode).

Koneco et al.[18] also studied the electrochemical reaction of CO_2 at a Cu electrode in a methyl based electrolyte using lithium supported salts as LiCl, LiBr, LiI, $LiClO_4$ and CH_3COOLi at low temperature (-30 °C). The main products were methane, ethylene, carbon monoxide and formic acid. A maximum faradic efficiency of methane was 71.8% in $LiClO_4$/methanol based electrolyte at -3.0 V vs Ag/AgCl saturated KCl. This group also studied the electrochemical reaction of CO_2 in water+methanol (8:2) with copper-electro-deposited nickel electrode. Under optimum experimental conditions, 20.2% methane Faradaic efficiency and 7.5% ethylene were obtained by electrochemical reduction of CO_2. A reaction mechanism of the electrochemical reduction of CO_2 at copper modified Ni electrode in water + methanol solution was proposed.

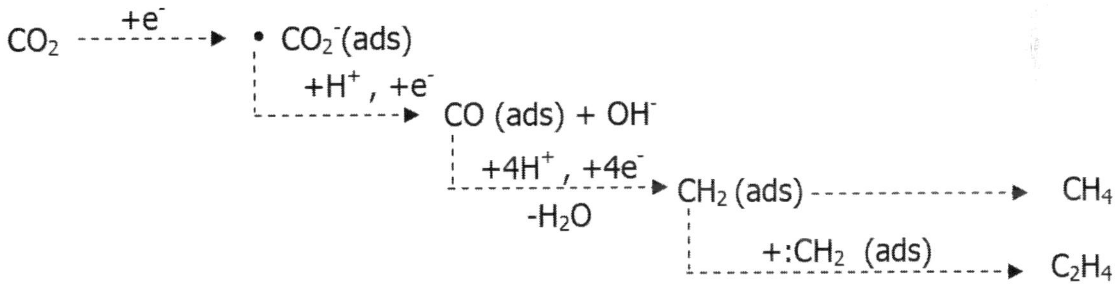

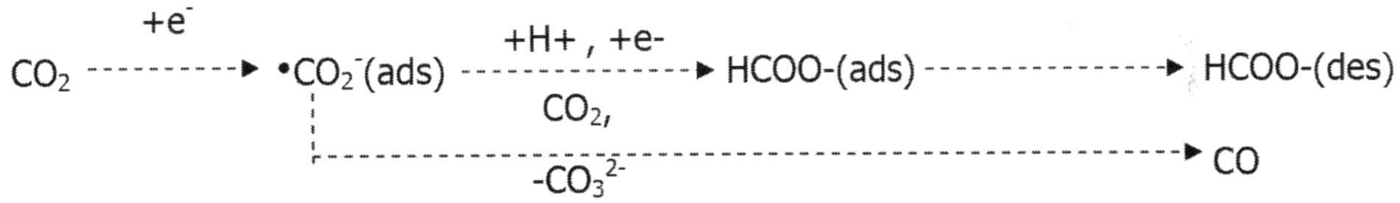

Ogura et al.[19] studied the electrolysis of CO_2 at the three phase interface on a copper mesh electrode in acidic halide solutions containing cupric ions. A selective formation of ethylene with a maximum Faradaic efficiency of 73.5% vs Ag/AgCl in a KBr solution was obtained. The probable reaction scheme is considered as: CO_2 (g) is first reduced to CO (g) at the three-phase interface, CO absorbs on a newly formed copper surface, and the electrochemical hydrogenation of copper to $:CH_2$ and the coupling of $:CH_2$ to C_2H_4 were followed.[20]

2.3. Plasma process

Yang et al.[4] proposed that a plasma technology could be more effective approach for the conversion carbon dioxide. Plasma process is a simple and fast process compared to other process. In the plasma process molecules are entered into electric fields where they are excited and ionized by collision with accelerated electrons to generate various species such as atoms, electrons, ions, free radicals etc. The mixtures of these species are plasma. There are two types of plasmas: low energy (1-20 eV) sometimes termed glow discharge, low temperature or cold plasmas and high temperature plasmas where the mean temperature of the gas can range from hundreds to millions of degrees. In the thermal plasma all the species, i.e free electrons, free radicals, neutral molecules and ions, are very hot (5,000-50,000 °C). On the other hand, in cold plasma only free electrons are hot, while heavier ions and neutral molecules remain close to room temperature. The activated species generated by plasma bombardments can recombine to form new products. The bond dissociation of CO_2 can be achieved at lower energy level in the plasma process.

Liu et al.[22] have proposed a non-thermal plasma approach for the formation of lower olefins from methane and CO_2:

$$2CO_2 + 2CH_4 \longrightarrow 2CO + C_2H_4 + 2H_2O \qquad (10)$$

At 800 °C, equilibrium yields of ethylene were fairly high at 57%.[18] However, the experimental yield over metal oxide catalysts was not sufficiently high (less than 9%). This group found that CO_2 plasmas can generate many active species. Some of them can be utilized to produce other valuable chemicals such as ethylene, propylene and oxygenates. Their experimental results showed that CO_2 is an excellent "catalyst" for the conversion of low alkanes to alkenes (especially ethylene and propylene).

Yun et al.[23] studied the thermal decomposition of CO_2. Their results showed higher conversion of CO_2 when it was reacted with other reactants such as CH_4 and H_2 compared to the direct decomposition of CO_2. The method of quenching or cooling the hot product stream in thermal plasma is often a vital factor to determine the final product distribution during thermal plasma studies. For many plasma reactions, the final product distribution depends on the rates of competing reactions during the quenching step. Therefore, the quenching technology in controlling temperature is very important in the thermal plasma process. They have indicated that the quenching is not effective if the speed of quenching is less than 105 K/s.

Yang et al.[4] reported the formation of different organic products using plasma method. In this method, carbon dioxide and counterpart molecules were mixed in reaction chamber and the plasma excitation was utilized to trigger chemical reactions. Depending on the magnitude of the input power and molecular structure of counter parts, the final products were polymer, oligo-polymer or low molecular weight small molecules. They reported that the conversion of carbon dioxide and chemical structures of the resulting products are strongly dependent on the selection of the counterpart molecules. They claimed that through this plasma process carbon dioxide can be converted to useful materials such as plastic or fuels.

Among the effective plasma systems, dielectric barrier discharge (DBD) plasma which is characterized by the presence of one or more insulating layers in the current path has received significant attention. But the plasma discharge energy is still too high to meet economic requirements. Further improvements are needed to get a better understanding of the plasma kinetics, better discharge control, and the simultaneous use of catalysts.

2.4. Photoelectric process

Photocatalytic reaction is one of the most available reactions at low temperature, because it can promote thermodynamically unfavorable reaction under mild conditions.[24] Photocatalysis is the acceleration of a photo reaction in the presence of a catalyst. In the catalyzed photolysis, light is absorbed by an absorbed substrate. Generally, photon reaction is the primary source of energy for electron hole-pair formation at ambient temperature. In photo-generated catalysis the photocatalytic activity (PCA) depends on the ability of the catalyst to create electron-hole-pairs, which generate free radicals (hydroxyl radicals: ·OH) able to undergo secondary reactions. Photocatalysis occurs under relatively mild conditions with lower energy input, especially where the reaction is activated by solar energy or other easily obtained light source. The schematic of the photocatalysis reaction of CO_2 is given below.

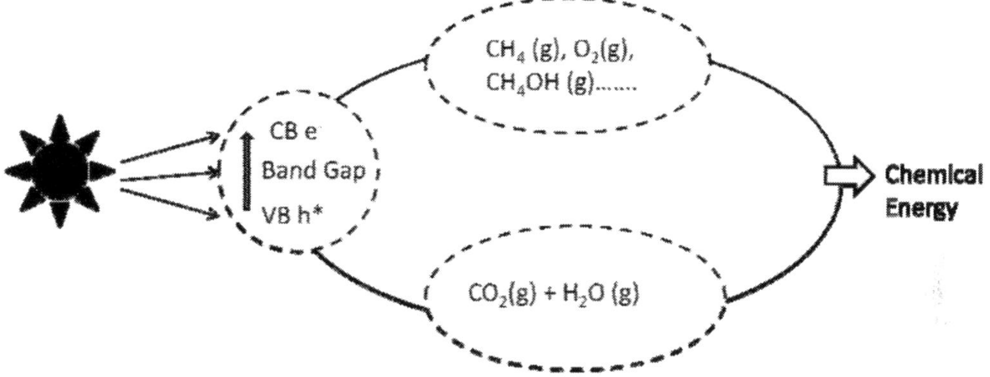

Figure 1. The schematic of the photo catalysis reaction of CO_2

Photocatalysis reduction of CO_2 was first demonstrated by Honda and co-worker[25] where they have reported to produce formaldehyde (HCHO), formic acid (HCOOH), methyl alcohol (CH_3OH) and trace of amount of methane from this reaction using different semiconductors such as tungsten trioxide (WO_3), titanium dioxide (TiO_2), zinc oxide (ZnO), cadmium sulfide (CdS), gallium phosphide (GaP) and silicon carbide. In their work, xenon- and mercury lamp irradiation was used to activate these semiconductors. Literature indicates that the product distribution of this reaction can be changed by using specific semiconductors.[26] The production of methane, C_2H_5OH and C_2H_6 were also reported in the literature.[26] Photoreduction of CO_2 can be performed using different reductant such as water, H_2, propanol, isopropyl alcohol, CH_4 etc dependent on the photocatalyst. The photoreduction of CO_2 by water is readily available and inexpensive. This process is also very promising and can form a useful carbon cycle. There are two important species such as H· (hydrogen atom) and $.CO_2^-$ (carbon dioxide anion radical) are mainly involved in CO_2 photoreduction. These species are produced by electron transfer according to the following reactions.[27]

$$H^+ + e^- \longrightarrow H^\cdot \quad (11)$$

$$CO_2 + e^- \longrightarrow {}^\cdot CO_2^- \quad (12)$$

In photogenerated catalysis, the photocatalytic activity (PCA) depends on the ability of the catalyst to create electron-hole-pairs, which generate free radicals (hydroxyl radicals: ·OH) which undergo secondary reactions. These radicals then form other stable substances. It should be noted that the stability of CO_2 in water is low and CO_2

photoreduction process competes with H_2 and H_2O_2 formation which consumes H^+ and e^-. Due to this limitation, water has been replaced by other reductants by some researchers in order to improve the desired product yields.

Both low[28] and high temperature[29] photo-catalytic reaction have been reported in the literature. Sometimes high temperature is used to increase the reaction rate by increasing collision frequency and diffusion rate. Saladin et al.[30] reported that high temperatures are beneficial to the rate of thermally activated steps such as desorption, which occurs after photochemical reaction. The maximum temperature in their work was 427 °C. On the other hand, Kohno et al.[31] noted that the photo excitation step of CO_2 with hydrogen over ZrO_2 can be deactivated with increased temperature. On the hand, Gokon et al.[29] indicated that the photo catalytic reaction cannot be induced when temperature is greater than 800 °C in the presence of ZnO due to the dominating effect of carbothermal reduction of ZnO. However, they have observed enhanced CO evaluation rate when the experiment was performed at 600 °C. Photocatalytic reaction of CO_2 at room temperature with gaseous water on powder TiO_2 was also reported.[32,33] But the CH_4 production was very low. Literature indicates that small TiO_2 particles have higher activity[29] and the highly dispersed TiO_2 exhibits a higher reductivity than bulk catalyst.[32] Kaneco et al.[14] stated that the formation of methane increased during photocatalysis reduction of CO_2 using TiO_2 suspension in isopropyl alcohol solution at 28 atm. On the other hand, Adachi et al.[34] indicated that ethylene cannot be produced at relatively high pressure because due to the accelerated formation of methane.

Photocatalysis is a potentially economical and environmental CO_2 removal process. It is very important to understand the parameters that influence the photocatalysis reaction to develop this for process for practical use. Improvement can be done by choosing semiconductor with appropriate band-gap energy, by developing reductants and by optimizing operating conditions such as temperature, pressure, light intensity and operating wavelength.

2.5. Enzymatic/biochemical process

Recently biocatalytic reduction of carbon has been gained significant attention because it requires mild reaction conditions and does not require high purity reactant. Biocatalytic reaction is capable to handle many impurities such as sulfur compounds which are common in flue gas. Biochemical reduction of carbon dioxide has been used to produce formate, phenol, ethanol, fuels, bi carbonate and menthol among them methanol production has received more attention because of its use in wide variety of application such as production of formaldehyde, in automotive anti-freeze, in a variety of chemical synthesis, as a general solvent, as an activation fuel (for water injection), as a denaturant for ethyl alcohol and a dehydrator for natural gas.[35] Certain microorganisms such as E-coli, formalose, CLEC Antarctica Lipase B, carbonic anhyhydrase can reduce CO_2 into valuable organic compounds via several steps. The enzymatic reduction of CO_2 requires an electron donar, which can be a coenzyme such as nicotinamide adenite dinucleotide (NADH) or a synthetic chemical like methyl viologen, depending on the specificity of the enzymes. The NADH-mediate reduction of CO_2 for production of methanol can be described as follows.

$$CO_2 + NADH + H^+ \xrightarrow{FateDH} HCOOH + NAD^+ \quad (13)$$

$$HCOOH + NADH + H^+ \xrightarrow{FaldDH} CH_2O + NAD^+ + H_2O \quad (14)$$

$$CH_2O + NADH + H^+ \xrightarrow{ADH} CH_3OH + NAD^+ \quad (15)$$

But the actual process is much more complicated because the solvation of CO_2 involves the formation of several species including CO_3^{2-}, HCO_3^-, and H_2CO_3. The thermodynamic equilibrium and the concentration distribution of these species depend on the pH and the other physicochemical properties of the solution. The enzyme can take all of these species as substrates for methanol synthesis other than the free state CO_2 only.[36]

The enzyme carbonic anhydrase is known as the fastest CO_2 catalyst. Conventionally, enzymatic reactions are performed in aqueous solutions, but recently, supercritical reactions in organic solvents and supercritical CO_2 have received extensive attention. Mobilized and immobilized both enzyme can be used for enzymatic reaction. Immobolized enzyme is easy to separate and reusable. But most immobilized enzymes show lower activity than mobilized enzyme.[37] Wu et al.[38] studied CO_2 enzymatic conversion in methanol using formate dehydrogenase ($F_{ate}DH$), formaldehyde dehydrogenase ($F_{ald}DH$) and alcohol dehydrogenase (ADH) co-encapsulated in silica gel prepared by modified silica process as catalysts. In their work TEOS was used as the precursor and NADH was used as the electron donar. They indicated that each dehydrogenase had its own optimum reaction temperatures. For example, for $F_{ate}DH$ and $F_{ald}DH$, the optimum temperature was 37 °C whereas for ADH, the optimum temperature was 25 °C. The maximum methanol yield was 92.1% at the temperature, pressure and pH of 37 °C, 0.3 MPa and 7, respectively. A wide variety of enzymes retain their characteristics reactivities and chemical functions when they are confined within the pores of the silica sol-gel derived matrix. The porosity of silica gel glasses allows small molecules and ion to diffuse into the matrix while the enzymes remain physically trapped in the pores, and thus resulting in an enhanced probability of reaction due to an increase in local concentration of substrates within the nanopores.[39]

3. Conclusion

Carbon dioxide reforming of methane is very attractive with a view to enhance natural resources of CO_2 and CH_4 into valuable chemicals. However, the reaction for syngas ($CO+H_2$) formation is processed at high temperature using various catalysts but the high temperature and the catalyst makes the process very costly. Catalytic conversion of CO_2 to methane or methanol is also attractive but the methane production requires large consumption of hydrogen per unit CO_2 and the methanol selectivity is very low even in the presence of catalyst. Electrochemical, photo electric, low temperature plasma or enzyme process could be effective methods for the conversion and fixation of CO_2.

4. References

1. Feld, L. (2000),"Bulletin of the United States Energy Information Administration", U.S. Department of Energy, http://www.eia.doe.gov/emeu/cabs/usa.html"
2. Buesseler, O. K., Doney S. C., Karl D. M., Boyd ,P. W., Caldeira, K., Ch_{ai}, F., Coale, K. H., de Baar,J. W. H., Falkowski, P. G., Johnson, K. S., Lampitt, R. S., Michaels, A. F., Naqvi, W. A., Smetacek, V., Takeda, S., and Watson, A. J. (2008), "Ocean Iron Fertilization—Moving Forward in a Sea of Uncertainty," Science, January, pp.162.
3. Varghese, O. K., Paulose, M., LaTempa, T. J. and Grimes, C. A. (2009), "High-Rate Solar Photocatalytic Conversion of CO_2 and Water Vapor to Hydrocarbon Fuels," Nano Lett., 9(2), pp. 731–737.
4. Yang, A. C.-M., Chang, Yi-H., and Chang, C.-C. (2009) "Conversion of Carbon Dioxide into Useful Organic Products by Using Plasma technology," IPC8 Class: AB01J1908FI.
5. Gallego, S. G., Mondragon, F., Tatibouet, J.-M., Barrault, J., and Batiot-Dupeyrat, C. (2008), "Carbon Dioxide Reforming of Methane over La_2NiO_4 as Catalyst Precursor—Characterization of Carbon Deposition", Catal. Today, 133-155, pp. 200–209.
6. Tomishige, K.; Himeno, Y.; Mastuo, Y.; Yoshinaga and Fujimoto K. (2000), "Catalytic Performance and Carbon Deposition ; Behavior of a NiO- MgO Solid Solution in Methane Reforming with Carbon Dioxide under Pressurized Conditions", Ind. Eng. Chem. Res., 39(6), pp.1891-1897.
7. Song, C., Srinivas, S. T., Sun, L. and Armor, J. N. (2000), "Comparison of High-Pressure and Atmospheric-Pressure Reactions for CO_2 Reforming of CH_4 over Ni/Na-Y and Ni/Al_2O_3 Catalysts. Am. Chem. Soc. Div. Petrol. Chem. Prep., 45 (1), pp. 143-148.
8. Shamsi, A. and Johnson, C D. (2007), "The Effect of CO_2 Reforming of Methane and the Carbon Deposition Route Using Noble Metal Catalyst", ACS Symposium Series, Vol. 959, Chapter 8, pp. 87-101.
9. Olah, G. A., Prakash, G. K. S. (2008), "Conversion of Carbon Dioxide to Methanol and/or Dimethyl Ether using Bi-Reforming of Methane or Natural Gas", IPC8 Class: AC07C2700F1, USPC Class: 518700.
10. Itkulova, S. S., Zhunsova, K. Z. and Zakumbaeva, G. D. (2005), "CO_2 Reforming of Methane over CO-Pd/Al_2O_3 Catalysts", Bull. Korean Chem. Soc., 26, pp. 2017-2022.
11. Inui, T. (1996), "Highly Effective Conversion of Carbon Dioxide to Valuable Compounds on Composite Catalysts," Catalysis Today, 29, pp. 329-337.
12. Inui, T., Funabiki, M., and Takegami, Y., (1980), "Simultaneous Methanation of CO and CO_2 on Supported Ni-Based Composite Catalysts", Ind. Eng. Chem. Prod. Res. Dev., 19, pp. 385–388
13. Halmann, M. M. (1993), "Chemical Fixation of Carbon Dioxide: Mrthods of Recycling CO_2 into useful Products", CRC Press, Florida, pp. 67-127.
14. Kaneco, S., Iwao, R., Iiba, K., Ohta, K., and Mizuno, T. (1998), "Electrochemical Conversion of Carbon Dioxide to Formic Acid on Pb in KOH/Methanol Electrolyte at Ambient Temperature and Pressure," Energy, 23(12), pp. 1107-1112.
15. Hori, Y., Kikuchi, K., and Sizuki, S. (1985), "Production of CO and CH_4 in Electrochemical Reduction of CO_2 at Metal Electrodes in Aqueous Hydrogencarbonate Solution", Chemistry Letters, 11, pp. 1695-1698.

16. Lide, D. R. (eds) (1984), "Handbook of Chemistry and Physics", 72nd Ed. CRC Press, Boston, pp. 4-6.
17. Mizuno, T., Kawamoto, M., and Koneco, M. (1998), "Electrochemical Reduction of Carbon Dioxide at Ti and Hydrogen-Storing Ti Electrodes in KOH-Methanol", Electrochem. Acta, 43(8) pp. 899-907.
18. Kaneco, S., Hiei, N., Xing, Y., Katsumata, H., Ohnishi, H., Suzuki, T., and Ohta, K. (2002), "Electrochemical Conversion of Carbon Dioxide to Methane in Aqueous $NaHCO_3$ Solution at Less than 273 K," Electrochem. Acta , 41(21), pp. 51-55.
19. Ogura, K. , Yano, H., and Shirai, F. (2003), "Selective Conversion of CO_2 to Ethylene by the Electrocatalysis at a Three-Phases (Gas/Liquid/Solid) Interface in an Acidic Solution Containing Cupric Ions", Fuel. Chem. Div. Pre., 48(1), pp. 264-265.
20. Huang, H. Y., Padin, J., and Yang, R. T. (1999) "Comparison of π-Complexations of Ethylene and Carbon Monoxide and Cu^+ and Ag^+", Ind. Eng. Chem. Res., 38, pp. 2720-2725.
21. Chapman, B. (1980) "Glow Discharge Processing", Wiley, New York.
22. Liu,C-J., Xu, G.-H. and Wang, T. (1999), "Non-thermal Plasma Approaches in CO_2 Utilization", Fuel Processing Technology, 58, pp. 119-134.
23. Yun, S.-H., Kim, G.-J., and Park, D.-W. (1997), "Decomposition and Conversion of Carbon Dioxide into Synthesis Gas using Thermal Plasma", J. Ind. Eng. Chem. 3(4), pp. 293-297.
24. Shi, X. D., Feng, Q. Y., and Zhong, H. S. (2005), "Photocatalytic Synthesis of Oxygenates from Gaseous CO_2 and CH_4 over Semiconductor", Chi. Chem. Lett. 16(5), pp. 685-687.
25. Inoue, T., Fujishima, A., Konishi, S., and Honda, K. (1979) "Photoelectric Catalytic Reduction of Carbon Dioxide in Aqueous Suspensions of Semiconductor Powders", Natur, 277, pp. 637-638.
26. Subrahmanyam, M., Kaneco, S. and Alonso-Vante, N. (1999), "A Screening for the Photo Reduction of Carbon Dioxide Supported on Metal Oxide Catalysts for C_1 -C_3 Selectivity", Appl. Catal. B: Environ., 23(2-3), pp. 169-174.
27. Usubharatana, P., McMartin, D., Veawab, A. and Tontiwachwuthikul, P. (2006), "Photocatalytic Process for CO_2 Emission Reduction from Industrial Flue Gas Streams", Ind. Eng. Chem. Res., 45, pp. 2558-2568.
28. Guan G., Kida T., and Yoshida A. (2003), "Reduction of Carbon Dioxide with Water under Concentrated Sunlight using Photocatalyst Combined with Fe-based Catalyst", 41, pp. 387-396.
29. Gokon, N., Hasegawa, N., Kaneko, H., and Aoki, H., Tamaura, Y., Kitamura, M. (2003), "Photocatalytic Effect of ZnO on Carbon Gasification with CO_2 for High-Temperature Solar Thermochemistry", Sol. Energy Mater. Sol. Cells, 80, pp. 335-341.
30. Saladin, F., and Alxneit, I. (1997), "Temperature Dependence of the Photosynthetic Reduction of CO_2 with H_2O at the Solid/Gas Interface of TiO_2", Chem. Soc. Faraday Trans, 93 (23), PP. 4159-4163.
31. Kohno, Y., Hayashi, H., Takenaka, S., Tanaka, T., Funabiki, T., and Yoshida, S. (1999), "Photoenhanced Reduction of Carbon Dioxide with Hydrogen over Rh/TiO_2", J. Photochem. Photobiol. A, 126, pp. 117-123.

32. Anpo, M., and Takeuchi, M. (2003), "The Design and Development of Highly Reactive Titanium Oxide Photocatalysts Operating Under Visible Light Irradiation", J. Catal., 216, pp. 505-516.
33. Yamashita, H., Nishiguchi, H., Kamada, N., and Anpo, M. (1994), "Photocatalytic Reduction of CO_2 with H_2O on TiO_2 and Cu/TiO_2 Catalysts", Res. Chem. Intermed., 20(8), pp. 815-823.
34. Adachi, K., Ohta, K., and Mizuno, T. (1994), "Photocatalytic Reduction of Carbon Dioxide to Hydrocarbon using Copper-Loaded Titanium Dioxide", Sol. Energy, 53(2), pp. 187-190.
35. Dave, C. B., Rao, S. M., and Burt, C. M. (2007), "Conversion of Carbon Dioxide to Methanol in Silica Sol-Gel Matrix", US patent, US2007/0042479AO, January.
36. Baskaya, F. S., Zhao X., Flickinger C. M., and Wang, P. (2010), "Thermodynamic Feasibility of Enzymatic Reduction of Carbon Dioxide to Methanol", Appl Biochem Biotechnol, September pp. 8758-8772.
37. Obert, R., and Dave, B. (1999), "Enzymatic Conversion of Carbon Dioxide to Methanol: Enhanced Methanol Production in Silica Sol-Gel Matrices", J. Am. Chem. Soc., 121, pp. 12192-12193.
38. Wu, H., Jiang Y. Z., Xui, W. S., and Huang, F. S. (2003) "A New Biochemical Way for Conversion of CO_2 to Methanol via Dehydrogenases Encapsulated in SiO_2 Matrix", Chinese Chemical Letters, 14(4), pp. 423 – 425.
39. Liu, C., Mallinson, R., Aresta, M., Jiang, Z., Wu, H., Xu, S., and Huang, S. (2002), "Enzymatic Conversion of Carbon Dioxide to methanol by Dehydrogenases Encapsulated in Sol-Gel Matrix", Fuel Chemistry Division Preprints, 47(1), pp. 306

CH_4 DRY REFORMING WITH CO_2: CHARACTERIZATION OF SURFACE CARBON SPECIES FORMED ON CeO_x-ZrO_y BASED Ni, Ni-Co AND Ni-Rh CATALYSTS

László Guczi[1], Györgyi Stefler[1], Alain Kiennemann[2], Agnieszka Pietraszek[2]

[1]Surface Chemistry and Catalysis, Institute of Isotopes, HAS, P. O. Box 77. H-1525 Budapest, Hungary, email: guczi@mail.kfki.hu
[2] LMSPC, Laboratoire de Matériaux, Surfaces et Procédés pour la Catalyse, UMR 7515 du CNRS-ECPM-ULP, 25 Rue Becquerel, 67087 Strasbourg Cedex 2, France

ABSTRACT

Methane dry reforming with CO_2 is greatly affected by the presence of various form of surface carbon (carbides, graphite, carbon nanotubes (CNT), etc). To characterize them two methods have been chosen: (i) temperature programmed oxidation (TPO) and (ii) transmission electron microscopy (TEM). CeZrNi, CeZrNiCo, CeZrNiRh and 5%Ni/CeZr catalysts were prepared by sol-gel and impregnated method, respectively. These catalysts are efficient in dry reforming of methane reaction. The surface carbon obtained during the reforming process was generally characterized by T_{max} in TPO. On sol/gel CeZrNi(s/g) and impregnated Ni/CeZr(imp.) catalysts the T_{max} appeared at 400°C and at 600°C, respectively, which shows different carbon structures. Simultaneously, TEM image pointed out the extensive CNT formation on the impregnated Ni catalyst along with an increase in Ni particle size while for the sol/gel sample the Ni particles are stabilized in the form of Ni nanoparticles. On bimetallic CeZrNiCo(s/g) and CeZrNiRh(s/g) catalysts two kinds of surface carbon species coexist. These characterized by T_{max} at 360°C was slowly transformed to those at higher temperature peak (600°C). The amount of carbon is considerably higher on the Co containing sample than for one with Rh. We can speculate that on CeZrNiCo sample the carbidic (CoC_x or NiC_x) species are predominantly formed and after the changing of catalyst structure the CNT and graphitic species become prevailing.

INTRODUCTION

In methane dry reforming using CO_2 there are two approaches to avoid substantial carbonaceous deposit formation: the first is to choose appropriate support and the second one to modify the most frequently used nickel with a second metal. In our previous paper (1) we studied the behaviour of nickel catalyst supported on $MgAl_2O_4$ spinel. It was established that three different types of carbon are formed, such as: Ni-carbide, graphite and carbon nanotubes and it was proved that the crystallized carbon nanotubes are responsible for the decrease in catalyst activity. The addition of small amount of gold decreased the deactivation process with a great extent.

Recently ceria has been used to modify the support. The activity of 13.5Ni-2K/10CeO_2-Al_2O_3 catalyst was tested in dry reforming of methane (2). The amount of coke deposited on the catalyst at different time intervals was estimated by thermogravimetric analysis (TGA). Results suggested that both CH_4 cracking and CO disproportionation contribute in coke deposition. Inconsiderable deactivation was observed for the catalyst over 60-hrs run showing that the amount of accumulated coke did not affect the activity. It indicates that some carbon species formed on the surface of the catalyst must be involved in the reaction to produce CO. TEM results displayed that a large part of the graphitic carbon deposited on the catalyst surface was of the filamentous form with nickel set on the top of these carbon filaments. This form of graphitic carbon is more active in the reforming reaction of methane, probably because of its close interaction with nickel particles.

Zirconium-based mixed oxides (3–5) were proved to be also a good support for nickel catalyst in methane dry reforming. The catalysts containing 5 wt% of Ni were very effective in CO_2 reforming reaction of methane and exhibited stable performance up to 50 hrs of reaction. The obtained results showed that the addition of promoters to the zirconia (e.g. Ce, La, K) improved both the activity and stability of the catalyst.

In the present paper we wish to study the Ce-Zr oxide based metal catalysts prepared by sol-gel or impregnation method and gather information about its structure, activity and the different carbonaceous species formed during the CH_4 dry reforming with CO_2.

EXPERIMENTAL

2. 1. Catalyst preparation

The monometallic CeZrNi(s/g) and bimetallic CeZrNiCo(s/g) and CeZrNiRh(s/g) catalysts were prepared by pseudo sol-gel method based on thermal decomposition of metallic propionates (6, 7). The metal precursors used for synthesis were: cerium (III) acetate sesquihydrate, zirconium (IV) acetylacetonate, rhodium (II) acetate dimer, nickel (II) acetate hydrate and cobalt (II) acetate hydrate. The salts were dissolved in the excess of propionic acid. The solutions were mixed and heated under reflux. The solvent was evaporated to obtain gel which was than calcined in air at 750°C during 4 hours. The Ni/CeZr catalyst was prepared by wet impregnation method in the excess of water which was then removed during heating up to 160°C under the stirring. The obtained dry residue was dried overnight and calcined in air at 750°C during 4 hours. The metal precursor used in this case was nickel (II) nitrate hexahydrate.

2. 2. Sample characterization

The specific BET surface area was determined by N_2 adsorption-desorption measurements at -196°C in a Tristar gas adsorption system. The crystalline structure of mixed oxides was determined by powder X-ray diffraction using Brucker D8 Advanced apparatus with Cu Kα radiation. The reducibility of the catalysts was studied by thermo-programmed reduction (TPR) using 0.05 g sample heated from room temperature to 900°C with a slope 15°C/ min. under hydrogen/argon flow. The major parameters are summarized in Table 1.

The amount of carbon on the surface was estimated from the amount of CO_2 evolved in TPO experiments carried out in 5 vol. % O_2 and 1 vol.% Ar in He. To detect CO_2 formation the mass spectrometer was employed. Using the correlation between the intensity and flow ratio of CO_2 to Ar the flow of CO_2 can be determined and then integrated through the time, which results in the amount of CO_2 formed (1).

2.3. Catalytic activity

The catalytic activity measurements were performed in a fixed bed quartz reactor. Prior to CO_2/CH_4 reaction, 0.1g of catalyst was reduced in 5% H_2-95%Ar at 750°C for 4 hours (0.0025L/min). After the pre-treatment, a feed mixture consisted of $CH_4/CO_2/Ar$ 10/10/80 was introduced into reactor. The total space velocity was equal to 30 $L.h^{-1}.g^{-1}$. The reforming tests were studied with an increase of temperature from 550°C to 800°C with the stabilization of 2 hours at each temperature.

Table 1. The major parameters of the various samples

Catalysts	abbreviation	Ni nominal content wt.(%)	Me nominal content wt.(%)	BET surface area (m^2/g)	Lattice parameter (Å)	Support particle size (nm)
5%Ni/Ce_2Zr_2	Ni/CeZr(imp.)	5	-	17	5.27	6
$Ce_{0.6}Zr_{2.97}Ni_{0.43}$	CeZrNi(s/g)	5	-	15	5.17	7
$Ce_2Zr_{1.51}Ni_{0.49}Co_{0.29}$	CeZrNiCo(s/g)	3	2	11	5.30	7
$Ce_2Zr_{1.51}Ni_{0.49}Rh_{0.03}$	CeZrNiRh(s/g)	4.5	0.5	21	5.30	6

RESULTS AND DISCUSSION

The XRD patterns (Fig.1 a-d) showed the Ni/CeZr, CeZrNi(s/g), CeZrNiRh(s/g) and CeZrNiCo(s/g) catalysts after calcination in air at 750°C for 4 hours. It can be clearly seen that the patterns match well to face-centered cubic structure of $Ce_{0.6}Zr_{0.4}O_2$ (JCPDS 38-1439). Nevertheless the small, broad diffraction patterns at 2Θ= 37.3° and 43.3° indexed as NiO, with (001) and (200) planes can also be observed. This phenomenon might be explained by the small rejection of Ni from the host Ce-Zr structure in the case of the catalysts prepared by sol gel method. For the bimetallic CeZrNiMe catalysts besides diffraction patterns characteristic of fluorite structure and NiO species no evidence of Co or Rh oxides was observed. In the case of Rh catalyst it could be anticipated to low Rh loading (~ 0.5wt.%) thus for CeZrNiCo(s/g) to different ionic radii than that of Ni (Ni^{2+} 69 pm and Co^{2+} 75pm) and lower loading (2 wt.% instead of 3wt.%) what can facilitate the insertion of these cations into the Ce-Zr structure. Furthermore the high dispersion of the metals could be also taken into account.

The catalytic tests (Fig. 2) demonstrate the temperature dependence on catalytic activity for CeZrNi and CeZrNiMe based catalysts. CH_4 conversion is lower than CO_2 one due to the presence of reverse water gas-shift reaction (RWGS) occurring simultaneously with CO_2 reforming of CH_4 (8). The highest activity in both CO_2 and CH_4 conversion was observed over catalyst where the part of Ni was substituted by Rh. The Ni/CeZr catalyst with Ni impregnated on it has also very interesting activity, lower than CeZrNiRh but higher than sol-gel prepared and CeZrNiCo ones.

In order to study the surface carbonaceous deposits formed during the dry reforming reactions temperature programmed oxidation was performed. The results are summarized in Fig. 3. It is obvious that on the sample containing impregnated nickel the amount of carbon is the highest.

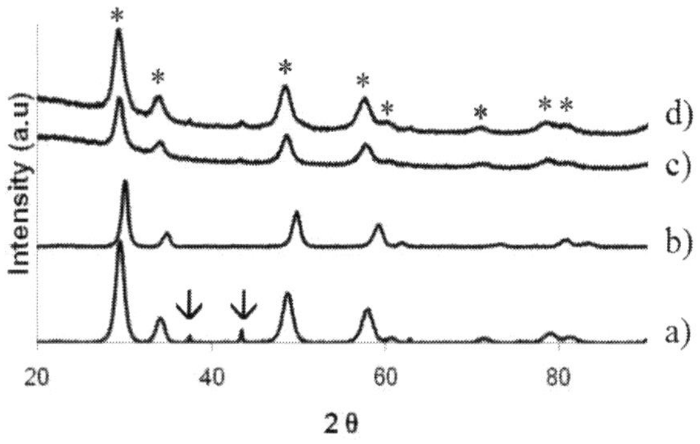

Figure 1. XRD of freshly prepared catalyst (a) Ni/CeZr(impr.); (b) CeZrNi(s/g); (c) CeZrNiCo(s/g); (d) CeZrNiRh(s/g). * - face-centered cubic, fluorite structure, ·- NiO species

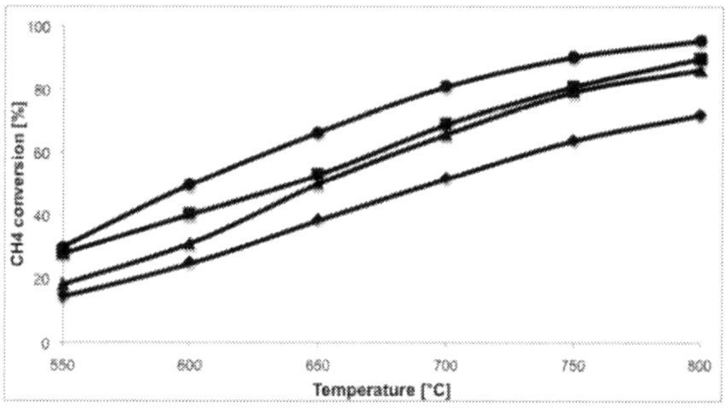

Figure 2. (■) Ni/CeZr(impr.); (▲) CeZrNiCo(s/g); (♦) CeZrNi(s/g) and (•) CeZrNiRh(s/g). Reaction conditions: $CH_4/CO_2/Ar$ 10/10/80; total space velocity: 30 $L.h^{-1}.g^{-1}$.

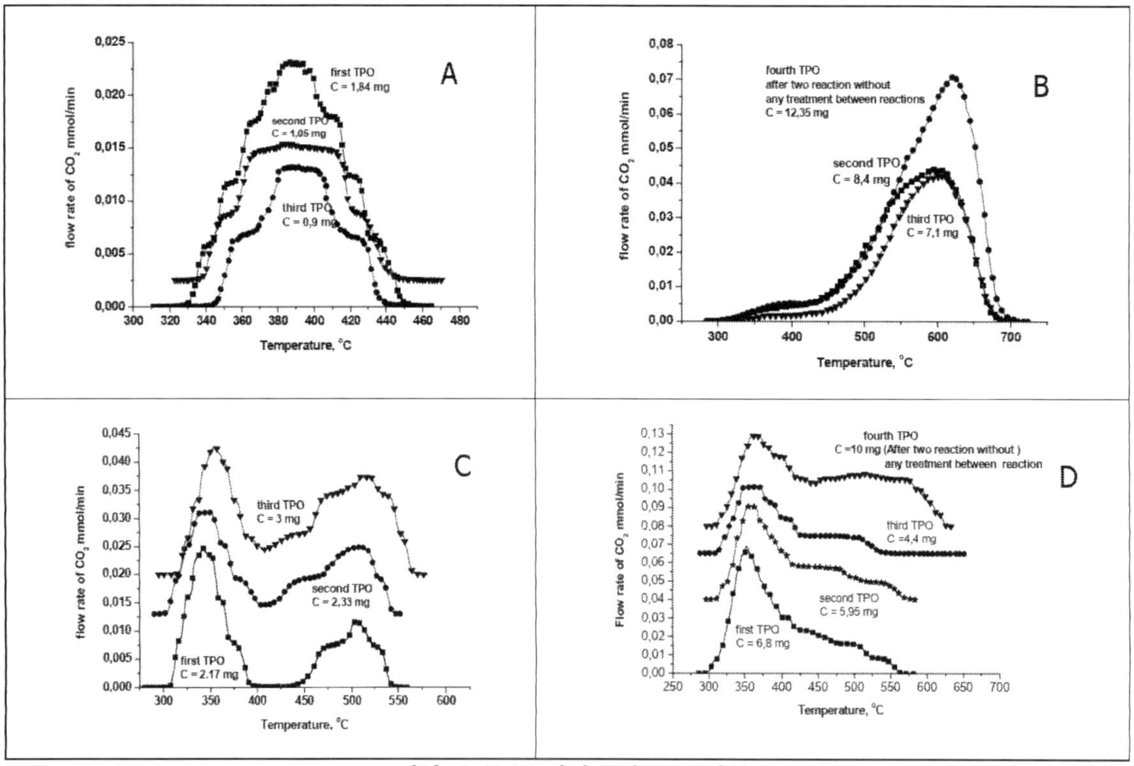

Fig. 3. TPO measurements on (A) CeZrNi; (B) Ni/CeZr; (C) CeZrNiCo and (D) CeZrNiRh samples

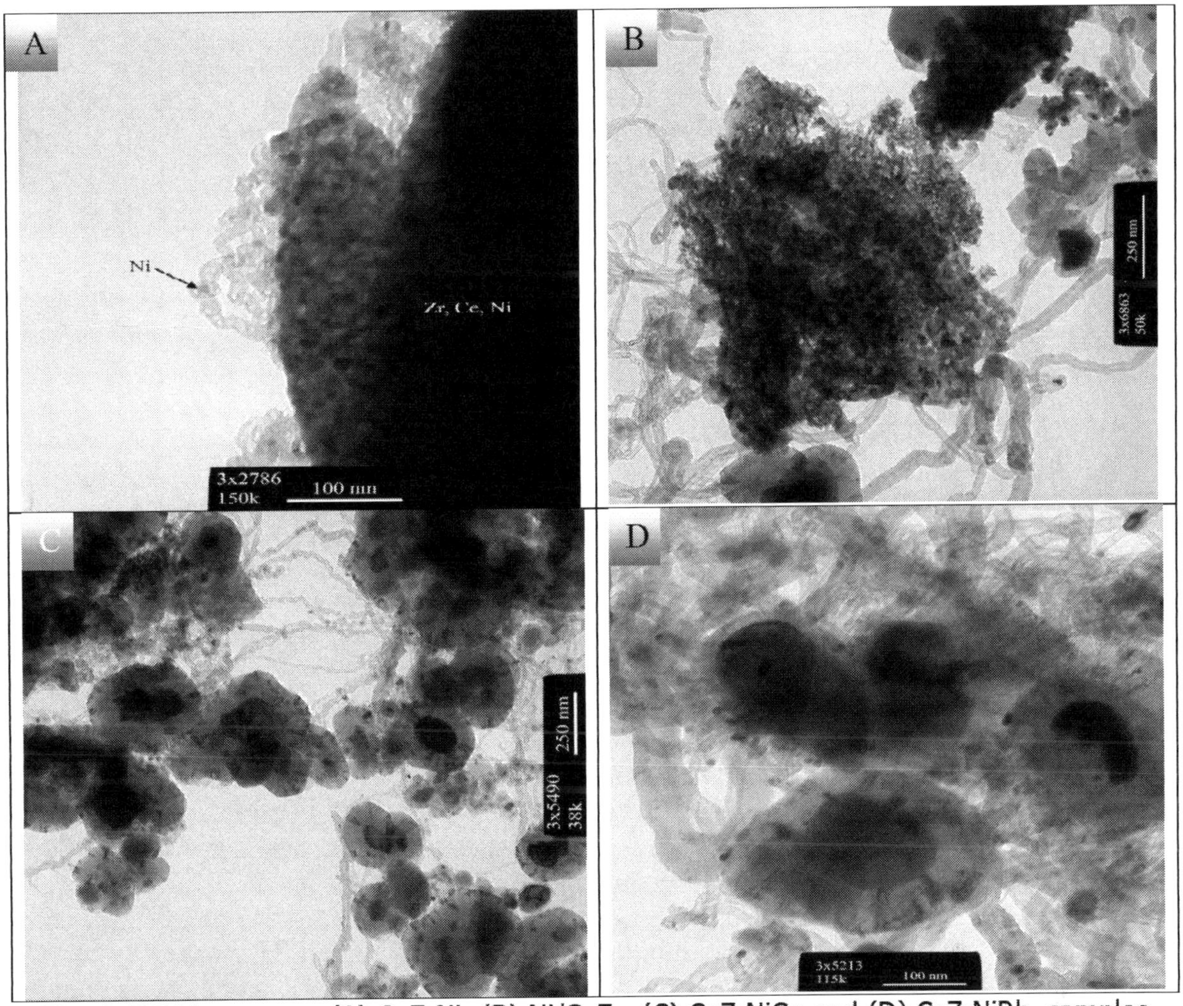

Fig. 4. TEM images on (A) CeZrNi; (B) Ni/CeZr; (C) CeZrNiCo and (D) CeZrNiRh samples

As shown by TEM (Fig. 4 A-D) the catalysts behave in different manner. The nickel being incorporated in a solid oxide composite material (CeZrNi) is rather compact and the size of the nickel atoms are in nano range and those are not involved in nanotube formation (C/Ni is negligible) (Fig. 4A). The Ni particles impregnated on CeZr support (Fig. 4B) are involved in filament like nanotubes growing originating from the surface nickel sites. The carbon/nickel ratio determined by EDS is rather small (0.7) as the surface carbon is involved in this process. It is interesting that in both case of CeZrNiCo (Fig.4C) and CeZrNiRh (Fig. 4D) the metal particles are detached from the bulk surface and wrapped in nanotubes. The EDS results show a large amount of carbon substantially more than in the sample A and B.
It can be noticed that TPR profiles of both catalysts are totally different (Fig. 5 A and B). In the sol-gel catalyst (Fig. 5A) the presence of strong metal-support interactions results in higher temperature of reduction and smaller Ni particles due to the incorporation of the metal inside the fluorite structure, what is in accordance with TEM images. The

presence of small particles leads to the formation of carbon species oxidised at lower temperatures (Fig. 3A). However, the Ni impregnated (Fig. 5B) is reduced in the lower temperatures due to the weak interactions between Ni and support resulting in the presence of big particles. These particles are responsible for the formation of carbon nanotubes which are oxidised at higher temperatures (Fig. 3B).

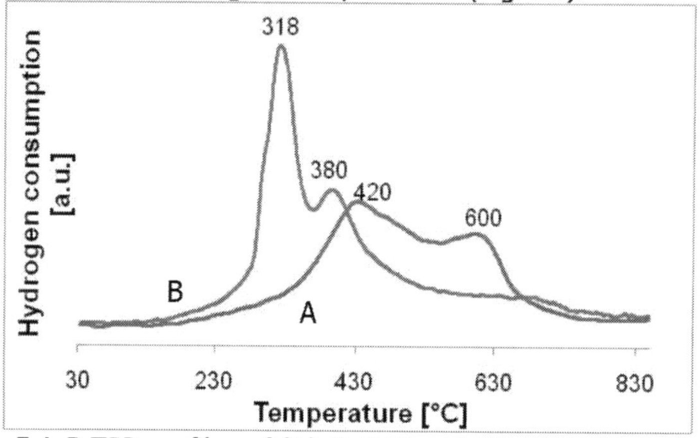

Fig. 5 A-B TPR profiles of (A) CeZrNi and (B) Ni/CeZr catalysts

The TPO results also support the results. The carbon formed on sample A (Fig. 3A) can be more easily removed from the surface (T_{max} is around 400°C) than those produced on sample B (Fig. 3B) with the T_{max} around 600°C. This is because carbon nanotubes are rather stable and only a part can be oxidized emerging in the gas phase. On the other hand there are both types of carbon species on samples C and D (Fig. 3C and 3D) what can be observed by the appearance of two CO_2 peaks at 400 and 600°C. This is in agreement with TPR profiles (not shown here) where the addition of Rh or Co resulting in shifting the Ni reduction temperature due to the formation of metal-H bond on the second metal. As Ni is reduced at lower temperatures we will have beside small particles, some bigger ones explaining the presence of two kinds of carbon species.
Nevertheless, the best catalyst is sample D because the interaction of Rh with nickel results in an optimum structure of surface carbonaceous deposit being responsible for the deactivation.

ACKNOWLEDGEMENT

The authors are grateful of the support from ERA-CHEMISTRY.

REFERENCES

1. L. Guczi, G. Stefler, O. Geszti, I. Sajó, Z. Pászti, A. Tompos, Z. Schay, Methane dry reforming with CO2: A study on surface carbon species, Applied Catalysis A (2010), doi:10.1016/j.apcata.2009.12.040
2. N.A. Pechimuthu, K.K. Pant, S.C. Dhingra, Deactivation studies over Ni-K/CeO_2-Al_2O_3 catalyst for dry reforming of methane, Industrial and Engineering Chemistry Research, 46 (2007) 1731-1736

3. M. Rezaei, , S.M. Alavi, S. Sahebdelfar, X. Liu, P. Bai, Z.-F. Yan, CO_2 reforming of CH_4 over nanocrystalline zirconia-supported nickel catalysts, Applied Catalysis B 77 (2008) 346–364.
4. S. Therdthianwong, , A. Therdthianwong, , Ch. Siangchin, S. Yongprapat, , Synthesis gas production from dry reforming of methane over Ni/Al_2O_3 stabilized by ZrO_2, International Journal of Hydrogen Energy 33 (2008) 991–999.
5. M. Rezaei, S.M. Alavi, S. Sahebdelfar, Z.-F. Yan, Mesoporous nanocrystalline zirconia powders: A promising support for nickel catalyst in CH_4 reforming with CO_2, Materials Letters 61 (2007) 2628–2631.
6. J. C. Vargas, E. Vanhaecke, A.C. Roger, A. Kiennemann, Hydrogen production by ethanol steam reforming. Study of cobalt-doped Ce-Zr catalysts, Studies in Surface Science and Catalysis 147 (2004) 115-120.
7. F. Romero-Sarria, J. C. Vargas, A. C. Roger, A. Kiennemann, Hydrogen production by steam reforming of ethanol. Study of mixed oxide catalysts $Ce_2Zr_{1.5}Me_{0.5}O_8$: Comparison of Ni/Co and effect of Rh, Catalysis Today 133-135 (2008) 149-153
8. M. C. J. Bradford, M. A. Vannice, CO_2 Reforming of CH_4, Catalysis Reviews Science and Engineering 41 (1999) 1

(134d) Heterogeneous Catalytic Systems for the Synthesis of Organic Carbonates

Jose Antonio Lopez-Sanchez, Chemistry Department, Cardiff Catalysis Institute, Cardiff, United Kingdom

Ceri Hammond, Chemistry Department, Cardiff Catalysis Institute, Cardiff, United Kingdom

Mohd Hasbi Ab Rahim, Chemistry Department, Cardiff Catalysis Institute, Cardiff, United Kingdom

Nikolaos Dimitratos, Chemistry Department, Cardiff Catalysis Institute, Cardiff, United Kingdom

Graham J. Hutchings, Chemistry Department, Cardiff Catalysis Institute, Cardiff, United Kingdom

We have shown that glycerol carbonate can successfully be synthesised at atmospheric pressure via the indirect carbonylation of glycerol with urea, and demonstrated that higher yields and selectivities of glycerol carbonate can be obtained through the use of newly developed heterogeneous catalysts, as opposed to previously reported homogeneous metal sulphates. The production of glycerol carbonate with this method has recently gained much popularity due to the low cost of the starting materials, and the ability of the overall process to successfully utilise a CO_2 carrier during the reaction. This environmentally friendly, 'green' process involves the reaction of glycerol and urea at mild temperatures (ca.100 °C). The favourability of this reaction not only lies upon its ability to synthesise glycerol carbonate from two readily available and cheap starting materials, but also the possibility of re-converting the ammonia that is released during the reaction back to urea with the Bosch-Meiser process, overall resulting in a chemical cycle that utilises CO_2.Organic carbonates are used as solvents, lubricants, plasticizes and monomers as well as in methylation and carbonylation processes for preparing isocyanates, polyurethanes and polycarbonates. Organic carbonates have been added to gasoline resulting in an improved octane value. Furthermore, their addition to diesel effectively reduces carbon particulates and their application as fuel additives potentially has a very large market. The formation of the organic carbonates via the reaction of alcohols with urea has the advantage that it can be utilised to carbonylate a wide range of alcohols as starting materials. However, current catalysts cited in the patent and commercial literature are based on metal sulfates and other catalysts that suffer from low intrinsic activity and selectivity. Although sulfate catalysts form the desired carbonate, the reaction is slow and requires very elevated quantities of the active metal, which, in addition also suffer from intense leaching, i.e. loss of active catalyst components during reaction, so that the reaction actually occurs catalytic homogeneous process instead of in a highly desired heterogeneous mechanism where the catalyst can easily be recovered as a solid. We have identified suitable heterogeneous catalysts for this reaction and also demonstrated the ease of purification for the product, with a simple vacuum-distillation yielding a product of ca. 98% purity. In this work we report the catalytic activity of heterogeneous catalysts based on gallium, zinc, silver, palladium and gold nanoparticles supported onto a range of oxides and zeolitic materials. The addition of each component to an acidic zeolite support (ZSM5) causes a dramatic increase in the selectivity of the reaction towards glycerol carbonate, while the effect on the conversion of glycerol strongly depends on the choice of oxide support. A major outcome of our research has been the unforeseen discovery that gold is a highly effective metal for the transformation of glycerol to glycerol carbonate. The postulated mechanism suggests that glycerol carbonate forms via the cyclization of a glycerol urethane intermediate.

See more of this Session: CO2 Conversion

See more of this Group/Topical: Topical 6: 10th Topical Conference on Gas Utilization

(145a) Modeling the U.S. Natural Gas Pipeline Network in the National Infrastructure Simulation and Analysis Center (NISAC) Agent-Based Laboratory for Economics (N-ABLE)

Greg E. Mackey, Infrastructure & Economic Systems Analysis, Sandia National Laboratories, Albuquerque, NM

Mark A. Ehlen, Infrastructure & Economic Systems Analysis, Sandia National Laboratories, Albuquerque, NM

Sue Downes, Infrastructure & Economic Systems Analysis, Sandia National Laboratories, Albuquerque, NM

Natural gas is an important component in chemical supply chains as it is a necessary feedstock to many chemical production processes. This paper describes how Sandia National Laboratories (Sandia) implements a model of the U.S. natural gas market and pipeline network within the National Infrastructure Simulation and Analysis Center (NISAC) Agent-Based Laboratory for Economics (N-ABLE). In this case, N-ABLE is used to simulate natural gas flows from suppliers to consumers, particularly during disruptive events such as hurricanes or earthquakes. The U.S. natural gas market is represented as a single natural gas marketer that brokers transactions between suppliers and consumers and enforces capacity constraints on the pipeline network using a maximum flow algorithm. Explicitly modeling the interactions between markets and flows allows Sandia to compare how normal conditions and disruptive events affect the supply, demand, and availability of natural gas. The paper concludes with an example disruption scenario.

See more of this Session: Professor Cedomir M. Sliepcevich Memorial Session Five - LNG Simulation

See more of this Group/Topical: Topical 6: 10th Topical Conference on Gas Utilization

Validation of Air Recirculation by CFD on the Multi-pass Air Heater for LNG Vaporization

Joseph H. Cho, Ph.D., P.E.
Technology Director

Dong M. Shin, Ph.D.
CFD Specialist

Jae-Young Yang, Ph.D.
Principal Process Engineer

Houston Technology Center
SK Engineering & Construction
Houston, Texas

AIChE Spring Meeting, March 2010
10th Topical Conference on Natural Gas Utilization
San Antonio, TX, March 21-25, 2010

ABSTRACT

In order to comply with the LNG industry's requirements for minimizing life cycle costs, the ambient air heater vaporization in association with closed loop heat transfer fluid system has been developed. This vaporization system has been proved and is a new trend in the LNG industry. However, one of the biggest concerns is to estimate the air recirculation across the air heater. Lower temperatures at the air heater inlet can cause a reduction in heat extraction from the air as a re-useable heat source. Understanding the mechanism for the temperature decrease is beneficial for designing the facility in a better way considering adequate design margins.

This paper presents effect of key parameters and variables on designing the ambient air heater with a proposal for optimization process of the ambient air heater vaporization. The paper also discusses the effective usage of Computational Fluid Dynamics (CFD) modeling for optimizing the air heater vaporization system. Models are through the interaction of the multi-pass air heat exchange tubes with the meteorological data at the given site plot.

INTRODUCTION

Two predominant vaporization technologies have been historically used: Open Rack Vaporizer (ORV) and Submerged Combustion Vaporizer (SCV). However, some concerns have arisen in the selection of vaporization technology. These are not only vaporization system's economics and availability, but also the impact of effluent seawater on the ecosystem and air emission (NOx, CO, CO_2, etc). The selection of the vaporization system is constrained by economic justification and site specific environmental issues.

This selection also directly affects LNG terminal availability, capital expenditure, and operating costs.

For the selection of LNG vaporization technology, the distinctive new trend is to use ambient air-heated vaporization system (AHV) in the LNG regasification industry, where reusable air is used as a heat source. This seems to be a commercially proven technology as operation at Dahej terminal since 2004, and Freeport LNG terminal in 2008 are successfully running with AHV technology. It is obvious that the trend is towards AHV technology. Recently, Trunkline LNG terminal is under commissioning after completion of revamping work, which replaced the existing SCV's with the same type of AHV used in Dahej terminal.

This paper briefly discusses concerns of vaporization technologies from engineering point of view and presents the design features of the ambient air-heated vaporizers. This paper also includes the optimization methodology of the AHV process. Technical features and economic advantages of each vaporization process are also addressed.

CONVENTIONAL VAPORIZATION TECHNOLOGY

Open Rack Type Vaporizer (ORV)

ORV's use seawater as their source of heat. ORV system consists of a vertical rack with an overhead distributor which allows seawater to fall over aluminum panels to the outer surface of long finned tube panels and vaporizes the LNG inside the tubes. The cooled seawater is then collected at the bottom from where it is routed and discharged back to the sea. One of the important design criteria is to limit the cooling temperature of the discharged water which should be suitable to the marine life. Figures 1 illustrates seawater distribution on each tube panel.

ORV's are widely used in Asia and Europe, and are well proven in base load LNG regasification service. However, the use of ORV's in the US terminals has been curtailed because of adverse impact on the marine life due to mechanical, thermal and chemical (residual biocide) pollution.

Great care must be taken to assure a continuous and uniform flow of seawater to the panels to avoid formation of ice and the subsequent loss of performance. Some other issues that must be addressed are:

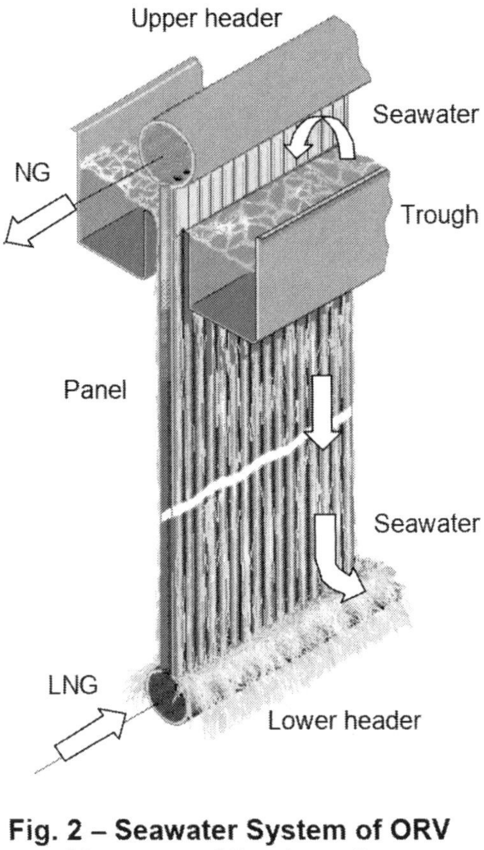

Fig. 2 – Seawater System of ORV
[Courtesy of Sumitomo]

- Return water discharge temperature: A site specific environmental study to determine the thermal pollution impact on aquatic life must be carried out [1]. Supplementary heat sources may be required depending on send-out temperature and seawater temperature during winter months.
- Return water de-chlorination and aeration to reduce chemical pollution: This will require an environmental assessment.
- High seawater suction and discharge velocity (mechanical pollution) will impact the ecosystem in the seawater.
- Seawater quality: Contaminated seawater can affect the ORVs' heat exchange surface coating. Suspended solids, mainly silt, should be limited, since it also contributes to the erosion of the coated surface. Cu and Hg ions are extremely hazardous on the heat exchange tubes made of aluminum alloy.
- Regular surface cleaning with water jet is required. After three to four (3-4) years of operation, it is normally required to do periodic zinc surface coating. It needs a shut down of the ORV for a short duration (it may take about a week per unit).
- Turn down is about 10:1, but it requires high attention to the ramp up of the LNG flow in order to avoid any uneven liquid distribution from the lower header. The uneven distribution will cause bowing of the heat exchange tubes. Even icing heights need to be confirmed before increasing the flow rate. The ramp up time from zero LNG flow to the designed capacity may vary but generally, it takes about 1 hour.

Submerged Combustion Vaporizer (SCV)

These vaporizers burn the natural gas taken from the send-out gas stream and pass the hot combustion gases into a water bath that contains the heating tubes for LNG (see Figure 2).

Typically, the largest single SCV units installed are for a gas send-out rate of approximately 150 MMScfd.

The SCV's are indirect fired heat exchangers with the burner and the coil all contained within a single vessel. They are highly efficient systems used to vaporize incoming liquid natural gas from a temperature of -160°C to gas at 5-10°C. Each SCV contains a burner unit, a tube bundle, and a weir assembly. A vaporizer tank encloses these components.

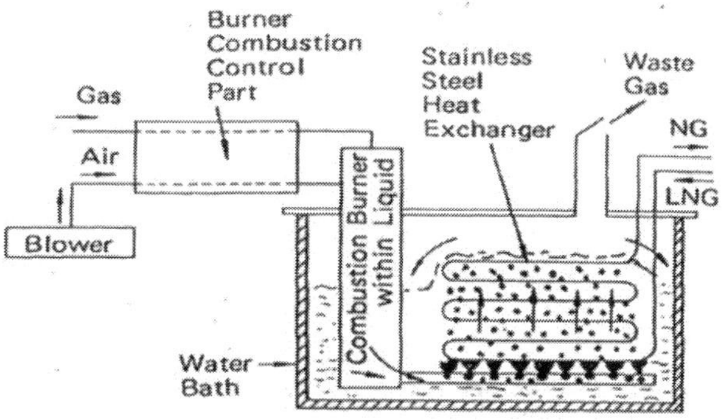

Figure 2 - Submerged Combustion Vaporizer

The weir assembly contains a water bath in which the tube bundle is submerged. The burner unit, which is partially submerged in water, combusts natural gas and routes the product directly into the water bath causing a rising froth flow. This rising flow is confined within the weir assembly, which surrounds the tube bundle. The water falls back over the weir assembly to contact the partially submerged burner before recirculation occurs. Because the combustion products directly contact the water bath, the submerged combustion vaporizer systems are able to reach efficiencies between 95 and 99% based on high heating values (HHV) of fuel gas [2].

The SCV's have a proven track record, especially as backup for ORV. Their turn-down ratio is approximately 8~10:1.

Limitations of SCV applications are:

- Emissions of NOx and CO: The available technology can reduce the NOx and CO emissions from the SCV's below 30 ppmv of each component. However, these emission levels do not satisfy the emission restrictions in some environmentally sensitive areas.
- Water quality: High chloride contents (over 30 ppm) in makeup water will cause pitting corrosion on the heat exchanger tube surface. The quality of initial fill and any subsequent make-up water, if required, needs to be carefully measured and controlled.
- Plot space of SCV: Generally, a SCV requires larger plot space than an ORV for the same thermal capacity.
- Blowdown of water from the SCV to control pH and treatment.

AMBIENT AIR HEATED VAPORIZERS

Natural Draft Ambient Air Vaporizer (AAV)

The natural draft Ambient Air Vaporizer (AAV) uses air as the heating medium, which is a free heat source. The AAV relies on wind and natural convective currents to move air over the tubes and fins of the vaporizer unit. As warm air contacts the tubes containing LNG, the air cools and becomes dense, causing it to flow downwards to the bottom of the vaporizer unit. This causes warm ambient air from the surrounding to be drawn through the top of the unit. This is sometimes called: "reverse chimney effect".

Such vaporizers are in service for peak shaving plants at locations where the LNG vaporization duty and flow rate are relatively small. At this smaller scale, plot space is not a major consideration. These systems are based on using single or multiple units in banks with common interconnecting pipes. Traditionally, these vaporizers have been designed with half of the parallel units on-line while the other half are off-line for defrosting.

Recently, Cheniere Energy has installed a pilot test train, consisting of 11 cells, to serve as a pilot testing facility at Sabine LNG Terminal. A designed full AAV train comprises 33 cells with expected vaporization capacity of 180 MMscfd [3]. However, its AAV requires considerable plot space and generates dense fog around the AAV.

Some other issues that must be addressed are:

A large number of small units in operation will have much more fog generation than fewer large units. Therefore, dense fog dispersion must be studied (See Fig. 3). When the air is calm, the fog will tend to remain in the vaporization area;

Fig. 3 – Dense Fog around the AAV for LN_2 Service [Courtesy of Cryoquip]

- The large plot space demand has to be included in the life cycle cost;
- De-icing becomes difficult when the ambient air temperature drops below freezing. In this case a back-up heat source for de-icing is required.
- Experience and risk have to be included in the evaluation because these ambient air draft vaporizers have only been used for peak shaving facilities (scale up issue);
- Fog and no-wind conditions control the number of vaporizer trains. Back-up facilities need to be included, depending on the specific meteorological conditions and terminal availability requirements [4].

Forced Draft Air Heater Vaporizer (AHV)

This scheme uses an intermediate fluid between a shell-and-tube LNG vaporizer (STV) and the forced draft air heater, which is intrinsically the same as the conventional air coolers to reheat the Heat Transfer Fluid (HTF) by ambient air (Fig 4). The HTF is supplied to the vaporizer after it is heated by the air heater. Air flow direction from top to bottom is recommended to minimize cold air recirculation. If the heat extracted from air is not enough for LNG vaporization in winter season, the system is fully backed up by fired heaters. If ambient air temperature is high enough, the back-up facility can be waived, resulting in lower Capital Expenditure (Capex) and Operating Expenditure (Opex) [4].

The forced draft ambient air vaporizer would significantly lower operating costs compared to SCV's or slightly lower than ORV's with tropical condition because the recovered heat from ambient air would save on fuel consumption for LNG vaporization. However, it should be noted that the operation and utilization rate of the force draft ambient air heaters are site specific.

The air heater vaporization technology with intermediate fluid has now been considered as a proven technology. As AHV technology uses the basic equipment such as air coolers and STV's which are existing quite some time in the LNG industry, a complete AHV system has been constructed and operated since early 2004 at Dahej LNG Terminal, India, and no specific problems have been reported (Fig. 5). Newly developed Kochi LNG terminal, India, has also adopted the same type of AHV system. The Freeport LNG terminal has operated STV's combined with the air heat recovery system since 2008. In fact, Federal Energy Regulatory Commission (FERC) recommended using AHV system at the Gulf of Mexico area as it does not permit ORV system because of seawater pollution. Trunkline LNG terminal has revamped vaporization system from SCV's to AHV system in 2009 and kept the existing SCV's as back-up facility. With many LNG vaporization operators choosing this technology, AHV system seems to be the new technology trend.

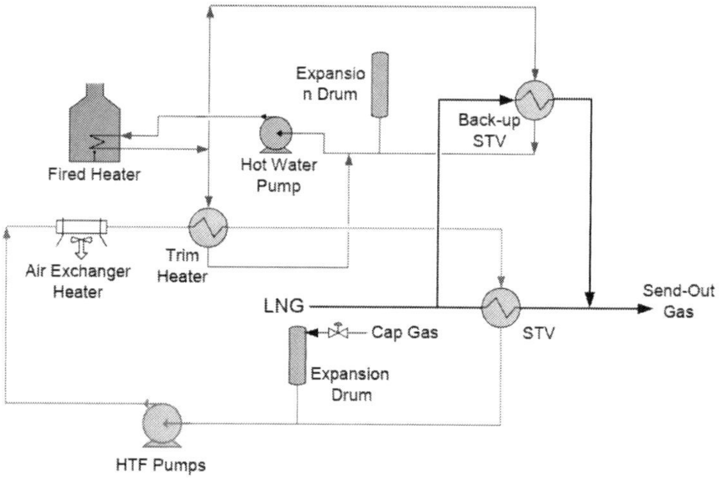

Fig. 4 – Forced Draft Air Heater Vaporizer

One of major concerns is fogging around the air heater since the air is cooled and sometimes it falls below saturation temperature, which will theoretically generate fog, which is generally formed around nuclei when the air is near or below the dew point. Low or no wind conditions, and low atmospheric pressure conditions are important parameters in generating fog. Salts, dust, and pollution particles can be nuclei. If the AHV is located near shore line, the air heater will generate fog all round the year. However, the forced draft air heater will not generate fog because of the following:

Fig. 5 – Air Heater Operated at Dahej Terminal
[Courtesy of Petronet]

- High velocities from the forced draft air fan
- Exit air temperature from the AHV is higher than AAV
- Exit air is less dense than AAV, resulting in easily mixing with adjacent air

- Tube fins designed to release condensate easily (downward forced draft removes condensate to grade)

If the AHV generates fog, it will be normally confined within a couple of meters of the AHV boundary. About 10 days of fogging yearly have been reported at Dahej terminal and it is easily dissipated when the sun rises [5,6].

OPTIMIZATED DESIGN OF THE AHV

Gathering Key Parameters

Two important things to keep in mind are that the vaporization system is quite site specific and the meteorological data vary all the time. In order to enable the AHV economics, identifying key parameters is required, which directly affect Capex and Opex of the AHV system. This includes the following:

- Main wind direction
- Equipment and facility arrangement plan (before establishing plot)
- Hourly meteorological data (wind velocity and direction, dry & wet bulb temp)

Size of large equipment, LNG tanks, and facilities (and buildings) have to be determined. Their initial locations will be confirmed or relocated based on AHV optimization results. Specially, LNG tank location and adjacent buildings have to be identified because these directly impact air flow path. For meteorological data analysis, minimum 10 years' hourly data are required.

Design of the Air Heater and Vaporizer System

Shell and Tube Vaporizer (STV)

The shell and tube heat exchangers have been used for wide range of services for long time. More than 140 units of STV's have been used for LNG service since 1974.

The temperature of LNG entering the vaporizer is between - 140 to -160 deg C during normal operation. However, its inlet HTF temperature is between 10 to 20 deg C. In order to increase safe operation without freezing of HTF, a co-current flow is recommended. (or combination of co-current and counter current depending on process requirements). The film temperature needs to be carefully checked for each baffle space to ensure it is significantly above the HTF freezing temperature [7].

The time it takes to freeze HTF without flow in the vaporizer needs to be checked. Normally, using HTF having a lower freezing point does not significantly increase the freeze time. Moreover, HTF having a lower freezing point has less effective thermal characteristics. HTF with a lower freezing point requires more heat exchange surface area than HTF with relatively high freezing point, which results in less cost effectiveness (high Capex and Opex). If the THF is not flowing into the vaporizer, adopting other safety systems and devises, such as window baffle, nitrogen ejection of the HTF, will enhance the safe operation of the vaporizer more than using HTF with lower freezing point.

The STV has higher turn down ratio (zero to 100%). Technically, it depends on flow controllability by the LNG flow control valve. Since an expansion joint is incorporated into the vaporizer, it allows quicker ramp up time (less than 15 min) than the ORV, which generally requires more than an hour. The special evaporated vapor mixing device is inserted into each heat exchange tube. The special device ensures no cold liquid break through.

Heat Transfer Fluid (HTF)

The selection of a HTF for LNG vaporization and the design temperature directly affects the equipment sizing, reliability of the vaporization system, operating cost and environmental impact. A low-viscosity HTF at low temperature decreases both heat exchange surface area of the air heater and the vaporizer such that the total installed cost can be reduced. The low-viscosity HTF also reduces HTF pumping power, resulting in low operating costs.

Generally, HTF available in market are:

- Ethylene glycol water solution
- Propylene glycol water solution
- Potassium based water solution

Ethylene glycol has been widely used for LNG vaporization industry. Though Propylene glycol is environmentally friendly it has a high viscosity at low temperature. Ethylene glycol is toxic. Potassium based HTF has good thermal characteristics and is environmentally friendly.

Air Heater Design

The air-fin heaters use fans mounted above the units to force air across banks of finned tubes. As the air cools, water condenses from the air and supplies a significant amount of the heating duty.

The water is collected in a basin below the air-fins. The water can be used as firewater or disposed of as required. It is a mineral-free water with pH same as the rainwater and need not require treatment prior to disposal. There may even be byproduct utilization with the generated water. However, if the HTF used in the AHV system which is toxic, HTF leak detection system and the condensate treatment system should be considered.

Heat is extracted by heat exchange from air to HTF through the heat exchange tubes. While air is cooled at the surface of the heat exchange tubes, the HTF is warmed up. During this process, moisture in the air is condensed at the surface of the tubes, results in using both sensible and latent heat. In order to calculate the heat extracted from the air, the following mechanism is modeled for an indirect heat exchange:

- Sensible Heat Transfer: from ambient air – fins – tube wall – HTF (1)
- Latent Heat Transfer: from condensing vapor – fins – tube wall – HTF (2)
- Sensible Heat Transfer: from ambient air – film – fins – tube wall – HTF (3)

The sensible heat transfer is modeled either in direct contact in the non-wetted surface area of the tube surfaces, or in the condensate film on the tube surface.

AHV System Design

There are two variables affecting the amount of heat that can be recovered from ambient air for a particular air heater design: ambient air temperature and dew point temperature (or Relative Humidity: RH). In order to determine a practical design for the air heater system, an optimization process must be performed in which total installed cost (TIC) and Opex are balanced to offer a favorable net present value (NPV) for the investment.

The first step is to set up the design air temperatures. If the selected design temperature is too low, then the air heater will be oversized. Conversely, if design air temperature is too high, the available heat recovery from air in winter season may not be enough, resulting in adding a supplementary heat source. A supplementary heat source may not be necessary under the tropical temperature conditions. However, where ambient temperature in winter is not high enough to provide vaporization heat, a supplementary heat source or back up vaporization level can be determined based on availability and economic study. A supplementary heat source is linked to the HTF heat as shown in Fig. 4, while a back up vaporization facility is to be used for the balance of LNG vaporization capacity.

The minimum threshold temperature to stop operation of the air heater should also be determined. The power consumption and an even point of heat gain from the air heaters can be used to determine the turn-off temperature of the air fan. However, continuous fan operation is recommended unless ambient air temperature drops below 5°C. During winter season, it is expected that the fans may be turned off for a considerable duration depending on the site conditions.

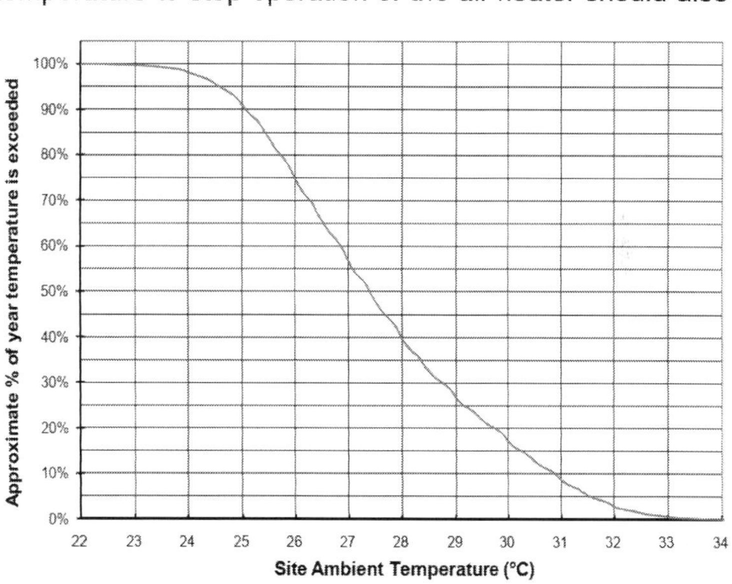

Fig. 6 - Temperature Exceed Curve (Tropical Case)

The ambient temperature exceedance curve at a hypothetical LNG regasification project site is shown in Fig. 6. To determine the design air temperature is a sort of iteration process to find out optimization point. The heart recovery and capital expenditure is a strong function of ambient temperature and relative humidity (RH).

Dew point temperature is an important consideration in the design of air heater units because it can be related to the amount of heat that an air heater can recover from latent heat as a result of condensation of water from the air. The amount of heat recovered from latent heat is significant in that it comes as an additional benefit over the amount of heat recovered from sensible heat. The relationship between design relative humidity and capital cost of the air heater is shown in Fig. 7.

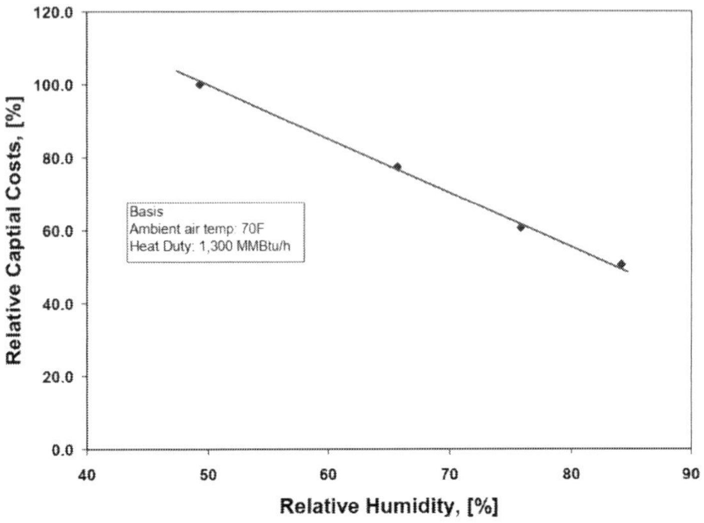

Fig. 7 – Relationship between and Capital Costs

A higher design RH decreases capital costs, but frequently requires operation of the supplementary heat source. Determining the appropriate design RH value also requires optimization based on the life cycle costs along with frequency analysis of RH data [8].

Optimization Process

At the beginning of the optimization process, the air heater design temperature should be selected based on engineering experience. A sound engineering background for approach air temperature of the air heater and expected rate of air recirculation will reduce iteration optimization process, which results in reliable design of the AHV system. The difference between the inlet temperature of the fluid providing the heat, and the outlet temperature of the fluid being heated, is called the approach temperature. The achievable approach temperature for a heat exchanger is a function of parameters such as the available heat exchanger surface area, resistance in heat transfer between the two fluids, and fluid mechanics (flow rate, viscosity, thermal characteristics, Reynolds number, etc).

During optimization process, economic and technical justification will be performed with the various design parameters. This includes wind directions, velocities, RH, and ambient temperature. Air path and its constraints due to large structures (LNG Tanks and buildings) have also to be reviewed. Figure 8 illustrates brief AHV optimization process.

Optimization of the air heater design consists of several steps. First, a judgment must be made as to the desired ambient temperature percentile at which the air heater will recover sensible heat from air temperature drop and latent heat from vapor condensation (moisture in air). The relevant range data of RH and ambient air temperature can be obtained from the analysis of the ambient air data. For the sake of this example, assume a reasonable percentile such as 90%. In other words, assume a design for the air heater to recover the heat for LNG vaporization is fully met during 90%

of the year without operation of a supplementary heat. The remaining 10% of the year only allow operation of back up vaporizers or supplementary heat sources.

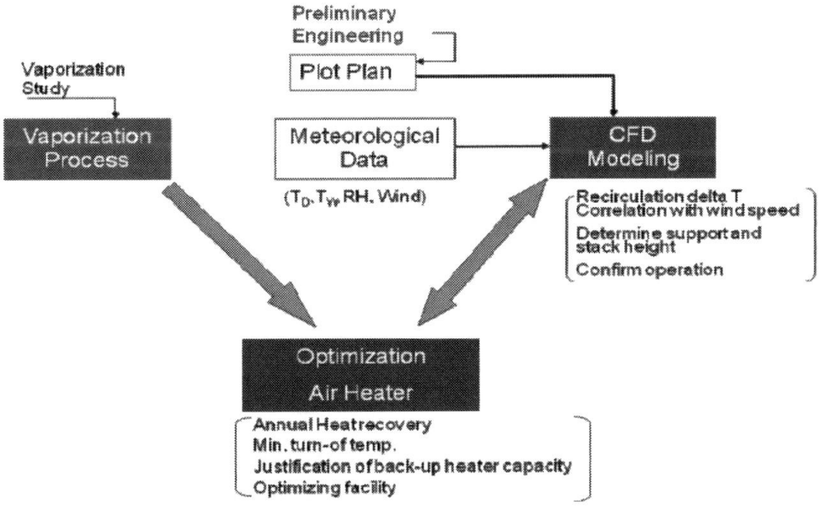

Fig. 8 – AHV Optimization Process

At the assumed ambient temperature percentile, determine a design for the heaters for different ambient temperatures. For example, determine a design for the air heaters at design ambient temperatures of 23°C, 25°C, and 27°C for this hypothetical case. Establish the TIC for each case. Next, calculate the TIC and Opex of the air heaters for each case based on power consumption of the air heater system and fuel gas for supplemental heating (when required).

In other words, based upon the design point for the ambient air temperature, each case will require more or less power consumption and more or less supplemental heating in a given year in comparison to the other cases depending on the actual site ambient conditions.

Higher water condensing results in the design of a smaller air heater size, resulting in low capital cost, but requires a large supplementary heat supply throughout the entire year when RH is lower than that of design case. Alternatively, a lower or no water condensing results in the design of a larger air heater size, which in turn will result in high capital cost. This is the iteration process to determine the optimum level of RH (or dew point).

A net present value (NPV) for each of the design cases considered is to be determined based on TIC and Opex. An ambient air temperature at which to design the air heater is to be determined based on the most attractive NPV. Re-investigation is required for confirming the confidence level of the dew point latent heat recovery percentile at which to design the air heater using the selected ambient air temperature design point. Repeat of the previous optimization exercise until the most attractive NPV can be obtained, but needs to keep the ambient air temperature design point constant and selecting new dew point percentile design cases. This iteration process is required to confirm the confidence level of the air data analysis.

CFD Modeling

Once the feasibility of air heater units has been established for particular site conditions, preliminary engineering may begin in conjunction with evaluation of site meteorological data. After preliminary engineering of the units has been completed, Modern Computational Fluid Dynamics (CFD) modeling should be developed in order to predict the behavior of the effluent air stream and resulting reduction of inlet air temperature to the air heater. At this point, further development of the air heater units becomes an iterative process with the CFD modeling playing a key role in the optimization of the air heater design.

Entrainment of cold fan exhaust into ambient air heater fan intakes (also referred to as cold air recirculation) can reduce the efficiency and operability of air heater units and must be taken into consideration in the air heater design. The colder air exhausted by air heaters has a higher density than that of warmer surrounding air and tends to settle downwards. This settling effect often has an impact on the performance of the air heater unit. Wind conditions also have an impact on cold air recirculation effects, as both wind direction and wind velocity can have a significant influence on the efficiency and operability of ambient air heaters.

The projection of recovered heat as a function of site conditions requires developing correlations of the re-circulated air temperatures with wind direction, wind velocity, and associated air paths around surrounding structures (LNG tanks, building, etc). The recirculated air temperatures under a given wind velocity and direction should be taken into account in estimating the recovered heat. High ambient air temperatures are usually not a cause for concern and in most cases are favorable. However, low ambient air temperatures must be considered in order to ensure that the required duty is achievable with the projected air heater size.

Correlations should be produced based on the results of the CFD models to determine sensitivity of the given design and plot plan to differing wind conditions. The correlations will give an indication as to the average inlet air temperature that can be expected depending upon wind conditions. The results of this evaluation should indicate that the inlet temperatures used to determine heat recovery of the air heater units cannot be based on actual ambient air temperature.

As such, the temperatures used to determine heat recovery must be adjusted to account for cold air recirculation. Site meteorological data in conjunction with the developed correlations can be used to determine corrected fan inlet air temperatures specified on an hourly basis. Based on this adjustment, heat recovery of the air heaters can then be calculated and used to determine operating costs.

The CFD modeling analysis is performed to predict heat exchanger recirculation behavior under a variety of wind conditions. Figs. 9 & 10 illustrate the air recirculation at 10 m/s of wind velocity and air stream path through the air heater.

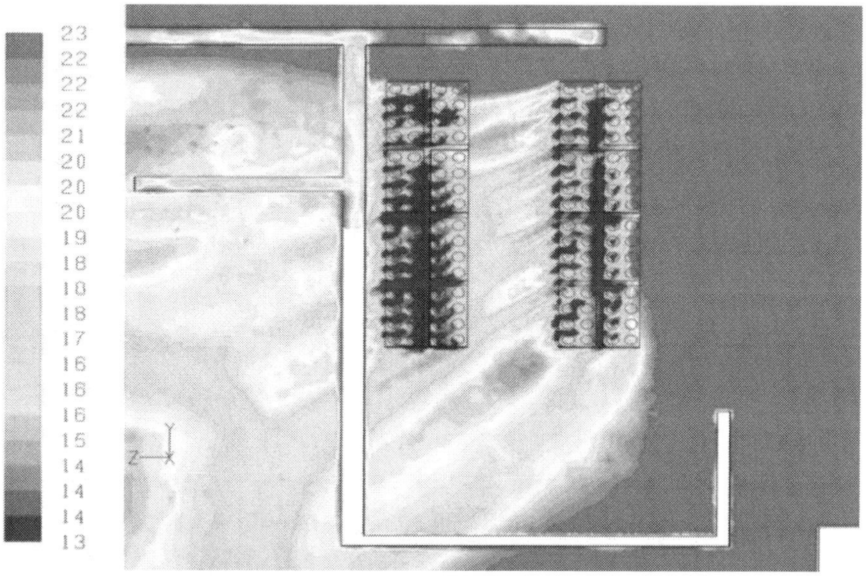

Fig. 9 – Air Flow around the Air Heater

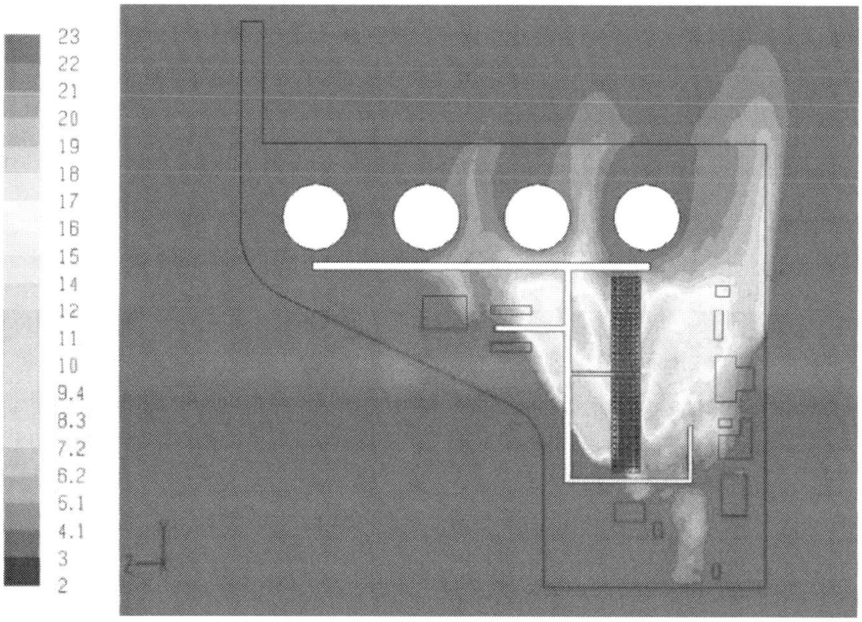

Fig 10 – Air Flow Impact due to Large Structure

The air recirculation needs to be carefully checked with the different air directions and wind velocities. The simulation results reveal that the parallel installation of the air heater is more severe in air recirculation (Fig. 9) than the installation in series (Fig. 10).

Figure 10 illustrates large structure (LNG tank) and adjacent building can also impact air path.

The main objective of the CFD modeling is to confirm the air recirculation is within the design point set by the process engineering with the various wind direction and velocities. If the air heater layout with the facility plot brings more temperature drop due to air recirculation, the air heater and plot need to be revised until the CFD modeling results are satisfied with the design conditions. If this cannot be achieved because of design restriction or site limitation, the most severe recirculated air temperature shall be applied to the calculation of heat recovery available.

CONCLUSION

The ORV's, a conventional vaporization technology, have been widely used in Asian countries. However, this vaporization technology recently brought concerns in seawater usage: mechanical, chemical and thermal pollution.

Newly developed air heater vaporization (AHV) technology has been operated at Dahej since 2004 without any specific operation issues. Other terminals are also operating and commissioning the same vaporization technology. Well proven equipment, such as the air coolers and the shell and tube vaporizer (STV) have been used in this technology. This new trend in LNG vaporization technology is environmentally friendly and has lower Capex and Opex than the conventional vaporization without pollutions.

More than 140 units of STV's for LNG services are under operation since 1974. The STV operation is reliable and flexible in response to send-out load fluctuation. The required ramp up time from zero to 100% load is faster than the ORV.

The selection of HTF is an important step of the AHV system design. Lowering freezing point of a HTF may not significantly enhance safe operation of the vaporizer. Instead, it requires more heat exchange surface area, results in less cost effectiveness. The suitable freezing temperature of the HTF ensures optimization of the AHV system with reliable safe operation.

Selection of design ambient temperature and design relative humidity requires various costing processes with the different meteorological conditions. This includes total installed costs and operating cost of the AHV system.

It should be noted that the air heater design requires complicated iteration process. Optimization involves extensive studies of local meteorological data, plot plan and the site location. CFD modeling can be utilized to predict the behavior of the effluent air stream, resulting in reduction in suction temperature of inlet air. Air recirculation needs to be checked with various wind directions and wind velocities in order to confirm that the recirculation is within the design point.

REFERENCES

1. World Bank Group: "Table 4, Liquid Effluent from Onshore Oil and Gas Production, Oil and Gas Development," Pollution Prevention and Abatement Handbook (1998).
2. "Louisiana Department of Environmental Quality (LDEQ) Office of Environmental Services, Agency Interest Number 119267, http://www.deq.louisiana.gov/apps/pubNotice/pdf/SPSOB11-9-06.pdf.
3. "Sabine Pass Registration Statement Form S-4/A",http://google.brand.edgar-online.com/EFX_dll/EDGARpro.dll?FetchFilingHTML1?ID=5081465&SessionID=PO26Wvigikg9247
4. Cho, J.H., Marthur, G., Kotzot, H., and Durr, C.A.: "Availabilities and Limitations of LNG-Vaporization Process", LNG Journal p. 24-26, June 2005.
5. Email communication with Mr. Steve Turner, 29 Jan. 2010.
6. Ahuja, M.M., Gupta, A.K., and Sunder, S.: "A Case Study – LNG Regasification Using Ambient Heat at Dahej, Gujarat, India", presented at Gas Tech 2006, Abu Dhabi, 4-7 Dec. 2006.
7. Cho, J.H., Kotzot, H., Vega, F., and Durr, C.A.: "Three-Shell Vaporizer to eb Cost Effective Air Heater Vaporization System", LNG Journal p. 48-51, November 2006.
8. Cho, J.H., Kotzot, H., Vega, F., and Durr, C.A.: "LNG Vaporization Needs Low-Cost and Environmentally-Friendly Solutions", LNG Journal p.28-30, March 2006.

BIOGRAPHY of PRESENTER

Joseph H. Cho is a Technology Director of Houston Technology Center at SK Engineering and Construction Co., Ltd. He is a registered professional engineer and has been in the LNG industry for twenty two years. He was involved in process design of LNG liquefaction plants and import terminals for various regions as a process lead engineer and technology manager. He graduated from the University of Oklahoma with a Ph.D. degree in Petroleum Engineering.

Email: jcho22@skec.com

(145c) Uncertainties in Modeling LNG Vapor Dispersion with CFD Codes

Ruifeng Qi, Chemical Engineering, Mary Kay O'Connor Process Safety Center, College Station, TX

Dedy Ng, Chemical Engineering, Mary Kay O'Connor Process Safety Center, College Station, TX

Simon P. Waldram, Chemical Engineering, Texas A&M University at Qatar, Doha, Qatar

M. Sam Mannan, Chemical Engineering, Mary Kay O'Connor Process Safety Center, College Station, TX

Computational fluid dynamics (CFD) models have been widely proposed and developed to predict LNG vapor dispersion behavior and to determine safe separation distances around LNG production storage and transportation facilities. This is largely due to the ability of CFD codes to describe fully the relevant physical processes in complex geometrical circumstances. However, like other consequence models, CFD models for LNG vapor dispersion entail a number of simplifying assumptions, approximations, empirical relations and constants. Therefore uncertainties in predicted results are inevitable. These uncertainties are generally associated with either the numerical solution methods or an incomplete modeling of the physics involved. In this paper, a review of factors which can impact the results of CFD modeling is provided. A sensitivity analysis was conducted to estimate the scale of uncertainties for key parameters. In addition, some recommended methods are introduced to control or reduce the magnitude of uncertainties to a prescribed confidence level.

See more of this Session: Professor Cedomir M. Sliepcevich Memorial Session Five - LNG Simulation
See more of this Group/Topical: Topical 6: 10th Topical Conference on Gas Utilization

CFD Simulation of Vapor Dispersion from LNG Jetting and Flashing Releases

Filippo Gavelli [a,1], *Scott Davis* [a], *Olav Roald Hansen* [a], *Mathieu Ichard* [b]

[a] *GexCon US,* [b] *GexCon AS*

Abstract

Accidental releases of liquefied natural gas (LNG) from high-pressure piping have recently become a matter of growing interest. In the context of U.S. onshore LNG facility sitin, for example, LNG terminal developers have been required to demonstrate that the flammable vapor cloud (to the ½ LFL concentration) due to "jetting" and "flashing" from pressurized LNG releases will not extend beyond the property boundaries.

The physical phenomena that occur as pressurized LNG flows from a breach are rather complex; as the release typically begins as a flashing jet that forms a gas-aerosol mixture as it expands to atmospheric pressure, which is followed by the rain-out or evaporation of the liquid droplets. The resulting vapor cloud, therefore, is likely to have very different characteristics from the "traditional" vapor cloud from a spill into an impoundment.

Integral models are available that can calculate the formation and dispersion of the vapor cloud from a high-pressure release in the absence of obstacles or significant terrain features. In many practical cases, however, obstacles are likely to be present along the path of the two-phase jet (e.g., flange guards, nearby equipment, pipe racks, etc.) or along the path of the vapor cloud (e.g., buildings, vapor barriers, etc.). In order to properly account for the effect of these obstacles, a more sophisticated modeling approach is necessary.

This paper discusses the simulation of high-pressure LNG releases using the computational fluid dynamics (CFD) model FLACS. FLACS can overcome the limitations of simpler models by: 1) accounting for the effects of obstacles and complex geometry on the flow of air and vapors; 2) simulating the behavior of the two-phase jet release from the source location. Case studies are presented to demonstrate the application of FLACS to different high-pressure LNG release scenarios.

Background

The analysis of LNG vapor dispersion hazards, particularly in the context of onshore facility siting, is typically focused on the consequences of atmospheric pressure releases, such as would result from a guillotine break to a storage tank withdrawal line. However, other types of leaks are possible in an LNG facility and their consequences need to be evaluated to ensure they do not affect the plant siting requirements. A scenario that has recently been brought up within the permitting process for several U.S. facilities is that of a high-pressure LNG release, for example,

[1] Corresponding author: fgavelli@gexcon.com

from the sendout line to the LNG vaporizers. In those instances, LNG terminal developers have been required to demonstrate that the flammable vapor cloud (to the ½ LFL concentration) due to "jetting" and "flashing" from pressurized LNG releases will not extend beyond the property boundaries.

The physical phenomena that occur as pressurized LNG flows from a breach are rather complex and include:

- Flashing (if the pressurized liquid is at temperatures above the atmospheric boiling point);
- Atomization of the liquid stream as it expands to atmospheric pressure;
- Entrainment of air by the liquid particle jet;
- Droplet evaporation; and
- Rainout (i.e., deposition of liquid droplets that reach the ground or hit objects before evaporating).

Semi-empirical correlations have been developed from experiments involving high-pressure releases [1, 2], so that quantities such as the flashing rate, air entrainment rate, vaporization and rainout can be predicted. These quantities can be used to calculate the source term for vapor dispersion in models, such as SLAB [3] or PHAST [4]. This type of approach is generally adequate provided that:

- The leak forms a unidirectional jet;
- The high-pressure jet can expand and fully evaporate without encountering any obstacles;
- The resulting vapor cloud can disperse downwind without encountering any obstacles.

In many practical cases, however, obstacles are likely to be present along the path of the two-phase jet (e.g., flange guards, nearby equipment, pipe racks, etc.) or along the path of the vapor cloud (e.g., buildings, vapor barriers, etc.). In order to properly account for the effect of these obstacles, a more sophisticated modeling approach is necessary.

This paper discusses the simulation of high-pressure LNG releases using the computational fluid dynamics (CFD) model FLACS. FLACS provides the following advantages when applied to the simulation of vapor dispersion from high-pressure LNG releases:

1. It is a CFD model capable of simulating the effects of obstacles and complex geometry on the flow of air and vapors;
2. It has been validated for LNG vapor dispersion; and
3. It includes a two-phase module capable of simulating the high-pressure jetting and flashing release from the source.

The paper will discuss the validation of the two-phase jet module for FLACS and will include examples to demonstrate the applicability of FLACS to "jetting and flashing" scenarios.

FLACS

The FLACS CFD model was developed in the early 1980s to simulate gas explosions in offshore platforms. The conservation equations are solved on a Cartesian grid using a finite volume method. The k-ε turbulence model is used to close the equations for fluid motion, with some modifications made for atmospheric boundary layer turbulence and dispersion applications [5-7]. A distributed porosity concept is implemented in FLACS to handle complex geometries, and turbulence production terms are parameterized for sub-grid objects[8]. At the upwind boundary of the domain, vertical profiles of wind speed and direction, temperature, turbulent kinetic energy and eddy dissipation rate are imposed according to the atmospheric stability class or to the Monin-Obukhov and surface roughness lengths. The vapor dispersion capabilities of FLACS have been validated against several field experiments involving both dense gas and passive gas releases, including CO_2 releases in an array of flat ''billboard''-shaped obstructions (the Kit Fox experiments) and tracer gas releases in an array of rectangular storage containers in the Mock Urban Setting Test (MUST) experiments [5]. More recently, FLACS was also validated against the 33 experimental tests included in the Model Validation Database for LNG vapor dispersion models [9].

Two alternative models can be used with FLACS to simulate the vapor cloud dispersion from a pressurized flashing jet release. The simpler method consists of a "flash" utility, which allows the user to define a pseudo-release location, where the release can be assumed to start as a single-phase jet (after flashing), for which the effects of droplet rain-out and evaporation, and air entrainment have been included [10]. At the single-phase pseudo-source location, the jet is a mixture of gas and air at a temperature below that of the source liquid, due to the droplets' evaporative cooling effect. The jet mass flow rate is larger than the leak flow rate, due to air entrainment, and the area of the pseudo-source is also increased. The "flash" utility, however, assumes that there is no impingement on objects during the jet evaporation and expansion process (which can take place over distances on the order of 10-20 m). When obstacles are closer than the "flash" utility allows, a different approach may be followed. The alternative approach consists of a "two-phase module", which allows the aerosol phase of the high-pressure release to be tracked within the three-dimensional computational domain, as the aerosol spreads and vaporizes. The depressurization and flashing of the high-pressure release are first calculated, and a pseudo-source for the two-phase (gas and aerosol) atmospheric-pressure jet is defined. The two-phase pseudo-source is located very close to the actual leak, effectively eliminating the restriction on object impingement that applies to the "flash" utility. The two-phase, atmospheric pressure jet is then used as the source for the CFD simulations. An accurate representation of the jet dispersion requires the behavior of the evaporating droplets to be modeled. In FLACS, the droplet evaporation and jet cooling effects that characterize this phase are implemented through the homogeneous equilibrium model [11]. This includes accounting for rain-out, liquid pool spreading and evaporation.

The two-phase flashing jet dispersion model was validated against experimental data from the INERIS experiments [12] as well as from the Desert Tortoise test series [13]. In the Desert Tortoise simulations, for example, rain-out predictions by the model were found to be within 30% of the observations in three of the four tests. Statistical Performance Measures to assess the performance of the model reported an over-prediction of 25% in the maximum gas concentration at different sensor locations, with 94% of all predictions falling within a factor of two of the experimental data.

Case Studies

Two high-pressure LNG release scenarios are examined in this paper. The first scenario consists of a guillotine break to a small-diameter horizontal high-pressure pipe connection, for the purpose of demonstrating the effect of obstructions on the dispersion of the flammable vapor cloud from a jetting and flashing release. The second scenario consists of a circumferential flange leak in a horizontal high-pressure pipe, for the purpose of demonstrating the application of FLACS' two-phase module to a multidirectional pressurized LNG release.

Scenario 1: Guillotine break

The leak scenario consists of a guillotine break to a 5 cm diameter horizontal connection to a 30 cm diameter pipe carrying LNG. The pipe runs horizontally at 10 m elevation; the pipe pressure is 10 barg and the LNG is at -150°C. The leak is assumed to continue at the nominal flow rate (i.e., with no drop in pressure inside the pipe) for 180 s and the simulation is continued until 200 s. Ambient conditions are assumed to be as follows: 20°C temperature, 2.0 m/s wind speed at 10 m elevation, and stable atmosphere (Pasquill-Gifford class "F"); the effects of humidity are neglected.

The two-phase flashing jet emerging from the breach is allowed to expand, entrain air and evaporate without encountering any obstructions. The scenario is solved by first calculating the characteristics of the two-phase jet, as well as those of a pseudo-source at the location where the liquid droplets have completely evaporated (i.e., where the two-phase jet has become single-phase). The flashing jet entrainment and evaporation calculations are performed using the "flash" utility included in FLACS, whose output is summarized in Table 1. The single-phase jet pseudo-source provides the inlet boundary condition for the three-dimensional vapor cloud dispersion simulation.

Table 1. Flashing jet and vaporized jet characteristics

	Flashing jet	Vaporized jet
Distance from breach	0 m	6.76 m
Flow area	19.6 cm^2	4116 cm^2
Mass flow rate	37.84 kg/s	135.51 kg/s
Temperature	-150°C	-171.1°C
Mass fraction of flashed vapor	6.87%	-
Gas-air equivalence ratio	∞	6.68

Two different cases are considered for the dispersion of the vapor cloud from the single-phase jet source:

1. No obstacles;
2. A vapor barrier (4 m tall and 200 m wide) is placed perpendicular to the wind direction, approximately 150 m downwind of the high-pressure release.

Figures 1 through 3 compare the size of the vapor cloud (to the 50% of LFL concentration) for the unobstructed dispersion (top) and obstructed dispersion (bottom) at different times. Figure 1 shows that, early in the simulation, the presence of the vapor barrier has no noticeable influence on the growth of the vapor cloud. As the cloud approaches the barrier, the momentum of the vapor cloud causes it to impinge against the barrier with limited lateral spread (Figure 2), unlike the typical behavior of low-momentum vapor clouds from atmospheric pressure releases. However, even though portion of the vapor cloud flows over the barrier, the presence of the vapor barrier reduces the maximum downwind distance reached by the ½-LFL cloud from approximately 310 m (unobstructed case) to approximately 165 m (obstructed case).

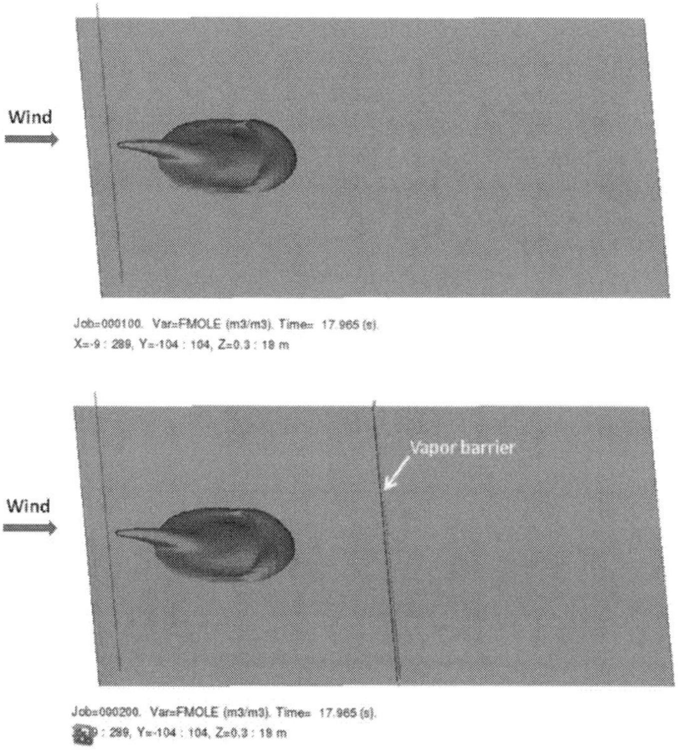

Figure 1. ½-LFL vapor cloud dispersion from high-pressure release, after approximately 18 s: unobstructed case (top) vs. obstructed case (bottom).

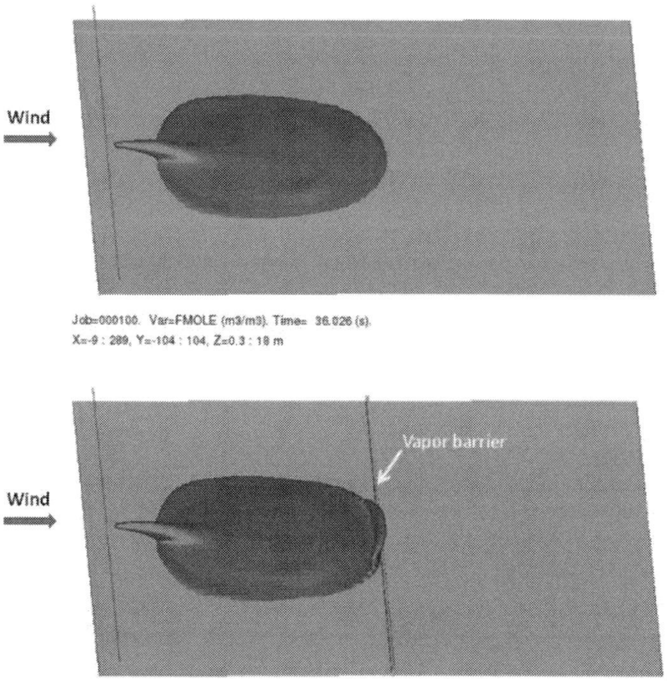

Figure 2. ½-LFL vapor cloud dispersion from high-pressure release, after approximately 36 s: unobstructed case (top) vs. obstructed case (bottom).

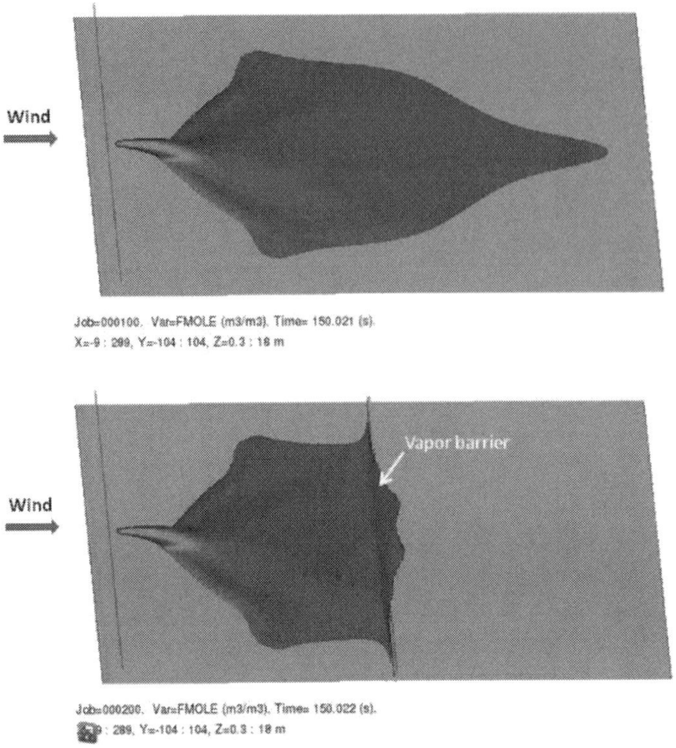

Figure 3. ½-LFL vapor cloud dispersion from high-pressure release, after approximately 150 s: unobstructed case (top) vs. obstructed case (bottom).

This example demonstrated how coupling the one-dimensional flashing jet calculations with a three-dimensional CFD model allows one to account for and benefit from the effect of obstacles (including passive mitigation features) on vapor cloud dispersion.

Scenario 2: Flange leak

The leak scenario assumes a complete gasket loss in a flange for a 30 cm diameter pipe carrying LNG. The pipe runs horizontally at 10 m elevation; the pipe pressure is 10 barg and the LNG is at -150°C. The leak is assumed to continue at the nominal flow rate (i.e., with no drop in pressure inside the pipe) for 180 s and the simulation is continued until 200 s. Ambient conditions are assumed to be as follows: 20°C temperature, 2.0 m/s wind speed at 10 m elevation, and stable atmosphere (Pasquill-Gifford class "F"); the effects of humidity are neglected.

The complete loss of a gasket results in a radial leak around the circumference of the flange, with a total flow area of approximately 29.9 cm^2. The resulting flow rate is approximately 92 kg/s, with a liquid volume fraction of approximately 4.5%, and is assumed to remain constant for 180 s. This case has been simulated with the FLACS's two-phase module and Figures 4 and 5 show the volume fraction isocontours for the liquid (top) and the vapor (bottom), respectively, 1 s and 30 s after the beginning of the leak.

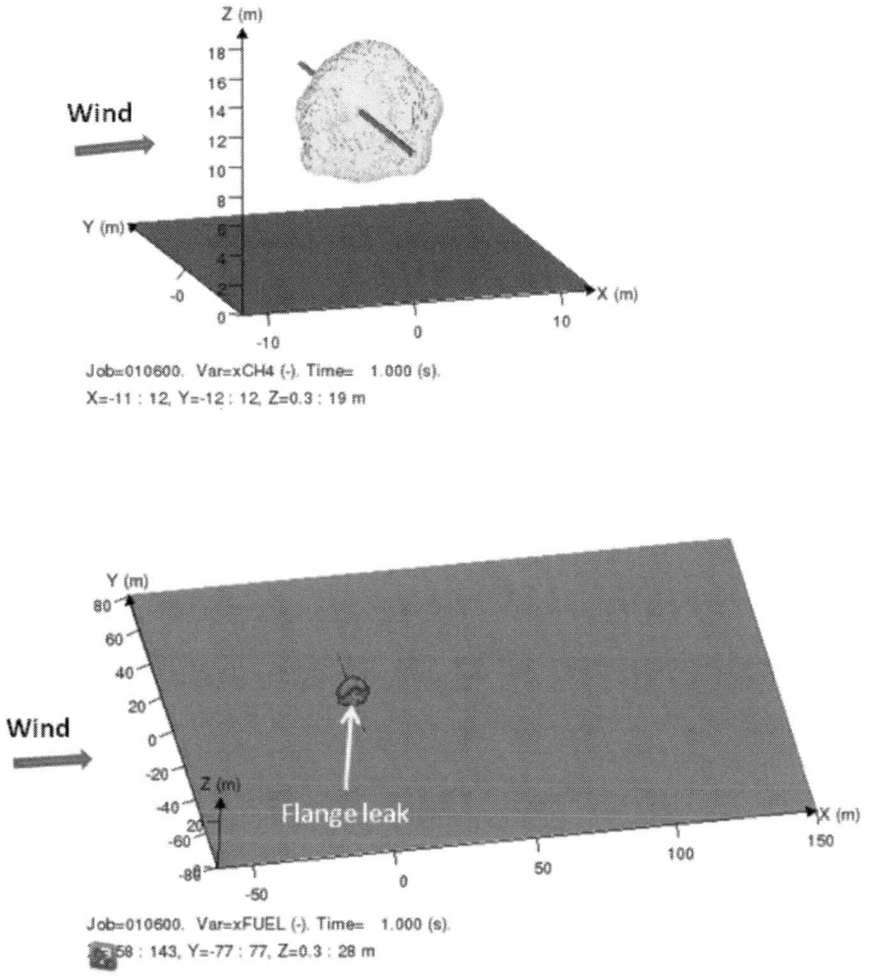

Figure 4. Liquid (top; 0.01% volume fraction, which after evaporation would correspond to an equivalent vapor cloud at approximately LFL) and vapor (bottom; 2.5% volume fraction) volume fraction from high-pressure flange leak, after approximately 1 s. Note that the top image is zoomed to the leak location to show the liquid volume fraction distribution more clearly.

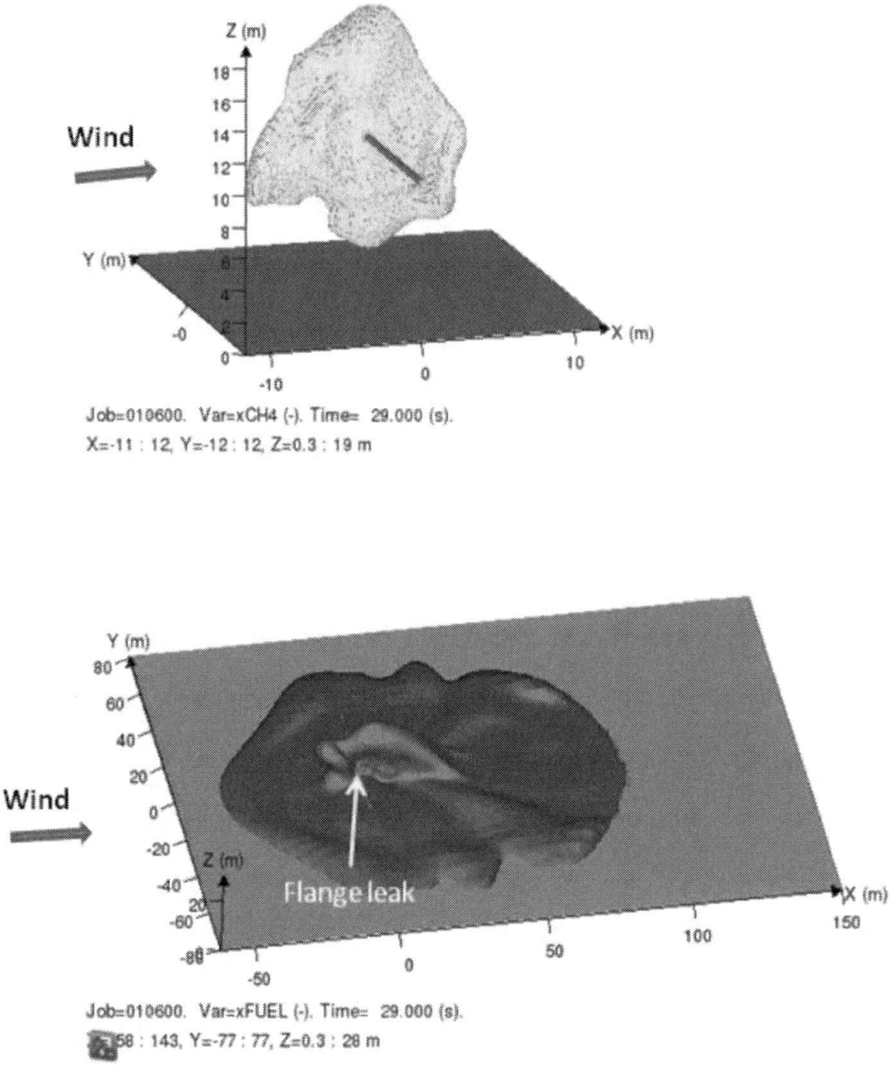

Figure 5. Liquid (top; 0.01% volume fraction, which after evaporation would correspond to an equivalent vapor cloud at approximately LFL) and vapor (bottom; 2.5% volume fraction) volume fraction from high-pressure flange leak, after approximately 30 s. Note that the top image is zoomed to the leak location to show the liquid volume fraction distribution more clearly.

The flashing jet correlations typically used for high-pressure releases (as in Example 1 above) cannot be applied to the radial leak pattern in this scenario without drastic and potentially very conservative assumptions. For example, if the vapor cloud dispersion is calculated using an integral model, only a single, unidirectional jet can be simulated: an intuitive, but not necessarily conservative approach would be to split the radial flow into two horizontal components – upwind and downwind – and to calculate the vapor cloud dispersion only for the downwind component.

In Figure 6, the vapor cloud dispersion from a single, horizontal flashing jet at half the leak flow rate (calculated as in example 1) is compared with the more realistic radial leak pattern simulated using FLACS' two-phase module. In the radial leak simulation, the two-phase jet emerges with zero net momentum in the downwind direction, which results in a wider and slower moving vapor cloud that dissipates below ½-LFL within a shorter distance than in the horizontal jet simulation.

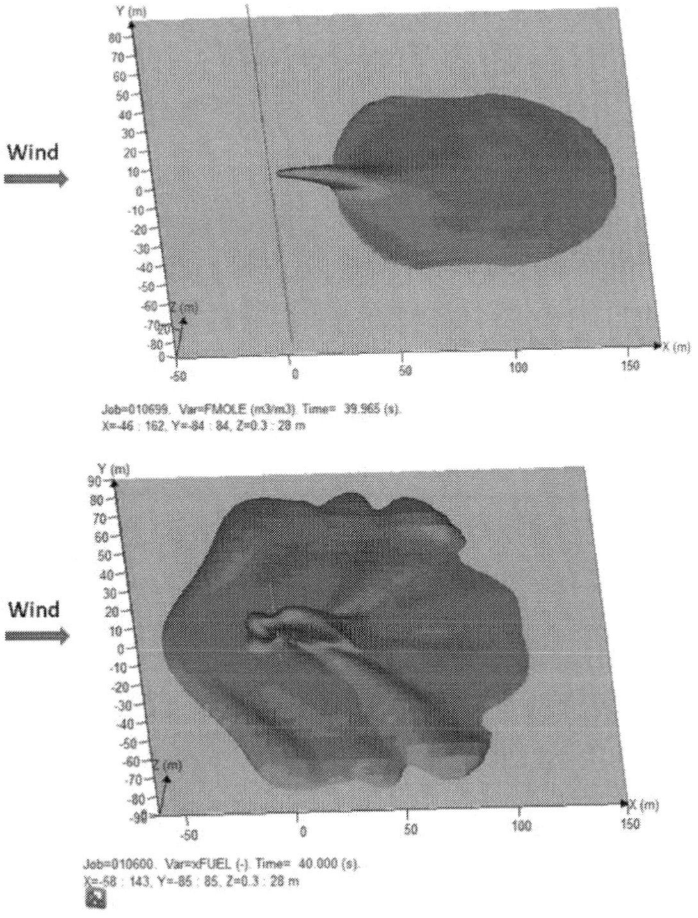

Figure 6. ½-LFL vapor cloud dispersion from high-pressure flange leak, after approximately 40 s: horizontal flashing jet (top) vs. two-phase radial leak (bottom). Note that the horizontal flashing jet (top image) represents only half of the total LNG spill flow rate.

Conclusions

This paper demonstrates the application of the CFD model FLACS to high-pressure releases of LNG, for the purpose of determining the vapor dispersion hazard distances for facility siting. FLACS provides the following advantages when applied to the simulation of vapor dispersion from high-pressure LNG releases:

1. It is a CFD model capable of simulating the effects of obstacles and complex geometry on the flow of air and vapors;
2. It has been validated for LNG vapor dispersion; and

3. It includes a two-phase module capable of simulating the high-pressure jetting and flashing release from the source.

The examples provided in this paper show the potentially significant benefits of performing vapor dispersion calculations from high-pressure releases using a CFD tool like FLACS, in terms of reducing the size of exclusion zones by performing more realistic simulations or by taking proper credit for passive mitigation features (e.g., vapor barriers).

References

[1] Witlox, H.; Harper, M.; Bowen, P. & Cleary, V., "*Flashing Liquid Jets and Two-Phase Droplet Dispersion. I. Overview (Literature Survey, Experiments and Model Implementation / Validation)*", 8th Annual Symposium, Mary Kay O'Connor Process Safety Center, 2005

[2] Cleary, V.; Bowen, P. & Witlox, H., "*Flashing Liquid Jets and Two-Phase Droplet Dispersion. II. Experiments for derivation of droplet atomisation correlations*", 8th Annual Symposium, Mary Kay O'Connor Process Safety Center, 2005

[3] Ermak, D. L., "*User's Manual for SLAB: an Atmospheric Dispersion Model for Denser-than-Air Releases*", Lawrence Livermore National Laboratory, 1990, UCRL-MA-105607

[4] Witlox, H. M. & Holt, A., "*A unified model for jet, heavy and passive dispersion including droplet rainout and re-evaporation*", CCPS, 1999

[5] Hanna, S. R.; Hansen, O. R. & Dharmavaram, S., "*FLACS CFD air quality model performance evaluation with Kit Fox, MUST, Prairie Grass, and EMU observations*", Atmospheric Environment, 2004, 38, 4675-468

[6] Hanna, S. R.; Hansen, O. R.; Ichard, M. & Strimaitis, D., "*CFD model simulation of dispersion from chlorine railcar releases in industrial and urban areas*", Atmospheric Environment, 2009, 43, 262-270

[7] Dharmavaram, S., Hanna, S.R. & Hansen, O.R, "*Consequence analysis – using a CFD model for industrial sites*", Process Safety Progress, 2005, 24, 316–327.

[8] Arntzen, B.A., "*Modelling of turbulence and combustion for simulation of gas explosions in complex geometries*", PhD thesis, 1998, NTNU, Trondheim, Norway. ISBN 82-471-0358-3.

[9] Hansen, O.R., Ichard, M. & Davis, S.G., "*Validation of FLACS for Vapor Dispersion from LNG Spills: Model Evaluation Protocol*", Proc. 2009 Mary Kay O'Connor Process Safety Symposium

[10] Salvesen, H.C., "*Modeling Release of Liquefied Gas Under High Pressure*", CMR-95-F20062, Christian Michelsen Research, 1995, 33 pp.

[11] Kukkonen, J., Kulmala, M., Nikmo, J., Vesala, T., Webber, D.M. & Wren, T., "*The homogeneous equilibrium approximation in models of aerosol cloud dispersion*", Atmospheric Environment, 1994, Vol. 25, No. 17, pp. 2763-2776

[12] Ichard, M., Hansen, O. R. & Melheim, J. A., "*Modeling of flashing releases around buildings*", 90th AMS annual meeting, Phoenix, AZ, USA, 2009

[13] Ichard, M., Hansen, O. R. & Melheim, J. A., "*Releases of pressurized liquefied gases: Simulations of the Desert Tortoise test series with the CFD model FLACS*", 91st AMS annual meeting, Atlanta, GA, USA, 2010

Speaker's Bio

Dr. Filippo Gavelli is the Head of the Dispersion Consulting group at GexCon US, Inc. and specializes in the analysis of heat transfer and fluid flow phenomena, including multiphase flows and cryogenic fluids. He applies his engineering and CFD modeling expertise to the atmospheric dispersion of hazardous gaseous releases, and has extensive experience modeling Liquefied Natural Gas (LNG) vapor cloud dispersion. Dr. Gavelli is responsible for GexCon US' dispersion modeling activities, which include risk assessments and consequence modeling for chemical and petrochemical facilities, offshore installations, hazardous materials transportation and various other applications. He is the lead author of several LNG safety-related papers and has contributed to numerous LNG-related technical committees and expert panels.

A Unified Model for LNG Pool Spread and Vapor Dispersion: Is Wind Scooping Really A Factor?

Filippo Gavelli[1], Scott Davis, Olav Roald Hansen

GexCon US, 7735 Old Georgetown Rd, Suite 1010, Bethesda, MD 20814

Abstract

U.S. regulations for the siting of LNG facilities require LNG spills to be collected into impounded areas, and the impoundments must be sited in such a way that LNG vapor clouds formed as a result of defined spills dissipate to below ½ LFL before reaching a property line that can be built upon. The traditional method to verify facility compliance with these regulations, which coupled the SOURCE5 model to calculate the LNG vaporization rate and the DEGADIS model to calculate vapor cloud dispersion, has recently been the subject of criticism.

In this paper, the vapor generation and vapor cloud dispersion scenario is simulated using an alternative tool: GexCon's computational fluid dynamics (CFD) model FLACS. FLACS incorporates a two-dimensional shallow water-based model for the simulation of LNG pool spreading and vaporization, and a three-dimensional model for the simulation of LNG vapor cloud dispersion. As such, FLACS provides a unified environment in which the entire LNG spill and dispersion scenario can be simulated efficiently and accurately. In fact, FLACS was successfully validated against the entire Model Evaluation Protocol database for LNG vapor dispersion.

The paper first compares FLACS' unified pool spread and vapor generation model with the traditional, (SOURCE5) method for LNG spills into a sump. The paper then utilizes the FLACS model to evaluate the effect of wind-driven turbulence (the "wind scooping" effect) on vapor dispersion from sumps, to demonstrate that the low wind speed requirement (2 m/s) in current regulations does indeed represent the worst-case scenario for vapor dispersion distances.

Background

U.S. regulations for the siting of LNG facilities require LNG spills to be collected into impounded areas, and the impoundments must be sited in such a way that LNG vapor clouds formed as a result of defined spills dissipate to below 50% of the lower flammable limit (LFL) before reaching a property line that can be built upon [1].

Traditionally, LNG terminal applications have verified the compliance of an impoundment with the flammable vapor cloud dispersion requirement by using the combination of the following models: SOURCE5 to calculate the LNG pool spread and vaporization rate within the impoundment; and DEGADIS to calculate the vapor cloud dispersion.

---

[1] Corresponding author: fgavelli@gexcon.com

Over the last few years, several concerns were raised about this approach, particularly in relation to the assumptions built into the SOURCE5 model. For example, it was pointed out that SOURCE5 assumes that the vapor cloud formed by vaporization of the LNG pool does not mix with air and remains at the boiling temperature until the entire sump volume has been filled [2]. In particular, the model was criticized for not including the "wind scooping" effect, that is, turbulence induced by the air flowing over the sump "scooping" LNG vapors out of the sump. Some have argued that wind scooping may result in longer vapor dispersion distances for higher wind speeds than the regulatory value of 2 m/s. More recently, other incorrect assumptions were identified in the treatment of the LNG pool spread within SOURCE5 [3], leading regulatory authorities in the United States to question its applicability to LNG spills into impoundments and to seek alternative models to evaluate these scenarios. Finally, it must be pointed out that SOURCE5, as most other pool spreading and vaporization models, only accounts for heat transfer from the impoundment floor and not from the walls in calculating the vapor generation rate. While this assumption is reasonable for spills in large and shallow impoundments, it neglects a significant contribution to vapor generation in smaller, deeper sumps.

This paper presents an alternative tool to calculate the vapor generation and vapor cloud dispersion from an LNG spill into a sump: GexCon's computational fluid dynamics (CFD) model FLACS. The model will be applied to hypothetical LNG spill scenarios and the results compared with the SOURCE5 vapor generation predictions. Additionally, FLACS will be used to evaluate the effect of wind scooping on vapor dispersion distances as wind speed increases above 2 m/s.

FLACS

GexCon's CFD model FLACS incorporates a fully two-dimensional shallow water-based model for the simulation of LNG pool spreading and vaporization, and a three-dimensional model for the simulation of LNG vapor cloud dispersion. As such, FLACS provides a unified environment in which the entire LNG spill and dispersion scenario can be simulated efficiently and accurately.

Pool Spread Model

FLACS incorporates a pool model that allows for the formation and spreading of the pool, with local evaporation rates based on the heat transfer from the ground or water, radiation, local wind speeds and turbulence levels, and the local vapor pressure above the pool [4]. Figure 1 shows a schematic of the physical processes modeled in the FLACS pool spreading model.

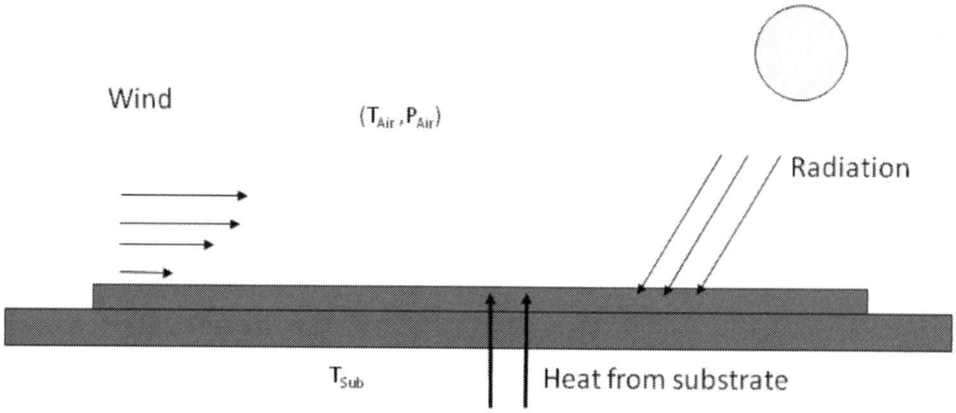

Figure 1. Schematic of the physical processes modeled in FLACS

Vapor Dispersion Model

FLACS solves the Reynolds Averaged Navier-Stokes equations by using the k-ε model with the standard set of constants taken from Launder & Spalding [5] for the turbulent closure. Buoyancy effects are taken into account in the turbulent equations. The atmospheric boundary layer is modeled by forcing profiles for the velocity, temperature and the turbulence parameters on inlet boundaries. Wind inlet profiles are imposed according to the Monin-Obukhov length L and the atmospheric roughness length z_0 for the given atmospheric stability class.

Validation

The FLACS CFD tool has been used to simulate all the experimental tests of the Model Validation Database of the Fire Protection Research Foundation's Model Evaluation Protocol (MEP) [6]. The MEP is used to assess the suitability of dispersion models for predicting hazard ranges associated with large spills of LNG. The models need to be validated against key experimental data which are provided in the Model Validation Database, which consists of both large-scale field trials as well as wind tunnel tests. The 33 tests of the Model Validation Database have been simulated with FLACS and the model predictions matched well with the experimental data, meeting the statistical performance requirements in the MEP [7].

LNG Spill and Vapor Dispersion Scenarios

Example 1 – SOURCE5 and Wall Heat Transfer

The first example presented in this paper addresses the LNG vaporization rate as calculated by the following methods:

1. The SOURCE5 model;

2. The pool spread and vapor generation model built in to FLACS, neglecting LNG heat transfer from the walls to the LNG pool;
3. The unified pool spread and vapor dispersion model built in to FLACS, accounting for LNG vapor formation by heat transfer from the walls.

The scenario consists of a 10-minute LNG spill with a flow rate of 0.547 m^3/s (8,670 gpm) into a concrete-lined sump measuring 10 m by 10 m by 3.3 m deep. The terrain is assumed to be flat, with 0.01 m surface roughness, and no obstructions are present other than the sump cavity (Figure 2). The ambient temperature is assumed to be 20°C (68 F), the wind speed is 2.0 m/s (4.5 mph) and the atmospheric stability class is F (stable).

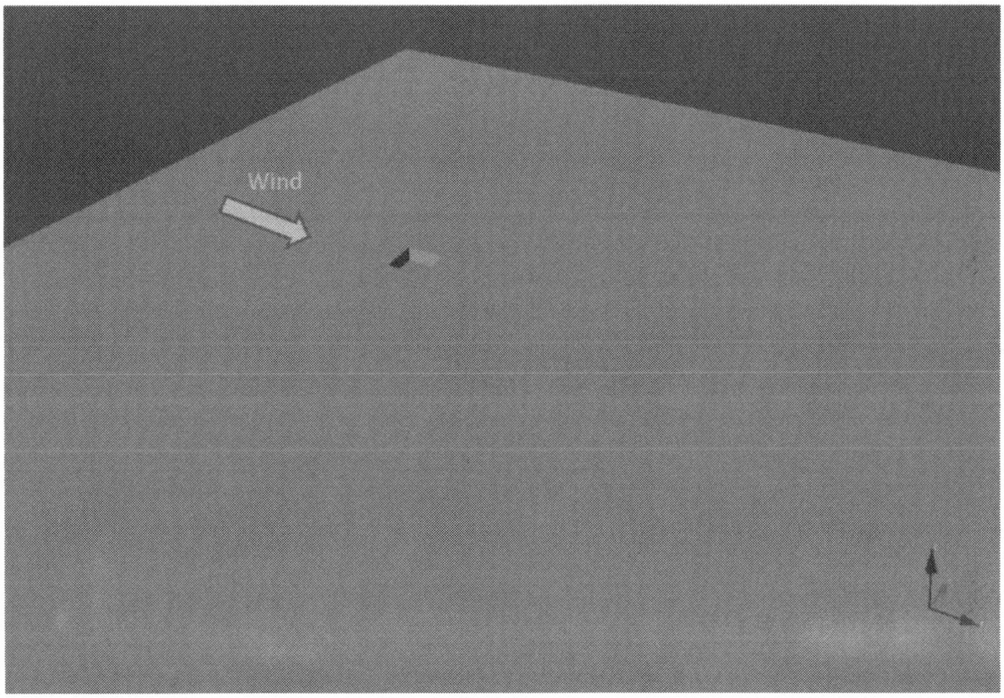

Figure 2. Geometry for scenario LNG sump simulation.

Figure 3 shows the LNG vapor generation rates as obtained following each of the methods above. The vapor generation rates during the floor wetting phase are significantly different between SOURCE5 and FLACS. This is due to the assumption made in SOURCE5, where the vapor generation rate is equal to the LNG spill flow rate until the entire sump floor has been wetted. This is clearly a non-physical assumption, as it would be impossible for the LNG pool to spread on the sump floor if the entire spill flow rate is immediately converted to vapor. Additionally, this assumption makes the SOURCE vapor generation term independent of the sump substrate material until the sump floor is wetted – another non-physical approximation that prevents designers from taking proper credit when utilizing insulating materials as substrate. The FLACS pool spreading model, on the contrary, provides a more realistic estimate of the vapor generation by calculating heat transfer from the floor to the pool based on the distribution of the LNG pool as well as the substrate material of the sump. When only heat transfer from the sump floor is considered, the highest vapor generation rates are obtained at the beginning of the spill, as the pool spreads to wet the sump floor. Once the sump floor is wet, the surface cools down and heat transfer to the pool is rapidly reduced.

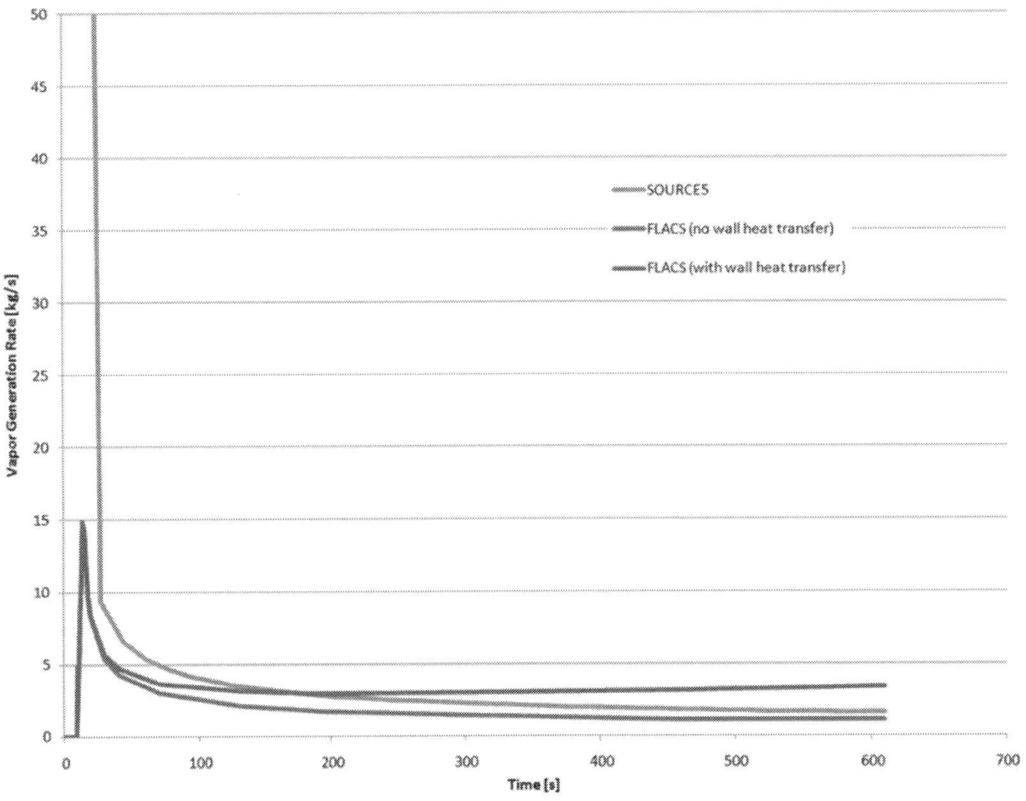

Figure 3. Vapor generation rates for a design spill into a sump. The peak vapor generation rate for the SOURCE5 model, in this example, is approximately 241 kg/s (the plot range is limited to 50 kg/s to better resolve the difference between the curves after the initial phase).

The comparison in Example 1 also demonstrates the effect of heat transfer from the sump walls to the LNG pool, once the floor of the sump has been wetted: as long as LNG is spilled into the sump and the liquid pool depth increases, new portions of the vertical walls continue to be wetted and, as a result, the heat transfer from the walls to the pool increases with time. The contribution of wall heat transfer is responsible for the difference between the red and blue traces in Figure 3. The vapor generation from the walls contributes to the growth of the vapor cloud within the sump and, therefore, results in a larger cloud emerging from the sump, as shown in Figure 4 (approximately 150 s into the spill, which started at time = 10 s) and in a longer vapor cloud dispersion distance, as shown in Figure 5 (approximately the maximum downwind cloud distance).

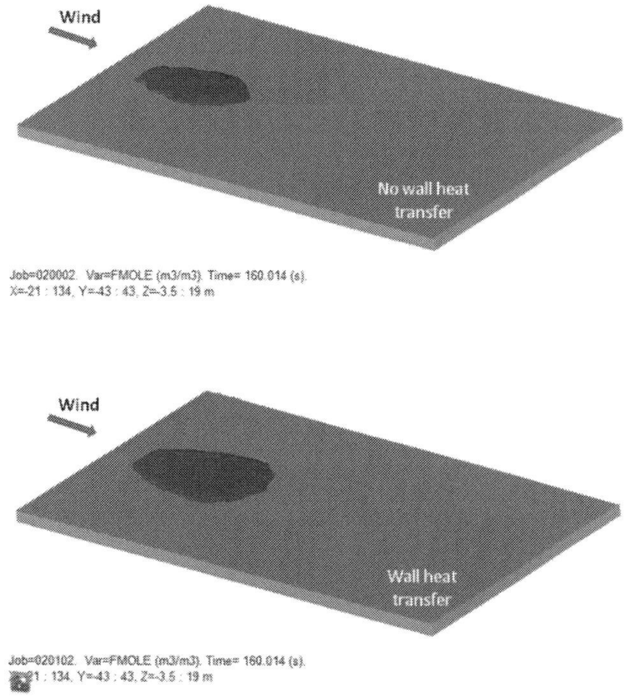

Figure 4. ½-LFL gas concentration isocontours for simulations without (top) and with wall heat transfer (bottom), approximately 150 s into the spill.

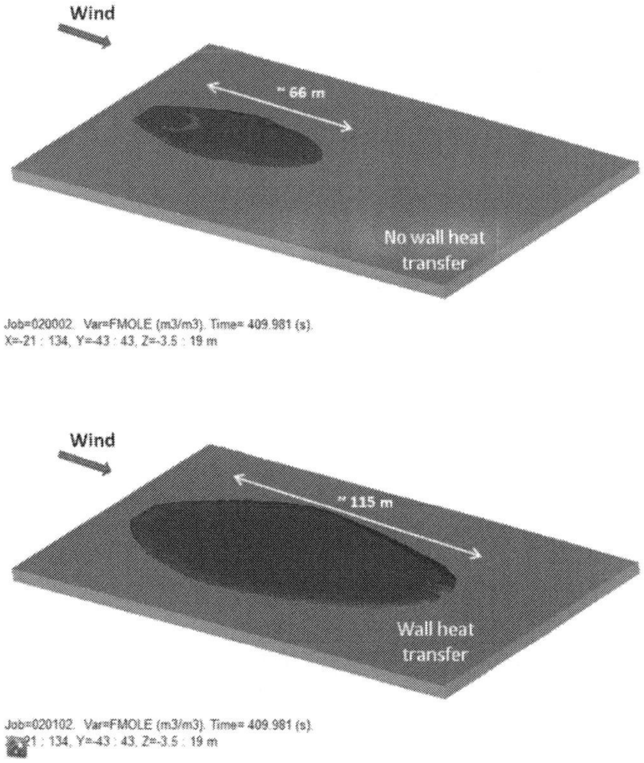

Figure 5. ½-LFL gas concentration vapor cloud at approximately the maximum cloud size, for simulations without (top) and with wall heat transfer (bottom).

The maximum vapor dispersion hazard distances for the FLACS simulations, with and without the contribution of sump wall heat transfer are summarized in Table 1.

Table 1. Maximum distances to 1/2-LFL for FLACS simulations with and without sump wall heat transfer

Solution Method	Max. Distance to ½-LFL
1) FLACS without wall contribution	66 m
2) FLACS with wall contribution	115 m

Example 2 – Wind Scooping

The second example presented in this paper addresses the so-called "wind scooping" effect. The purpose is to evaluate whether the current wind speed requirements in NFPA 59A and in U.S. federal regulations (i.e., a wind speed of 2.0 m/s) represent the worst-case ambient conditions, or a higher wind speed may indeed result in increased hazard to the public.

The scenario consists of a 10-minute LNG spill with a flow rate of 0.547 m^3/s (8,670 gpm) into a concrete-lined sump. The terrain is assumed to be flat and no other obstructions are present. The ambient temperature is assumed to be 20 C (68 F). The wind speed is varied parametrically from 2 to 5 m/s (4.5 to 11.2 mph) and the most stable atmospheric stability class compatible with each wind speed [turner ref.] is assumed: F stability for wind speeds of 2 and 3 m/s, E stability for wind speed of 4 m/s; D stability for wind speed of 5 m/s.

The wind scooping effect is expected to be different for different sump dimensions, particularly as the aspect ratio (defined as the ratio of streamwise length over depth) changes. In order to evaluate the implications on vapor dispersion hazard distances, the same wind speed parameterization is performed for two different sumps, with the same total volume but different aspect ratios:

1. Deep sump: 10 m by 10 m by 3.3 m deep (L/D = 3);
2. Shallow sump: 12.5 m by 12.5 m by 2.1 m deep (L/D = 6).

Figure 6 shows the vapor generation rate for the two sump geometries. The different sump floor areas result in different vapor generation rates between the two geometries: a higher vaporization rate is initially obtained from the shallow sump, due to its larger floor area. The contribution to vapor generation from the sump walls becomes more significant in the deeper sump, as the liquid depth increases at a faster rate, resulting in a slight increase in vapor generation at later times.

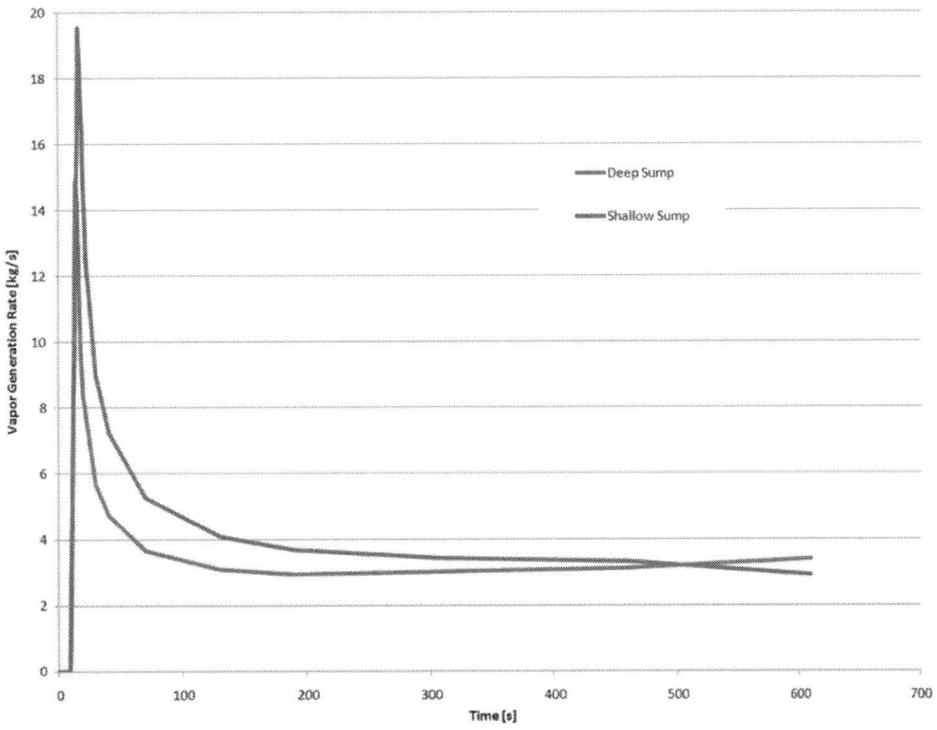

Figure 6. Vapor generation rates for the two sump geometries.

Figure 7 and Figure 8 show the ½-LFL vapor cloud isocontours for the deep sump scenario, respectively 60 and 340 seconds into the spill (which started at time = 10 s). Even though the vapor generation rates are the same for all four cases, higher wind speeds "scoop" a larger volume of LNG vapors (at concentrations equal to or greater than ½-LFL) from the sump, as shown in Figure 7. However, the LNG vapors emerging from the sump are dispersed to below ½-LFL at a faster rate as the wind speed increases, resulting in progressively shorter hazard distances, as shown in Figure 8.

Figure 9 and Figure 10 show the ½-LFL vapor cloud isocontours for the shallow sump scenario, respectively 45 and 320 seconds into the spill (which started at time = 10 s). Similar conclusions can be drawn for this scenario as for the deep sump: higher wind speeds "scoop" a larger volume of LNG vapors (at concentrations equal to or greater than ½-LFL) from the sump (see Figure 9), but the LNG vapors emerging from the sump are dispersed at a faster rate, resulting in progressively shorter hazard distances as the wind speed increases (Figure 10).

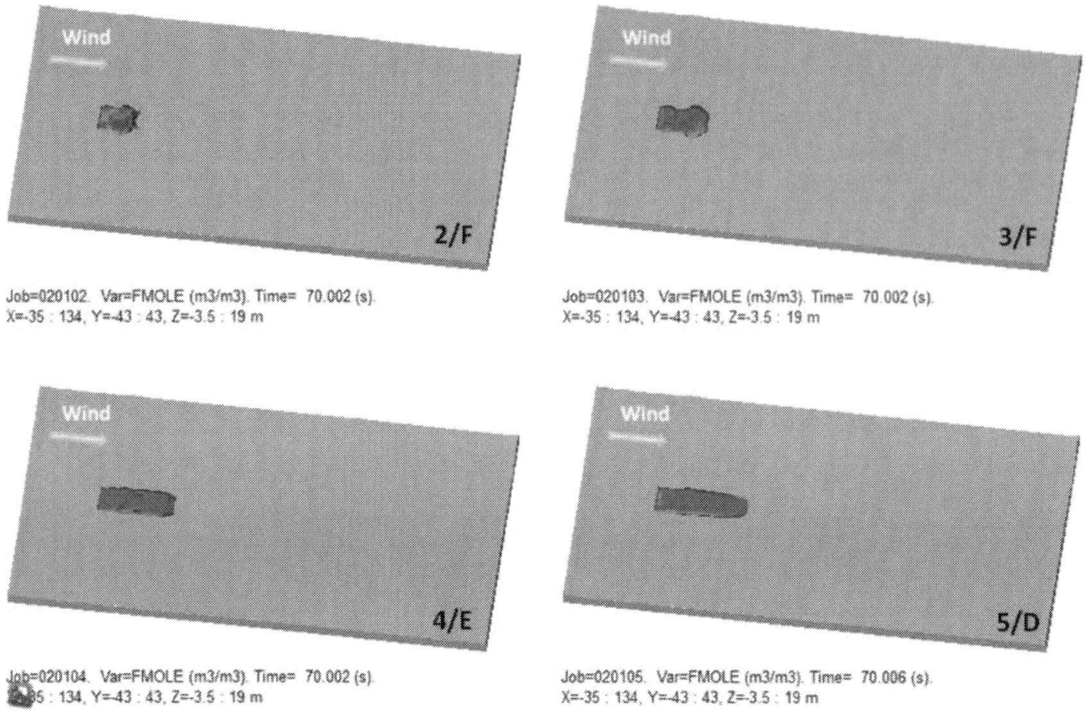

Figure 7. ½-LFL gas concentration isocontours for the deep sump under different wind / stability conditions, approximately 60 s into the spill.

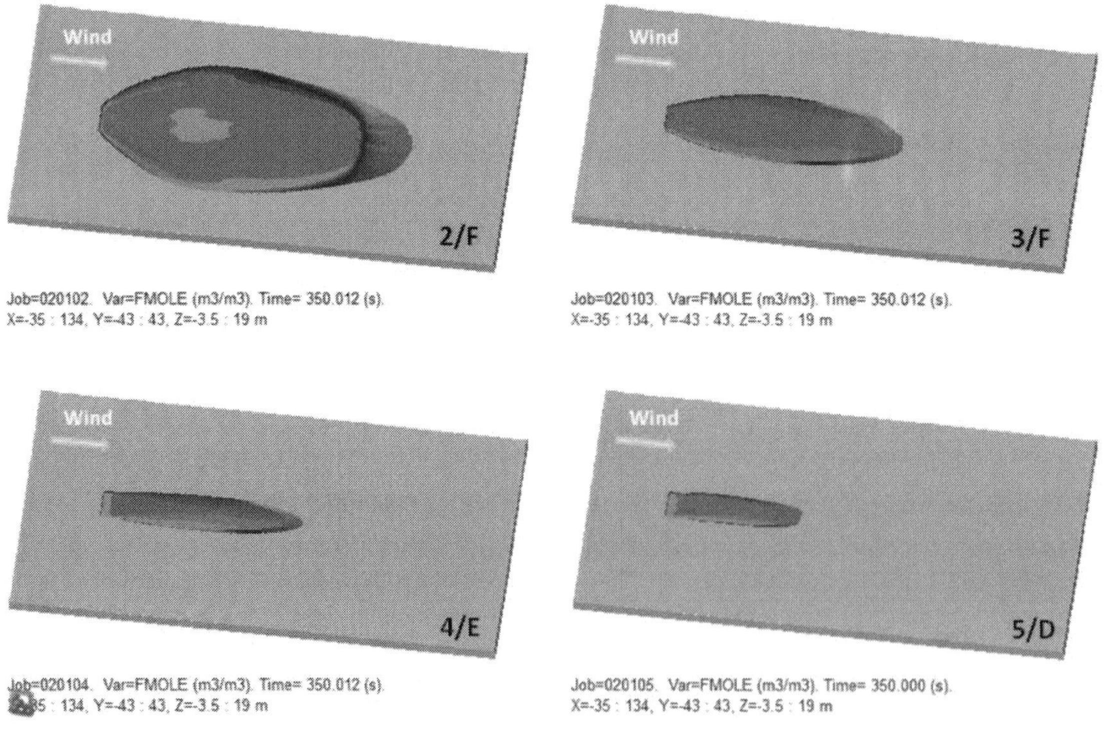

Figure 8. ½-LFL gas concentration isocontours for the deep sump under different wind / stability conditions, approximately 340 s into the spill.

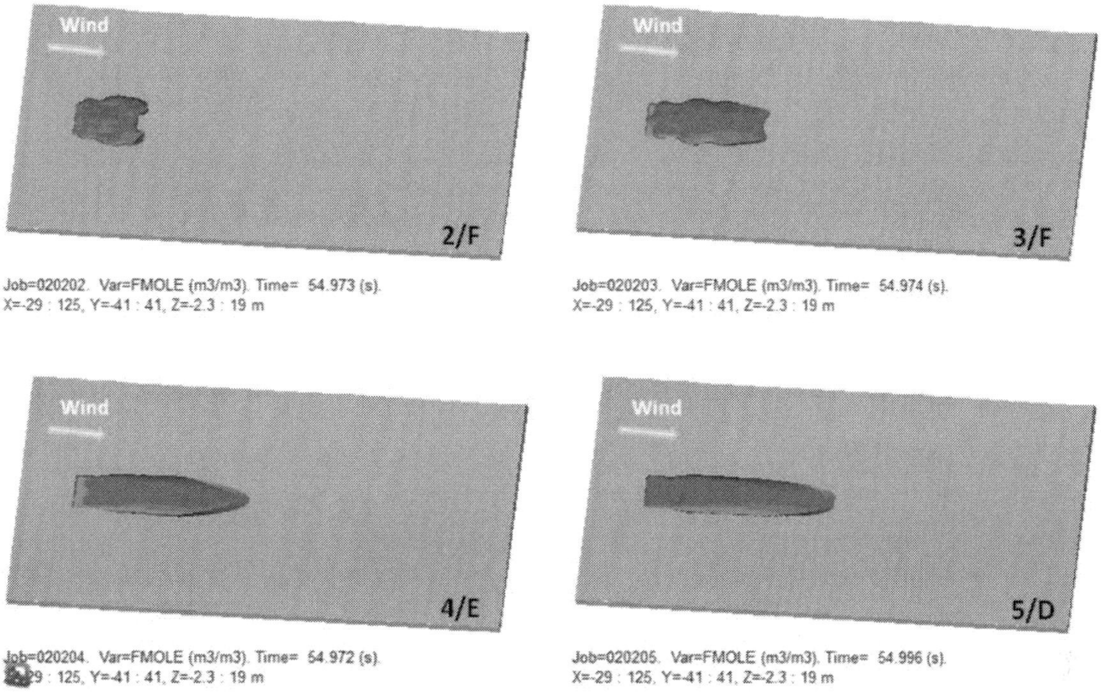

Figure 9. ½-LFL gas concentration isocontours for the shallow sump under different wind / stability conditions, approximately 45 s into the spill.

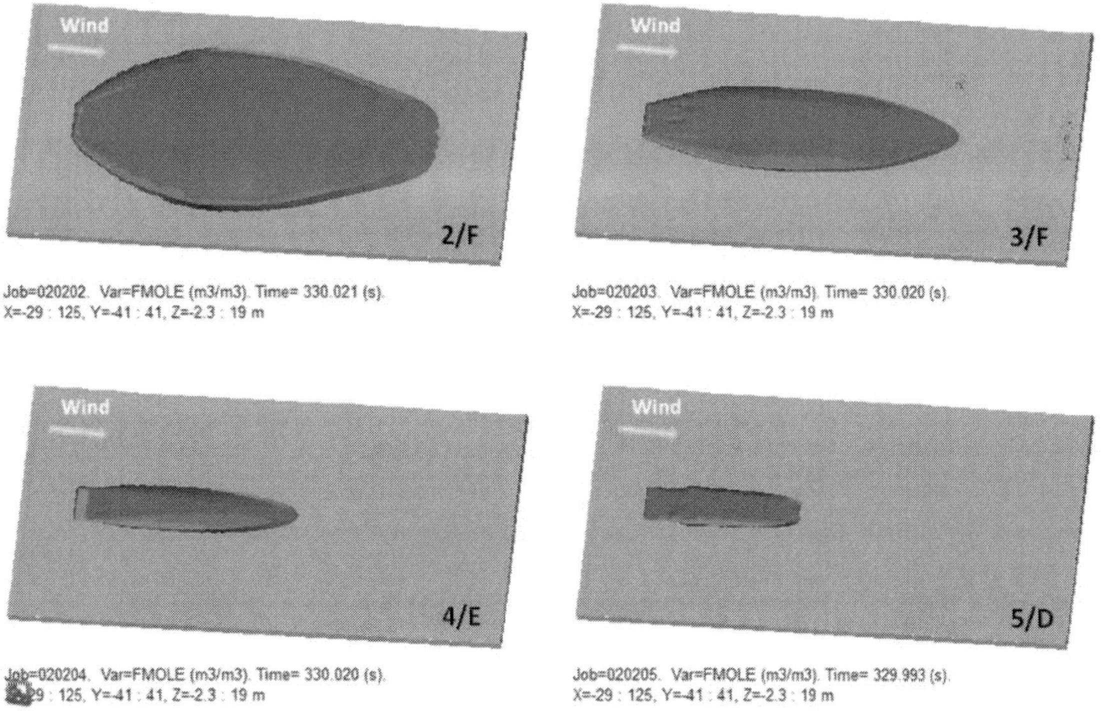

Figure 10. ½-LFL gas concentration isocontours for the shallow sump under different wind / stability conditions, approximately 320 s into the spill.

The maximum vapor dispersion hazard distances, calculated for an LNG spill under different wind conditions and different sump geometries, are summarized in Table 2. The results indicate that the maximum distance to ½-LFL for the vapor cloud from an LNG spill into a sump is reached for the low wind speed case of 2 m/s, as currently required by the NFPA 59A standard and by U.S. federal regulations.

Table 2. Maximum distances to 1/2-LFL for different wind speeds and sump geometries

Wind Speed / Atmospheric Stability	Deep Sump (L/D = 3.0)	Shallow Sump (L/D = 6.0)
2 / F	115 m	123 m
3 / F	86 m	97 m
4 / E	71 m	79 m
5 / D	46 m	66 m

Conclusions

This paper presented an alternative tool to the traditional SOURCE5 / DEGADIS method to calculate the vapor generation and vapor cloud dispersion from an LNG spill into a sump. GexCon's computational fluid dynamics (CFD) model FLACS, which provides an unified source term (i.e., pool spreading and vapor generation) and dispersion model, was applied to the typical scenario that all LNG facilities are required to simulate: a 10-minute LNG spill into a sump.

The pool spreading and vapor generation rate calculated by FLACS for an LNG spill into a sump were compared with the SOURCE5 calculations. Due to the assumptions built into the SOURCE5 model, the vapor generation rate calculated using SOURCE5 appears unphysically large during the sump floor wetting phase of the spill. The effect of heat transfer from the sump walls to the LNG pool was also demonstrated to be significant, at least for sumps with a relatively small footprint and tall walls. This finding should be taken into consideration when evaluating pool spreading and vaporization models in alternative to SOURCE5.

Finally, the wind scooping effect was then evaluated, by simulating the dispersion of an LNG spill into a sump, under increasing wind speeds (from 2 to 5 m/s). The results indicated that increasing wind speed above the current regulatory requirement of 2 m/s does not lead to longer vapor dispersion hazard distances. The CFD simulations showed that while "wind scooping" increases with wind speed, turbulent mixing also increases and at a faster rate, such that the net effect of an increasing wind speed is a reduction in vapor dispersion hazard distances.

References

[1] Title 49 Code of Federal Regulations, Part 193.2059

[2] Havens, J. & Spicer, T., "*United States regulations for siting LNG terminals: Problems and potential*", Journal of Hazardous Materials, *140 (3),* 2006, pp. 439-443.

[3] Fire Protection Research Foundation, "*LNG Source Term Models for Hazard Analysis: a Review of the State-of-the-Art and an Approach to Model Assessment*", March 2009.

[4] Hansen, O.R., Melheim, J.A., Storvik, I.E., 2007. "*CFD-Modeling of LNG Dispersion Experiments*". AIChE Spring National Meeting, 7th Topical Conference on Natural Gas Utilization, Houston, USA.

[5] Launder, B.E., Spalding, D.B., 1974, "*The numerical computation of turbulent flows*". Comp. Meth. In App. Mech. And Eng., **3**, 269-289.

[6] Ivings M J, Jagger S F, Lea C J and Webber D M, 2007, "*Evaluating vapor dispersion models for safety analysis of LNG facilities*", The Fire Protection Research Foundation, May 9 2007.

[7] Hansen, O.R., Ichard, M. and Davis, S.G., "*Validation of FLACS for Vapor Dispersion from LNG Spills: Model Evaluation Protocol*", Proc. 2009 Mary Kay O'Connor Process Safety Symposium

Speaker's Bio

Dr. Filippo Gavelli is the Head of the Dispersion Consulting group at GexCon US, Inc. and specializes in the analysis of heat transfer and fluid flow phenomena, including multiphase flows and cryogenic fluids. He applies his engineering and CFD modeling expertise to the atmospheric dispersion of hazardous gaseous releases, and has extensive experience modeling Liquefied Natural Gas (LNG) vapor cloud dispersion. Dr. Gavelli is responsible for GexCon US' dispersion modeling activities, which include risk assessments and consequence modeling for chemical and petrochemical facilities, offshore installations, hazardous materials transportation and various other applications. He is the lead author of several LNG safety-related papers and has contributed to numerous LNG-related technical committees and expert panels.

Process Simulator for Operator Training and Control Studies at Al-Khaleej Gas Project

Abdulla Al-Subaey
Ras Gas Company Limited
P.O. Box 24200
Doha, State of Qatar
asalsubaey@rasgas.com.qa

John Womack
Markham Dynamics
32 Markham Rd
Princeton, NJ 08540
jwwomack@aol.com

Ahmet Pekediz
Honeywell Process Solutions
300-250 York Street
London, Ontario, Canada
ahmet.pekediz@honeywell.com.
Tel: (519)640-6613

Prepared for Presentation for the 10th Topical Conference on Natural Gas Utilization, AIChE Spring National Meeting, March 21-25, 2010, San Antonio, Texas.

Abstract

Al-Khaleej Gas, a joint venture project, is operated by Rasgas Company Limited (Rasgas) in the State of Qatar. Al-Khaleej Gas has been developed into two phases, AKG-1 and AKG-2, to process a total of 2000 MMSCFD of sour natural gas. Up to 800 MMSCFD of this demand is "uninterruptable", therefore, the AKG plants are integrated with RasGas' LNG facilities to provide emergency gas supplies. During the AKG-1 project, a dynamic process simulator was employed for operator training as well as for control check-out. This process model was later updated and significantly expanded for a study aimed at improving the sales gas header controls. Those controls were revamped on the basis of experience and of the study results. During the second phase of the project, a detailed, multi-unit, AKG-2 model was developed and connected to the existing sales gas header model. The combined AKG-1, AKG-2 and sales has header model was integrated with the actual plant controls and plant protection logic system. The combined model was made available for plant operator training for conducting plant startup and emergency shutdown scenarios. Final Control Studies were carried with the plant control database to test response of sales gas header controls. The presentation will discuss the size of the model(s), some of the problems which they were used to solve, and some of the difficulties encountered. This work might be considered an example of the use of a life-cycle model for the AKG facility.

Extended Abstract

The Rasgas complex at Ras Laffan in Qatar is among the largest gas processing facilities in the world. Projects now nearing completion are the Train-6/7 LNG expansion and the Al-Khaleej Gas Sales expansion. These projects have increased the gas supply into the complex from offshore production wells by some 4 BSCFD.

Consistent with their established operating philosophy, RasGas purchased two new Operator Training Systems (OTS) for pre-start-up and continuing training of panel operators for these new projects. Each OTS includes Honeywell Unisim process (dynamic) simulation models connected to emulated Invensys I/A panel operator consoles. Each OTS was delivered to site several months before start-up, and included the pre-commissioning versions of complete contol system configuration of Distributed Control System (DCS) database, Protective Emergency Shutdown (PESD) PLC configuration as well as Sequence Code configurations for the OTS-process scope.

The LNG Train-6 OTS model includes the UCARSOL Acid Gas Treating Unit and associated Sulfur Recovery Unit as well as Gas Dehydration and Selexol Units, the NGL extraction unit and the new APX-LNG Unit (C3, MR, and N2 refrigeration sections), and associated NGL Fractionation and C3/C4 Liquid Treating Units. The control system for this large process model manipulates some 700 emergency shutdown and DCS control valves. There are approximately 35000 I/O points delivered with the Honeywell Unisim process model.

The AKG Sales Gas System OTS process model includes the UCARSOL Acid Gas Treating Unit, Sulfur Recovery Unit, Gas Dehydration and Selexol Units, as well as two NGL Extraction Units (AKG-1 and AKG-2) and the associated Fractionation and Treating Units which produce sales-quality C2, C3 and C4 as by-products. The model also includes the sales gas system, which can supply up to 2000 MMSCFD, and demand models for the five gas sales system customers. An important constraint is that three major customer are contractually entitled uninterruptible gas supplies of up to 800 MMSCFD. Therefore training operations personnel becomes even more crucial. The control system for this large process model manipulates some 1070 emergency shutdown and control valves.

Each of the Unisim process simulation models runs on a single (quad core) XP based simulation computer which is networked to it's own DCS/PESD system and to Instructor and Trainee stations.

As with the new super size LNG Trains, the simulation models have also become bigger with more details added to the dynamic simulation model compared to previous OTS models delivered to Rasgas.

RasGas began operations at Ras Laffan in 1996 and has made heavy use of OTS systems from that time onwards. Before the present expansion projects there were two OTS systems in their Training Centre with process models and I/A consoles used to model LNG Trains1, 3 & 4 and the first phase of the AKG gas sales project.

These older systems also use actual plant configurations for DCS, PESD and Sequence Code. The major difference between the older OTS systems and the new ones is the size of the process models. For example, the Train-1 OTS has five different models (one for each process unit) to handle an LNG process configuration which is essentially similar to the Train-6 process, but with approximately one-third of the LNG capacity.

The process scope for the single OTS model for the full Al Khaleej Gas system is shown in Figure 1. The model represents all process equipment and controls which impact on operation of the sales gas header system.

The AKG Operator Training Simulator is designed for use in training for complete or partial process start-up; for operator training during normal/smooth running of the plant; and for training in handling of severe emergency situations. One key training objective is to provide the control-room panel operators with experience in meeting gas supply guarantees during major upsets.

Final testing of the OTS, and it's on-site use for pre-startup training also provided final shakedown tests for the DCS controls, PESD and sequence codes. This testing led to several DCS configuration improvements being made before plant start-up, preventing hours of trouble-shooting during the (expensive) "real world" start-up of the facilities.

In contrast to the older (smaller scope) OTS models, the AKG OTS clearly shows the effects of interaction between process units. For example, a machinery trip in the AKG-2 NGL Extraction unit causes a swift reduction of gas flow into that unit which causes a pressure surge within the UCARSOL unit. This rapidly caused a high-differential pressure trip of the main amine circulation pumps, leading to a total stoppage of gas flow from the AKG-2 units. Advance warning of this interaction allowed RasGas to develop a strategy to avoid the problem during operation of the facility.

The Dynamic Simulation Study and Experience with the OTS Model

The importance of sales gas header pressure control was recognized very early in the AKG Expansion project because initial control performance, with only the AKG-1 project in operation, was only marginally satisfactory. A dynamic simulation study (DSS) is the obvious tool with which to evaluate alternative control strategies. However, DSS results were needed swiftly because any additional control valves (long delivery due to size and pressure rating) had to be installed during a planned shut-down whose timing was already determined.

Because of the time constraint it was decided to expand the existing (AKG-1) OTS model (which already included a partial-model of the sales gas header) for this purpose rather than add this task to the DSS work of the main engineering contractor.
Since data for the AKG Expansion (AKG-2) NGL-unit was not yet available the DSS study used a scaled-up version of the observed responses for the AKG-1 NGL-unit to represent dynamics of the new unit. This preliminary model was partially validated against observed responses to an AKG-1 process trip which were retrieved from DCS-system archives. After that the model was used to test control-system alternatives with several major supply and demand upsets. The study developed a simple multi-controller feedback pressure control strategy which is combined with feed-forward elements (for help with major upsets in gas supply or demand), see Figure 2. These controls were implemented by the Expansion-project DCS/PESD.

Later in the project, after engineering was complete and equipment data was available, the initial DSS model was expanded by inclusion of fully detailed models of the AKG-2 process units. This model was used for the OTS Factory Acceptance Test and for pre-start-up panel operator training. After delivery of the OTS another (small) DSS was completed to re-evaluate performance of the sales gas header controls.

Conclusion

Dynamic simulation models were successfully employed by Rasgas for training panel operators for startup and shutdown scenarios prior to plant start-up. The previously delivered AKG1 model was successfully integrated with the new AKG2 process model to simulate sales gas header. In addition, the dynamic model was employed for improving sales gas header control.

For several years process simulation advocates have discussed "life-cycle modeling", the concept that a single process model might evolve from process design through the construction and start-up phases to be updated for use during operation, debottlenecking, and the entire life-cycle of a process plant. There have been, and remain, significant administrative difficulties with this idea; among them the proprietary knowledge positions of different process licensors and of engineering contractors.

Using a high-quality Operator Training Simulator system as the starting point for such a model enables an Operating company to by-pass some of these problems. If (a) the OTS process models are updated after start-up by systematically using operating data to improve their accuracy, and (b) the OTS's DCS/PESD system is kept in synchronization with the real-world DCS/PESD system by including OTS-updates in the Management-of-Change Procedures, then the Operating Company could expect to capture benefits in development of debottlenecking planning, turn-around planning, and in better evaluation of control system or equipment change proposals.

RasGas' experience with these latest OTS systems suggests that the concept of life-cycle models is beneficial and it should be considered by operating companies.

Figure-1: AKG OTS Scope Overview

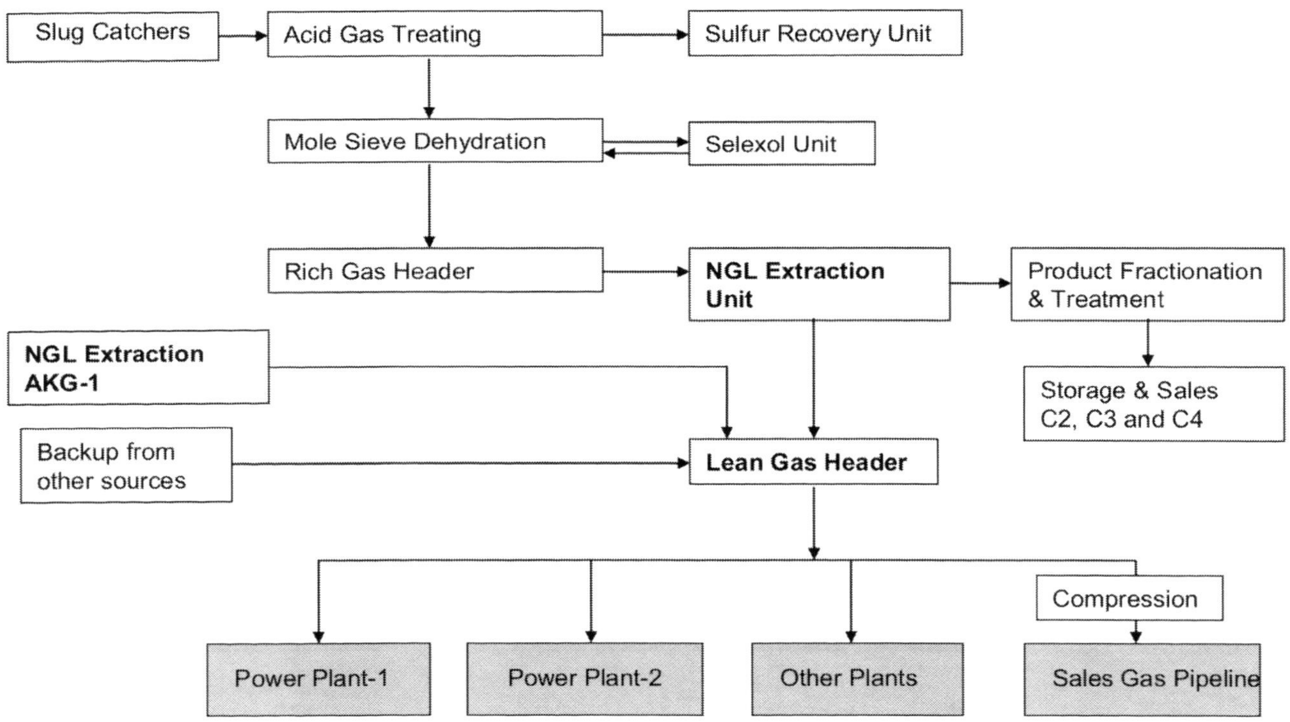

Figure-2: Control Scope Overview

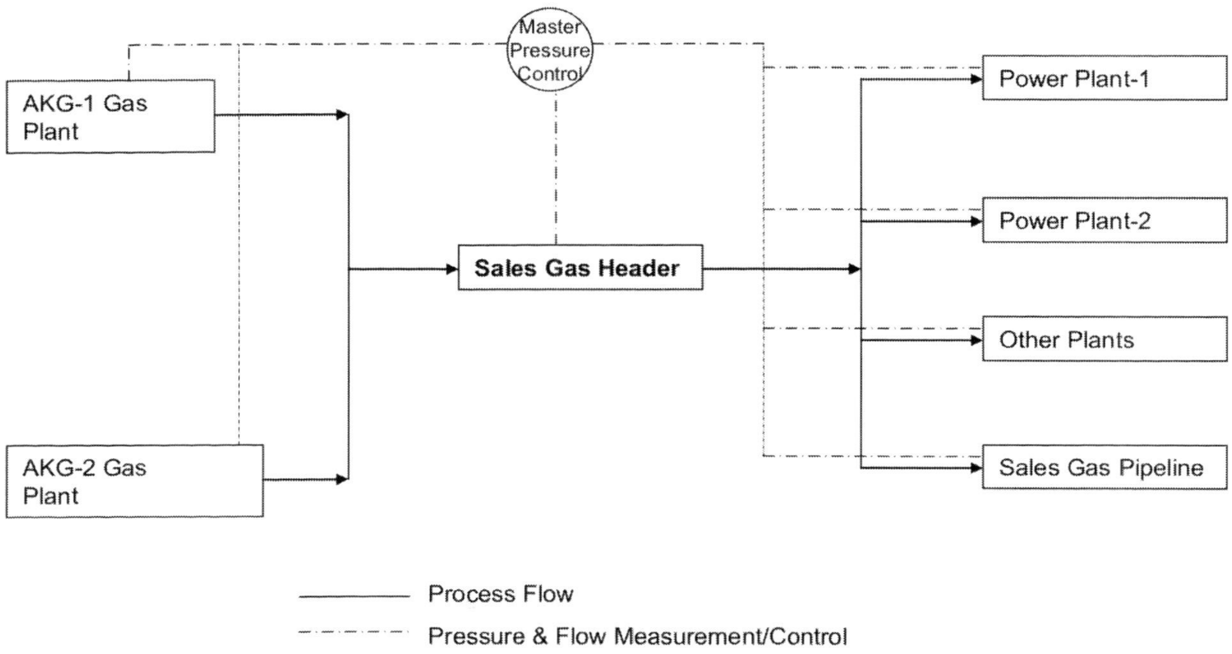

Author Biographies

 **Abdulla Al-Subaey**	 **John Womack**	 **Ahmet Pekediz**
Abdulla is Deputy Commissioning Manager for Rasgas Train 6/7, AKG2 and COP projects. A process engineer, he holds a Mechanical Engineering degree and has over 10 years of LNG and Gas Processing operating experience with Rasgas Company Ltd.	John has many years of dynamic simulation experience and worked with Mobil R&DC for 23 years. He has consulted with Rasgas to support their simulation and OTS needs for over 10 years. John holds a DEngSc in Chemical Engineering	Ahmet is Senior Application Analyst with Honeywell Process Solutions and has over 12 years experience in dynamic simulation of LNG and Gas Processing plants. DCS and PESD skills derive from his Lead Engineer role in many Operator Training Systems. Ahmet holds a PhD in Chemical Engineering.

DESIGNING FOR AN LNG LOADING ARM FAILURE

Over the last few decades, LNG has vaulted onto the world stage as a viable candidate for future energy needs. Scores of studies have been done to illustrate the potential consequences of an accidental release of LNG. Some excellent studies have been reported, but uncertainties still exist because of randomness of events, unconfirmed extrapolation, and other factors that could affect an actual outcome

One area where an accidental release of significant magnitude could pose serious problems is during the loading operations of large LNG carriers in service today. With loading rates exceeding 12,000 cubic meters per hour, a failure in the loading delivery system could cause embrittlement of a ship, supporting loading structures, possible RPTs, potentially large vapor dispersion zones, and/or a confined vapor cloud explosion.

Typically, and in the codes, 10 minutes is considered a "worst case scenario" for duration of an event. By any account, a 16" loading arm failure represents a potentially catastrophic event.

A suggested approach to determining the consequences to such a release is provided, along with a system that can detect a spill or fire, activate shutdowns, and mitigation systems to stop the accidental release in less than one minute and minimize the potential effects of such an event.

Issues such as reliability/availability, code compliance, hydraulic transient analysis, and actual design considerations are discussed.

Hazard Identification

WorleyParsons is a leading global provider of professional services to the resources and energy sectors, and the complex process industries. With 120 offices in 37 countries and over 30,000 employees, WorleyParsons covers the full asset spectrum, both in size and lifecycle, from the creation of new assets, to services that sustain and improve operating assets.

Zero harm is the corporate vision for health, safety, and the environment. One Way is the enterprise wide integrity management system which lays down corporate expectations for achieving the Zero Harm vision, and in turn provides links to the policies, standards, and processes that are to be followed in daily operations to meet the expectations.

SEAL, Safe and Sustainable Engineering for Asset Lifecycle, is the program that provides the framework for implementation of the WorleyParsons' vision. Task Sheet EPP-9009 provides the overview of the work processes and necessary links to relevant documents to be used by Engineering.

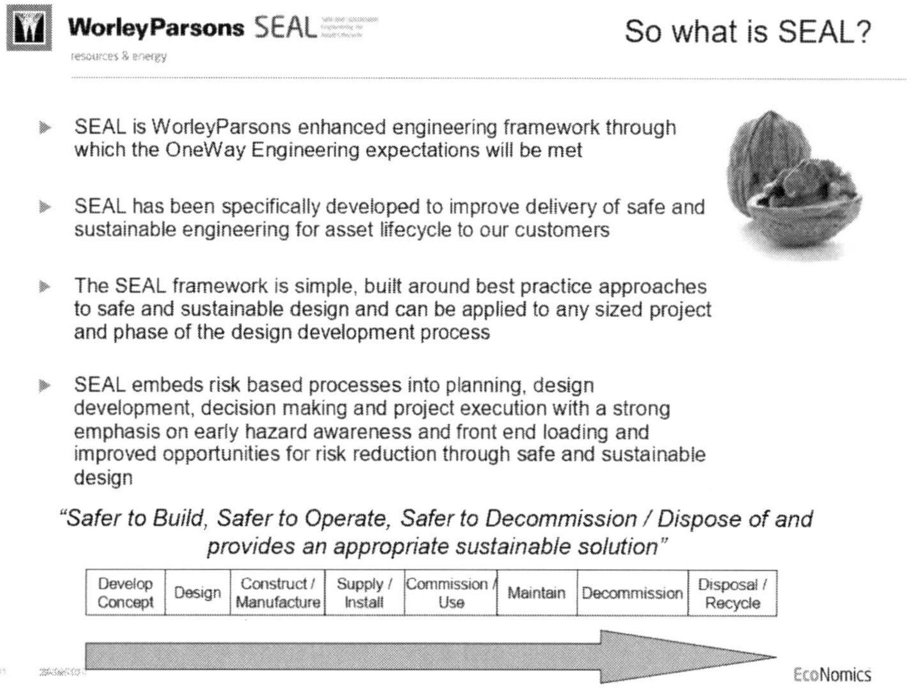

Early detection of hazards with a strategy for managing that hazard is a corner stone of the SEAL process and development of Discipline Context statements are required where each discipline identifies risks to design elements within their scope of responsibility.

WorleyParsons has a long history of developing and supporting world scale LNG facilities. During an early hazard identification exercise, the subject of this paper was identified as a potential hazard and steps were taken to properly quantify and manage the risk inherent in the process of loading/unloading at LNG marine terminals.

Properties of LNG

LNG (Liquefied Natural Gas) is produced by chilling natural gas to approximately -162°C at atmospheric pressure. It is stored in insulated tanks which maintain the vapor pressure slightly above atmospheric pressure.

LNG occupies less than one six-hundredth the volume of natural gas at 15°C and atmospheric pressure, making it economically feasible to transport large volumes of gas by ship.

LNG is a cryogenic liquid, that is, it is so cold that it can cause extreme thermal effects. It can freeze living tissue exposed to it for more than a few seconds, and can cause the embrittlement and subsequent failure of structural materials such as carbon steel.

LNG does not pose a hazard unless it is accidentally released in an uncontrolled manner from its specially designed cryogenic containment system. When accidentally released into the environment, it absorbs heat from its surroundings and starts to revert to the gaseous state, producing a cold, heavier-than-air, pancake-shaped cloud of vapor. In the initial phase of vapor cloud dispersion, gravity forces spread the heavy, dense cloud. During this phase, the cloud may spread both upwind and downwind. As the cold natural gas cloud moves along a surface, it entrains air and becomes warmed and diluted. Portions of the cloud can be flammable.

Methane is flammable when the methane-air mixture is in the range of 5% to 15% by volume. In the second phase of gas cloud dispersion, turbulent forces at the edge and upper surface of the cloud tend to dilute the methane-air mixture to concentrations below the lower flammability limit, shrinking the size of the flammable cloud as it moves downwind, until a point is reached at which no potentially flammable vapor remains.

Although methane is invisible (and odorless), the cold methane vapor causes condensation of water vapor in the air, thereby revealing the presence of the methane vapor by a dense cloud of fog. Under most temperature and humidity conditions, the extent of the visible fog exceeds the extent of the flammable vapor cloud. The boundaries of such a fog can be considered a rough indicator of the area of potentially hazardous conditions.

An LNG spill can result in a fire, either at the moment of a spill or after a vapor cloud has formed (if an ignition source is present). Flame front speed of methane is slow, resulting in a relatively lazy progression back to the initial source of the vapor cloud, a leak or pool, depending on circumstances. The size of the fire is directly related to the surface area of the spill.

However, LNG vapor confined in a building, conduit or other enclosed space can explode, if combined in the right proportions with air. Those explosions can be catastrophic.

Confined Methane Explosions

Over the years, there have been a number of catastrophic explosions involving confined methane vapor. Recently, the Chemical Safety Board (CSB) released the results of their investigations into confined methane explosions. One of those was CSB's ongoing federal investigation into the June 9, 2009, natural gas explosion at the ConAgra Slim Jim production facility in Garner, North Carolina, which caused four deaths, three critical life-threatening burn injuries, and other injuries that sent a total of 67 people to the hospital.

CSB investigators determined that the catastrophic explosion resulted from the accumulation of significant amounts of natural gas that had been purged indoors.

Other incidents examined by the CSB include: a 1999 explosion at a Ford power plant in Dearborn, Michigan, killing six, injuring 38, and causing a $1 billion property loss; a 2008 explosion at a Hilton Hotel under construction in San Diego, California that injured fourteen people; a 2005 school explosion in Porterville, California, and an explosion at a hotel in Cheyenne, Wyoming, in 2007

LNG Shipping Activities

International LNG shipments began in the 70's, with the subsequent development of LNG carriers that continues to grow in size and range. A trend from the beginning has been to employ economies of scale, from the liquefaction plants, storage, and shipping perspective. This has also resulted in increasingly larger carriers as well as greatly increased loading and unloading rates.

Currently, there are two main ship types; a dual membrane of Gaz Transport or TechniGaz (GTT) design or a single wall spherical design by Moss. Both the membrane and spherical design afford significant physical barriers between the cargo and the external environment.

All LNG carriers (both currently in service and planned) provide double hull construction, providing a significant measure of protection against LNG release due to external damage, such as could occur in groundings, collisions, or allisions (i.e., striking a fixed object).

Also, LNG carriers are generally guided into ports and restricted waterways by local pilots with extensive experience in the port area. This reduces the likelihood of navigational failures or human errors by normal tanker crews.

Docking and departing of LNG tankers generally involve the use of harbor tugs to assist in maneuvering the LNG carrier. In the United States, LNG carrier arrivals and departures are subject to operational plans developed and enforced by the cognizant USCG Marine Safety Office (MSO). These plans generally mandate safety zones and restrict traffic in the area of any LNG tanker.

The LNG shipping industry has an unusually good safety record in terms of cargo loss. There have been between 80,000 and 100,000 loaded port transits with no significant loss of containment failures.

LNG vessels have experienced a similar number of events to other forms of shipping related to minor collisions, small leaks and fires, but none have resulted in a containment failure of release of cargo.

Catastrophic Events involving LNG Shipping

Numerous studies have been conducted on potential ship collisions, terrorist attacks, strikings, and other speculated means of opening large holes in the carrier and releasing large quantities of LNG with catastrophic results.

Some studies have merit, although strong differences of opinion exist among experts as to the validity of large, almost instantaneous, releases and their catastrophic consequences.

For example, one incident involving a collision between *Pacific Ares*, a 15,000 deadweight ton bulk carrier and *Yoko Maru No. 10*, a combination refrigerated LPG/oil products carrier, in Tokyo Bay, resulted in a naphtha fire and ignition of venting LPG. To protect other ships, the *Yoko Maru No. 10* was towed out to sea and it was decided to sink her. After being bombarded by more than 200 five inch shells for some four hours, the *Yoko Maru No. 10* finally sunk.

It is noteworthy that despite the initial collision, and resultant fire from naphtha and LPV vapors, no rupture or explosion took place and there was no release of the liquefied gas cargo either into the hull or into the sea until the final bombardment and sinking.[1]

Another study concluded that it was necessary to put these hypothetical large hole releases into the context of what is most probable from historical records of LNG shipping and they suggest that a large scale catastrophic release is unlikely to occur in the foreseeable future of the LNG trade in the U.S.[2]

Safety Rules and Regulations

There are many rules and regulations regarding LNG tankers. The result of these various rules and regulations is that most safety aspects of the design, construction, and operation of LNG tankers are adequately addressed.

LNG carriers are unique in that LNG is a cryogenic liquid; accordingly, specialized materials, construction methods, and operating procedures are needed to safely handle LNG.

The general rules and regulations that govern ships at sea do not address the particular concerns of LNG; therefore, specific rules and regulations have been developed in order to ensure the safety of LNG tankers and their ports of call.

The IMO, through its signatory nations, provides the basis for the framework of international regulations for shipping. The IMO publishes documents that outline the international requirements for vessel design and construction, operation, safety security, and crewing. There are several IMO publications and or conventions regarding design, construction, and operation of LNG tankers that are allowed to operate in U.S. and international waters. These conventions are:

- Convention on the International Regulations for Preventing Collisions at Sea
- Safety of Life at Sea (SOLAS)
- Code for the Construction and Equipment of Ships Carrying Liquefied Gases in Bulk (Gas Carrier Code)
- International Code for Ships Carrying Liquefied Gases in Bulk (IGC Code)
- 1994/1996 Amendments to the IGC (replaced the Gas Carrier Code)
- International Convention on Standards of Training, Certification and Watchkeeping (STCW) for Seafarers
- International Management Code for the Safe Operation of Ships and for Pollution Prevention (ISM Code) — adopted by IMO Resolution A.741 (18) in 1994

IGC Code

Probably the most important of these international requirements for new LNG carrier design is the IGC Code. This provides the basic requirements that flag states and classification societies then supplement with more detailed requirements.

The IGC Code requires remotely controlled emergency shutdown valves on the ship for stopping liquid and vapor transfers between the ship and shore. Cargo pumps and compressors must be arranged to shut down automatically if the emergency shutdown system closes the emergency shutdown valves. In addition, there must be a remotely operated shutdown valve at each cargo hose connection used in transfers.

The emergency shutdown control system must be capable of being actuated by a single control in either of two locations on the ship. It must also be actuated by fusible links that will respond in the event of a fire in locations, including tank domes and loading stations.

Per the IGC Code, LNG carriers require a firewater main system and hydrants/hose stations with the ability to supply at least two jets of water to any part of the deck in the cargo area and parts of the cargo containment and tank covers above the deck. The firewater supply system also has to supply water for a water spray system that provides water for cooling, fire prevention, and crew protection in specific areas. LNG carriers also require a fixed dry chemical powder-type extinguishing system to allow the crew to fight fires on the deck in the cargo

These systems, along with the pressure relief and vacuum pressure protection systems, provides a major role in preventing LNG cargo tank upsets or human errors from damaging the tank or resulting in a large scale release.

The IGC Code also requires a permanently installed gas detection system with audible and visual alarms for areas, including cargo compressor rooms, motor rooms, cargo control rooms, enclosed spaces in the cargo area, specific ventilation hoods and gas ducts in the propulsion machinery space for the fuel gas system, and air locks. These detectors are required to activate alarms when 30% of the LFL is reached. Ships also are required to have at least two sets of portable gas detectors.

IGC Loading Terminal Safety Systems

LNG unloading systems and docks are equipped with LNG vapor detection, fire detection, and associated safety shutdown systems that shut down pumping operations and close valves to isolate the transfer lines. The shutdown systems can be actuated by the ship's crew or LNG terminal personnel. In most cases, these systems also respond automatically to any detection of LNG in the atmosphere by shutting down pumping operations and closing valves to isolate the LNG transfer lines. The systems can also be actuated by the ship's crew or LNG terminal personnel. This serves to limit the amount of LNG that would be released in a leak that occurred during the unloading process.

Current designs for LNG terminals have emergency release couplings that are fitted between the ship's cargo manifold and the receiving station. These couplings are designed to release if vessel movement exceeds predetermined limits. If the couplings release, the result is a complete shutdown of loading operations.

Operational safety has continued to improve through training, procedures, regulations, security measures, and operational experience to the point where the earlier suggested scenarios involving LNG transporter's experiencing events which result in very large openings that resulted in instantaneous releases and catastrophic consequences are not considered credible by many experts.

The Current Project's Marine Terminal

The marine terminal itself was a conventional design. Concrete deck with retention curb around the loading arms designed to hold relatively small LNG spills, two 75 foot towers at each end of the dock area, and numerous mooring points, both on the dock itself and at both ends to secure the vessel prior to loading operations commencing.

Loadings arms are conventional design, with slewing limit switches, extension limit switches, selector valves, hydraulic power unit, PERC quick disconnect, and remote control, were integrated into a Safety Instrumented System (SIS), located in the marine terminal building located on the dock.

A secondary drainage system was intended to channel a major spill away from the dock and ship. However modeling revealed that a major loading arm failure could basically allow liquid LNG to impact the ship, and dock structure and overflow into the water immediately adjacent to the ship, rather than follow the designed path.

Marine Terminal Hazard Analysis

During the initial WorleyParsons' hazard review, a question was raised about the possibility of a failure of an LNG loading arm system. Given that there were loading rates in excess of 12,000 m^3/hr, it was suggested that the loading rate itself could pose a potentially catastrophic spill scenario.

The challenge brought out in the WorleyParsons' hazard review was the possibility of a loading arm system failure that allowed LNG to be discharged out of a 16" loading arm at a rate of 6,000 m^3 plus (the loss of backpressure on the LNG pumps would result in a higher flow than the design case) onto the side of the ship, onto the loading dock, and into the water where rapid vaporization might take place, including RPTs.

Confined Vapor Cloud Explosion

It was further suggested that, although an explosion of unconfined methane is highly unlikely, if LNG liquid and vapors were to collect between the ship and dock, and be confined underneath the dock, an explosion might take place, resulting in extensive damage to the loading terminal and ship.

In the Layers Of Protective Analysis (LOPA)[3] following the Hazard assessment, it was decided that further analysis was required before a decision could be made regarding an adequate level of protective layers and their SIL requirements.

Scenario Definitions

A loading arm failure scenario had been examined in an earlier study[4], and modeling using PHAST software had indicated significant consequences if the methane vapors were confined.

Earlier, a DOE study by Quest Consultants, Inc., of Norman Oklahoma[5] had examined the possible results of a large LNG spill on water, and introduced a method to factor in the effects of wave action. Although criticized at the time, experts now recognize that, in fact, wave action can play a significant role in LNG spread, vaporization, and potential consequences of large scale releases.

Because WorleyParsons' modelers had not factored the effects of wave action on spreading of LNG vapors previously, it was decided to start from the beginning and perform a series of dispersion models to determine if such a scenario had merit.

Using the software available from Quest Consultants, CANARY, studies were undertaken to perform an analysis of the results of a loading arm failure at various elevations (base, joint of articulation point, and at the ship connection), and an analysis of an ignition of LNG vapors collected between the ship, loading dock and underneath the dock. Once the initial scenarios were developed, additional modeling was used to determine and how the potential consequences might be mitigated.

Release Models

Canary Model, version 4.3, was used in the consequence analysis for the modeling of dispersion, pool fires, and explosions. Canary is a proprietary model, (from Quest Consultants), composed of sophisticated, state-of-the-art chemical engineering, fluid dynamics, and dispersion sub-models which are all based on peer-reviewed, public-domain technical work available in the literature. Canary is a comprehensive computer package which has algorithms to account for the following material behaviors in simulating a release: the release rate, the liquid-vapor flash, any liquid pool formation and vaporization, aerosol formation and evaporation, momentum-jet dispersion, dense-cloud dispersion, passive or neutral density dispersion (Gaussian), and radiant heat from fires. The program includes a thermodynamic sub-model for multi-component streams, such as the feed gas and LNG streams.

Two weather conditions were used in the dispersion modeling, one for day conditions and one for night / calm conditions ("D" stability with 5 m/s wind speed and "F" stability with 2 m/s wind speed, respectively). For radiant heat modeling from pool fires, conservative values for wind speed of 5 and 10 m/s were used.

The Vapor Cloud Explosion (VCE) prediction methodology that was used in the analysis is based on the Baker-Strehlow method, using the Baker-Strehlow-Tang (BST) blast curves[5] . The explosion model that was used is VCEC, (Vapor Cloud Explosion Contours), which was developed by Wilfred Baker Engineering for the Blast Technology Cooperative. VCEC incorporates the BST curves, as well as other factors such as cloud expansion and energy density factor. The program estimates the flame speed based on the level of congestion, confinement, material reactivity, and vent length. Based on the energy within the flammable vapor cloud and the flame speed, the program estimates side-on overpressures.

The congestion and confinement determination for a potential VCE were based on the preliminary layouts developed as part of the pre-FEED package. The VCE analysis assumed that the area beneath the entire marine terminal area is filled with a flammable cloud mixture, at stoichiometric concentration.

Modeling Results

As expected, the amount of liquid LNG that reached the water varied, based on the height of the release point selected. At release rates of 6,000 m^3 per hour (one loading arm failure), the retention curb around the loading arm area quickly filled and overflowed into the water.

Also as expected, wind and waves played a significant role in how the LNG vaporized, but the physical presence of the large tanker made migration under the marine terminal a definite possibility.

It was further determined that LNG vapors, if trapped between the ship and marine terminal, and at a stochiometric concentrations, could result in significant overpressures generated, possibly resulting in severe damage to both ship and terminal.

Discussions were held regarding expert opinions on how such an event could occur, and how likely it was, but it was decided not to pursue a quantitative analysis to determine the likelihood, but rather, based on the potential consequences, including personnel, equipment, and business interruption loss, to proceed directly into developing a means to detect and mitigate the event if it did occur.

In almost all scenarios, (various height release points) vaporized LNG could be expected to overwhelm the relatively small retention basin located around the loading arms and fall to the water immediately below the dock.

Accidental Release Detection System

The first step was to identify a means of detection. That analysis, based on the modeling of the accidental release, was to place optical gas detection devices on the two elevated 75' towers located on each side of the loading platform.

This would provide an unobstructed view of the loading arms, from their base up to the connections on the side of the ship. There were also gas detection sensors located on the marine terminal loading arm bases to detect smaller spills.

Because a complete shut down of loading facilities was planned if a major spill occurred, continuous low temperature thermistors were located around the perimeter of the retention basin to provide confirmation of a major spill.

The logic behind the addition of spot gas detectors and low temperature detectors at the base of the loading arms was data that indicates the majority of catastrophic events occurs from smaller releases that damage equipment and result in larger releases/explosions with catastrophic results. The detection system was intended to detect both small and large spill events.

For that same reason, optical fire detectors were located on the 75' towers, and a pneumatic pilot head system was located around the periphery of the retention basin.

These detectors, along with manual pull stations, and alarms, were hard wired back to an NFPA 72, UL listed, fire and gas detection/alarm panel, located in the marine terminal building. **This was the easy part;**

SIL Classification

The team had decided that the potential consequences required SIL 2 rated systems to mitigate the risk. Basically, this meant that both the FGS and SIS would need to have a minimum SIL 2 classification as defined by ISA 84.

Differences between Prevention and Mitigation

An excellent paper by Paul Gruhn [5] describes very succinctly the differences between a Safety Instrumented System (SIS), a preventive system, and the applicability of SIL levels, and Fire and Gas Detection/alarm systems (FGS), which, at best, are mitigation systems, and attempts to assign and verify SIL classifications to FGS.

The SIS, which will be an integral part of the final safety system, can be reliably classified as a SIL 3 and verified as such. A TMR-based SIS, with inputs from the FGS, ship, plant operations, and marine terminal activities, to monitor and shut down of all loading operations in case of an emergency was planned.

The FGS controller, logic, and detection devices can be classified as a SIL 2 rated system, but the ability to prove that this system can demonstrate the same reliability to detect and mitigate a release is highly controversial. The FGS may, or may not, detect the release, and, depending on a multitude of variables, may not be able to mitigate the consequences.

The use of different types of detection devices and events (LNG, vapor, fire) was an attempt to provide the highest likelihood of detection possible. The mitigation portion was much more involved.

Automatic/remote controlled Oscillating Monitors

To protect the ship and loading arms from cryogenic temperatures or fire, a 1500 gpm monitor was located at the top of each of the two 75' towers at the ends of the marine terminal. The monitors had adjustable nozzles, set on a semi-fog pattern, adjusted to spray water over both the ship's side and the loading arms.

The fire water monitors were set on automatic water-driven oscillating bases that were designed to oscillate the monitor back and forth over the ship side and loading arms. Deluge valves, with adjustable opening rates to prevent water hammer, located on the marine terminal deck could be activated by the SIS, FGS, manual pull stations, or the pilot head system.

The intent was to activate the two monitors in case of an LNG release, thus resulting in increased air entrainment to dilute and vaporize the LNG, and provide some degree of protection against embrittlement of exposed carbon steel on the ship and towers. The benefits, in case of a fire, are obvious, and intended to help minimize heat flux damage.

Shutdown of Loading Operations

The next step was to analyze the shutdown of the loading operation. This was complicated because of the large capacity LNG pumps which were located inside the LNG storage tanks and the ability to generate damaging hydraulic transients capable of causing failure of the piping delivery system.

Hydraulic Transient Analysis

A hydraulic transient analysis was performed to simulate pipeline pressures and pipe thrust loads during selected events that could occur when loading LNG from storage tanks to ships and a shutdown occurs.

The Pre-FEED level analysis was based on preliminary piping design information. It was intended to identify objectionable pressures and thrust loads resulting from hydraulic transients and identify mitigating measures to correct them.

The analysis precedes the detailed verification analysis, which must be performed to evaluate the system once the final design and control system logic have been established during the EPC phase of the project.

Study Assumptions

1) Each simulation used a time step value of 0.002 seconds, which resulted in pipe section lengths of approximately 2 meters, depending on the wave speed in the pipe.
2) The duration of each run allowed for several wave reflections throughout the system following the completion of the changes in state of each component comprising the scenario, ensuring that the severest conditions of the simulation were captured.
3) 24-inch XV's should be closed no faster than full stroke in 24 seconds
4) 30-inch XV's should be closed no faster than full stroke in 30 seconds

5) Interlocks must be provided to assure that the 30 inch shoreline block valve cannot close without the dock valves closing first. (This is to avoid severe column separation and rejoining in the jetty piping.)

6) When executing shutdown, operating pumps must stay on line for at least 30 seconds after the dock valves are finished closing

7) Interlocks must be provided to assure that the 30-inch shoreline block valve closes 30 seconds after the dock valves are finished closing

8) There should be a 30-second delay following loss of power to all pumps before any open pipeline valves begin to close

9) Concurrent loss of more than one pump should trigger shutting off all remaining pumps immediately to avoid excessive pipe loadings

10) Pump flow control valves should be equipped with operators incapable of operating them faster than full stroke in 30 seconds

SCENARIOS INVESTIGATED

Of primary initial concern were **Scenarios 1 through 4,** where an Emergency Shut Down (typical of that which would occur prior to an emergency separation of the loading arm coupling) is initiated when the system is operating at 50 percent of the maximum flow, assuming two loading arms were operating.

The initial interest in these first four scenarios was because at the lower flow, the hydraulic grade line in the system fell to such an elevation that the liquid was at vapor pressure in the tank down-corners and for approximately 50 meters in the loading line at the point where it broke from horizontal and headed to the jetty.

The concern was whether a valve closure at the downstream end would cause vapor cavities to collapse and create severe pressure rises and pipe thrusts. These scenarios were chosen over the maximum design flow scenarios because they posed a greater potential for column rejoining, and the likelihood was considered significant that a system upset could initiate an emergency shut down as the system is ramping up to or down from full flow.

Scenarios 5 through 7 examined pressure rise as a result of the same Emergency Shut Down when the system operated at its maximum design flow of 12,000 m3/h, with a total of eight pumps on line.

RESULTS

It was found that closing the loading-arm-valves during the 50-percent flow cases resulted in column rejoining and severe pipe thrust loads. Scenarios 1 and 4 showed that if the pumps were left on during these events, portions of the piping could be subjected to pressures well over 1,000 psi and pipe thrust loads well over 1,000 kN. Turning off the pumps at the same time as initiating the valve closure (Scenarios 2 and 3) resulted in reducing both loads and pressures, but thrust loads from column rejoining still exceeded 500 kN in some pipe sections.

Emergency shut down at full design capacity (12,000 m3/h), assuming a 5-second PERC valve closure (Scenario 5), produced column separation and severe transient conditions if the pumps were shut down at the same time as the valves initiated their closing. High-pressures approached 800 psi at the downstream end of the system, and 1,000 psi the tanks. Leaving the pumps operating (Scenario 6) resulted in more manageable pressure rise (approximately 440 psi) but was still in excess of allowable transient pressure rise for Class 150 pipe.

Slowing down the PERC valve speed to close over a 10-second period (Scenario 7) rather than a 5-second period produced pressures and loads that were acceptable. Pressures were less than 350 psi, and thrust loads were less than 180 kN.

DISCUSSION AND RECOMMENDATIONS

Results showed that the system design must include measures to mitigate the high pressures and thrust loads that could occur when operating the system. Two separate phenomena were identified that needed to be addressed: Design Condition 1: Low pressure and resulting cavity formation during low flow steady state operation of the system. Design Condition 2: The pressure rise resulting from rapid closure of valves at the downstream end of the system.

Design Condition 1: When the Loading Arms' Powered Emergency Release Coupling (PERC) valves close at a rate of 5 seconds under maximum design flow of 12,000 m3/h, the pressure upstream from them rises to over 400 psi.

Design Condition 2: It was also found that vapor cavities formed in the tank down-legs and in the loading line when system flow was less than approximately 6,000 m^3/h. This results in column rejoining and severe pressure rises and pipe thrust during emergency shut-down conditions.

The hydraulic transient analysis indicated that the safety shut down valve, located at the ship/shore interface, could not be closed faster than one second per inch of line size. In this particular case, that meant that once a shutdown signal had been generated from the SIS, the safety valve closure rate could not be faster than one second per inch of valve diameter.

In addition, because the large LNG pumps could generate significant transients when shutting down, the SIS would have to stagger the shutdown signals to them as well. Again, hydraulic analysis was used to arrive at the optimal timing sequence.

Recommendations

From study results, it was recommended that:

1) At a minimum, piping anchorages should be designed to withstand the loadings listed in Att. 6 for those Scenarios that are pertinent to the final design.
2) 16-inch PERC valves should be closed no faster than full stroke in 10 seconds if no pressure relief measures are employed. If a 5-second closure rate is required, the mitigating measures described above must be taken to protect against high pipe pressure (Design Condition 1).
3) Pressure sustaining devices should be added to the 30" loading line immediately upstream from the loading arm manifold to protect against Design Condition 2.
4) Design Condition l's pressure rise during PERC valve closure could be effectively reduced by providing timed pressure relief valves that dump LNG into a receiving drum at or near the Loading Dock.. Alternatively, pipe wall and flange class ratings could be increased to withstand the 440-psi maximum surge pressures.
5) Design Condition 2 could be addressed by installing upstream-pressure-controlling valves immediately upstream from the loading arm manifold to sustain the hydraulic grade line at an elevation above the vapor line elevation throughout the system. This would eliminate the conditions that lead to column rejoining at the lower flows.

Summary

As with many things, the closer the hazard investigation examined the possible consequences of a loading arm failure, the more issues surfaced. As it turns out, the design of the detection/alarm/shutdown system was fairly straightforward, given the codes and standards available to provide guidance.

A detailed examination of the possible hydraulic transients on the loading system during emergency shutdown however, revealed a potential for overpressuring the loading system itself, resulting in a possible catastrophic failure. The detailed hydraulic analysis provided the information necessary to develop the final design and logic for safe shutdowns during operations and emergencies.

REFERENCES

1) Cornwell, John B., Modeling LNG Spills in Boston Harbor, Quest Consultants, Inc., Norman, Oklahoma

2) Pitblado, R.M., Baik, J., Hughes, G.J., Ferro, C., Shaw, S.J., Consequences of Marine Incidents, CCPS Conference June 29, 2004

3) Dowell, A. M., Chair, Layers of Protective Analysis, AIChE CCPS, 3 Park Avenue, New York, NY

4) Pitblado, R.M., Baik, J., Hughes, G.J., Ferro, C., Shaw, S.J., Consequences of Marine Incidents, CCPS Conference June 29, 2004

5) Cornwell, John B., Modeling LNG Spills in Boston Harbor, Quest Consultants, Inc., Norman, Oklahoma

6) Pierorazio, A.J., Thomas, J.K., Baker, Q.A., Ketchum, D.E., An Update to the Baker-Strehlow-Tang Vapor Cloud Explosion Prediction Methodolgy Flame Speed Table, 38th Loss Prevention Symposium, AIChE, 2004

7) NFPA 72, National Fire Alarm Code, National Fire Protection Association, Quincy, MA,

8) Taylor, D.W., The Role of Consequence Modeling in LNG Facility Siting, The Journal of Hazardous Materials, Vol.142, Issue 3, Pages 776-785, April 11, 2007

9) Taylor, D.W., The Use of Gas Dispersion Modeling for Gas Sensor Location., AIChE, Process Plant Symposium, March 1994

10) IEC 61508, Functional Safety of Electrical/electronic/Programmable electronic safety-related systems.

11) IEC 61511, Safety Instrumented Systems for the Process Industry Sector

12) EN 54, Fire Detection and fire alarm systems.

13) Lees, F.P., Loss Prevention in the Process Industries, Hazard Identification, Assessment and Control, Second Edition, Butterworth-Heinemann, Oxford, 1996.

14) Mercx, W.P.M., Modelling and Experimental Research Into Gas Explosions, final report of the MERGE project, carried out for the Commission of the European Communities (CEC) under contract STEP-CT-0111.

15) Andrews, G.E. and D. Bradley, "Determination of Burning Velocities: A Critical Review," Combustion and Flame, 18: 133-53, 1972.
16) NFPA, Guide for Venting of Deflagrations, NFPA 68, National Fire Protection Association, Quincy, MA,
17) Gtinther, R. and G. Janisch, "Measurements of Burning Velocity in a Flat Flame Front," Combustion and Flame, 19: 49-53, 1972.
18) Baker, Q. A., M.J. Tang, E.A. Scheier and G.J. Silva, "Vapor Cloud Explosion Analysis," 28th Annual Loss Prevention Symposium, 1994.

A brief bio of the author;
Dennis W. Taylor, CFPS #1376, is the Technical Director, Process Safety/Fire Protection, for WorleyParsons, located in Houston, Texas. He has served on five API COSFP committees, and currently serves on NFPA 15 and NFPA 750. He has authored and presented over a dozen papers to API, TSA, AIChE, VPPPA, Petroleum Refiners Association, and ILTA. He has served as a peer reviewer for the Texas Institute for Advancement of Chemical Technology, and The Journal of Hazardous Materials.
With over 40 years experience in Loss Prevention, he was nominated to become a Bechtel Fellow in 2003.
Dennis holds AS, BS, and MBA degrees, and is an NFPA Board-certified Fire Protection Specialist.

(160b) Consequence Modeling on Subsea Release Using PLUMERISE and Numerical Model

Anna Qiao, Solution, Det Norske Veritas (U.S.A.), Inc., Houston, TX

As more offshore installations are developed in deep and ultra-deep water, subsea release, including a range of accident scenarios like the release of hydrocarbon inventory from topside process equipment, subsea pipelines, risers, or blowout events, has raised more concerns. Subsea release will result in dispersion of the hydrocarbon as it rises to the sea surface, and potentially a pool or pool fire could form on the surface depending on components and conditions of the released material. In this presentation, the author reviews the previous methodologies on vapor dispersion and pool fire modeling for subsea release, and then PLUMERISE model developed by DNV is introduced for subsea release modeling. This program calculates the boil area on the sea surface and the gas concentration above the sea surface from underwater blowout. The gas dispersion in the sea and the initial mixing with air just above the sea level are calculated as the source term to other dispersion modeling on the dispersion in the air. In these calculations the program can take into account the effect of seawater current. The pool spread on sea surface was simulated by a comprehensive numerical model. The results were compared with the common methodologies used for pool/pool fire consequence modeling. The tools developed for subsea release modeling has facilitated the offshore installation layout design, like the location section on living quarters, and provided the risk profile for subsea operations decisions like subsea pipeline routing.

See more of this Session: Professor Cedomir M. Sliepcevich Memorial Session Six - LNG Safety

See more of this Group/Topical: Topical 6: 10th Topical Conference on Gas Utilization

A medium-scale field test on expansion foam application – key findings of LNG pool fire suppression on land

Geunwoong Yun, Dedy Ng, M. Sam Mannan[1]

Mary Kay O'Connor Process Safety Center,

Artie McFerrin Department of Chemical Engineering, Texas A&M University System,

College Station, TX 77843-3122, USA

[1]Corresponding author: mannan@tamu.edu

AIChE Spring Meeting, March 2010

10th Topical Conference on Natural Gas Utilization

San Antonio March 21-25, 2010

1. Introduction

The Liquefied Natural Gas (LNG) is one of the fastest growing energy sources in U.S. to diversify the energy portfolio and fulfill the increasing energy demands. In order to meet the increasing LNG demand, many LNG facilities including LNG importation terminals are operating currently. Moreover, many projects for LNG terminals have been proposed to fill the gap between supply and demand of LNG in North America. However, LNG is a highly flammable gas which may result in fire or explosion when it is discharged into undesired conditions. In order to secure LNG safety, several safety guidelines, such as NFPA 59 A [1] and 49 CFR, are available. For controlling spilled LNG vapors and suppressing LNG fires, these guidelines recommend several safety measures, such as expansion foam application.

Expansion foam is one of the important safety measures to control vapor dispersion and LNG pool fires. The expansion foam is also used for fire-extinguishing systems for interior or exterior hazards in other industries such as refineries and petrochemicals. Fire-fighting foam, within the scope of this research, is an aggregation of small bubbles of lower density than LNG or oil that can cover horizontal surfaces of

LNG. Air foam, which is generally used in LNG facilities, is made by mixing air into a water solution, containing a foam concentrate, by foam generators. The expansion foam flows freely over a volatile liquid or burning liquid surface and forms a continuous blanket that seals volatile flammable vapors from access to air. It is also capable of resisting disruption from wind or flame attack so as to keep continuous blanketing. According to NFPA 11 [2], foam can be subdivided into three ranges of expansion having foam-to-solution volume ratios: (1) Low-expansion foam – expansion up to 20; (2) Medium-expansion foam – expansion from 20 to 200; (3) High-expansion foam – expansion from 200 to approximately 1000.

It has been recognized that the effectiveness of expansion foam may be achieved by: (1) reducing the LNG vapor density to make it positively buoyant by increasing its temperature adding heat from water in the foam to LNG vapors while penetrating through the foam; (2) reducing radiant and convective heat input from surroundings to the LNG pool surface; (3) reducing the fire growth by lowering the oxygen concentration in the vicinity of the burning liquid. In other words, heat transfer between water in the foam and LNG vapors can play an important role to disperse LNG vapors and suppress LNG fires.

Expansion foam has been identified as one of the promising safety measures for LNG pool fire suppression. However, little is known concerning the mechanisms and effectiveness of foam in suppressing LNG fire. Thus, there is a need to perform outdoor LNG spill tests for emergency preparedness and response training in case of accidental release. In this work, a medium-scale field tests observing the effects of expansion foam dispersed onto LNG pool fire were conducted on a concrete pit (6.40 m $\times$ 10.06 m $\times$ 1.22 m) by Mary Kay O'Connor Process Safety Center (MKOPSC) in 2009. LNG was released into the pit up to 0.20 m (8 inch) thickness and then the vapor cloud was ignited to form a pool fire. Subsequently, high expansion foam was applied to control the fire. Several parameters were measured during the LNG spill tests, such as temperatures of foam and fire, radiant heat fluxes, and fire sizes (height and area). In addition, mass burning rate was measured using thermocouples that are installed at different elevations around the concrete pit.

2. Results

When expansion foam was applied on LNG pool fire, initially the fire grew for around 5 seconds and the intensity of radiant heat flux reached the maximum value. After that, when expansion foam started to suppress the fire (around 5 seconds after foam application), the fire length decreased gradually, accordingly heat flux also decreased. Results showed that at the heat flux level of 5 kW/m^2, the hazard distances decrease from 19.72 m to 5.49 m which is about 72 % reduction by foam application. The radiant heat flux contours around the pit were also obtained from the radiometer measurements at a certain distance from fire. It was clear that the heat flux contour is significantly dependent on the prevailing wind direction. During December test in 2009, It was attempted that one radiometer was placed in the middle of fire to measure heat flux of fire itself. At the same location and position, two thermocouples were also installed to simulate heat flux using a theoretical equation [3] with temperature information and the result was compared with measured data by the radiometer. The comparison shows that calculated result gives larger values than measured one and it is believed that fire is not perfect grey body, in other words, it is moving fluid, not a body. Thus, these kinds of assumptions in calculation can cause uncertainty. The measured radiant heat flux in the fire was used to make heat flux contour around the LNG pit to obtain thermal hazard zone.

In addition, it was observed that expansion foam reduces fire height around 58 %. The fire was recorded by two video cameras and the fire heights over time were obtained by photographical analysis using Image J software. Then, the fire height profiles were compared with radiant heat flux profiles measured by radiometers placed in the same wind direction (i.e. a video and two radiometers were placed in the same crosswind direction). The results show that radiant heat flux profiles exactly follow the fire length profile during the whole burning period. It implies that radiant heat flux is strong function of fire height, thus this result can be used for LNG pool fie modeling. In addition, the correlation between LNG fire length and diameter were developed by previous researchers, but the correlation with expansion foam application has not been studied. Thus, in this paper, the correlation with foam was developed using obtained fire profiles and modifying the Thomas correlation [4], which has been widely accepted in the LNG fire-diameter correlation. We believe that it will be useful for LNG industry to design or site facilities with expansion foam system. Moreover, for the calculation of thermal hazard zone, when considering to give some credits to LNG industry for using safeguards such as expansion foam system, this correlated equation can be used.

Using thermocouples, mass burning rate of LNG pool applied by 1.22 m foam depth was measured as 0.082 ± 0.002 kg/m^2s. Based on temperature data of expansion foam during fire test, the minimum effective foam depth was determined as 0.61 m, where foam temperatures stayed below 100 ^oC to maintain the shape of bubbles and provide heat insulation to LNG pool. In addition, a schematic model of expansion foam and LNG system was proposed to understand the heat transfer and the underlying mechanisms of fire control and suppression. The model suggests that expansion foam can have three layers; thin frozen layer, non-frozen layer, and foam breaking layer resulted from the evaporation of water in foam at above 100 ^oC. One of the interesting facts is that even after fire test ice plates were observed. It is believed that when expansion foam was applied on the LNG surface some portion of water droplets in the foam drained to the LNG surface and it became ice plates due to heat loss from LNG, even though LNG vapors were still burning. Additionally, we believe that the ice plates can be methane hydrate, which is frozen ice bulk containing methane gases, because during LNG vaporization some amount of evaporated methane could be captured by frozen foam layers and became ice plates.

Another interesting finding is that paper labels attached to thermocouples in LNG for LNG level measurement were intact after pool fire test. It can be explained that initially those paper labels were in LNG, thus they could not be burnt. When LNG level decreased, the paper labels were covered by expansion foam and they were not burnt because oxygen was deficient for combustion process in the foam layer. This fact implies that expansion foam can reduce fire size and block radiant heat from fire back to LNG surface, thus resulting in less evaporation of LNG.

3. Conclusion

Finally, these findings can be used to validate the effectiveness of expansion foam on LNG fire control and provide design guidelines of foam system to the LNG industry. The understanding of LNG fire and its mitigation techniques will improve the fire and emergency response services and help to save lives and protect properties in the LNG industry.

References

1. National Fire Protection Association, *Standard for the Production, Storage, and Handling of Liquefied Natural Gas (LNG)*. 2009. **59A**.
2. National Fire Protection Association (NFPA), *Standard for Low-, Medium-, and High-Expansion foam*. 2002. **11A**.
3. Babrauskas, V., *Estimating large pool fire burning rates*. Fire Technology, 1983. **Volume 19, Number 4**: p. pp 251-261.
4. Raj, P.K., *LNG fires: A review of experimental results, models and hazard prediction challenges*. Journal of Hazardous Materials, 2007. **140**(3): p. 444-464.

Biography of speaker

The Department of Chemical Engineering / Mary Kay O'Connor Process Safety Center
Texas A&M University, 3122 TAMU, College Station, TX 77843-3122
Phone: (979) 862-1310, Fax: (979) 458-1493

E-mail: g0y0859@chemail.tamu.edu

Yun, Geunwoong is a PhD candidate of the Chemical Engineering Department of Texas A&M University advised by Dr. Sam Mannan and a research assistant of the Mary Kay O'Connor Process Safety Center. He attended SungKyunKwan University and received his BS in chemical engineering, and also received a MS degree in mechanical engineering from the YonSei University in Rep. of Korea. Moreover, He obtained a MS degree in chemical engineering from the Texas A&M University in U.S.

Mr. Yun has worked for the Korea Gas Safety Corporation for 9 years as a safety engineer and research scientist for gas related facilities (including LNG terminals) and equipment. He was a national secretary for 5 years for ISO standardization in the field of gas cylinders and bursting discs.

Mr. Yun has the Certified Welding Inspector Certification issued by AWS (American Welding Society) and Design Corrosion Control Certification by NACE (The Corrosion Society of America). He received the Certification of LOPA Training as well as SIL (Safety Integrity Level) Verification by SIS-TECH. Furthermore, he has Gas Safety Expert Certificate and Occupational Safety Expert Certificate by the Human Resources Development Service of Korea.

(160d) Vapor Cloud Explosion Siting Issues for LNG Facilities

Jeffrey D. Marx, Quest Consultants Inc., Norman, OK

John B. Cornwell, Quest Consultants Inc., Norman, OK

As building siting studies become increasingly important for newly-designed facilities, the primary hazard is often vapor cloud explosion (VCE) overpressures. This is certainly true for liquefied natural gas (LNG) facilities, which typically do not have any acutely toxic materials. At these facilities, VCEs and fires are the hazards that define the potential impacts, and the risk, to both onsite and offsite populations. While the approach for facility layout with regard to fires is generally straightforward and handled with fire protection measures, the approach for VCE modeling is not always clear. VCE modeling must demonstrate at what locations potential explosion impacts may affect offsite populations or onsite buildings. This is often accomplished using building damage levels which are correlated to the predictions of produced by a VCE model. This paper discusses some of the challenges involved in using VCE modeling to predict explosion overpressures, specifically in regard to LNG facilities. Both import and export facilities are evaluated in order to demonstrate the important issues with LNG projects.

See more of this Session: Professor Cedomir M. Sliepcevich Memorial Session Six - LNG Safety

See more of this Group/Topical: Topical 6: 10th Topical Conference on Gas Utilization

The effect of turbulence on the evaporation of cryogenic liquid spills on water

Timothy L. Morse, Ph.D.
Harri K. Kytömaa, Ph.D., P.E.
Exponent, Inc.
Natick, MA

The present study provides new measurements of the rate of evaporation of a cryogenic liquid floating on a water surface with different levels of turbulence intensity. The turbulent water surface is generated with an upward-pointing submerged jet with controlled jet velocity, an approach which has often been used in studies of free-surface turbulence. The rate of evaporation of the cryogenic liquid layer (liquid nitrogen in the present tests) is measured experimentally. Tests were carried out for different initial layer thicknesses and turbulence intensities of the water surface. These tests reveal not only a strong dependence of the evaporation rate on the turbulence intensity, but also a strong dependence on the thickness of the cryogenic liquid layer above the water surface. Models of liquefied natural gas (LNG) spills on water currently use a single rate of evaporation; these findings show that this approach is inadequate. Future models should incorporate the relative speed between the spill and the water below, the water turbulence intensity, and the LNG spill thickness for improved accuracy.

1. Introduction

Natural gas demand has been rapidly growing worldwide over the past few years, leading to a large number of new LNG terminals, particularly in Europe and North America and the rapid construction of additional global LNG liquefaction capacity. There has also been an associated increase in the development of offshore terminals as well as a rising interest in offshore operations. An important element of the overall risk of near-shore and offshore LNG operations is the behavior of spills on water: how the spill first interacts with the water at the point of entry, how the liquid pool first forms and spreads, and how the emanating vapors disperse.

Several models exist for estimating the behavior of an LNG spill on water, which all include a number of simplifying assumptions. One important component of these models, necessary for the successful prediction of the formation, growth, and dispersion of vapor clouds, is the rate of evaporation of LNG. The evaporation rate has often been assumed to be constant over the entire pool, and independent of the speed of the pool relative to the water it is traveling over. However, it is known that in a spill the point of entry as well as the leading edge will both exhibit elevated levels of turbulence and elevated evaporation rates. There have been some studies which have analyzed the effect of turbulence on evaporation rates, for example Hissong (2007) calculated empirical "turbulence factors" which adjusted the evaporation rate. Also, some previous tests conducted by Exponent in a recirculating water channel demonstrated that relative speed between the water and the liquid nitrogen poured on top influence the evaporation rate (Kytömaa et al. 2008).

The purpose of the present study is to provide new data regarding the rate of evaporation of a cryogenic liquid on a water surface with a carefully controlled turbulence intensity. Our measurements also reveal that the thickness of the cryogenic liquid on the water surface has a considerable effect on the rate of evaporation. This result was not initially expected, but may be as important as the effect of turbulence for the accurate modeling of LNG spills, depending on the specifics of the spill characteristics. The details regarding our experimental approach are described in the next section, followed by brief comments regarding the water surface turbulence characteristics. Our main findings regarding the thickness of the cryogenic liquid layer and water turbulence intensity are discussed in sections 4 and 5, followed by our conclusions.

2. Experimental details

The evaporation tests in this study were performed in a vertical cylinder with a submerged turbulent jet, shown in Figure 1. The vertical cylinder is double-walled (consisting of two transparent acrylic tubes) to minimize heat transfer through the cylinder walls. Nitrogen gas is piped into the gap space between the walls to remove water vapor and prevent condensation. The cylinder is partially filled with water, with a submerged jet entering from the bottom of the cylinder at the center. The free surface is 3.6 cylinder diameters above the jet nozzle, and is located above the point where the widening jet meets the pipe wall. The jet is generated by a centrifugal pump, with the flow rate controlled by a ball valve downstream of the pump. The turbulence intensity at the water surface is determined by the jet velocity at the nozzle and thus is controlled by the valve. Water recirculates back to the pump from the bottom of the cylinder, so that the volume of water in the cylinder is constant. The cylinder has an inner diameter of 17.1 cm, the jet nozzle diameter is 1.3 cm. The water level is held constant at a height of 62.2 cm above the nozzle (or 3.6 cylinder diameters). The fluctuations in height of the turbulent water surface at the center of the cylinder are measured optically as a measure of the turbulence intensity.

Once the turbulent surface has been established and characterized, a layer of a cryogenic liquid is poured onto the surface. The liquid is poured down a tube (to minimize the amount of mixing during the pouring procedure) which passes through a valve at the top of the cylinder. The tube is then removed and the valve is closed. Once the pouring valve is closed, the escaping vapor flows out of an orifice at the top of the cylinder (different orifice sizes are used depending on the approximate expected evaporation rate). The pressure in the cylinder and the temperature of the escaping vapor are measured and used with the equation for compressible flow out of an orifice to calculate the exit velocity and thus the evaporation rate. The present tests were performed using only liquid nitrogen. Further investigation with this facility will use liquefied natural gas.

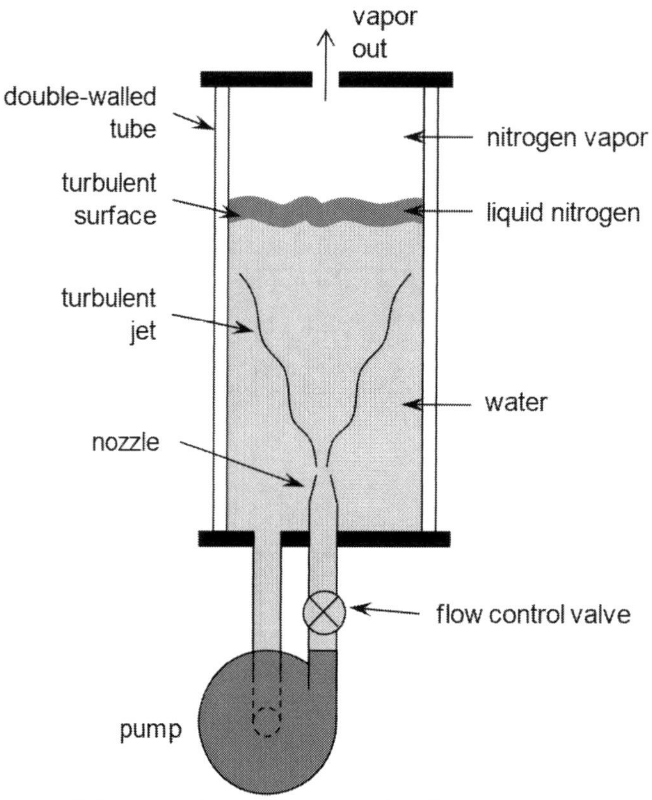

Figure 1. Schematic diagram of the experimental setup. The flow rate of the submerged jet, and thus the turbulence intensity at the surface, is controlled by a valve. Liquid nitrogen is poured on from above.

The bulk water temperature throughout the tests remained relatively constant, changing by only a few degrees Celsius. At the interface with the liquid nitrogen there was a small amount of ice formation, confined to a few millimeters at the edges of the cylinder. A lower jet velocity (and hence lower surface turbulence intensity) leads to more ice formation. This effectively provided a lower bound on the minimum turbulence intensity that could be tested. If the ice thickness, measured from the cylinder wall, ever grew to more than 0.5 cm, the test results were discarded. Of course, the presence of ice may be an important factor for some LNG spills (see for example the discussion of the effects of ice on LNG spills by Vesovic, 2006); however, in this study, we only considered cases with negligible ice.

3. Water surface turbulence characteristics

The submerged jet apparatus used in this study is similar to the setup used by Sonin et al. (1986), Hu (1993), and Law et al. (1999), although in the case of Hu (1993), the jet was inverted. These prior studies (among others) discuss details regarding the turbulent characteristics of submerged jets, and the resulting free surface. They show that for distances more than about 3 cylinder diameters from the nozzle, the flow is essentially isotropic, with equal horizontal and vertical velocity fluctuations, and has essentially zero mean flow. In the present study, the water surface is 3.6 cylinder diameters from the nozzle. Very near the surface (within about 0.1 diameters), Law et al. (1999) using detailed PIV measurements of the flow showed that the vertical velocity fluctuations start to become somewhat damped as kinetic energy is transferred to the horizontal velocity components, which show a corresponding increase in variations near the surface.

In the present study, the water surface height in the center of the cylinder was used as a measure of the turbulence intensity. The magnitude of velocity fluctuations may be estimated by balancing kinetic and potential energy changes:

$$\langle u^2 \rangle = 2g\sqrt{\langle \Delta h^2 \rangle},$$

where Δh is the variation in height of the water surface, g is the gravitational acceleration and $\langle u^2 \rangle$ is the mean-square of the velocity fluctuation ($u_{\mathrm{rms}} = \sqrt{\langle u^2 \rangle}$).

The turbulence intensity considered in this study is the value calculated for the water surface before any cryogenic liquid is poured on. The presence and evaporation of a cryogenic liquid on top of the water would be expected to cause changes in the turbulence intensity. The water surface height variations with liquid nitrogen were also measured, however they cannot be directly compared to the water surface height variations without liquid nitrogen, since the upper boundary conditions are different. Even so, the height variation with liquid nitrogen does provide a qualitative assessment of the intensity of the turbulence at the water surface.

4. Evaporation rate dependence on cryogenic liquid thickness

Our main intent with this study was to document the effect of turbulence intensity on the rate of evaporation of liquid nitrogen. Our expectation was that for a constant level of turbulence intensity the measured evaporation rate would be relatively constant over time as the thickness of liquid nitrogen decreased, then drop off rapidly once the liquid nitrogen was gone. Instead, we found that the evaporation rate decreased gradually over time, which may be seen in the example test run shown in Figure 2. For this case, the evaporation rate starts at about 0.5 kg/s m^2, and then gradually decreases to zero over the course of about three minutes. In addition to this broad trend, there are several intermittent bursts in the evaporation rate. These evaporation bursts occur roughly at times with increased fluctuations of the water surface height, i.e. at times with increased turbulence (shown in Figure 2b). The total amount of LN2 evaporated during this test was 0.89 kg, yielding an average evaporation rate of about 0.2 kg/s m^2.

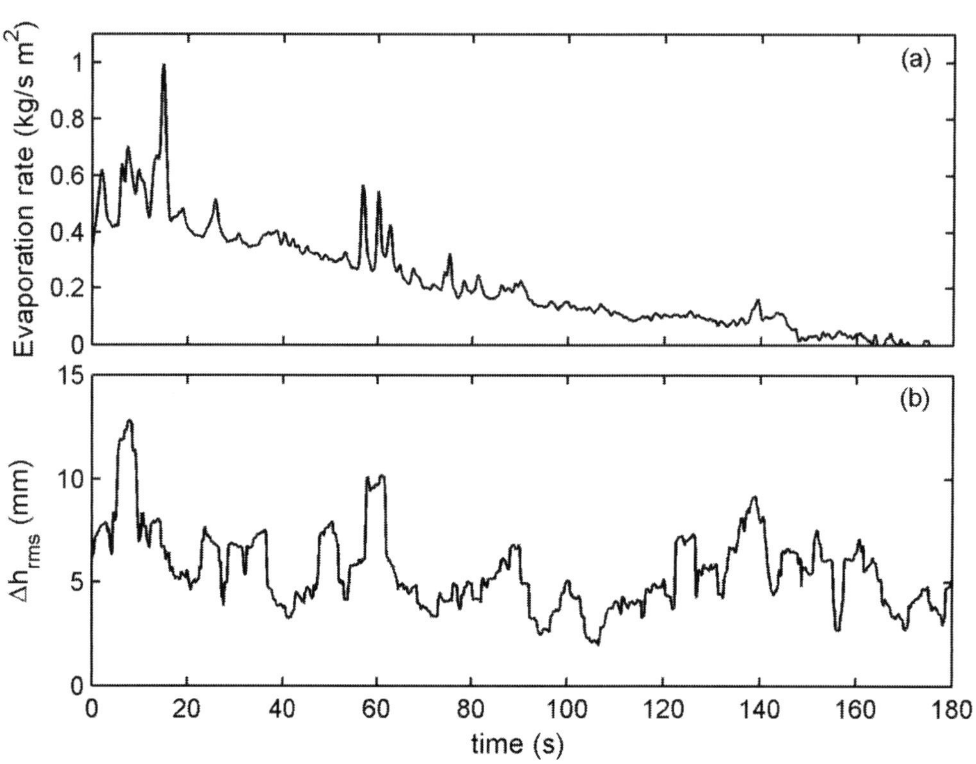

Figure 2. Evaporation rate (a) and root-mean-square of the water surface height variations (b) for an evaporation test. The height variation root-mean-square at each time was calculated over a four second moving window. For this test u_{rms} = 0.12 m/s (when no liquid nitrogen is present).

In addition to the gradual decrease in evaporation rate over time for any one particular test, we also observed a significant dependence on the initial amount of liquid nitrogen that was poured on. The evaporation rate over time is shown for three different runs in Figure 3, each with a different initial amount of liquid nitrogen, but with the same average turbulence intensity, in this case u_{rms} = 0.12 m/s. The initial mass of nitrogen was determined by integrating the evaporation rate over the entire run. A higher initial amount of liquid nitrogen consistently yielded a higher initial evaporation rate. To explain the results shown in Figure 3, we plot the evaporation rate as a function of the thickness of liquid nitrogen remaining, shown in Figure 4. The three cases nearly collapse onto a single curve of evaporation rate vs. thickness of liquid nitrogen remaining. Although, in all three cases, there are fluctuations in the evaporation rate including instances of large bursts of evaporation as described above.

Several additional evaporation runs were performed keeping the water surface turbulence intensity constant (i.e. without changing the control valve position). The initial average evaporation rate (i.e. the evaporation rate averaged over the first 60 seconds for each test) varies approximately linearly with the initial thickness of liquid nitrogen, at least over the range of initial thicknesses tested, as may be seen in Figure 5. We would not necessarily expect the evaporation rate to continue to rise linearly at very high liquid nitrogen layer thicknesses.

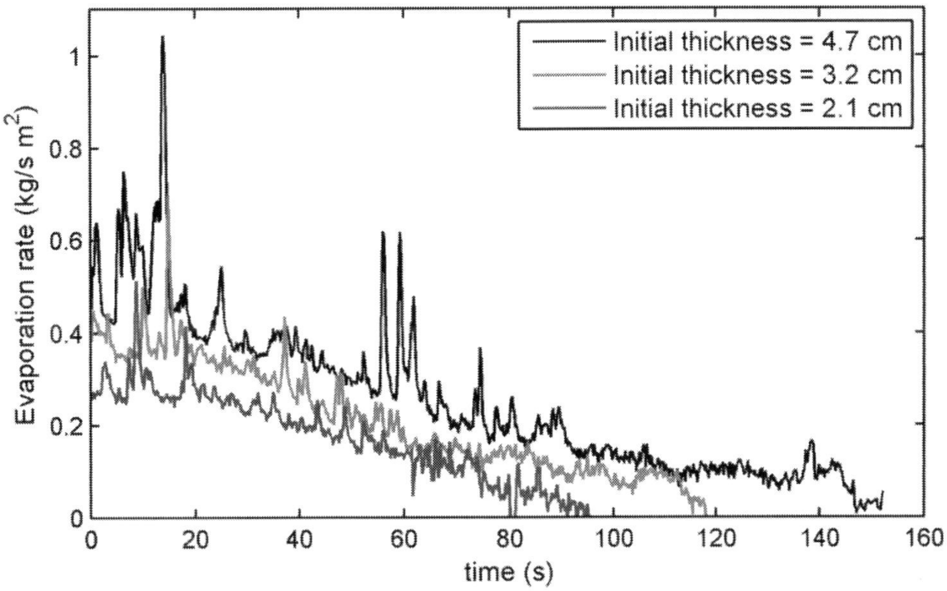

Figure 3. Evaporation rate over time for three different initial thickness of liquid nitrogen. For all cases u_{rms} = 0.12 m/s (when no liquid nitrogen is present).

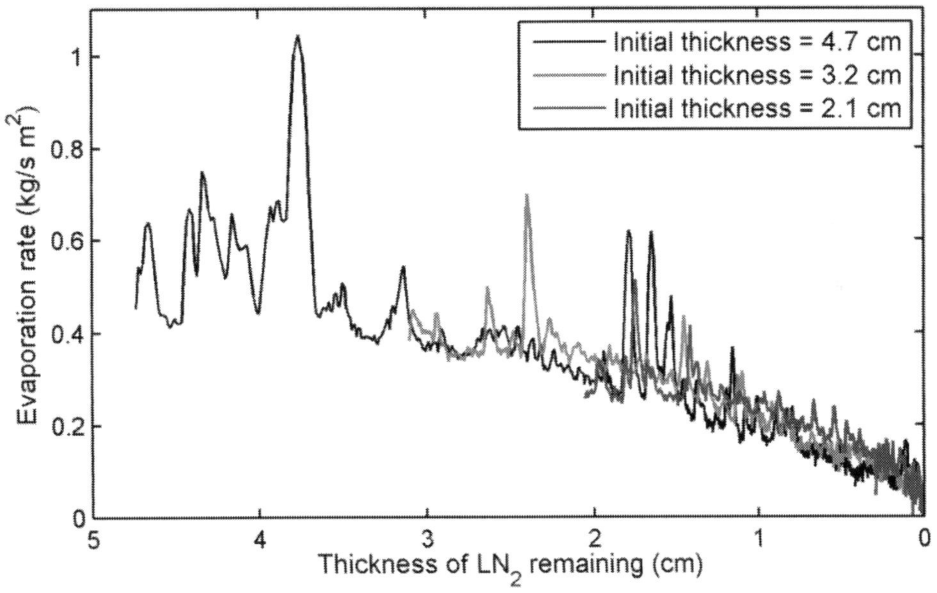

Figure 4. Evaporation rate vs. thickness of liquid nitrogen remaining for the same three runs as in Figure 3. For all cases u_{rms} = 0.12 m/s (when no liquid nitrogen is present).

The reason for this thickness dependence is not fully explained at this time. The rate of evaporation of a cryogenic liquid is dependent on the heat transfer between the cryogenic liquid and the water, which depends strongly on whether there is nucleate boiling, film boiling, or some transitional state. We expect that the added mass of liquid nitrogen affects the boiling dynamics, but we have not examined precisely how or why this occurs.

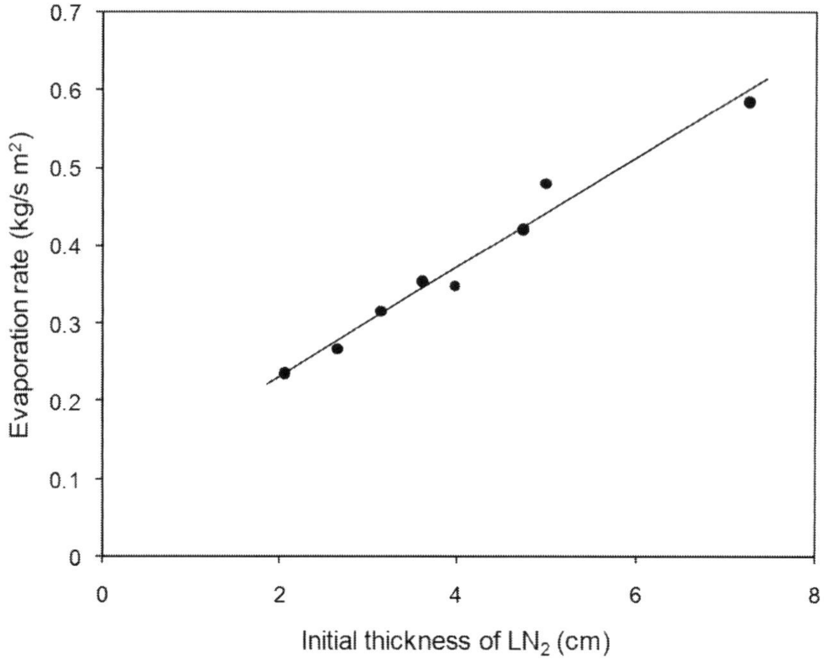

Figure 5. Evaporation rate for different initial thicknesses of liquid nitrogen above the water surface. The evaporation rate shown is the average over the first 60 seconds of each test.

We considered the possibility that this apparent effect was simply due to heat transfer through the side walls of the cylinder. Since a thicker layer of liquid nitrogen results in a larger surface area against the cylinder walls, this at first seems plausible. However, with the double-walled design, the heat transfer through the side walls would be expected to be much lower than the heat transfer at the water interface. To confirm this, we conducted a test with the pump switched off, which caused ice to quickly form and cover the entire water surface. In this case the evaporation rate was essentially constant as the thickness of liquid nitrogen decreased gradually. This leads us to conclude that the thickness of liquid nitrogen affects the evaporation rate by influencing the dynamics of boiling at the water interface.

5. Evaporation rate dependence on turbulence intensity

The main goal of this study was to identify the dependence of the evaporation rate for a cryogenic liquid on the turbulence intensity of the water surface. In light of the dependence on thickness of the cryogenic liquid layer described above, we conducted a series of tests with similar initial thicknesses of liquid nitrogen (about 3.5 cm), but with varying turbulence intensity. The results from these tests are shown in Figure 6. As in Figure 5, the evaporation rate for each case is the average rate over the first 60 seconds of the test. The turbulence intensity shown is measured in the absence of liquid nitrogen. The scatter above and below the best-fit curve provides an indication of the variability in evaporation rate, which is about 10%. The effect of turbulence intensity is considerable, the rate of evaporation doubles for about a four-fold increase in turbulence intensity.

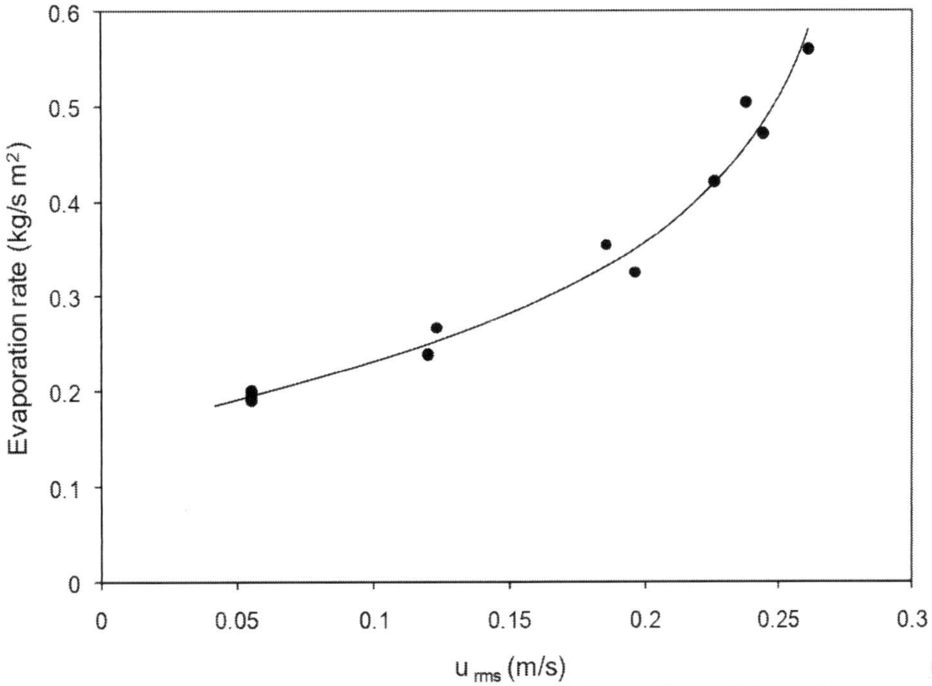

Figure 6. Evaporation rate for different water surface turbulence intensities. The evaporation rate shown is the average over the first 60 seconds of each test.

The cluster of three points at u_{rms} = 0.055 m/s are for three runs with the same valve position (and thus same mean turbulence intensity) but different orifice diameters at the top of the cylinder. Thus the pressure in the cylinder was different for these three cases, varying from about 0.06 psig to 0.5 psig. The close agreement in the calculated evaporation rate indicates that our measurement technique is consistent, and that pressure in the cylinder does not affect the evaporation rate (at least up to 0.5 psig). This has implications for our understanding of the effect of liquid nitrogen thickness on evaporation. To achieve a pressure of 0.5 psig, the equivalent thickness of liquid nitrogen would need to be 43 cm, yet that pressure alone does not apparently affect the rate of evaporation. Thus the thickness of liquid nitrogen affects the evaporation rate not primarily because of the increased pressure at the interface, but instead by some other mechanism.

6. Conclusions

The evaporation tests performed in this study demonstrate that the rate of evaporation of a cryogenic liquid on a water surface is not constant. It depends significantly on both the turbulence intensity of the water surface, and on the thickness of the cryogenic liquid layer on the water surface. The water turbulence characteristics in the case of an LNG spill would depend on several factors such as the geometry of the spilling stream of LNG, the physical characteristics of the waterway of interest (channel, harbor, open ocean), as well as the current weather conditions (wind, precipitation), and the spreading rate of the spill (a faster spreading rate would increase the turbulence intensity at the interface). In addition to the important factor of turbulence intensity, the thickness of the cryogenic liquid layer is also quite important, a finding that we did not expect initially. This study shows that for an increase in liquid layer thickness from 2 to 7 cm, the evaporation rate of liquid nitrogen increases from 0.23 to 0.57 kg/s m^2.

The phenomena observed in this study for liquid nitrogen are expected to carry over to liquefied natural gas, although the precise values of the evaporation rate will be different. Therefore, these results suggest that future efforts to model LNG spills should include both turbulence intensity and liquid layer thickness as important factors. Further tests with LNG are needed to quantitatively determine the effects of these factors on evaporation rates.

References

Hissong, D.W. 2007. Keys to modeling LNG spills on water. *Journal of Hazardous Materials*. 140:465-477.

Hu, W.K. 1993. Particle ejection and suspension due to isotropic turbulence. M.S. Thesis. Massachusetts Institute of Technology.

Kytomaa, H.K., Chernovsky, M.K., Gavelli, F.P. 2008. A new experimental study on the spreading of liquid nitrogen over water. Offshore Technologies Conference. OTC 19561.

Law, C.N.S., Khoo, B.C., Chew, T.C. 1999. Turbulence structure in the immediate vicinity of the shear-free air-water interface induced by a deeply submerged jet. *Experiments in Fluids*. 27:321-331.

Sonin, A.A., Shimko, M.A., Chun, J.H. 1986. Vapor condensation onto a turbulent liquid-I. The steady condensation rate as a function of liquid-side turbulence. *International Journal of Heat and Mass Transfer*. 29:1319-1332

Vesovic, V. 2007. The influence of ice formation on vaporization of LNG on water surfaces. *Journal of Hazardous Materials*. 140: 518-526.

Biography of the Speaker

Dr. Harri Kytömaa is Corporate Vice President and the Director of the Thermal Science Practice at Exponent. He specializes in mechanical engineering and the analysis of thermo-fluid phenomena, and performs research in cryogenic fluid mechanics, heat transfer and phase change phenomena. He has worked on numerous LNG projects with the industry in the US and internationally and has published extensively in this area. He is currently a member of the ISO TC67, WG10 working group that is developing a guidance document on the major hazards associated with the planning and design of onshore LNG facilities and associated marine activities. Dr. Kytömaa has held several positions, including that of Associate Professor of Mechanical Engineering at the Massachusetts Institute of Technology.

AUTHOR INDEX

AUTHOR INDEX